Communication Technology Update and Fundamentals: 17th Edition

Communication Technology Update and Fundamentals, now in its 17th edition, has set the standard as the single best resource for students and professionals looking to brush up on how communication technologies have developed, grown, and converged, as well as what's in store for the future.

The book covers the fundamentals of communication technology in five chapters that explain the communication technology ecosystem, its history, theories, structure, and regulations. Each chapter is written by experts who provide a snapshot of an individual field. The book also dives into the latest developments in electronic mass media, computers, consumer electronics, networking, and telephony. Together, these updates provide a broad overview of these industries and examine the role communication technologies play in our everyday lives. In addition to substantial updates to each chapter, the 17th edition includes the first-ever chapter on Artificial Intelligence; updated user data in every chapter; an overview of industry structure, including recent and proposed mergers and acquisitions; and sidebars exploring sustainability and relevance of each technology to Gen Z.

Communication Technology Update and Fundamentals continues to be the industry-leading resource for both students and professionals seeking to understand how communication technologies have developed and where they are headed.

Dr. August E. ("Augie") Grant is J. Rion McKissick Professor of Journalism and Director of the Center for Teaching Excellence at the University of South Carolina. Grant is a technology futurist who specializes in research on new media technologies and consumer behavior. His teaching and research combine the study of traditional and emerging media, with emphases on media management, organizational structure, integrated communication, and consumer behavior. He's been editor of *Communication Technology Update and Fundamentals* since 1992. His work in media convergence includes founding and serving for five years as Executive Editor of *The Convergence Newsletter*, chairing twelve national conferences on convergent journalism since 2002, and directing the Newsplex Summer Seminar program to train academics in convergent journalism.

Jennifer H. Meadows is Professor in the Department of Media Arts, Design, and Technology at California State University, Chico. Her teaching and research integrate practical and theoretical concerns related to television production, virtual reality, digital signage, the relationships between media and users, and changing media behavior in the new media marketplace. She has been working on the *Communication Technology Update and Fundamentals* since 1995 and has been co-editor since 1998.

Technology Futures, Inc. (TFI) provides top-quality custom studies and technology forecasts on key trends in communications, artificial intelligence, and other high-tech fields. Founded in 1978, TFI's specialties include technology forecasting, valuation, innovation, and business leadership. TFI brings to the table a futures focus, in-depth experience, a wide range of proven forecasting and analytic methods, and an excellent track-record of making valuable forecasts over several decades. www.tfi.com.

Communication Technology Update and Fundamentals

17th Edition

Editors
August E. Grant
Jennifer H. Meadows
In association with **Technology Futures, Inc.**

Routledge
Taylor & Francis Group

NEW YORK AND LONDON

Editors:
August E. Grant
Jennifer H. Meadows

Technology Futures, Inc.
Production & Graphics Editor:
Helen Mary V. Marek

Publisher: Margaret Farrelly
Editorial Assistant: Priscille Biehlmann
Production Editor: Alfred Symons

Seventeenth edition published 2021
by Routledge
52 Vanderbilt Avenue, New York, NY 10017

and by Routledge
2 Park Square, Milton Park, Abingdon, Oxon, OX14 4RN

Routledge is an imprint of the Taylor & Francis Group, an informa business

Publisher's note: This book has been prepared from camera-ready copy provided by the editors.
Typeset in Palatino Linotype by H.M.V. Marek, Technology Futures, Inc.

[First edition published by Technology Futures, Inc. 1992]
[Sixteenth edition published by Focal Press 2018]

Library of Congress Cataloging-in-Publication Data
A catalog record has been requested for this book

ISBN: 9780367420130 (hbk)
ISBN: 9780367420161 (pbk)
ISBN: 9780367817398 (ebk)

Visit the Communication Technology Update and Fundamentals website: http://www.tfi.com/ctu/

Table of Contents

Glossary and Updates can be found on the
Communication Technology Update and Fundamentals website: http://www.tfi.com/ctu/

Preface

As this book is going to press in early April 2020, the COVID-19 virus is in the early stages of spreading throughout the world. While we don't have a crystal ball that will show us how this pandemic will change things, history teaches us that crises like these always offer an opportunity for proliferation of new communication technologies.

Over the past year, the two dozen authors whose work appears in this book have studied the history, recent developments, and current status of these individual technologies. Although they finished their work before the virus started having its severe effects, they have provided you the foundation you need to understand the changes that are underway in the communication industries, putting you in a position to apply these lessons as you develop your career plans, no matter what career you are pursuing.

As you can clearly see, studying these technologies is like following a moving target. Consider that this is the 17th edition of *Communication Technology Update and Fundamentals*, which has tracked the evolution of the communication industries since 1992. In addition to updating every chapter with the latest developments, we have a first-time chapter exploring Artificial Intelligence (Chapter 25) and a few chapters that have been rewritten from scratch to provide a more contemporary discussion, including Regulation (Chapter 5), Automotive Telematics (Chapter 12), eHealth (Chapter 17), and The Internet (Chapter 21).

One thing shared by all of the contributors to this book is a passion for communication technology. In order to keep this book as current as possible we asked the authors to work under extremely tight deadlines. Authors begin working in late 2019, and most chapters were submitted in January or February 2020 with the final details added in March 2020. Individually, the chapters provide snapshots of the state of the field for individual technologies, but together they present a broad overview of the role that communication technologies play in our everyday lives. The efforts of these authors have produced a remarkable compilation, and we thank them for all for their hard work in preparing this volume.

The most remarkable process is the system created by our Production and Graphics Editor extraordinaire, TFI's Helen Mary V. Marek, who moved all 27 chapters from draft to camera-ready in weeks, as she produced on-demand graphics, adding visual elements to help make the content more understandable. Our editorial and marketing team at Routledge, including Margaret Farrelly and Priscille Biehlmann, ensured that production and promotion of the book were as smooth as ever.

Our spouses, Diane Grant and Floyd Meadows, supported us even more than usual as we have been forced to work remotely, alternating between an unhealthy focus on a screen and too many conferences to resolve editorial issues. They have watched us create this book every two years, but this year we have needed—and appreciate—their support more than ever!

You can keep up with developments on technologies discussed in this book by visiting our companion website, where we use the same technologies discussed in the book to make sure you have the latest information. The companion website for the *Communication Technology Update and Fundamentals*: www.tfi.com/ctu. The complete Glossary for the book is on the site, where it is much easier to find individual entries than in the paper version of the book. We have also moved the vast quantity of statistical data on each of the communication technologies that were formerly printed in Chapter 2 to the site. As always, we will periodically update the website to supplement the text with new information and links to a wide variety of information available over the internet.

Your interest and support is the reason we do this book every two years, and we listen your suggestions so that we can improve the book after every edition. You are invited to send us updates for the website, ideas for new topics, and other contributions that will inform all members of the community. You are invited to communicate directly with us via e-mail, snail mail, social media, or voice.

Thank you for being part of the CTUF community!

Augie Grant and *Jennifer Meadows*

April 3, 2020

Augie Grant
School of Journalism and Mass Communications
University of South Carolina
Columbia, SC 29208
Phone: 803.777.4464
augie@sc.edu
Twitter: @augiegrant

Jennifer H. Meadows
Dept. of Media Arts, Design, and Technology
California State University, Chico
Chico, CA 95929-0504
Phone: 530.898.4775
jmeadows@csuchico.edu
@mediaartsjen

Section I

Fundamentals

The Communication Technology Ecosystem

August E. Grant, Ph.D.*

Communication technologies are the nervous system of contemporary society. These technologies are critical to commerce, essential to entertainment, and intertwined in our interpersonal relationships. Because these technologies are so vitally important, any change in communication technologies has the potential to impact virtually every area of society.

The starting point for studying these technologies is identifying the wide range of factors related to the technology, which will be referred to throughout this book as the Communication Technology Ecosystem. As discussed below, it is not enough to study technologies as "things," but rather to examine a much wider range of factors, starting with some history.

One of the hallmarks of the industrial revolution was the introduction of new communication technologies as mechanisms of control that played an important role in almost every area of the production and distribution of manufactured goods (Beniger, 1986). These communication technologies have evolved throughout the past two centuries at an increasingly rapid rate. This evolution shows no signs of slowing, so an understanding of this evolution is vital for any individual wishing to attain or retain a position in business, government, or education.

The economic and political challenges faced by the United States and other countries since the beginning of the new millennium clearly illustrate the central role these communication systems play in our society. Just as economic prosperity is often credited to advances in technology, major downturns in the technology sector can also create challenges in the general economy. Today, communication technology is seen by many as a tool for making more efficient use of a wide range of resources including time and energy.

* J. Rion McKissick Professor of Journalism, School of Journalism and Mass Communications, University of South Carolina (Columbia, South Carolina).

Communication technologies play as critical a part in our private lives as they do in commerce and control in society. Geographic distances are no longer barriers to relationships thanks to the bridging power of communication technologies. We can also be entertained and informed in ways that were unimaginable a century ago thanks to these technologies—and they continue to evolve and change before our eyes.

This text provides a snapshot of the state of technologies in our society. The individual chapter authors have compiled facts and figures from hundreds of sources to provide the latest information on more than two dozen communication technologies. Each discussion explains the roots and evolution, recent developments, and current status of the technology as of mid-2020. In discussing each technology, we address them from a systematic perspective, looking at a range of factors beyond hardware.

The goal is to help you analyze emerging technologies and be better able to predict which ones will succeed and which ones will fail. That task is more difficult to achieve than it sounds. Let's look at an example of how unpredictable technology can be.

The Alphabet Tale

As this book goes to press in mid-2020, Alphabet, the parent company of Google, is the most valuable media company in the world in terms of market capitalization (the total value of all shares of stock held in the company). To understand how Alphabet attained that lofty position, we have to go back to the late 1990s, when commercial applications of the internet were taking off. There was no question in the minds of engineers and futurists that the internet was going to revolutionize the delivery of information, entertainment, and commerce. The big question was how it was going to happen.

Those who saw the internet as a medium for information distribution knew that advertiser support would be critical to its long-term financial success. They knew that they could always find a small group willing to pay for content, but the majority of people preferred free content. To become a mass medium similar to television, newspapers, and magazines, an internet advertising industry was needed.

At that time, most internet advertising was banner ads—horizontal display ads that stretched across most of the screen to attract attention, but took up very little space on the screen. The problem was that most people at that time accessed the internet using slow, dial-up connections, so advertisers were limited in what they could include in these banners to about a dozen words of text and simple graphics. The advertisers' dream was to use rich media, including full-motion video, audio, animation, and every other trick that makes television advertising so successful.

When broadband internet access started to spread, advertisers were quick to add rich media to their banners, as well as create other types of ads using graphics, video, and sound. These ads were a little more effective, but many internet users did not like the intrusive nature of rich media messages.

At about the same time, two Stanford students, Sergey Brin and Larry Page, had developed a new type of search engine, Google, that ranked results on the basis of how often content was referred to or linked from other sites, allowing their computer algorithms to create more robust and relevant search results (in most cases) than having a staff of people indexing web content. What they needed was a way to pay for the costs of the servers and other technology.

According to Vise & Malseed (2006), their budget did not allow the company, then known as Google, to create and distribute rich media ads. They could do text ads, but they decided to do them differently from other internet advertising, using computer algorithms to place these small text ads on the search results that were most likely to give the advertisers results. With a credit card, anyone could use this "AdWords" service, specifying the search terms they thought should display their ads, writing the brief ads (less than 100 characters total—just over a dozen words), and even specifying how much they were willing to pay every time someone clicked on their ad. Even more revolutionary, the Google team decided that no one should have to pay for an ad unless a user clicked on it.

For advertisers, it was as close to a no-lose proposition as they could find. Advertisers did not have to pay unless a person was interested enough to click on the ad. They could set a budget that Google computers could follow, and Google provided a control panel for advertisers that gave a set of measures that was a dream for anyone trying to make a campaign more effective. These measures indicated not only the overall effectiveness of the ad, but also the effectiveness of each message, each keyword, and every part of every campaign.

The result was remarkable. Google's share of the search market was not that much greater than the companies that had held the #1 position earlier, but Google was making money—lots of money—from these little text ads. Wall Street investors noticed, and, once Google went public, investors bid up the stock price, spurred by increases in revenues and a very large profit margin. Today, Google's parent company, renamed Alphabet, is involved in a number of other ventures designed to aggregate and deliver content ranging from text to full-motion video, but its little text ads on its Google search engine are still the primary revenue generator.

In retrospect, it is easy to see why Google was a success. Their little text ads were effective because of context—they always appeared where they would be the most effective. They were not intrusive, so people did not mind the ads on Google pages, and later, on other pages that Google served through its "content network." Plus, advertisers had a degree of control, feedback, and accountability that no advertising medium had ever offered (Grant & Wilkinson, 2007).

So, what lessons should we learn from this story? Advertisers have their own set of lessons, but there are a separate set of lessons for those wishing to understand new media. First, no matter how insightful, no one is ever able to predict whether a technology will succeed or fail. Second, success can be due as much to luck as to careful, deliberate planning and investment. Third, simplicity matters—there are few advertising messages as simple as the little text ads you see when doing a Google search.

The Alphabet tale provides an example of the utility of studying individual companies and industries, so the focus throughout this book is on individual technologies. These individual snapshots, however, comprise a larger mosaic representing the communication networks that bind individuals together and enable them to function as a society. No single technology can be understood without understanding the competing and complementary technologies and the larger social environment within which these technologies exist. As discussed in the following section, all of these factors (and others) have been considered in preparing each chapter through application of the "technology ecosystem." Following this discussion, an overview of the remainder of the book is presented.

The Communication Technology Ecosystem

The most obvious aspect of communication technology is the hardware—the physical equipment related to the technology. The hardware is the most tangible part of a technology system, and new technologies typically spring from developments in hardware. However, understanding communication technology requires more than just studying the hardware. One of the characteristics of today's digital technologies is that most are based upon computer technology, requiring instructions and algorithms more commonly known as "software."

In addition to understanding the hardware and software of the technology, it is just as important to understand the content communicated through the technology system. Some consider the content as another type of software. Regardless of the terminology used, it is critical to understand that digital technologies require a set of instructions (the software) as well as the equipment and content.

The hardware, software, and content must also be studied within a larger context. Rogers' (1986) definition of "communication technology" includes some of these contextual factors, defining it as "the hardware equipment, organizational structures, and social values by which individuals collect, process, and exchange information with other individuals" (p. 2). An even broader range of factors is suggested by Ball-Rokeach (1985) in her media system dependency theory, which suggests that communication media can be understood by analyzing dependency relations within and across levels of analysis, including the individual, organizational, and system levels. Within the system level, Ball-Rokeach identifies three systems for analysis: the media system, the political system, and the economic system.

These two approaches have been synthesized into the "Technology Ecosystem" illustrated in Figure 1.1. The core of the technology ecosystem consists of the hardware, software, and content (as previously defined). Surrounding this core is the organizational infrastructure: the group of organizations involved in the production and distribution of the technology. The next level moving outwards is the system level, including the political, economic, and media systems, as well as other groups of individuals or organizations serving a common set of functions in society. Finally, the individual users of the technology cut across all of the other areas, providing a focus for understanding each one. The basic premise of the technology ecosystem is that all areas of the ecosystem interact and must be examined in order to understand a technology.

(The technology ecosystem is an elaboration of the "umbrella perspective" (Grant, 2010) that was explicated in earlier editions of this book to illustrate the elements that need to be studied in order to understand communication technologies.)

Figure 1.1

The Communication Technology Ecosystem

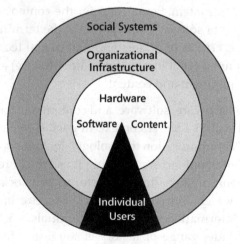

Source: A. E. Grant

Adding another layer of complexity to each of the areas of the technology ecosystem is also helpful. In order to identify the impact that each individual characteristic of a technology has, the factors within each area of the ecosystem may be identified as "enabling," "limiting," "motivating," and "inhibiting" depending upon the role they play in the technology's diffusion.

Enabling factors are those that make an application possible. For example, the fact that the coaxial cable used to deliver traditional cable television can carry hundreds of channels is an enabling factor at the hardware level. Similarly, the decision of policy makers to allocate a portion of the radio frequency spectrum for cellular telephony is an enabling factor at the system level (political system). One starting point to use in examining any technology is to make a list of the underlying factors from each area of the technology ecosystem that make the technology possible in the first place.

Limiting factors are the opposite of enabling factors; they are those factors that create barriers to the adoption or impacts of a technology. A great example is related to the cellular telephone illustration in the previous paragraph. The fact that the policy makers discussed above initially permitted only two companies to offer cellular telephone service in each market was a system level limitation on that technology. The later introduction of digital technology made it possible for another four companies to compete for mobile phone service. To a consumer, six telephone companies may seem to be more than is needed, but to a start-up company wanting to enter the market, this system-level factor represents a definite limitation. Again, it is useful to apply the technology ecosystem to create a list of factors that limit the adoption, use, or impacts of any specific communication technology.

Motivating factors are a little more complicated. They are those factors that provide a reason for the adoption and use of a technology. Technologies are not adopted just because they exist. Rather, individuals, organizations, and social systems must have a reason to take advantage of a technology. The desire of local telephone companies for increased profits, combined with the fact that growth in providing local telephone service is limited, is an organizational factor motivating the telcos to enter the markets for new communication technologies. Individual users desiring information more quickly can be motivated to adopt electronic information technologies. If a technology does not have sufficient motivating factors for its use, it cannot be a success.

Inhibiting factors are the opposite of motivating ones, providing a disincentive for adoption or use. An example of an inhibiting factor at the organizational level might be a company's history of bad customer service. Regardless of how useful a new technology might be, if customers don't trust a company, they are not likely to purchase its products or services. One of the most important inhibiting factors for most new technologies is the cost to users. Each potential user must decide whether the cost is worth the service, considering their budget and the number of competing technologies. Competition from other technologies is one of the biggest barriers any new (or existing) technology faces. Any factor that works against the success of a technology can be considered an inhibiting factor. As you might guess, there are usually more inhibiting factors for most technologies than motivating ones. And if the motivating factors are more numerous and stronger than the inhibiting factors, it is an easy bet that a technology will be a success.

All four factors—enabling, limiting, motivating, and inhibiting—can be identified at the individual user, organizational, content, and system levels. However, hardware and software can only be enabling or limiting; by themselves, hardware and software do not provide any motivating factors. The motivating factors must always come from the messages transmitted or one of the other areas of the ecosystem.

The final dimension of the technology ecosystem relates to the environment within which communication technologies are introduced and operate. These factors can be termed "external" factors, while ones relating to the technology itself are "internal" factors. In order to understand a communication technology or be able to predict how a technology will diffuse, both internal and external factors must be studied.

Applying the Communication Technology Ecosystem

The best way to understand the communication technology ecosystem is to apply it to a specific technology. One of the fastest diffusing technologies discussed later in this book is the "personal or virtual assistant" such as Amazon Alexa or Google Home—devices that use this technology such as the Amazon Echo provide a great application of the communication technology ecosystem.

Let's start with the hardware. Most personal assistants are small or medium-sized units, designed to sit on a shelf or table. Studying the hardware reveals that the unit contains multiple speakers, a microphone, some computer circuitry, and a radio transmitter and receiver. Studying the hardware, we can get clues about the functionality of the device, but the key to the functionality is the software.

The software related to the personal assistant enables conversion of speech heard by the microphone into text or other commands that connect to another set of software designed to fulfill the commands given to the system. From the perspective of the user, it doesn't matter whether the device converts speech to commands or whether the device transmits speech to a central computer where the translation takes place—the device is designed so that it doesn't matter to the user. The important thing that becomes apparent is that the hardware used by the system extends well beyond the device through the Internet to servers that are programmed to deliver answers and content requested through the personal assistant.

So, who owns these servers? To answer that question, we have to look at the organizational infrastructure. It is apparent that there are two distinct sets of organizations involved—one set that makes and distributes the devices themselves to the public and the other that provides the back-end processing power to find answers and deliver content. For the Amazon devices that use Alexa, Amazon has designed and arranged for the manufacture of the device. (Note that few companies specialize in making hardware; rather, most communication hardware is made by companies that specialize in manufacturing on a contract basis.) Amazon also owns and controls the servers that interpret and seek answers to questions and commands. But to get to those servers, the commands have to first pass through cable or phone networks owned by other companies, with answers or content provided by servers on the Internet owned by still other companies. At this point, it is helpful to examine the economic relationships among the companies involved. The users'

Internet Service Provider (ISP) passes all commands and content from the home device to the cloud-based servers, which are, in turn, connected to servers owned by other companies that deliver content.

So, if a person requests a weather forecast, the servers connect to a weather service for content. A person might also request music, finding themselves connected to Amazon's own music service or to another service such as Spotify or Sirius/XM. A person ordering a pizza will have their message directed to the appropriate pizza delivery service, with the only content returned being a confirmation of the order, perhaps with status updates as the order is fulfilled.

The pizza delivery example is especially important because it demonstrates the economics of the system. The servers used are expensive to purchase and operate, so the company that designs and sells personal assistants has a motivation to contract with individual pizza delivery services to pay a small commission every time someone orders a pizza. Extending this example to multiple other services will help you understand why some services are provided for free but others must be paid, with the pieces of the system working together to spread revenue to all of the companies involved.

The point is that it is not possible to understand the personal assistant without understanding all of the organizations implicated in the operation of the device. And if two organizations decide not to cooperate with each other, content or service may simply not be available.

The potential conflicts among these organizations can move our attention to the next level of the ecosystem, the social system level. The political system, for example, has the potential to enable services by allowing or encouraging collaboration among organizations. Or it can do the opposite, limiting or inhibiting cooperation with regulations. (Net neutrality, discussed in Chapter 5, is a good example of the role played by the political system in enabling or limiting capabilities of technology.) The system of retail stores enables distribution of the personal assistant devices to local retail stores, making it easier for a user to become an "adopter" of the device.

Studying the personal assistant also helps illustrate the enabling and limiting functions. For example, the fact that Amazon has programmed the Alexa app to accept commands in dozens of languages from Spanish to Klingon is an enabling factor, but the fact that there are dozens of other languages that have not been programmed is definitely a limiting factor.

Similarly, the ease of ordering a pizza through a personal assistant is a motivating factor, but having your device not understand your commands is an inhibiting factor. Finally, examination of the environment gives us more information, including competitive devices, public sentiment, and general economic environment.

All of those details help us to understand how personal assistants work and how companies can profit in many different ways from their use. But we can't fully understand the role that these devices play in the lives of their users without studying the individual user. We can examine what services are used, why they are used, how often they are used, the impacts of their use, and much more.

Applying the Communication Technology Ecosystem thus allows us to look at a technology, its uses, and its effects using a multi-level perspective that provides a more comprehensive insight than we would get from just examining the hardware or software.

Each communication technology discussed in this book has been analyzed using the technology ecosystem to ensure that all relevant factors have been included in the discussions. As you will see, in most cases, organizational and system-level factors (especially political factors) are more important in the development and adoption of communication technologies than the hardware itself. For example, political forces have, to date, prevented the establishment of a single world standard for television production and transmission. As individual standards are selected in countries and regions, the standard selected is as likely to be the product of political and economic factors as of technical attributes of the system.

Organizational factors can have similar powerful effects. For example, as discussed in Chapter 4, the entry of a single company, IBM, into the personal computer business in the early 1980s resulted in changes in the entire industry, dictating standards and anointing an operating system (MS-DOS) as a market leader. Finally, the individuals who adopt (or choose not to adopt) a technology, along with their motivations and the manner in which they use the technology, have profound impacts on the development and success of a technology following its initial introduction.

Perhaps the best indication of the relative importance of organizational and system-level factors is the number of changes individual authors made to the chapters in this book between the time of the initial chapter submission in January 2020 and production of the final, camera-ready text in April 2020. Very little new information was added regarding hardware, but numerous changes were made due to developments at the organizational and system levels.

To facilitate your understanding of all of the elements related to the technologies explored, each chapter in this book has been written from the perspective of the technology ecosystem. The individual writers have updated developments in each area to the extent possible. Obviously, not every technology experienced developments in each area of the ecosystem, so each chapter is limited to areas in which relatively recent developments have taken place.

Why Study New Technologies?

One constant in the study of media is that new technologies seem to get more attention than traditional, established technologies. There are many reasons for the attention. New technologies are more dynamic and evolve more quickly, with greater potential to cause change in other parts of the media system. Perhaps the reason for our attention is the natural attraction that humans have to motion, a characteristic inherited from our most distant ancestors.

There are a number of other reasons for studying new technologies. Maybe you want to make a lot of money—and there is a lot of money to be made (and lost!) on new technologies. If you are planning a career in the media, you may simply be interested in knowing how the media are changing and evolving, and how those changes will affect your career.

Or you might want to learn lessons from the failure of new communication technologies so you can avoid failure in your own career, investments, etc. Simply put, the majority of new technologies introduced do not succeed in the market. Some fail because the technology itself was not attractive to consumers (such as the 1980s' attempt to provide AM stereo radio). Some fail because they were far ahead of the market, such as Qube, the first interactive cable television system, introduced in the 1970s. Others failed because of bad timing or aggressive marketing from competitors that succeeded despite inferior technology.

The final reason for studying new communication technologies is to identify patterns of adoption, effects, economics, and competition so that we can be prepared to understand, use, and/or compete with the next generation of media. Virtually every new technology discussed in this book is going to be one of those "traditional, established technologies" in a few short years, but there will always be another generation of new media to challenge the status quo.

Overview of Book

The key to getting the most out of this book is therefore to pay as much attention as possible to the reasons that some technologies succeed and others fail. To that end, this book provides you with a number of tools you can apply to virtually any new technology that comes along. These tools are explored in the first five chapters, which we refer to as the *Communication Technology Fundamentals*. You might want to skip over these to get to the latest developments that are making an impact today, but you will be much better equipped to learn lessons from these technologies if you are armed with these tools.

The first of these is the "technology ecosystem" discussed previously that broadens attention from the technology itself to the users, organizations, and system surrounding that technology. To that end, each of the technologies explored in this book provides details about all of the elements of the ecosystem.

Of course, studying the history of each technology can help you find patterns and apply them to different technologies, times, and places. The next chapter, A History of Communication Technologies, provides an overview of most of the technologies discussed later in the book, allowing you to compare the year introduced, growth rate, number of users, etc. This chapter highlights commonalties in the evolution of individual technologies, as well as presents the "big picture" before we delve into the details. By focusing on the number of users over time, this chapter also provides a useful basis of comparison across technologies.

Another useful tool in identifying patterns across technologies is theories related to new communication technologies. By definition, theories are general statements that identify the underlying mechanisms for adoption and effects of these new technologies. Chapter 3 provides an overview of a wide range of these theories that you can apply to both the technologies in this book and any new technologies that follow.

Chapter 4 explores the structure of communication industries, examining organizational relationships and differentiating between the companies that make the technologies and those that sell the technologies. The most important force at the system level of the ecosystem, regulation, is introduced in Chapter 5.

These introductory chapters provide a structure and a set of analytic tools that define the study of communication technologies. Following this introduction, the book then addresses the individual technologies.

The technologies discussed in this book are organized into three sections: Electronic Mass Media, Computers & Consumer Electronics, and Networking Technologies. These three are not necessarily exclusive; for example, Digital Signage could be classified as either a mass medium or a computer technology. The decision regarding where to put each technology was made by determining which set of current technologies most closely resemble the technology. Thus, Digital Signage was classified with electronic mass media. This process also locates the discussion of an entertainment technology—DVRs—in the Home Video chapter in the Computers & Consumer Electronics section.

Each chapter is followed by a brief bibliography that represents a broad overview of literally hundreds of books and articles that provide details about these technologies. It is hoped that the reader will not only use these references but will examine the list of source material to determine the best places to find newer information since the publication of this *Update*.

To provide a broader perspective, each chapter also includes two sidebars, with one examining global and sustainability factors related to the technologies and the other providing a Generation Z perspective.

Most of the technologies discussed here continue to evolve. As this book was completed, many developments were announced but not released, corporate mergers were under discussion, and regulations had been proposed but not passed. Our goal is for the chapters in this book to establish a basic understanding of the structure, functions, and background for each technology, and for the supplementary website to provide brief synopses of the latest developments for each technology. (The address for the website is www.tfi.com/ctu.)

The final chapter returns to the "big picture" presented in this book, exploring the process of starting a company to exploit or profit from these technologies. Any text such as this one can never be fully comprehensive, but ideally this text will provide you with a broad overview of the current developments in communication technology.

Bibliography

Ball-Rokeach, S. J. (1985). The origins of media system dependency: A sociological perspective. *Communication Research, 12* (4), 485-510.

Beniger, J. (1986). *The control revolution*. Cambridge, MA: Harvard University Press.

Grant, A. E. (2010). Introduction to communication technologies. In A. E. Grant & J. H. Meadows (Eds.) *Communication Technology Update and Fundamentals (12th ed)*. Boston: Focal Press.

Grant, A. E. & Wilkinson, J. S. (2007, February). Lessons for communication technologies from Web advertising. Paper presented to the Mid-Winter Conference of the Association of Educators in Journalism and Mass Communication, Reno.

Rogers, E. M. (1986). *Communication technology: The new media in society*. New York: Free Press.

Vise, D. & Malseed, M. (2006). *The Google story: Inside the hottest business, media, and technology success of our time*. New York: Delta.

2

A History of Communication Technology

Yicheng Zhu, Ph.D. & Mengqi Wang, B.A.*

The other chapters in this book provide details regarding the history of one or more communication technologies. However, one needs to understand that history works, in some ways, like a telescope. The closer an observer looks at the details, i.e. the particular human behaviors that changed communication technologies, the less they can grasp the *big picture*.

This chapter attempts to provide the *big picture* by discussing recent advancements along with a review of happenings "before we were born." Without the understanding of the collective memory of the trailblazers of communication technology, we will be "children forever"

when we make interpretations and implications from history records. (Cicero, 1876).

The print, electronic, and digital era will be discussed in this chapter. To provide a useful perspective, numerical statistics of adoption and use of these technologies are compared across time. To that end, this chapter follows patterns adopted in previous summaries of trends in U.S. communications media (Brown & Bryant, 1989; Brown, 1996, 1998, 2000, 2002, 2004, 2006, 2008, 2010, 2012, 2014; Zhu & Brown, 2016; Zhu, 2018). Non-monetary units are reported when possible, although dollar expenditures appear as supplementary measures. A notable exception is the de facto standard

* Zhu is an Assistant Professor in the School of Journalism and Communication at Beijing Normal University (Beijing, China). Wang is a master student in mass communication at Beijing Normal University (Beijing, China). (The authors and the editors acknowledge the contributions of the late Dan Brown, Ph.D., who created the first versions of this chapter and the related figures and tables.)

of measuring motion picture acceptance in the market: box office receipts.

Government sources are preferred for consistency in this chapter. However, they have recently become more volatile in terms of format, measurement and focus due to the shortened life circle of technologies (for example, some sources do not distinguish laptops from tablets when calculating PC shipments). Readers should use caution in interpreting data for individual years and instead emphasize the trends over several years. One limitation of this government data is the lag time before statistics are reported, with the most recent data being a year or more old. The companion website for this book (www.tfi.com/ctu) reports more detailed statistics than could be printed in this chapter.

Communication technologies are evolving at a much faster pace today than in the past, and the way in which we differentiate technologies is more about concepts than products. For example, audiocassettes and compact discs seem doomed in the face of rapid adoption of newer forms of digital audio recordings. But what fundamentally changed our daily experience is the surge of individual power brought by technological convenience: digitized audio empowered our mobility and efficiency both at work and at play. Quadraphonic sound, CB radios, 8-track audiotapes, and 8mm film cameras ceased to exist as standalone products in the marketplace, and we exclude them, not because they disappeared, but because their concepts were converted or integrated into newer and larger concepts. This chapter traces trends that reveal clues about what has happened and what may happen in the use of respective media forms.

To illustrate the growth rates and specific statistics regarding each technology, a large set of tables and figures have been placed on the companion website for this book at www.tfi.com/ctu. Your understanding of each technology will be aided by referring to the website as you read each section.

The Print Era

Printing began in China thousands of years before Johann Gutenberg developed the movable type printing press in 1455 in Germany. Gutenberg's press triggered a revolution that began an industry that remained stable for another 600 years (Rawlinson, 2011).

Printing in the United States grew from a one-issue newspaper in 1690 to become the largest print industry in the world (U.S. Department of Commerce/International Trade Association, 2000). This enterprise includes newspapers, periodicals, books, directories, greeting cards, and other print media.

Newspapers

Publick Occurrences, Both Foreign and Domestick was the first newspaper produced in North America, appearing in 1690 (Lee, 1917). Table 2.1 and Figure 2.1 from the companion website (www.tfi.com/ctu) for this book show that U.S. newspaper firms and newspaper circulation had extremely slow growth until the 1800s. Early growth suffered from relatively low literacy rates and the lack of discretionary cash among the bulk of the population. The progress of the industrial revolution brought money for workers and improved mechanized printing processes. Lower newspaper prices and the practice of deriving revenue from advertisers encouraged significant growth beginning in the 1830s. Newspapers made the transition from the realm of the educated and wealthy elite to a mass medium serving a wider range of people from this period through the Civil War era (Huntzicker, 1999).

The Mexican and Civil Wars stimulated public demand for news by the mid-1800s, and modern journalism practices, such as assigning reporters to cover specific stories and topics, began to emerge. Circulation wars among big city newspapers in the 1880s featured sensational writing about outrageous stories. Both the number of newspaper firms and newspaper circulation began to soar. Although the number of firms would level off in the 20th century, circulation continued to rise.

Figure 2.A

Communication Technology Timeline

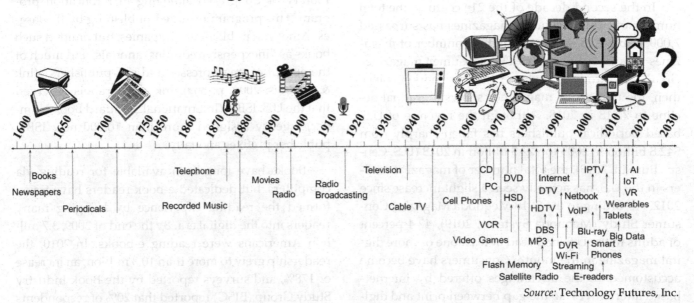

Source: Technology Futures, Inc.

The newspaper industry shrunk as we entered the 21st century when new technologies such as the internet became popular outlets for advertising. Newspaper publishing revenue declined from $29 billion in 2010 to $23 billion in 2017 (U.S. Census Bureau, 2016). In the meantime, percentage of revenue generated from online newspapers rose from 5.6% to 15.2%. (U.S. Census Bureau, 2016). Advertising as a source of revenue for newspaper publishers dropped by 27% from $20.4 billion to $14.9 billion (U.S. Census Bureau, 2017).

Periodicals

"The first colonial magazines appeared in Philadelphia in 1741, about 50 years after the first newspapers" (Campbell, 2002, p. 310). Few Americans could read in that era, and periodicals were costly to produce and circulate. Magazines were often subsidized and distributed by special interest groups, such as churches (Huntzicker, 1999). The *Saturday Evening Post*, the longest running magazine in U.S. history, began in 1821 and became the first magazine to both target women as an audience and to be distributed to a national audience. By 1850, nearly 600 magazines were operating.

By early in the 20th century, national magazines became popular with advertisers who wanted to reach wide audiences. No other medium offered such opportunity. However, by the middle of the century, many successful national magazines began dying in the face of advertiser preferences for the new medium of television and the increasing costs of periodical distribution. Magazines turned to smaller niche audiences that were more effectively targeted. Table 2.2, Figure 2.2, and Figure 2.3 on the companion website (www.tfi.com/ctu) show the number of American periodical titles by year, revealing that the number of new periodical titles nearly doubled from 1958 to 1960.

Single copy magazine sales were mired in a long period of decline in 2009 when circulation fell by 17.2%. However, subscription circulation fell by only 5.9%. In 2010, the Audit Bureau of Circulation reported that, among the 522 magazine titles monitored by the Bureau, the number of magazine titles in the United States fell by 8.7% (Agnese, 2011).

In 2010, 20,707 consumer magazines were published in North America, reaching a paid circulation of $8.8 billion. Subscriptions accounted for $6.2 billion (71%) of that circulation. During that year, 193 new North American magazines began publishing, but 176 magazines closed. Many print magazines were also available in digital form, and many had eliminated print circulation in favor of digital publishing. In 2009,

81 North American magazines moved online, but the number of additional magazines that went online in 2010 dropped to 28 (Agnese, 2011).

In the second decade of the 21st century, the total number of print consumer magazines has surpassed 7,000. Table 2.15 shows the highest number of magazines was in 2012, when 7,390 consumer magazines circulated throughout the American market. Since then, the number of magazines shows a gradual decline to 7,218 in 2018. Although the revenue of U.S. based periodical publishers has fallen sharply from $42.8 billion in 2005 to $22.3 billion in 2018 (U.S. Census Bureau, 2019), the total number of magazine readers in the U.S. has actually seen a slight increase since 2012. According to the 2018 Mequoda Magazine Consumer Study (Mequoda Systems, 2019), 42.4 percent of adults in the U.S. had read at least one or more digital magazine in the month. Consumers have become accustomed to the advantages offered by internet-based alternatives, so the gap between print and digital magazine has narrowed.

Books

Stephen Daye printed the first book in colonial America, *The Bay Psalm Book*, in 1640 (Campbell, 2002). Books remained relatively expensive and rare until after the printing process benefited from the industrial revolution. Linotype machines developed in the 1880s allowed for mechanical typesetting. After World War II, the popularity of paperback books helped the industry expand. The U.S. book publishing industry includes 87,000 publishers, most of which are small businesses. Many of these literally operate as "mom-and-pop desktop operations" (Peters & Donald, 2007, p. 11).

Table 2.3 and Figures 2.3 and 2.4 from the companion website (www.tfi.com/ctu) show new book titles published by year from the late 1800s through 2016. While times of war negatively affected the book industry, the number of book titles in the U.S. has been generally increasing with short-lived fluctuations like those in 1983-1986 and 1997-1999. The U.S. Bureau of the Census reports furnished data based on material from R. R. Bowker, which changed its reporting methods beginning with the 1998 report. Ink and Grabois (2000) explained the increase as resulting from the change in the method of counting titles "that results in a more accurate portrayal of the current state of American book publishing" (p. 508). The older counting process included only books included by the Library of Congress Cataloging in Publication program. This program included publishing by the largest American publishing companies, but omitted such books as "inexpensive editions, annuals, and much of the output of small presses and self-publishers" (Ink & Grabois, 2000, p. 509). Ink and Grabois observed that the U.S. ISBN (International Standard Book Number) Agency assigned more than 10,000 new ISBN publisher prefixes annually.

Books have long been available for reading via computers, but dedicated e-book readers have transformed the reading experience by bringing many readers into the digital era. By the end of 2009, 3.7 million Americans were reading e-books. In 2010, the readership grew to more than 10.3 million, an increase of 178%, and surveys reported by the Book Industry Study Group (BISG) reported that 20% of respondents had stopped buying printed books in favor of e-books within a year. By July 2010, Amazon reported that sales of e-books surpassed that of print hardcover sales for the first time, with "143 e-books sold for every 100 print hardcover books" (Dillon, 2011, p. 5). From mid-December 2011 through January 2012, the proportion of Americans owning both e-book readers and tablet computers nearly doubled from 10% to 19%, with 29% owning at least one of the devices (Rainie, 2012). In January 2014, the e-book penetration rate in the U.S. reached 32% (Pew Research Center, 2014), and 51% of U.S. households owned a tablet in April 2015 (Nielsen, 2015a). However, e-book sales revenue in the United States reached its peak in 2014 ($1.6 billion) and continued to drop in 2015 ($1.4 billion) and 2016 ($1.1 billion) (Association of American Publishers, 2017).

The Electronic Era

The telegraph transitioned from the print era to a new period by introducing a means of sending messages far more rapidly than was previously possible. Soon, Americans and people around the world enjoyed a world enhanced by such electronic media as wired telephones, motion pictures, audio recording, radio, television, cable television, and satellite television.

Telephone

With the telephone, Alexander Graham Bell became the first to transmit speech electronically in 1876. By June 30, 1877, 230 telephones were in use, and the number rose to 1,300 by the end of August, mostly to avoid the need for a skilled interpreter of telegraph messages. The first switching office connected three company offices in Boston beginning on May 17, 1877, reflecting a focus on business rather than residential use during the telephone's early decades. Hotels became early adopters of telephones as they sought to reduce the costs of employing human messengers, and New York's 100 largest hotels had 21,000 telephones by 1909. After 1894, non-business telephone use became common, in part because business use lowered the cost of telephone service. By 1902, 2,315,000 telephones were in service in the United States (Aronson, 1977). Table 2.4 and Figure 2.4 on the companion website (www.tfi.com/ctu) document the growth to near ubiquity of telephones in U.S. households and the expanding presence of wireless telephones.

Wireless Telephones

Guglielmo Marconi sent the first wireless data messages in 1895. The growing popularity of telephony led many to experiment with Marconi's radio technology as another means for interpersonal communication. By the 1920s, Detroit police cars had mobile radiophones for voice communication (ITU, 1999). The Bell system offered radio telephone service in 1946 in St. Louis, the first of 25 cities to receive the service. Bell engineers divided reception areas into cells in 1947, but cellular telephones that switched effectively among cells as callers moved did not arrive until the 1970s. The first call on a portable, handheld cell phone occurred in 1973. However, in 1981, only 24 people in New York City could use their mobile phones at the same time, and only 700 customers could have active contracts. To increase the number of people who could receive service, the Federal Communications Commission (FCC) began offering cellular telephone system licenses by lottery in June 1982 (Murray, 2001). Other countries, such as Japan in 1979 and Saudi Arabia in 1982, operated cellular systems earlier than the United States (ITU, 1999).

The U.S. Congress promoted a more advanced group of mobile communication services in 1993 by creating a classification that became known as Commercial Mobile Radio Service. This classification allowed for consistent regulatory oversight of these technologies and encouraged commercial competition among providers (FCC, 2005). By the end of 1996, about 44 million Americans subscribed to wireless telephone services (U.S. Bureau of the Census, 2008).

The new century brought an explosion of wireless telephones, and phones morphed into multipurpose devices (i.e., smartphones) with capabilities previously limited to computers. By the end of 2016, wireless phone penetration in the United States reached 395.9 million subscribers (CTIA, 2017), it further trended up to 417.5 million by 2017 (FierceWireless, 2017). *CTIA-The Wireless Association* (CTIA, 2017) reported that more than half (50.8%) of all American households were wireless-only by the end of 2016, up from about 10% in 2006. By 2017, worldwide shipments of smartphones exceeded 1.5 billion units (IDC, 2017), five times the quantity shipped in 2010 (IDC as cited by Amobi, 2013). However, in 2018 and 2019, IDC reported fewer shipments at around 1.4 billion units (IDC, 2019). IDC's (2015a) prediction about the maturation of the Chinese smartphone market and its impact brought to the global shipment may be true for the years to come.

Smartphones have long gone beyond their original functions of helping people to make calls while on the move. Wireless communication technology standards such as 3G (which facilitated internet surfing), 4G (which facilitated streaming) and 5G (Internet of Things) have, in their own time, changed the role of wireless phones in our society and in people's daily lives. With 5G becoming commercially applied in 2019, wireless phones can continue to be a shaping force of human beings' social and political environment.

Motion Pictures

In the 1890s, George Eastman improved on work by and patents purchased from Hannibal Goodwin in 1889 to produce workable motion picture film. The Lumière brothers projected moving pictures in a Paris café in 1895, hosting 2,500 people nightly at their movies. William Dickson, an assistant to Thomas Edison, developed the kinetograph, an early motion picture camera, and the kinetoscope, a motion picture viewing system. A New York movie house opened in 1894, offering moviegoers several coin-fed kinetoscopes. Edison's Vitascope, which expanded the length of films over those shown via kinetoscopes and allowed larger audiences to simultaneously see the moving images, appeared in public for the first time in 1896. In France in that same year, Georges Méliès started the first motion picture theater. Short movies became part of public entertainment in a variety of American venues by 1900 (Campbell, 2002), and average weekly movie attendance reached 40 million people by 1922.

Average weekly motion picture theater attendance, as shown in Table 2.5 and Figure 2.6 on the companion website (www.tfi.com/ctu), increased annually from the earliest available census reports on the subject in 1922 until 1930. After falling dramatically during the Great Depression, attendance regained growth in 1934 and continued until 1937. Slight declines in the prewar years were followed by a period of strength and stability throughout the World War II years. After the end of the war, average weekly attendance reached its greatest heights: 90 million attendees weekly from 1946 through 1949. After the introduction of television, weekly attendance would never again reach these levels.

Although a brief period of leveling off occurred in the late 1950s and early 1960s, average weekly attendance continued to plummet until a small recovery began in 1972. This recovery signaled a period of relative stability that lasted into the 1990s. Through the last decade of the century, average weekly theatre attendance enjoyed small but steady gains.

Box office revenues, which declined generally for 20 years after the beginning of television, began a recovery in the late 1960s, then began to skyrocket in the 1970s. The explosion continued until after the turn of the new century. However, much of the increase in revenues came from increases in ticket prices and inflation, rather than from increased popularity of films with audiences, and total motion picture revenue from box office receipts declined during recent years, as studios realized revenues from television and videocassettes (U.S. Department of Commerce/International Trade Association, 2000).

As shown in Table 2.5 on the companion website (www.tfi.com/ctu), American movie fans spent an average of 12 hours per person per year from 1993 through 1997 going to theaters. That average stabilized through the first decade of the 21st century (U.S. Bureau of the Census, 2010), despite the growing popularity of watching movies at home with new digital tools. In 2011, movie rental companies were thriving, with *Netflix* boasting 25 million subscribers and *Redbox* having 32,000 rental kiosks in the United States (Amobi, 2011b). However, recent physical sales and rental of home entertainment content suffered from the rise of web streaming services and consumer behavior change (Digital Entertainment Group, 2017). *Redbox* kiosk rentals started to decline in 2013 ($1.97 billion) to $1.76 billion in 2015 (Outerwall, 2016).

The record-breaking success of *Avatar* in 2009 as a 3D motion picture triggered a spate of followers who tried to revive the technology that was a brief hit in the 1950s. *Avatar* earned more than $761 million at American box offices and nearly $2.8 billion worldwide.

In the United States, nearly 8,000 of 39,500 theater screens were set up for 3D by the end of 2010, half of them having been installed in that year. The ticket prices for 3D films ran 20–30% higher than that of 2D films, and 3D films comprised 20% of the new films released. Nevertheless, American audiences preferred subsequent 2D films to 3D competitors, although 3D response remained strong outside the United States, where 61% of the world's 22,000 3D screens were installed. In 2014, there were 64,905 3D screens worldwide, except for the Asian Pacific region (55% annual growth), the annual growth rates of 3D screen numbers have stabilized around 6–10% (HIS quoted in MPAA, 2015). Another factor in the lack of success of 3D in America might have been the trend toward viewing movies at home, often with digital playback. In 2010, home video purchases and rentals reached $18.8 billion in North America, compared with only $10.6 billion spent at theaters (Amobi, 2011b). U.S. home entertainment spending rose to $20.8 billion in 2017, with revenues in the physical market shrinking ($12 billion in 2016) and digital subscriptions to web streaming such as Netflix soaring (Digital Entertainment Group, 2018). In the meantime, the 2014 U.S. domestic box office slipped 1.4% to $10.44 billion (Nash

Information Services quoted in Willens, 2015) and remained at a similar level for the next three years (Digital Entertainment Group, 2017). The MPAA's (2019) survey showed that only 35% of the total motion picture revenue comes from offline box office, 22% comes from rental or purchase and the other 43% comes from streaming distribution.

Globally, the Asian Pacific region and Latin America have been the main contributors to global box office revenue since 2004 (MPAA, 2017). And the bloom of the Chinese movie market has been a major reason for the increase of global revenue. Chinese box office revenue continued to soar in recent years, even when other global sectors remained comparatively flat or trended downwards (MPAA, 2019).

Audio Recording

Thomas Edison expanded on experiments from the 1850s by Leon Scott de Martinville to produce a talking machine or phonograph in 1877 that played back sound recordings from etchings in tin foil. Edison later replaced the foil with wax. In the 1880s, Emile Berliner created the first flat records from metal and shellac designed to play on his gramophone, providing mass production of recordings. The early standard recordings played at 78 revolutions per minute (rpm). After shellac became a scarce commodity because of World War II, records were manufactured from polyvinyl plastic. In 1948, CBS Records produced the long-playing record that turned at 33-1/3 rpm, extending the playing time from three to four minutes to 10 minutes. *RCA* countered in 1949 with 45 rpm records that were incompatible with machines that played other formats. After a five-year war of formats, record players were manufactured that would play recordings at all of the speeds (Campbell, 2002).

The Germans used plastic magnetic tape for sound recording during World War II. After the Americans confiscated some of the tapes, the technology was adopted and improved, becoming a boon for Western audio editing and multiple track recordings that played on bulky reel-to-reel machines. By the 1960s, the reels were encased in plastic cases, variously known as 8-track tapes and compact audio cassettes, which would prove to be deadly competition in the 1970s for single song records playing at 45 rpm

and long-playing albums playing at 33-1/3 rpm. Thomas Stockholm began recording sound digitally in the 1970s, and the introduction of compact disc (CD) recordings in 1983 decimated the sales performance of earlier analog media types (Campbell, 2002). Tables 2.6 and 2.6A and Figures 2.7 and 2.7A on the companion website (www.tfi.com/ctu) show that total unit sales of recorded music generally increased from the early 1970s through 2004 and kept declining after that mainly because of the rapid decline in CD sales. Figure 2.7a shows trends in downloaded music.

The 21st century saw an explosion in new digital delivery systems for music. Digital audio players, which had their first limited popularity in 1998 (Beaumont, 2008), hit a new gear of growth with the 2001 introduction of the Apple iPod, which increased the storage capacity and became responsible for about 19% of music sales within its first decade. Apple's online iTunes store followed in 2003, soon becoming the world's largest music seller (Amobi, 2009). However, as shown in Table 2.6A, notwithstanding with the drop in CD sales, a new upward trend was prompted by the emergence of paid online music subscriptions, sound exchange, synchronization, etc. Revenue-wise, paid subscriptions generated $2.26 billion, while CD revenue declined to $1.17 billion. Again, we are witnessing another "technological dynasty" after the "CD dynasty", the "cassette dynasty" and the "LP/EP dynasty" of recorded music (RIAA, 2019).

Radio

Guglielmo Marconi's wireless messages in 1895 on his father's estate led to his establishing a British company to profit from ship-to-ship and ship-to-shore messaging. He formed a U.S. subsidiary in 1899 that would become the American Marconi Company. Reginald A. Fessenden and Lee De Forest independently transmitted voice by means of wireless radio in 1906, and a radio broadcast from the stage of a performance by Enrico Caruso occurred in 1910. Various U.S. companies and Marconi's British company owned important patents that were necessary to the development of the infant industry, so the U.S. firms, including AT&T formed the Radio Corporation of America (*RCA*) to buy the patent rights from Marconi.

The debate still rages over the question of who became the first broadcaster among KDKA in Pittsburgh (Pennsylvania), WHA in Madison (Wisconsin), WWJ in Detroit (Michigan), and KQW in San Jose (California). In 1919, Dr. Frank Conrad of Westinghouse broadcast music from his phonograph in his garage in East Pittsburgh. Westinghouse's KDKA in Pittsburgh announced the presidential election returns over the airwaves on November 2, 1920. By January 1, 1922, the Secretary of Commerce had issued 30 broadcast licenses, and the number of licensees swelled to 556 by early 1923. By 1924, RCA owned a station in New York, and Westinghouse expanded to Chicago, Philadelphia, and Boston. In 1922, AT&T withdrew from RCA and started WEAF in New York, the first radio station supported by commercials. In 1923, AT&T linked WEAF with WNAC in Boston by the company's telephone lines for a simultaneous program. This began the first network, which grew to 26 stations by 1925. RCA linked its stations with telegraph lines, which failed to match the voice quality of the transmissions of AT&T. However, AT&T wanted out of the new business and sold WEAF in 1926 to the National Broadcasting Company, a subsidiary of RCA (White, 1971).

The 1930 penetration of radio sets in American households reached 40%, then approximately doubled over the next 10 years, passing 90% by 1947 (Brown, 2006). Table 2.7 and Figure 2.8, on the companion website (www.tfi.com/ctu), show the rapid rate of increase in the number of radio households from 1922 through the early 1980s, when the rate of increase declined. The increases resumed until 1993, when they began to level off.

Although thousands of radio stations were transmitting via the internet by 2000, Channel1031.com became the first station to cease using FM and move exclusively to the internet in September 2000 (Raphael, 2000). Many other stations were operating only on the internet when questions about fees for commercial performers and royalties for music played on the web arose. In 2002, the Librarian of Congress set royalty rates for internet transmissions of sound recordings (U.S. Copyright Office, 2003). A federal court upheld the right of the Copyright Office to establish fees on streaming music over the internet (*Bonneville v. Peters*, 2001).

In March 2001, the first two American digital audio satellites were launched, offering the promise of hundreds of satellite radio channels (Associated Press, 2001). Consumers were expected to pay about $9.95 per month for access to commercial-free programming that would be targeted to automobile receivers. The system included amplification from about 1,300 ground antennas. By the end of 2003, about 1.6 million satellite radio subscribers tuned to the two top providers, XM and Sirius (Schaeffler, 2004). These two players merged soon before the 2008 stock market crisis, during which the new company, Sirius XM Radio, lost nearly all of its stock value. In 2011, the service was used by 20.5 million subscribers, with its market value beginning to recover (Sirius XM Radio, 2011). The company continues to attract new subscribers and reported its highest subscriber growth since 2007 in 2015 with a 30% growth rate to 29.6 million subscribers; this number continued to grow to more than 34.6 million in 2019 (Sirius XM Radio, 2016; 2018; 2019).

Television

Paul Nipkow invented a scanning disk device in the 1880s that provided the basis from which other inventions would develop into television. In 1927, Philo Farnsworth became the first to electronically transmit a picture over the air. Fittingly, he transmitted the image of a dollar sign. In 1930, he received a patent for the first electronic television, one of many patents for which RCA would be forced, after court challenges, to negotiate. By 1932, Vladimir Zworykin discovered a means of converting light rays into electronic signals that could be transmitted and reconstructed at a receiving device. RCA offered the first public demonstration of television at the 1939 World's Fair.

The FCC designated 13 channels in 1941 for use in transmitting black-and-white television, and the commission issued almost 100 television station broadcasting licenses before placing a freeze on new licenses in 1948. The freeze offered time to settle technical issues, and it ran longer because of U.S. involvement in the Korean War (Campbell, 2002). As shown in Table 2.8 on the companion website (www.tfi.com/ctu), nearly 4 million households had television sets by 1950, a 9% penetration rate that would escalate to 87% a decade later. Penetration has remained steady at about 98% since

1980 until a recent small slide to about 96% in 2014. Figure 2.8 illustrates the meteoric rise in the number of households with television by year from 1946 through 2015. In 2010, 288.5 million Americans had televisions, up by 0.8% from 2009, and average monthly time spent viewing reached 158 hours and 47 minutes, an increase of 0.2% from the previous year (Amobi, 2011a). In 2015, Nielsen estimated that 296 million persons age 2 and older lived in 116.3 million homes that have TV, which showed that "Both the universe of U.S. television homes and the potential TV audience in those homes continue to grow" (Nielsen, 2015b). In 2014, Americans were spending an average of 141 hours per month watching TV and paid more attention to online video streaming and Over-The-Top TV service such as Netflix (IDATE quoted in Statista, 2016).

By the 1980s, Japanese high-definition television (HDTV) increased the potential resolution to more than 1,100 lines of data in a television picture. This increase enabled a much higher-quality image to be transmitted with less electromagnetic spectrum space per signal. In 1996, the FCC approved a digital television transmission standard and authorized broadcast television stations a second channel for a 10-year period to allow the transition to HDTV. As discussed in Chapter 6, that transition made all older analog television sets obsolete because they cannot process HDTV signals (Campbell, 2002).

The FCC (2002) initially set May 2002 as the deadline by which all U.S. commercial television broadcasters were required to be broadcasting digital television signals. Progress toward digital television broadcasting fell short of FCC requirements that all affiliates of the top four networks in the top 10 markets transmit digital signals by May 1, 1999.

Within the 10 largest television markets, all except one network affiliate had begun HDTV broadcasts by August 1, 2001. By that date, 83% of American television stations had received construction permits for HDTV facilities or a license to broadcast HDTV signals (FCC, 2002). HDTV penetration into the home marketplace would remain slow for the first few years of the 21st century, in part because of the high price of the television sets.

Although 3D television sets were available in 2010, little sales success occurred. The sets were quite expensive, not much 3D television content was available, and the required 3D viewing glasses were inconvenient to wear (Amobi, 2011b).

During the fall 2011-12 television season, The Nielsen Company reported that the number of households with televisions in the United States dropped for the first time since the company began such monitoring in the 1970s. The decline to 114.7 million from 115.9 million television households represented a 2.2% decline, leaving the television penetration at 96.7%. Explanations for the reversal of the long-running trend included the economic recession, but the decline could represent a transition to digital access in which viewers were getting TV from devices other than television sets (Wallenstein, 2011). But the number of television households has since increased again, with Nielsen reporting 116.3 million U.S. television households in 2015 (Nielsen, 2015b).

Cable Television

Cable television began as a means to overcome poor reception for broadcast television signals. John Watson claimed to have developed a master antenna system in 1948, but his records were lost in a fire. Robert J. Tarlton of Lansford (Pennsylvania) and Ed Parsons of Astoria (Oregon) set up working systems in 1949 that used a single antenna to receive programming over the air and distribute it via coaxial cable to multiple users (Baldwin & McVoy, 1983). At first, the FCC chose not to regulate cable, but after the new medium appeared to offer a threat to broadcasters, cable became the focus of heavy government regulation. Under the Reagan administration, attitudes swung toward deregulation, and cable began to flourish. Table 2.9 and Figure 2.9 on the companion website (www.tfi.com/ctu) show the growth of cable systems and subscribers, with penetration remaining below 25% until 1981, but passing the 50% mark before the 1980s ended.

In the first decade of the 21st century, cable customers began receiving options including internet access, digital video, video on demand, DVRs, HDTV, and telephone services. By fall 2010, 3.2 million (9%) televisions were connected to the internet (Amobi, 2011b). The success of digital cable led to the FCC decision to eliminate analog broadcast television as of

February 17, 2009. However, in September 2007, the FCC unanimously required cable television operators to continue to provide carriage of local television stations that demand it in both analog and digital formats for three years after the conversion date. This action was designed to provide uninterrupted local station service to all cable television subscribers, protecting the 40 million (35%) U.S. households that remained analog-only (Amobi & Kolb, 2007).

Telephone service became widespread via cable during the early years of the 21st century. For years, some cable television operators offered circuit-switched telephone service, attracting 3.6 million subscribers by the end of 2004. Also by that time, the industry offered telephone services via voice over internet protocol (VoIP) to 38% of cable households, attracting 600,000 subscribers. That number grew to 1.2 million by July 2005 (Amobi, 2005).

The growth of digital cable in the first decade of the new century also saw the growth of video-on-demand (VOD), offering cable television customers the ability to order programming for immediate viewing.

Cable penetration declined in the United States after 2000, as illustrated in Figure 2.9 on the companion website (www.tfi.com/ctu). However, estimated use of a combination of cable and satellite television increased steadily over the same period (U.S. Bureau of the Census, 2008).

Worldwide, pay television flourished in the new century, especially in the digital market. From 2009 to 2014, pay TV subscriptions increased from 648 million households to 923.5 million households. (Statista, 2016). This increase was and is expected to be led by rapid growth in the Asia Pacific region and moderate growth in Latin America and Africa, while pay TV subscription slowly declined in North America and Eastern Europe (Digital TV Research, 2017). The number of pay television subscribers in the U.S. was falling from more than 95 million in 2012 to 94.2 million in 2015 (Dreier, 2016). This figure continued to drop as major pay-television providers lost about 1.7 million subscribers in the third quarter of 2019 (Leichtman Research Group, 2019). Statista (2019) also reported that number of Netflix subscribers may have already surpassed the number of subscribers to major U.S. cable providers for the first time in 2017.

Direct Broadcast Satellite & Other Cable TV Competitors

Satellite technology began in the 1940s, but HBO became the first service to use it for distributing entertainment content in 1976 when the company sent programming to its cable affiliates (Amobi & Kolb, 2007). Other networks soon followed this lead, and individual broadcast stations (WTBS, WGN, WWOR, and WPIX) used satellites in the 1970s to expand their audiences beyond their local markets by distributing their signals to cable operators around the U.S.

Competitors for the cable industry include a variety of technologies. Annual FCC reports distinguish between home satellite dish (HSD) and direct broadcast satellite (DBS) systems. Both are included as MVPDs (multi-channel video program distributors), which include cable television, wireless cable systems called multichannel multipoint distribution services (MMDS), and private cable systems called satellite master antenna television (SMATV). Table 2.10 and Figure 2.10 on the companion website for this book (www.tfi.com/ctu), show trends in home satellite dish, DBS, MMDS, and SMATV (or PCO, Private Cable Operator) subscribers. However, the FCC (2013a) noted that little public data was available for the dwindling services of HSD, MMDS, and PCO, citing SNL Kagan conclusions that those services accounted for less than 1% of MVPDs and were expected to continue declining over the coming decade.

In developed markets like the U.S., internet-based Over-the-Top TV services such as Netflix, Hulu, and SlingTV have grown substantially since 2015. SlingTV's subscriber totals grew from 169,000 in March 2015 to 523,000 by the end of 2015 (Ramachandran, 2016). The number of Netflix subscribers grew to 52.77 million in the U.S. by the third quarter of 2017 (Netflix, 2017) and its international subscribers are accumulating even faster, adding up to 109.25 million subscribers worldwide (Netflix, 2017). In the company's 2016 long term market view, Netflix reported: "People love TV content, but they don't love the linear TV experience, where channels present programs only at particular times on non-portable screens with complicated remote controls" (Netflix, 2016). It is possible that the concept of TV is again being redefined by people's need for cord-cutting and screen convergence. From 2014 to 2017, share of cord-cutters/nevers among all U.S. TV households grew from 18.8% to 24.6% (Convergence Consulting Group, 2017).

According to Nielsen, there were 119.9 million TV homes in the U.S. for the 2018-19 TV season. the percentage of total U.S. homes with televisions receiving traditional TV signals via broadcast, cable, DBS or Telco, or via a broadband internet connection connected to a TV set was 95.9% (Nielsen, 2018). Nielsen's Local Watch Report (Nielsen, 2019a) focused on the impact of streaming access to subscription video on-demand services and reported that 65% of homes had access to an internet-connected device or smart TV. Moreover, although the majority of U.S. homes had pay-TV service subscription, their adoption of other alternatives such as OTA (over-the-air) or SOVD (subscription video-on-demand) or vMVPD (virtual multichannel video programming distributor, a.k.a. "skinny bundles") seems to be on the rise, and, most importantly, are not mutually exclusive. According to Nielsen (2019b), U.S. TV owners are becoming accustomed to a self-prepared TV menu across technological platforms.

Home Video

Although VCRs became available to the public in the late 1970s, competing technical standards slowed the adoption of the new devices. After the longer taping capacity of the VHS format won greater public acceptance over the higher-quality images of the Betamax, the popularity of home recording and playback rapidly accelerated, as shown in Table 2.11 and Figure 2.11 on the companion website (www.tfi.com/ctu).

By 2004, rental spending for videotapes and DVDs reached $24.5 billion—far surpassing the $9.4 billion spent for tickets to motion pictures in that year. During 2005, DVD sales increased by 400% over the $4 billion figure for 2000 to $15.7 billion. However, the annual rate of growth reversed direction and slowed that year to 45% and again the following year to 2%. VHS sales amounted to less than $300 million in 2006 (Amobi & Donald, 2007).

Factors in the decline of VHS and DVD use included growth in cable and satellite video-on-demand

services, growth of broadband video availability, digital downloading of content, and the transition to DVD Blu-ray format (Amobi, 2009). The competing new formats for playing high-definition content was similar to the one waged in the early years of VCR development between the Betamax and VHS formats. Similarly, in early DVD player development, companies touting competing standards settled a dispute by agreeing to share royalties with the creator of the winning format. Until early 2008, the competition between proponents of the HD-DVD and Blu-ray formats for playing high-definition DVD content remained unresolved, and some studios were planning to distribute motion pictures in both formats. Blu-ray seemed to emerge the victor in 2008 when large companies (e.g., Time Warner, Walmart, Netflix) declared allegiance to that format. By July 2010, Blu-ray penetration reached 17% of American households (Gruenwedel, 2010), and 170 million Blu-ray discs shipped that year (Amobi, 2011b).

Digital video recorders (DVRs, also called personal video recorders, PVRs) debuted during 2000, and about 500,000 units were sold by the end of 2001 (FCC, 2002). The devices save video content on computer hard drives, allowing fast-forwarding, reversing, and pausing of live television; replay of limited minutes of previously displayed live television; automatic recording of all first-run episodes; automatic recording logs; and superior quality to that of analog VCRs. Multiple tuner models allow viewers to watch one program, while recording others simultaneously.

DVR providers generate additional revenues by charging households monthly fees, and satellite DVR households tend to be less likely to drop their satellite subscriptions. Perhaps the most fundamental importance of DVRs is the ability of consumers to make their own programming decisions about when and what they watch. This flexibility threatens the revenue base of network television in several ways, including empowering viewers to skip commercials. Advertiser responses included sponsorships and product placements within programming (Amobi, 2005).

Reflecting the popularity of the DVR, time shifting was practiced in 2010 by 107.1 million American households (up 13.2% from 2009). Time shifted viewing increased by 12.2% in 2010 from 2009 to an average of 10 hours and 46 minutes monthly (Amobi, 2011a). As shown in Table 2.11 on the companion website (www.tfi.com/ctu), its penetration rate also bounced back to 45.1% in 2014, however, it declined again in 2016 (Plastics Industry Association, 2017).

The Digital Era

The digital era represents a transition in modes of delivery of mediated content. Although the tools of using digital media may have changed, in many cases, the content remains remarkably stable. With other media, such as social media, the digital content fostered new modes of communicating. This section contains histories of the development of computers and the internet. Segments of earlier discussions could be considered part of the digital era, such as audio recording, HDTV, films on DVD, etc., but discussions of those segments remain under earlier eras.

Computers

The history of computing traces its origins back thousands of years to such practices as using bones as counters (Hofstra University, 2000). Intel introduced the first microprocessor in 1971. The MITS Altair, with an 8080 processor and 256 bytes of RAM (random access memory), sold for $498 in 1975, introducing the desktop computer to individuals. In 1977, Radio Shack offered the TRS80 home computer, and the Apple II set a new standard for personal computing, selling for $1,298. Other companies began introducing personal computers, and, by 1978, 212,000 personal computers were shipped for sale.

Early business adoption of computers served primarily to assist practices such as accounting. When computers became small enough to sit on office desktops in the 1980s, word processing became popular and fueled interest in home computers. With the growth of networking and the internet in the 1990s, both businesses and consumers began buying computers in large numbers. Computer shipments around the world grew annually by more than 20% between 1991 and 1995 (Kessler, 2007).

By 1997, the majority of American households with annual incomes greater than $50,000 owned a personal computer. At the time, those computers sold for about $2,000, exceeding the reach of lower income groups. By the late 1990s, prices dropped below $1,000 per system (Kessler, 2007), and penetration in American households passed 60% within a couple of years (U.S. Bureau of the Census, 2008).

Table 2.12 and Figure 2.12 on the companion website (www.tfi.com/ctu) trace the rapid and steady rise in American computer shipments and home penetration. After 2006, U.S. PC shipments started to decline and with penetration reaching 65.3% in 2015 (IDC, 2016). By 1998, 42.1% of American households owned personal computers (U.S. Bureau of the Census, 2006). After the start of the 21st century, personal computer prices declined, and penetration increased from 63% in 2000 to 77% in 2008 (Forrester Research as cited in Kessler, 2011). Worldwide personal computer sales increased by 34% from 287 million in 2008 to 385 million in 2011 (IDC as cited Kessler, 2011). However, this number suffered from a 3.2% decrease in 2012 (IDC, 2016). This downward trend continued, and in 2016 worldwide shipments were down to 260 million, lower than in 2008 (IDC, 2015b).

IDC (2017) also predicted that the downward trend would continue. However, while pointing out that shipments of desktops and slate tables would decline, it predicted that shipments of detachable tablets would increase from 21.5 million units in 2016 to 37.1 units in 2021. It is important to note that media convergence is present even with PCs, as an analyst from IDC pointed out: "a silver lining is that the industry has continued to refine the more mobile aspects of personal computers—contributing to higher growth in Convertible & Ultraslim Notebooks" (IDC, 2015c).

Internet

The internet began in the 1960s with ARPANET, or the Advanced Research Projects Agency (ARPA) network project, under the auspices of the U.S. Defense Department. The project intended to serve the military and researchers with multiple paths of linking computers together for sharing data in a system that would remain operational even when traditional communications might become unavailable. Early users—mostly university and research lab personnel—took advantage of electronic mail and posting information on computer bulletin boards. Usage increased dramatically in 1982 after the National Science Foundation (NSF) supported high-speed linkage of multiple locations around the United States. After the collapse of the Soviet Union in the late 1980s, military users abandoned ARPANET, but private users continued to use it, and multimedia transmissions of audio and video became possible once this content could be digitized. More than 150,000 regional computer networks and 95 million computer servers hosted data for internet users (Campbell, 2002).

Penetration & Usage

During the first decade of the 21st century, the internet became the primary reason that consumers purchased new computers (Kessler, 2007). Cable modems and digital subscriber line (DSL) telephone line connections increased among home users as the means for connecting to the internet, as more than half of American households accessed the internet with high-speed broadband connections.

Tables 2.13 and 2.14 and Figures 2.13 and 2.14 on the companion website (www.tfi.com/ctu) show trends in internet penetration in the United States. By 2008, 74 million (63%) American households had high-speed broadband access (Kessler, 2011). In June 2013, 91,342,000 fixed broadband subscriptions and 299,447,000 million American subscribers had wireless broadband subscriptions (OECD, 2013). In 2012, although the number of DSL+ users decreased, the numbers on other fronts are showing growth (FCC, 2013b). In 2013, however, subscription to all types of internet connections declined, and resulted in a 6% decline in total penetration (FCC, 2014).The penetration gradually climbed up to 76.2% by 2016(ITU, 2017).

Immersive Virtual Environments

Immersive Virtual Environments (IVEs) are more famously known or associated with virtual reality (VR) technologies (Biocca & Delaney, 1995). The creation of IVEs can be traced back to the 1930s when the first sensorama machine was built to provide immersive movie experiences (Jones & Dawkins, 2018). In 1989, NASA used an IVE to train astronauts (Cater & Huffman, 1995). Later, game companies such as Sega and Nintendo also created head-mounted-displays (HMDs) to enhance gaming experience (Rebenitsch, 2015).

Today, IVEs are created with an unprecedented verisimilitude: the lighter, clearer HMDs and faster internet connection worldwide facilitated this process. Consequentially, IVEs are now able to provide strong sense of physical and social presence to the user. People in IVEs can now not only communicate with the environment (e.g. a virtual tour), but also with other simulated or avatar characters inside the IVEs (e.g. virtual life gaming or first-person 360-degree movies). On the popular market, IVEs are also usually delivered with HMDs instead of a whole room of equipment in the past.

With the continuous advancement of technologies related to IVEs, the communication media which usually exists outside of our physical body is perhaps evolving into the next level: that we become a part of the medium with which we communicate with others. Almost 10% of all Americans engage with IVEs once a month in 2018 (eMarketer, 2019), and global shipments have been growing steadily since 2017. However, because of their relative novelty, the future of IVEs is uncertain.

Synthesis

Although this chapter has emphasized the importance of quantitative records of technology adoption, any understanding and interpretation of these numbers should consider the social contexts and structures in each historical period.

Early visions of the internet (see Chapter 21) did not include the emphasis on entertainment and information to the general public that has emerged. The combination of this new medium with older media belongs to a phenomenon called *convergence*, referring to the merging of functions of old and new media (Grant, 2009). By 2002, the FCC (2002) reported that the most important type of convergence related to video content is the joining of internet services. The report also noted that companies from many business areas were providing a variety of video, data, and other communications services.

The word *convergence* itself excludes the idea that technologies are monolithic constructs that symbolize the separation between two or more types of products. Rather, we see that older media, including both print and electronic types, have gone through transitions from their original forms into digital media.

In this way, many of the core assumptions and social roles of older media have converged with new desires and imaginations of society, and then take form in new media (Castells, 2011). From a media ecology perspective, media convergence at a given time in history symbolizes the era itself—the medium can be the message.

For example, print media, with the help of a complex writing system, largely protected the centrality and authority of government power in ancient China. When government announcements were replaced by folklore announcements (i.e. the press), the power of the government was impacted. Telephone and radio appeared in the 19ᵗʰ century, and they precisely fit into the Western social philosophy of that time: searching for certainty, control and communicative linearity. Nowadays, we are returning to the "tribal" methods of communication: that everywhere is a source and nowhere is an ending target. (McLuhan, 1994)

Just as media forms began converging nearly a century ago when radios and record players merged in the same appliance, media in recent years have been converging at a more rapid pace. As the popularity of print media generally declined throughout the 1990s, the popularity of the internet grew rapidly, particularly with the increase in high-speed broadband connections, for which adoption rates achieved comparability with previous new communications media. Consumer flexibility through the use of digital media became the dominant media consumption theme during the first decade of the new century.

Bibliography

Agnese, J. (2011, October 20). Industry surveys: Publishing & advertising. In E. M. Bossong-Martines (Ed.), *Standard & Poor's industrysSurveys, Vol. 2.*

Amobi, T. N. (2005, December 8). Industry surveys: Broadcasting, cable, and satellite industry survey. In E. M. Bossong-Martines (Ed.), *Standard & Poor's industry surveys, 173* (49), Section 2.

Amobi, T. N. (2009). Industry surveys: Movies and home entertainment. In E. M. Bossong-Martines (Ed.), *Standard & Poor's industry surveys, 177* (38), Section 2.

Amobi, T. N. (2011a). Industry surveys: Broadcasting, cable, and satellite. In E. M. Bossong-Martines (Ed.), *Standard & Poor's industry surveys, Vol. 1.*

Amobi, T. N. (2011b). Industry surveys: Movies & entertainment. In E. M. Bossong-Martines (Ed.), *Standard & Poor's industry surveys, Vol. 2.*

Amobi, T. N. (2013). Industry surveys: Broadcasting, cable, & satellite. In E. M. Bossong-Martines (Ed.), *Standard & Poor's industry surveys, Vol. 1.*

Amobi, T. N. & Donald, W. H. (2007, September 20). Industry surveys: Movies and home entertainment. In E. M. Bossong-Martines (Ed.), *Standard & Poor's industry surveys, 175* (38), Section 2.

Amobi, T. N. & Kolb, E. (2007, December 13). Industry surveys: Broadcasting, cable & satellite. In E. M. Bossong-Martines (Ed.), *Standard & Poor's industry surveys, 175* (50), Section 1.

Aronson, S. (1977). Bell's electrical toy: What's the use? The sociology of early telephone usage. In I. Pool (Ed.). *The social impact of the telephone.* Cambridge, MA: The MIT Press, 15-39.

Association of American Publishers. (2017). Publisher Revenue for Trade Books Up 10.2% in November 2016. http://newsroom.publishers.org/publisher-revenue-for-trade-books-up-102-in-november-2016/.

Associated Press. (2001, March 20). Audio satellite launched into orbit. *New York Times.* http://www.nytimes.com/aponline/national/AP-Satellite-Radio.html?ex=986113045& ei=1&en=7af33c7805ed8853.

Baldwin, T. & McVoy, D. (1983). *Cable communication.* Englewood Cliffs, NJ: Prentice-Hall.

Beaumont, C. (2008, May 10). Dancing to the digital tune As the MP3 turns 10. *The Daily Telegraph,* p. 19. LexisNexis Database.

Biocca, F., & Delaney, B. (1995). Immersive virtual reality technology. In F. Biocca & Mark. R. Levy (Eds.), *Communication in the age of virtual reality* (Vol. 15, pp. 57–124). Lawrence Erlbaum Associates.

Bonneville International Corp., et al. v. Marybeth Peters, as Register of Copyrights, et al. Civ. No. 01-0408, 153 F. Supp.2d 763 (E.D. Pa., August 1, 2001).

Brown, D. & Bryant, J. (1989). An annotated statistical abstract of communications media in the United States. In J. Salvaggio & J. Bryant (Eds.), *Media use in the information age: Emerging patterns of adoption and consumer use.* Hillsdale, NJ: Lawrence Erlbaum Associates, 259-302.

Brown, D. (1996). A statistical update of selected American communications media. In Grant, A. E. (Ed.), *Communication Technology Update* (5th ed.). Boston, MA: Focal Press, 327-355.

Brown, D. (1998). Trends in selected U. S. communications media. In Grant, A. E. & Meadows, J. H. (Eds.), Communication Technology Update (6th ed.). Boston, MA: Focal Press, 279-305.

Brown, D. (2000). Trends in selected U. S. communications media. In Grant, A. E. & Meadows, J. H. (Eds.), *Communication Technology Update* (7th ed.). Boston, MA: Focal Press, 299-324.

Brown, D. (2002). Communication technology timeline. In A. E. Grant & J. H. Meadows (Eds.), *Communication technology update* (8th ed.) Boston: Focal Press, 7-45.

Brown, D. (2004). Communication technology timeline. In A. E. Grant & J. H. Meadows (Eds.). *Communication technology update* (9th ed.). Boston: Focal Press, 7-46.

Brown, D. (2006). Communication technology timeline. In A. E. Grant & J. H. Meadows (Eds.), *Communication technology update* (10th ed.). Boston: Focal Press. 7-46.

Brown, D. (2008). Historical perspectives on communication technology. In A. E. Grant & J. H. Meadows (Eds.), *Communication technology update and fundamentals* (11th ed.). Boston: Focal Press. 11-42.

Brown, D. (2010). Historical perspectives on communication technology. In A. E. Grant & J. H. Meadows (Eds.), *Communication technology update and fundamentals* (12th ed.). Boston: Focal Press. 9-46.

Brown, D. (2012). Historical perspectives on communication technology. In A. E. Grant & J. H. Meadows (Eds.), *Communication technology update and fundamentals* (13th ed.). Boston: Focal Press. 9-24.

Brown, D. (2014). Historical perspectives on communication technology. In A. E. Grant & J. H. Meadows (Eds.), *Communication technology update and fundamentals* (14th ed.). Boston: Focal Press. 9-20.

Campbell, R. (2002). *Media & culture.* Boston, MA: Bedford/St. Martins.

Castells, M. (2011). The rise of the network society: The information age: Economy, society, and culture (Vol. 1). John Wiley & Sons.

Cater, J. P., & Huffman, S. D. (1995). Use of the Remote Access Virtual Environment Network (RAVEN) for Coordinated IVA—EVA Astronaut Training and Evaluation. *Presence: Teleoperators & Virtual Environments*, 4(2), 103–109.

Cicero, M. T. (1876). *Orator ad M. Brutum.* BG Teubner.

Convergence Consulting Group. (2017). The Battle for the North American (U.S./Canada) Couch Potato: Online & Traditional TV and Movie Distribution. http://www.convergenceonline.com/downloads/NewContent2017.pdf?lbisphpreq=1.

CTIA. (2017). Annual Wireless Industry Survey. https://www.ctia.org/industry-data/ctia-annual-wireless-industry-survey.

Digital Entertainment Group. (2017). DEG Year End 2016 Home Entertainment Report. http://degonline.org/portfolio_page/deg-year-end-2016-home-entertainment-report/.

Digital Entertainment Group. (2018). DEG Year End 2017 Home Entertainment Report. http://degonline.org/portfolio_page/deg-year-end-2017-home-entertainment-report/.

Digital TV Research. (2017). Digital TV Research's July 2017 newsletter. https://www.digitaltvresearch.com/press-releases?id=204.

Dillon, D. (2011). E-books pose major challenge for publishers, libraries. In D. Bogart (Ed.), *Library and book trade almanac* (pp. 3-16). Medford, NJ: Information Today, Inc.

Dreier, T. (2016, March 11). Pay TV industry losses increase to 385,000 subscribers in 2015. http://www.streamingmedia.com/Articles/Editorial/Featured-Articles/Pay-TV-Industry-Losses-Increase-to-385000-Subscribers-in-2015-109711.aspx.

eMarketer. (2019, March 27). Virtual and Augmented Reality Users 2019. https://www.emarketer.com/content/virtual-and-augmented-reality-users-2019.

Federal Communications Commission. (2002, January 14). *In the matter of annual assessment of the status of competition in the market for the delivery of video programming* (eighth annual report). CS Docket No. 01-129. Washington, DC 20554.

Federal Communications Commission. (2005). *In the matter of Implementation of Section 6002(b) of the Omnibus Budget Reconciliation Act of 1993: Annual report and analysis of competitive market conditions with respect to commercial mobile services* (10th report). WT Docket No. 05-71. Washington, DC 20554.

Federal Communications Commission. (2013a, July 22). *In the matter of annual assessment of the status of competition in the market for the delivery of video programming* (fifteenth annual report). MB Docket No. 12-203. Washington, DC 20554.

Federal Communications Commission. (2013b, December). Internet access services: Status as of December 31, 2012. http://hraunfoss.fcc.gov/edocs_public/attachmatch/DOC-324884A1.pdf.

Federal Communications Commission. (2014, October). Internet access services: Status as of December 31, 2013. https://transition.fcc.gov/Daily_Releases/Daily_Business/2014/db1016/DOC-329973A1.pdf.

FierceWireless. (2017). How Verizon, AT&T, T-Mobile, Sprint and more stacked up in Q3 2017: The top 7 carriers. https://www.fiercewireless.com/wireless/how-verizon-at-t-t-mobile-sprint-and-more-stacked-up-q3-2017-top-7-carriers.

Grant, A. E. (2009). Introduction: Dimensions of media convergence. In Grant, A E. and Wilkinson, J. S. (Eds.) *Understanding media convergence: The state of the field.* New York: Oxford University Press.

Gruenwedel, E. (2010). Report: Blu's household penetration reaches 17%. *Home Media Magazine, 32,* 40.

Hofstra University. (2000). *Chronology of computing history.* http://www.hofstra.edu/pdf/ CompHist_9812tla1.pdf.

Huntzicker, W. (1999). *The popular press, 1833-1865.* Westport, CT: Greenwood Press.

Ink, G. & Grabois, A. (2000). *Book title output and average prices: 1998 final and 1999 preliminary figures,* 45th edition. D. Bogart (Ed.). New Providence, NJ: R. R. Bowker, 508-513.

International Data Corporation. (2015a). *Worldwide Smartphone Market Will See the First Single-Digit Growth Year on Record, According to IDC* [Press release]. http://www.idc.com/getdoc.jsp?containerId=prUS40664915.

International Data Corporation. (2015b). *Worldwide PC Shipments Will Continue to Decline into 2016 as the Short-Term Outlook Softens, According to IDC* [Press release]. http://www.idc.com/getdoc.jsp?containerId=prUS40704015.

International Data Corporation. (2015c). *PC Shipments Expected to Shrink Through 2016 as Currency Devaluations and Inventory Constraints Worsens Outlook, According to IDC* [Press release]. http://www.idc.com/getdoc.jsp?containerId=prUS25866615.

International Data Corporation. (2016). *United States Quarterly PC Tracker.* http://www.idc.com/tracker/showproductinfo.jsp?prod_id=141.

International Data Corporation. (2017). Worldwide Quarterly Mobile Phone Tracker. https://www.idc.com/tracker/showproductinfo.jsp?prod_id=37.

International Data Corporation. (2019). Worldwide Quarterly Mobile Phone Tracker. https://www.idc.com/tracker/showproductinfo.jsp?prod_id=37.

International Telecommunications Union. (1999). *World telecommunications report 1999.* Geneva, Switzerland: Author.

ITU. (2017). Percentage of Individuals using the Internet. http://www.itu.int/en/ITU-D/Statistics/Documents/statistics/2017/Individuals_Internet_2000-2016.xls.

Jones, S., & Dawkins, S. (2018). The sensorama revisited: Evaluating the application of multi-sensory input on the sense of presence in 360-degree immersive film in virtual reality. In *Augmented reality and virtual reality* (pp. 183–197). Springer.

Kessler, S. H. (2007, August 26). Industry surveys: Computers: Hardware. In E M. Bossong-Martines (Ed.), *Standard & Poor's industry surveys, 175* (17), Section 2.

Kessler, S. H. (2011). Industry surveys: Computers: Consumer services and the Internet. In E. M. Bossong-Martines (Ed.), *Standard & Poor's industry surveys, Vol. 1.*

Lee, J. (1917). *History of American journalism.* Boston: Houghton Mifflin.

Leichtman Research Group. (2019, November 13). *Major Pay-TV Providers Lost About 1,740,000 Subscribers in 3Q 2019* [Press release]. https://www.leichtmanresearch.com/major-pay-tv-providers-lost-about-1740000-subscribers-in-3q-2019/

McLuhan, M.A. (1994). *Understanding media: The extensions of man.* MIT press.

Mequoda Systems. (2019) 2018 Mequoda Magazine Consumer Study. https://www.mequoda.com/free-reports/digital-magazines-report/

MPAA. (2015). *2014 Theatrical Market Statistics* [Press release]. http://www.mpaa.org/wp-content/uploads/2015/03/MPAA-Theatrical-Market-Statistics-2014.pdf.

MPAA. (2017). *2016 Theatrical Market Statistics* [Press release]. https://www.mpaa.org/wp-content/uploads/2017/03/2016-Theatrical-Market-Statistics-Report-2.pdf.

MPAA. (2019) Theme Report: A Comprehensive analysis and survey of the theatrical and home entertainment market environment (THEME) for 2018. https://www.motionpictures.org/wp-content/uploads/2019/03/MPAA-THEME-Report-2018.pdf

Murray, J. (2001). *Wireless nation: The frenzied launch of the cellular revolution in America.* Cambridge, MA: Perseus Publishing.

Nielsen. (2015a). *Q1 2015 Local Watch Report: Where You Live and Its Impact On Your Choices.* [Press release] http://www.nielsen.com/us/en/insights/reports/2015/q1-2015-local-watch-report-where-you-live-and-its-impact-on-your-choices.html.

Nielsen. (2015b). *Nielsen Estimates 116.3 Million TV Homes In The U.S., Up 0.4%.* [Press release] http://www.niel-sen.com/us/en/insights/news/2014/nielsen-estimates-116-3-million-tv-homes-in-the-us.html.

Nielsen. (2018). *Nielsen Estimates 119.9 Million TV Homes In The U.S. For The 2018-2019 TV Season.* [Press release] https://www.nielsen.com/us/en/insights/article/2018/nielsen-estimates-119-9-million-tv-homes-in-the-us-for-the-2018-19-season/.

Nielsen. (2019a, August). *Nielsen Local Watch Report & Audience Insights: TV Streaming Across Our Cities.* https://www.niel-sen.com/wp-content/uploads/sites/3/2019/08/local-watch-report-aug-2019.pdf.

Nielsen. (2019b, March). *Over-The-Air TV Is Booming In U.S. Cities.* https://www.nielsen.com/us/en/insights/article/2019/over-the-air-tv-is-booming-in-us-cities/.

Netflix. (2016). *Netflix's View: Internet TV is replacing linear TV.* http://ir.netflix.com/long-term-view.cfmOECD. (2013). Broadband and telecom. http://www.oecd.org/internet/broadband/oecdbroadbandportal.htm.

Netflix. (2017). Consolidated Segment Information. https://ir.netflix.com/financial-statements.

OECD. (2013). Broadband and telecom. http://www.oecd.org/internet/broadband/oecdbroadbandportal.htm.

Outerwall. (2016). Outerwall Annual Report to U.S. Security and Exchange Commission. http://d1lge852tjjqow.cloud-front.net/CIK-0000941604/a5b4e8c7-6f8f-428d-b176-d9723dcbc772.pdf?noexit=true .

Peters, J. & Donald, W. H. (2007). Industry surveys: Publishing. In E. M. Bossong-Martines (Ed.), *Standard & Poor's industry surveys. 175* (36). Section 1.

Pew Research Center. (2014, January 16). *E-Reading Rises as Device Ownership Jumps.* http://www.pewinternet.org/files/2014/01/PIP_E-reading_011614.pdf.

Pew Research Center. (2019). *Newspaper Fact Sheet.* https://www.journalism.org/fact-sheet/newspapers/

Plastics Industry Association. (2017). Watching: Consumer Technology Plastics' Innovative Chapter in The Consumer Technology Story. http://www.plasticsindustry.org/sites/plastics.dev/files/PlasticsMarketWatchConsumerTechnologyWebVersion.pdf.

Rainie, L. (2012, January 23). Tablet and e-book reader ownership nearly doubles over the holiday gift-giving period. *Pew Research Center's Internet & American Life Project.* http://pewinternet.org/Reports/2012/E-readers-and-tablets.aspx.

Ramachandran, S. (2016, February 18). Dish Network's Sling TV Has More Than 600,000 Subscribers. *Wall Street Journal.* http://www.wsj.com/article_email/dish-networks-sling-tv-has-more-than-600-000-subscribers-1455825689-

Raphael, J. (2000, September 4). Radio station leaves earth and enters cyberspace. Trading the FM dial for a digital stream. *New York Times.* http://www.nytimes.com/library/tech/00/09/biztech/articles/04radio.html.

Rawlinson, N. (2011, April 28). Books vs ebooks. *Computer Act!ve.* General OneFile database.

RIAA. (2019) *U.S. Sales Database.* https://www.riaa.com/u-s-sales-database/

Rebenitsch, L. (2015). Managing cybersickness in virtual reality. *XRDS: Crossroads, The ACM Magazine for Students, 22*(1), 46–51.

Schaeffler, J. (2004, February 2). The real satellite radio boom begins. *Satellite News, 27* (5). Lexis-Nexis.

Sirius XM Radio Poised for Growth, Finally. (2011, May 10). *Newsmax.*

Sirius XM Radio. (2016, February 2). SiriusXM Reports Fourth Quarter and Full-Year 2015 Results [Press release]. http://s2.q4cdn.com/835250846/files/doc_financials/annual2015/SiriusXM-Reports-Fourth-Quarter-and-Full-Year-2015-Results.pdf.

Sirius XM Radio. (2018, January 10). SiriusXM Beats 2017 Subscriber Guidance; Issues 2018 Subscriber and Financial Guidance [Press release]. http://investor.siriusxm.com/investor-overview/press-releases/press-release-details/2018/SiriusXM-Beats-2017-Subscriber-Guidance-Issues-2018-Subscriber-and-Financial-Guidance/default.aspx.

Sirius XM Radio. (2019, October 31). *SiriusXM Reports Third Quarter 2019 Results* [Press release]. http://s1.q4cdn.com/750174072/files/doc_news/SiriusXM-Reports-Third-Quarter-2019-Results-2019.pdf.

State Administration of Press, Publication, Radio, Film and Television of The People's Republic of China. (2016, January 3). *China Box Office Surpassed 44 Billion Yuan in 2015.* http://www.sarft.gov.cn/art/2016/1/3/art_160_29536.html.

Statista. (2016). *Average daily TV viewing time per person in selected countries worldwide in 2014 (in minutes).* http://www.statista.com/statistics/276748/average-daily-tv-viewing-time-per-person-in-selected-countries/.

Statista. (2019) *Video Streaming in the U.S.: An Industry in Flux.* https://www.statista.com/study/69365/us-video-streaming-market/

The Recording Industry Association of America. (2017). *U.S. Sales Database.* https://www.riaa.com/u-s-sales-database

U.S. Bureau of the Census. (2006). *Statistical abstract of the United States: 2006* (125th Ed.). Washington, DC: U.S. Government Printing Office.

U.S. Bureau of the Census. (2008). *Statistical abstract of the United States: 2008* (127th Ed.). Washington, DC: U.S. Government Printing Office. http://www.census.gov/compendia/statab/.

U.S. Bureau of the Census. (2010). *Statistical abstract of the United States: 2008* (129th Ed.). Washington, DC: U.S. Government Printing Office. http://www.census.gov/compendia/statab/.

U.S Census Bureau. (2016). 2015 Annual Services Report. http://www2.census.gov/services/sas/data/Historical/sas-15.xls.

U.S. Census Bureau. (2017). 2016 Annual Services Report. http://www2.census.gov/services/sas/data/table4.xls.

U.S. Census Bureau. (2019). 2018 Annual Services Report. https://www2.census.gov/programs-surveys/services/tables/time-series/sas-latest/table4.xlsx

U.S. Copyright Office. (2003). *106th Annual report of the Register of Copyrights for the fiscal year ending September 30, 2003.* Washington, DC: Library of Congress.

U.S. Department of Commerce/International Trade Association. (2000). *U.S. industry and trade outlook 2000.* New York: McGraw-Hill.

Wallenstein, A. (2011, September 16). Tube squeeze: economy, tech spur drop in TV homes. *Daily Variety,* 3.

White, L. (1971). *The American radio.* New York: Arno Press.

Willens, M. (2015, January 7). Home Entertainment 2014: U.S. DVD Sales and Rentals Crater, Digital Subscriptions Soar. *International Business Times.* http://www.ibtimes.com/home-entertainment-2014-us-dvd-sales-rentals-crater-digital-subscriptions-soar-1776440.

Zhu, Y. & Brown, D. (2016). A history of communication technology. In A. E. Grant & J. H. Meadows (Eds.), *Communication technology update and fundamentals (15th edition).* Taylor & Francis. 9-24.

Zhu, Y. (2018). A history of communication technology. In A. E. Grant & J. H. Meadows (Eds.), *Communication technology update and fundamentals (16th edition).* Taylor & Francis. 9-24.

3

Understanding Communication Technologies

Jennifer H. Meadows, Ph.D.*

Think back just 20 years ago. In many ways we were still living in an analog world. There were no smart watches or virtual assistants, no Netflix or Disney+, you couldn't just grab a Lyft. The latest communication technologies at that time were Palm Pilots and DVDs, and those were cutting edge. Communication technologies are in a constant state of change. New ones are developed while others fade. This book was created to help you understand these technologies, but there is a set of tools that will not only help you understand them, but also understand the next generation of technologies.

All of the communication technologies explored in this book have a number of characteristics in common, including how their adoption spreads from a small group of highly interested users to the general public (or not), what the effects of these technologies are upon the people who use them (and on society in general), and how these technologies affect each other.

For more than a century, researchers have studied adoption, effects, and other aspects of new technologies, identifying patterns that are common across dissimilar technologies, and proposing theories of technology adoption and effects. These theories have proven to be valuable to entrepreneurs seeking to develop new technologies, regulators who want to control those technologies, and everyone else who just wants to understand them.

* Professor, Department of Media Arts, Design, and Technology, California State University, Chico (Chico, California).

The utility of these theories is that they allow you to apply lessons from one technology to another or from old technologies to new technologies. The easiest way to understand the role played by the technologies explored in this book is to have a set of theories you can apply to virtually any technology you discuss. The purpose of this chapter is to give you these tools by introducing you to the theories.

The technology ecosystem discussed in Chapter 1 is a useful framework for studying communication technologies, but it is not a theory. This perspective is a good starting point to begin to understand communication technologies because it targets your attention at a number of different levels that might not be immediately obvious: hardware, software, content, organizational infrastructure, social systems, and, finally, the user.

Understanding each of these levels is aided by knowing a number of theoretical perspectives that can help us understand the different sections of the ecosystem for these technologies. Theoretical approaches are useful in understanding the origins of the information-based economy in which we now live, why some technologies take off while others fail, the impacts and effects of technologies, and the economics of the communication technology marketplace.

The Information Society & the Control Revolution

Our economy used to be based on tangible products such as coal, lumber, and steel. Now, though, information is the basis of our economy. Information industries include education; research and development; creating informational goods such as computer software, banking, insurance; and even entertainment and news (Beniger, 1986).

Information is different from other commodities like coffee and pork bellies, which are known as "private goods." Instead, information is a "public good" because it is intangible, lacks a physical presence, and can be sold as many times as demand allows without regard to consumption.

For example, if 10 sweaters are sold, then 10 sweaters must be manufactured using raw materials.

If 10 subscriptions to an online news service are sold, there is no need to create a different story for each user; 10—or 10,000—subscriptions can be sold without additional raw materials.

This difference actually gets to the heart of a common misunderstanding about ownership of information that falls into a field known as "intellectual property rights." A common example is the purchase of a digital music download. A person may believe that, because they purchased the music, they can copy and distribute that music to others. Just because the information (the music) was purchased doesn't mean they own the song and performance (intellectual property).

Several theorists have studied the development of the information society, including its origin. Beniger (1986) argues that there was a control revolution: "A complex of rapid changes in the technological and economic arrangements by which information is collected, stored, processed, and communicated and through which formal or programmed decisions might affect social control" (p. 52). In other words, as society progressed, technologies were created to help control information. For example, information was centralized by mass media.

In addition, as more and more information is created and distributed, new technologies must be developed to control that information. For example, with the explosion of information available over the internet, search engines were developed to help users find relevant information.

Another important point is that information is power, and there is power in giving information away. Power can also be gained by withholding information. At different times in modern history, governments have blocked access to information or controlled information dissemination to maintain power.

Adoption

Why do some technologies succeed while others fail? This question is addressed by a number of theoretical approaches including the diffusion of innovations, social information processing theory, critical mass theory, the theory of planned behavior, the technology acceptance model, and more.

Diffusion of Innovations

The diffusion of innovations, also referred to as diffusion theory, was developed by Everett Rogers (1962; 2003). This theory tries to explain how an innovation is communicated over time through different channels to members of a social system. There are four main aspects of this approach.

First, there is the innovation. In the case of communication technologies, the innovation is some technology that is perceived as new. Rogers also defines characteristics of innovations: relative advantage, compatibility, complexity, trialability, and observability.

So, if someone is deciding to purchase a new mobile phone, characteristics would include the relative advantage over other mobile phones; whether or not the mobile phone is compatible with the existing needs of the user; how complex it is to use; whether or not the potential user can try it out; and whether or not the potential user can see others using the new mobile phone with successful results.

Information about an innovation is communicated through different channels. Mass media is good for awareness knowledge. For example, each new iPhone has web content, television commercials, and print advertising announcing its existence and its features.

Interpersonal channels are also an important means of communication about innovations. These interactions generally involve subjective evaluations of the innovation. For example, a person might ask some friends how they like their new iPhones.

Rogers (2003) outlines the decision-making process a potential user goes through before adopting an innovation. This is a five-step process.

The first step is **knowledge**. You find out there is a new mobile phone available and learn about its new features. The next step is **persuasion**—the formation of a positive attitude about the innovation. Maybe you like the new phone. The third step is when you decide to accept or reject the innovation (**adoption**). Yes, I will get the new mobile phone. **Implementation** is the fourth step. You use the innovation, in this case, the mobile phone. Finally, **confirmation** occurs when you decide that you made the correct decision. Yes, the mobile phone is what I thought it would be; my decision is reinforced.

Another stage discussed by Rogers (2003) and others is "reinvention," the process by which a person who adopts a technology begins to use it for purposes other than those intended by the original inventor. For example, mobile phones were initially designed for calling other people regardless of location, but users have found ways to use them for a wide variety of applications ranging from alarm clocks to personal calendars and flashlights.

Have you ever noticed that some people are the first to have the new technology gadget, while others refuse to adopt a proven successful technology? Adopters can be categorized into different groups according to how soon or late they adopt an innovation.

The first to adopt are the **innovators**. Innovators are special because they are willing to take a risk adopting something new that may fail. Next come the **early adopters**, the **early majority**, and then the **late majority**, followed by the last category, the **laggards**. In terms of percentages, innovators make up the first 2.5% percent of adopters, early adopters are the next 13.5%, early majority follows with 34%, late majority are the next 34%, and laggards are the last 16%.

Adopters can also be described in terms of ideal types. Innovators are venturesome. These are people who like to take risks and can deal with failure. Early adopters are respectable. They are valued opinion leaders in the community and role models for others. Early majority adopters are deliberate. They adopt just before the average person and are an important link between the innovators, early adopters, and everyone else. The late majority are skeptical. They are hesitant to adopt innovations and often adopt because they pressured. Laggards are the last to adopt and often are isolated with no opinion leadership. They are suspicious and resistant to change. Other factors that affect adoption include education, social status, social mobility, finances, and willingness to use credit (Rogers, 2003).

Adoption of an innovation does not usually occur all at once; it happens over time. This is called the rate of adoption. The rate of adoption generally follows an S-shaped "diffusion curve" where the X-axis is time and the Y-axis is percent of adopters. You can note the different adopter categories along the diffusion curve.

Figure 3.1 shows a diffusion curve. See how the innovators are at the very beginning of the curve, and the laggards are at the end. The steepness of the curve depends on how quickly an innovation is adopted. For example, DVD has a steeper curve than VCR because DVD players were adopted at a faster rate than VCRs.

Figure 3.1

Innovation Adoption Rate

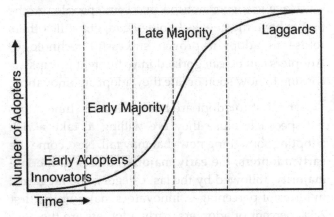

Source: Technology Futures, Inc.

Also, different types of decision processes lead to faster adoption. Voluntary adoption is slower than collective decisions, which, in turn, are slower than authority decisions. For example, a company may let its workers decide whether to use a new software package, the employees may agree collectively to use that software, or finally, the management may decide that everyone at the company is going to use the software. In most cases, voluntary adoption would take the longest, and a management dictate would result in the swiftest adoption.

Moore (2001) further explored diffusion of innovations and high-tech marketing in *Crossing the Chasm*. He noted there are gaps between the innovators and the early adopters, the early adopters and the early majority, and the early majority and late majority.

For a technology's adoption to move from innovators to the early adopters the technology must show a major new benefit. Innovators are visionaries that take the risk of adopting something new such as mixed reality headsets.

Early adopters then must see the new benefit of mixed reality headsets before adopting. The chasm between early adopters and early majority is the greatest of these gaps. Early adopters are still visionary and want to be change agents. They don't mind dealing with the troubles and glitches that come along with a new technology. Early adopters were likely to use a beta version of a new service or product.

The early majority, on the other hand, are pragmatists and want to see some improvement in productivity—something tangible. Moving from serving the visionaries to serving the pragmatists is difficult; hence Moore's description of "crossing the chasm."

Finally, there is a smaller gap between the early majority and the late majority. Unlike the early majority, the late majority reacts to the technical demands on the users. The early majority is more comfortable working with technology. So, the early majority would be comfortable using a virtual reality headset like Oculus Rift but the late majority is put off by the perceived technical demands. The technology must alleviate this concern before late majority adoption.

Figure 3.2

Product Lifecycle

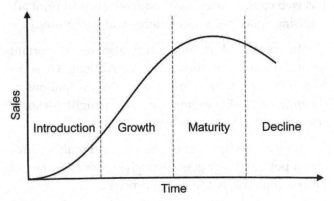

Source: Technology Futures, Inc.

Another perspective on adoption can be found in most marketing textbooks (e.g., Kottler & Keller, 2011): the product lifecycle. As illustrated in Figure 3.2, the product lifecycle extends the diffusion curve to include the maturity and decline of the technology. This perspective provides a more complete picture of a technology because it focuses our attention beyond the initial use of the technology to the time that the technology is in regular use, and ultimately, disappears from market. Remember laserdisc or Myspace?

Considering the short lifespan of many communication technologies, it may be just as useful to study the entire lifespan of a technology rather than just the process of adoption.

Other Theories of Adoption

Other theorists have attempted to explain the process of adoption as well. Among the most notable perspectives in this regard are the Theory of Planned Behavior (TPB), the Technology Acceptance Model (TAM) and the Unified Theory of Acceptance and Use of Technology. These models emerged from the need to identify factors that can help predict future adoption of a new technology when there is no history of adoption or use of the technology.

The Theory of Planned Behavior (TPB) (Ajzen, 1991; Ajzen & Fishbein, 1980; Fishbein & Ajzen, 1975) presumes that a suitable predictor of future behavior is "behavioral intention," a cognitive rather than behavioral variable that represents an individual's plans for adopting (or not adopting!) an innovation. Behavioral intentions are, in turn, predicted by attitudes toward the innovation and the innovators.

The Technology Acceptance Model (Davis, 1986; Davis et al., 1989) elaborates on TPB by adding factors that may predict attitudes toward an innovation and behavioral intentions, including perceived usefulness, perceived ease of use, and external variables.

The Unified Theory of Acceptance and Use of Technology attempts to combine elements of several theories of technology adoptions including the Theory of Planned Behavior, the Technology Acceptance model, and Diffusions of Innovations (Venkatesh et al., 2003). The authors argue that there are four core determinates of user acceptance and behavior. These are performance expectancy, effort expectancy, social influence, and facilitating conditions.

A substantial body of research has demonstrated the efficacy of these factors in predicting behavioral intentions (e.g., Jiang et al., 2000; Chau & Hu, 2002), but much more research is needed regarding the link between behavioral intentions and actual adoption at a later point in time.

Another theory that expands upon Rogers' diffusion theory is presented in Chapter 4. Grant's pre-diffusion theory identifies organizational functions that must be served before any consumer adoption of the technology can take place.

Critical Mass Theory

Have you ever wondered who had the first e-mail address or the first telephone? Who did they communicate with? Interactive technologies such as telephony and e-mail become more and more useful as more people adopt these technologies. There have to be some innovators and early adopters who are willing to take the risk to try a new interactive technology.

These users are the "critical mass," a small segment of the population that chooses to make big contributions to the public good (Markus, 1987). In general terms, any social process involving actions by individuals that benefit others is known as "collective action." In this case, the technologies become more useful if everyone in the system is using the technology, a goal known as "universal access."

Ultimately, universal access means that you can reach anyone through some communication technology. For example, in the United States, the landline phone system reaches almost everywhere, and everyone benefits from this technology, although a small segment of the population initially chose to adopt the telephone to get the ball rolling. There is a stage in the diffusion process that an interactive medium has to reach in order for adoption to take off. This is the critical mass. Interestingly, 40% of homes are mobile phone only but the important factor to remember is that even early mobile phone users could call anyone on a land line so the technology didn't have the same critical hurdle to overcome (CTIA, 2016).

Another conceptualization of critical mass theory is the "tipping point" (Gladwell, 2002). Here is an example. The videophone never took off, in part, because it never reached critical mass. The videophone was not really any better than a regular phone unless the person you were calling also had a videophone. If there were not enough people you knew who had videophones, then you might not adopt it because it was not worth it.

On the other hand, if most of your regular contacts had videophones, then that critical mass of users might drive you to adopt the videophone. Critical mass is an important aspect to consider for the adoption of any interactive technology. Think of the iPhone's FaceTime App. The only people who can use FaceTime are iPhone users. Skype, on the other hand, works on all mobile operating systems so it doesn't have the same limitations as FaceTime.

Another good example is facsimile or fax technology. The first method of sending images over wires was invented in the '40s—the 1840s—by Alexander Bain, who proposed using a system of electrical pendulums to send images over wires (Robinson, 1986). Within a few decades, the technology was adapted by the newspaper industry to send photos over wires, but the technology was limited to a small number of news organizations.

The development of technical standards in the 1960s brought the fax machine to corporate America, which generally ignored the technology because few businesses knew of another business that had a fax machine.

Adoption of the fax took place two machines at a time, with those two usually being purchased to communicate with each other, but rarely used to communicate with additional receivers. By the 1980s, enough businesses had fax machines that could communicate with each other that many businesses started buying fax machines one at a time.

As soon as the critical mass point was reached, fax machine adoption increased to the point that it became referred to as the first technology adopted out of fear of embarrassment that someone would ask, "What's your fax number?" (Wathne & Leos, 1993). In less than two years, the fax machine became a business necessity.

Social Information Processing

Another way to look at how and why people choose to use or not use a technology is social information processing. This theory begins by critiquing rational choice models, which presume that people make adoption decisions and other evaluations of technologies based upon objective characteristics of the technology. In order to understand social information processing, you first have to look at a few rational choice models.

One model, social presence theory, categorizes communication media based on a continuum of how the medium "facilitates awareness of the other person and interpersonal relationships during the interaction" (Fulk, et al., 1990, p. 118).

Communication is most efficient when the social presence level of the medium best matches the interpersonal relationship required for the task at hand. For example, most people would propose marriage face to face instead of using a text message.

Another rational choice model is information richness theory. In this theory, media are also arranged on a continuum of richness in four areas: speed of feedback, types of channels employed, personalness of source, and richness of language carried (Fulk, et al., 1990). Face-to-face communications is the highest in social presence and information richness.

In information richness theory, the communication medium chosen is related to message ambiguity. If the message is ambiguous, then a richer medium is chosen. In this case, teaching someone how to dance would be better with an online video that illustrates the steps rather than just an audio podcast that describes the steps.

Social information processing theory goes beyond the rational choice models because it states that perceptions of media are "in part, subjective and socially constructed." Although people may use objective standards in choosing communication media, use is also determined by subjective factors such as the attitudes of coworkers about the media and vicarious learning, or watching others' experiences.

Social influence is strongest in ambiguous situations. For example, the less people know about a medium, then the more likely they are to rely on social information in deciding to use it (Fulk, et al., 1987).

Think about whether you prefer an Apple or Android mobile phone. Although you can probably list objective differences between the two, many of the important factors in your choice are based upon subjective factors such as which one is owned by friends and coworkers, the perceived usefulness of the phone, and advice you receive from people who sell you the phone.

In the end, these social factors probably play a much more important role in your decision than "objective" factors such as the camera, processor speed, memory capacity, etc.

Impacts & Effects

Do video games make players violent? Do users seek out social networking sites for social interactions? These are some of the questions that theories of impacts or effects try to answer.

To begin, Rogers (1986) provides a useful typology of impacts. Impacts can be grouped into three dichotomies: desirable and undesirable, direct and indirect, and anticipated and unanticipated.

Desirable impacts are the functional impacts of a technology. For example, a desirable impact of social networking is the ability to connect with friends and family. An undesirable impact is one that is dysfunctional, such as bullying or stalking.

Direct impacts are changes that happen in immediate response to a technology. A direct impact of wireless telephony is the ability to make calls while driving. An indirect impact is a byproduct of the direct impact. To illustrate, laws against driving and using a handheld wireless phone are an impact of the direct impact described above.

Anticipated impacts are the intended impacts of a technology. An anticipated impact of text messaging is to communicate without audio. An unanticipated impact is an unintended impact, such as people sending text messages in a movie theater and annoying other patrons. Often, the desirable, direct, and anticipated impacts are the same and are considered first. Then, the undesirable, indirect, and unanticipated impacts are noted later.

Here is an example using e-mail. A desirable, anticipated, and direct impact of e-mail is to be able to quickly send a message to multiple people at the same time. An undesirable, indirect, and unanticipated impact of e-mail is spam—unwanted e-mail clogging the inboxes of millions of users.

Uses & Gratifications

Uses and gratifications research is a descriptive approach that gives insight into what people do with technology. This approach sees the users as actively seeking to use different media to fulfill different needs (Rubin, 2002). The perspective focuses on "(1) the social and psychological origins of (2) needs, which generate (3) expectations of (4) the mass media or other sources, which lead to (5) differential patterns of media exposure (or engagement in other activities), resulting in (6) needs gratifications and (7) other consequences, perhaps mostly unintended ones" (Katz, et al., 1974, p. 20).

Uses and gratifications research surveys audiences about why they choose to use different types of media. For example, uses and gratifications of television studies have found that people watch television for information, relaxation, to pass time, by habit, excitement, and for social utility (Rubin, 2002).

This approach is also useful for comparing the uses and gratifications between media, as illustrated by studies of the world wide web (www) and television gratifications that found that, although there are some similarities such as entertainment and to pass time, they are also very different on other variables such as companionship, where the web was much lower than for television (Ferguson & Perse, 2000). Uses and gratifications studies have examined a multitude of communication technologies including mobile phones (Wei, 2006), radio (Towers, 1987), satellite television (Etefa, 2005), and social media (Raacke & Bonds-Raacke, 2008).

Media System Dependency Theory

Often confused with uses and gratifications, media system dependency theory is "an ecological theory that attempts to explore and explain the role of media in society by examining dependency relations within and across levels of analysis" (Grant, et al., 1991, p. 774). The key to this theory is the focus it provides on the dependency relationships that result from the interplay between resources and goals.

The theory suggests that, in order to understand the role of a medium, you have to look at relationships at multiple levels of analysis, including the individual level—the audience, the organizational level, the media system level, and society in general.

These dependency relationships can be symmetrical or asymmetrical. For example, the dependency relationship between audiences and network television is asymmetrical because an individual audience member may depend more on network television to reach their goal than the television networks depend on that one audience member to reach their goals.

A typology of individual media dependency relations was developed by Ball-Rokeach & DeFleur (1976) to help understand the range of goals that individuals have when they use the media. There are six dimensions: social understanding, self-understanding, action orientation, interaction orientation, solitary play, and social play.

Social understanding is learning about the world around you, while self-understanding is learning about yourself. Action orientation is learning about specific behaviors, while interaction orientation is about learning about specific behaviors involving other people. Solitary play is entertaining yourself alone, while social play is using media as a focus for social interaction.

Research on individual media system dependency relationships has demonstrated that people have different dependency relationships with different media. For example, Meadows (1997) found that women had stronger social understanding dependencies for television than magazines, but stronger self-understanding dependencies for magazines than television.

In the early days of television shopping (when it was considered "new technology"), Grant et al. (1991) applied media system dependency theory to the phenomenon. Their analysis explored two dimensions: how TV shopping changed organizational dependency relations within the television industry and how and why individual users watched television shopping programs.

By applying a theory that addressed multiple levels of analysis, a greater understanding of the new technology was obtained than if a theory that focused on only one level had been applied.

Social Learning Theory/Social Cognitive Theory

Social learning theory focuses on how people learn by modeling others (Bandura, 2001). This observational learning occurs when watching another person model the behavior. It also happens with symbolic modeling, modeling that happens by watching the behavior modeled on a television or computer screen. So, a person can learn how to fry an egg by watching another person fry an egg in person or on a video.

Learning happens within a social context. People learn by watching others, but they may or may not perform the behavior. Learning happens, though, whether the behavior is imitated or not.

Reinforcement and punishment play a role in whether or not the modeled behavior is performed. If the behavior is reinforced, then the learner is more likely to perform the behavior. For example, if a student is successful using online resources for a presentation, other students watching the presentation will be more likely to use online resources.

On the other hand, if the action is punished, then the modeling is less likely to result in the behavior. To illustrate, if a character drives drunk and gets arrested on a television program, then that modeled behavior is less likely to be performed by viewers of that program.

Reinforcement and punishment is not that simple though. This is where cognition comes in—learners think about the consequences of performing that behavior. This is why a person may play *Grand Theft Auto* and steal cars in the videogame, but will not then go out and steal a car in real life. Self-regulation is an important factor. Self-efficacy is another important dimension: learners must believe that they can perform the behavior.

Social learning/cognitive theory, then, is a useful framework for examining not only the effects of communication media, but also the adoption of communication technologies (Bandura, 2001).

The content that is consumed through communication technologies contains symbolic models of behavior that are both functional and dysfunctional. If viewers model the behavior in the content, then some form of observational learning is occurring.

A lot of advertising works this way. A movie star uses a new shampoo and then is admired by others. This message models a positive reinforcement of using the shampoo. Cognitively, the viewer then thinks about the consequences of using the shampoo.

Modeling can happen with live models and symbolic models. For example, a person can watch another playing Just Dance, 2020, a videogame where the player has to mimic the dance moves of an avatar in the game. The other player considers the consequences of this modeling.

In addition, if the other person had not played with this gaming system, watching the other person play with the system and enjoy the experience will make it more likely that he or she will adopt the system. Therefore, social learning/cognitive theory can be used to facilitate the adoption of new technologies and to understand why some technologies are adopted and why some are adopted faster than others (Bandura, 2001).

Economic Theories

Thus far, the theories and perspectives discussed have dealt mainly with individual users and communication technologies. How do users decide to adopt a technology? What impacts will a technology have on a user?

Theory, though, can also be applied to organizational infrastructure and the overall technology market. Here, two approaches will be addressed: the theory of the long tail that presents a new way of looking at digital content and how it is distributed and sold, and the principle of relative constancy that examines what happens to the marketplace when new media products are introduced.

The Theory of the Long Tail

Former *Wired Magazine* editor Chris Anderson developed the theory of the long tail. While some claim this is not a "theory," it is nonetheless a useful framework for understanding new media markets.

This theory begins with the realization that there are not any huge hit movies, television shows, and albums like there used to be. What counts as a hit TV show today, for example, would be a failed show just 15 years ago.

One of the reasons for this is choice: 40 years ago viewers had a choice of only a few television channels. Today, you could have hundreds of channels of video programming on cable or satellite and limitless amounts of video programming on the internet.

New communication technologies are giving users access to niche content. There is more music, video, video games, news, etc. than ever before because the distribution is no longer limited to the traditional mass media of over-the-air broadcasting, newspapers, etc. or the shelf space at a local retailer.

The theory states that, "our culture and economy are increasingly shifting away from a focus on a relatively small number of 'hits' at the headend of the demand curve and toward a huge number of niches in the tail" (Anderson, n.d.).

Figure 3.3 shows a traditional demand curve; most of the hits are at the head of the curve, but there is still demand as you go into the tail. There is a demand for niche content and there are opportunities for businesses that deliver content in the long tail.

Both physical media and traditional retail have limitations. For example, there is only so much shelf space in the store. Therefore, the store, in order to maximize profit, is only going to stock the products most likely to sell. Digital content and distribution changes this.

Figure 3.3

The Long Tail

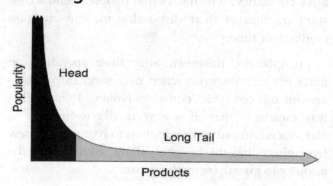

Source: Anderson (n.d.)

For example, Amazon and Netflix can have huge inventories of hard-to-find titles, as opposed to a Red Box kiosk, which has to have duplicate inventories at each location. All digital services, such as Spotify, eliminate all physical media. You subscribe and can download the content digitally, and there is no need for a warehouse to store DVDs and CDs.

Because of these efficiencies, these businesses can better serve niche markets. Taken one at a time, these niche markets may not generate significant revenue but when they are aggregated, these markets are significant.

Anderson (2006) suggests rules for long tail businesses. Make everything available, lower the price, and help people find it. Traditional media are responding to these services. For example, Disney premiered their streaming service, Disney+, in late 2019. Previously, Disney fans who wanted access to the large catalog of Disney content has to purchase DVDs and digital downloads (Sorrentino & Solsman, 2019).

The audience is changing, and expectations for content selection and availability are changing. The audience today, Anderson argues, wants what they want, when they want it, and how they want it.

The Principle of Relative Constancy

So now that people have all of this choice of content, delivery mode, etc., what happens to older media? Do people just keep adding new entertainment media, or do they adjust by dropping one form in favor of another?

This question is at the core of the principle of relative constancy, which says that people spend a constant fraction of their disposable income on mass media over time.

People do, however, alter their spending on mass media categories when new services/products are introduced (McCombs & Nolan, 1992). What this means is that, if a new media technology is introduced, in order for adoption to happen, the new technology has to be compelling enough for the adopter to give up something else.

For example, a person who signs up for Netflix may spend less money on movie tickets. A satellite radio user will spend less money purchasing music

subscriptions or CDs. So, when considering a new media technology, the relative advantage it has over existing service must be considered, along with other characteristics of the technology discussed earlier in this chapter. Remember, the money users spend on any new technology has to come from somewhere.

Critical & Cultural Theories

Most of the theories discussed above are derived from sociological perspectives on media and media use, relying upon quantitative analysis and the study of individuals to identify the dimensions of the technologies and the audience that help us understand these technologies. An alternate perspective can be found in critical and cultural studies, which provide different perspectives that help us understand the role of media in society.

Critical and cultural studies make greater use of qualitative analysis, including more analysis of macro-level factors such as media ownership and influences on the structure and content of media.

Marxist scholars, for example, focus on the underlying economic system and the division between those who control and benefit from the means of production and those who actually produce and consume goods and services.

This tradition considers that ownership and control of media and technology can affect the type of content provided. Scholars including Herman & Chomsky (2008) and Bagdikian (2000) have provided a wide range of evidence and analysis informing our understanding of these structural factors.

Critical theory of communication technology (Feenberg, 2009) examines these technologies contribution to democracy including bringing together like-minded publics around political issues. Community building and fragmentation using communication technologies such as social networking sites is discussed as well.

Feminist theory provides another set of critical perspectives that can help us understand the role and function of communication technology in society. Broadly speaking, feminist theory encompasses a broad range of factors including bodies, race, class, gender, language, pleasure, power, and sexual division of labor (Kolmar & Bartkowski, 2005).

New technologies represent an especially interesting challenge to the existing power relationships in media (Allen, 1999), and application of feminist theory has the potential to explicate the roles and power relationships in these media as well as to prescribe new models of organizational structure that take advantage of the shift in power relationships offered by interactive media (Grant et al., 2009).

Cultural studies also address the manner in which content is created and understood. Semiotics, for example, differentiates among "signs," "signifiers," and "signified" (Eco, 1979), explicating the manner in which meaning is attached to content. Scholars such as Hall (2010) have elaborated the processes by which content is encoded and decoded, helping us to understand the complexities of interpreting content in a culturally diverse environment.

Critical and cultural studies put the focus on a wide range of systemic and subjective factors that challenge conventional interpretations and understandings of media and technologies.

Conclusion

This chapter has provided a brief overview of several theoretical approaches to understanding communication technology. As you work through the book, consider theories of adoption, effects, economics, and critical/cultural studies and how they can inform you about each technology and allow you to apply lessons from one technology to others. For more in-depth discussions of these theoretical approaches, check out the sources cited in the bibliography.

Bibliography

Ajzen, I. (1991). The theory of planned behavior. *Organizational Behavior and Human Decision Processes*, 50, 179–211.

Ajzen, I. & Fishbein, M. (1980). *Understanding attitudes and predicting social behavior*. Prentice-Hall.

Allen, A. (1999). *The power of feminist theory*. Westview Press.

Anderson, C. (n.d.). *About me*. http://www.thelongtail.com/about.html.

Anderson, C. (2006). *The long tail: Why the future of business is selling less of more*. Hyperion.

Ball-Rokeach, S. & DeFleur, M. (1976). A dependency model of mass-media effects. *Communication Research*, 3, 1 3-21.

Bagdikian, B. H. (2000). *The media monopoly* (Vol. 6). Beacon Press.

Bandura, A. (2001). Social cognitive theory of mass communication. *Media Psychology*, 3, 265-299.

Beniger, J. (1986). The information society: Technological and economic origins. In S. Ball-Rokeach & M. Cantor (Eds.). *Media, audience, and social structure*. Sage, pp. 51-70.

Chau, P. Y. K. & Hu, P. J. (2002). Examining a model of information technology acceptance by individual professionals: An exploratory study. *Journal of Management Information Systems*, **18**(4), 191–229.

CTIA (2016). Your Wireless Life. CTIA. http://www.ctia.org/your-wireless-life/how-wireless-works/wireless-quick-facts.

Davis, F. D. (1986). *A technology acceptance model for empirically testing new end-user information systems: Theory and results*. Unpublished doctoral dissertation, Massachusetts Institute of Technology.

Davis, F. D., Bagozzi, R. P. & Warshaw, P. R. (1989). User acceptance of computer technology: A comparison of two theoretical models. *Management Science*, **35**(8), 982–1003.

Eco, U. (1979). *A theory of semiotics* (Vol. 217). Indiana University Press.

Etefa, A. (2005). *Arabic satellite channels in the U.S.: Uses & gratifications*. Paper presented at the annual meeting of the International Communication Association, New York. http://www.allacademic.com/meta/p14246_index.html.

Feenberg, A. (2009). Critical theory of communication technology: Introduction to the special section. *The Information Society*, 25: 77-83.

Ferguson, D. & Perse, E. (2000, Spring). The World Wide Web as a functional alternative to television. *Journal of Broadcasting and Electronic Media*. 44 (2), 155-174.

Fishbein, M. & Ajzen, I. (1975). *Belief, attitude, intention and behavior: An introduction to theory and research*. Reading, MA: Addison-Wesley.

Fulk, J., Schmitz, J. & Steinfield, C. W. (1990). A social influence model of technology use. In J. Fulk & C. Steinfield (Eds.), *Organizations and communication technology*. Sage, pp. 117-140.

Fulk, J., Steinfield, C., Schmitz, J. & Power, J. (1987). A social information processing model of media use in organizations. *Communication Research, 14* (5), 529-552.

Gladwell, M. (2002). *The tipping point: How little things can make a big difference*. Back Bay Books.

Grant, A. E., Guthrie, K. & Ball-Rokeach, S. (1991). Television shopping: A media system dependency perspective. *Communication Research, 18* (6), 773-798.

Grant, A. E., Meadows, J. H., & Storm, E. J. (2009). A feminist perspective on convergence. In Grant, A. E. & Wilkinson, J. S. (Eds.) *Understanding media convergence: The state of the field*. Oxford University Press.

Hall, S. (2010). Encoding, decoding1. *Social theory: Power and identity in the global era, 2*, 569.

Herman, E. S., & Chomsky, N. (2008). *Manufacturing consent: The political economy of the mass media*. Random House.

Jiang, J. J., Hsu, M. & Klein, G. (2000). E-commerce user behavior model: An empirical study. *Human Systems Management, 19*, 265–276.

Katz, E., Blumler, J. & Gurevitch, M. (1974). Utilization of mass communication by the individual. In J. Blumler & E. Katz (Eds.). *The uses of mass communication: Current perspectives on gratifications research*. Sage.

Kolmar, W., & Bartkowski, F. (Ed.). (2005). *Feminist theory: A reader*. McGraw-Hill Higher Education.

Kottler, P. and Keller, K. L. (2011) *Marketing Management (14th ed.)* Prentice-Hall.

Markus, M. (1987, October). Toward a "critical mass" theory of interactive media. *Communication Research, 14* (5), 497-511.

McCombs, M. & Nolan, J. (1992, Summer). The relative constancy approach to consumer spending for media. *Journal of Media Economics*, 43-52.

Meadows, J. H. (1997, May). *Body image, women, and media: A media system dependency theory perspective*. Paper presented to the 1997 Mass Communication Division of the International Communication Association Annual Meeting, Montreal, Quebec, Canada.

Moore, G. (2001). *Crossing the Chasm*. Harper Business.

Raacke, J. and Bonds-Raacke, J (2008). MySpace and Facebook: Applying the Uses and Gratifications Theory to exploring friend-networking sites. *CyberPsychology & Behavior*, 11, 2, 169-174.

Robinson, L. (1986). *The facts on fax*. Steve Davis Publishing.

Rogers, E. (1962). *Diffusion of Innovations*. Free Press.

Rogers, E. (1986). *Communication technology: The new media in society*. Free Press.

Rogers, E. (2003). *Diffusion of Innovations, 3rd Edition*. Free Press.

Rubin, A. (2002). The uses-and-gratifications perspective of media effects. In J. Bryant & D. Zillmann (Eds.). *Media effects: Advances in theory and research*. Lawrence Earlbaum Associates, pp. 525-548.

Sorrentino, M. and Solsman, J. (2019). Disney Plus: Everything you need to know about Disney's streaming service. *Cnet*. https://www.cnet.com/news/disney-plus-streaming-service-everything-to-know-mandalorian-baby-yoda/.

Towers, W. (1987, May 18-21). *Replicating perceived helpfulness of radio news and some uses and gratifications*. Paper presented at the Annual Meeting of the Eastern Communication Association, Syracuse, New York.

Venkatesh, V, Morris, M, Davis, G, and Davis, F. (2003). User acceptance of information technology: Toward a unified View. MSI Quarterly, 27, 3, 425-478.

Wathne, E. & Leos, C. R. (1993). Facsimile machines. In A. E. Grant & K. T. Wilkinson (Eds.). *Communication Technology Update: 1993-1994*. Technology Futures, Inc.

Wei, R. (2006). Staying connected while on the move. *New Media and Society, 8* (1), 53-72.

4

The Structure of the Communication Industries

August E. Grant, Ph.D.*

The first factor that many people consider when studying communication technologies is changes in the equipment and use of the technology. But, as discussed in Chapter 1, it is equally important to study and understand all areas of the technology ecosystem. In editing this *Communication Technology Update* for more than 25 years, one factor stands out as having the greatest amount of short-term change: the organizational infrastructure of the technology.

The continual flux in the organizational structure of communication industries makes this area the most

dynamic area of technology to study. "New" technologies that make a major impact come along only a few times a decade. New products that make a major impact come along once or twice a year. Organizational shifts are constantly happening, making it almost impossible to know all of the players at any given time.

Even though the players are changing, the organizational structure of communication industries is relatively stable. The best way to understand the industry, given the rapid pace of acquisitions, mergers, start-ups, and failures, is to understand its organizational functions. This chapter addresses the organizational structure and explores the functions of those industries,

* J. Rion McKissick Professor of Journalism, School of Journalism and Mass Communications, University of South Carolina (Columbia, South Carolina).

which will help you to understand the individual technologies discussed throughout this book.

In the process of using organizational functions to analyze specific technologies, you must consider that these functions cross national as well as technological boundaries. Most hardware is designed in one country, manufactured in another, and sold around the globe. Although there are cultural and regulatory differences addressed in the individual technology chapters later in the book, the organizational functions discussed in this chapter are common internationally.

What's in a Name? The AT&T Story

A good illustration of the importance of understanding organizational functions comes from analyzing the history of AT&T, one of the biggest names in communication of all time. When you hear the name "AT&T," what do you think of? Your answer probably depends on how old you are and where you live. If you live in Florida, you may know AT&T as your local phone company. In New York, it is the name of one of the leading mobile telephone companies. If you are older than 60, you might think of the company's old nickname "Ma Bell."

The Birth of AT&T

In the study of communication technology over the last century, no name is as prominent as AT&T. But the company known today as AT&T is an awkward descendent of the company that once held a monopoly on long-distance telephone service and a near monopoly on local telephone service through the first four decades of the 20th century. The AT&T story is a story of visionaries, mergers, divestiture, and rebirth.

Alexander Graham Bell invented his version of the telephone in 1876, although historians note that he barely beat his competitors to the patent office. His invention soon became an important force in business communication, but diffusion of the telephone was inhibited by the fact that, within 20 years, thousands of entrepreneurs established competing companies to provide telephone service in major cities. Initially, these telephone systems were not interconnected,

making the choice of telephone company a difficult one, with some businesses needing two or more local phone providers to connect with their clients.

The visionary who solved the problem was Theodore Vail, who realized that the most important function was the interconnection of these telephone companies. Vail led American Telephone & Telegraph to provide the needed interconnection, negotiating with the U.S. government to provide "universal service" under heavy regulation in return for the right to operate as a monopoly. Vail brought as many local telephone companies as he could into AT&T, which evolved under the eye of the federal government as a behemoth with three divisions:

- *AT&T Long Lines*—the company that had a virtual monopoly on long distance telephony in the United States.

- *The Bell System*—Local telephone companies providing service to 90% of U.S. subscribers.

- *Western Electric*—A manufacturing company that made equipment needed by the other two divisions, from telephones to switches. (Bell Labs was a part of Western Electric.)

As a monopoly that was generally regulated on a rate-of-return basis (making a fixed profit percentage), AT&T had little incentive—other than that provided by regulators—to hold down costs. The more the company spent, the more it had to charge to make its profit, which grew in proportion with expenses. As a result, the U.S. telephone industry became the envy of the world, known for "five nines" of reliability; that is, the telephone network was designed to be available to users 99.999% of the time. The company also spent millions every year on basic research, with its Bell Labs responsible for the invention of many of the most important technologies of the 20th century, including the transistor and the laser.

Divestiture

The monopoly suffered a series of challenges in the 1960s and 1970s that began to break AT&T's monopoly control. First, AT&T lost a suit brought by the Hush-a-Phone company, which made a plastic mouthpiece that fit over the AT&T telephone mouthpiece to make it easier to hear a call made in a noisy area (*Hush-a-phone v. AT&T*, 1955; *Hush-a-phone v. U.S.*, 1956). (The idea of a company having to win a lawsuit in order to sell such an innocent item seems frivolous today, but this suit was the first major crack in AT&T's monopoly armor.) Soon, MCI won a suit to provide long-distance service between St. Louis and Chicago, allowing businesses to bypass AT&T's long lines (*Microwave Communications, Inc.*, 1969).

Since the 1920s, the Department of Justice (DOJ) had challenged aspects of AT&T's monopoly control, earning a series of consent decrees to limit AT&T's market power and constrain corporate behavior. By the 1970s, it was clear to the antitrust attorneys that AT&T's ownership of Western Electric inhibited innovation, and the DOJ attempted to force AT&T to divest itself of its manufacturing arm. In a surprising move, AT&T proposed a different divestiture, spinning off all of its local telephone companies into seven new "Baby Bells," keeping the now-competitive long distance service and manufacturing arms. The DOJ agreed, and a new AT&T was born (Dizard, 1989).

Cycles of Expansion & Contraction

After divestiture, the leaner, "new" AT&T attempted to compete in many markets with mixed success; AT&T long distance service remained a national leader, but few bought overpriced AT&T personal computers. In the meantime, the seven Baby Bells focused on serving their local markets, with most named after the region they served. Nynex served New York and the extreme northeast states, Bell Atlantic served the mid-Atlantic states, BellSouth served the southeastern states, Ameritech served the midwest, Southwestern Bell served south central states, U S West served a set of western states, and Pacific Telesis served California and the far western states.

Over the next two decades, consolidation occurred among these Baby Bells. Nynex and Bell Atlantic merged to create Verizon. U S West was purchased by Qwest Communication and renamed after its new parent, which was, in turn, acquired by CenturyLink in 2010. As discussed below, Southwestern Bell was the most aggressive Baby Bell, ultimately reuniting more than half of the Baby Bells.

In the meantime, AT&T entered the 1990s with a repeating cycle of growth and decline. It acquired NCR Computers in 1991 and McCaw Communications (then the largest U.S. cellular phone company) in 1993. Then, in 1995, it divested itself of its manufacturing arm (which became Lucent Technologies) and the computer company (which took the NCR name). It grew again in 1998 by acquiring TCI, the largest U.S. cable TV company, renaming it AT&T Broadband, and then acquired another cable company, MediaOne. In 2001, it sold AT&T Broadband to Comcast, and it spun off its wireless interests into an independent company (AT&T Wireless), which was later acquired by Cingular (a wireless phone company co-owned by Baby Bells SBC and BellSouth) (AT&T, 2008).

The only parts of AT&T remaining were the long distance telephone network and the business services, resulting in a company that was a fraction of the size of the AT&T behemoth that had a near monopoly on telephony in the United States two decades earlier.

Under the leadership of Edward Whitacre, Southwestern Bell became one of the most formidable players in the telecommunications industry. With a visionary style not seen in the telephone industry since the days of Theodore Vail, Whitacre led Southwestern Bell to acquire Baby Bells Pacific Telesis and Ameritech (and a handful of other, smaller telephone companies), renaming itself SBC. Ultimately, SBC merged with BellSouth and purchased what was left of AT&T, then renamed the company AT&T, an interesting case comparable to a child adopting its parent. The "new" AT&T extended into other markets with the acquisition of DirecTV in 2015 and Time Warner (now Warner Media) in 2018.

Today's AT&T is a dramatically different company with a dramatically different culture than its parent, but the company serves most of the same markets in a much more competitive environment. The lesson is that it is not enough to know the technologies or the company names; you also have to know the history of both in order to understand the role that a company plays in the marketplace.

Functions within the Industries

The AT&T story is an extreme example of the complexity of communication industries. These industries are easier to understand by breaking their functions into categories that are common across most of the segments of these industries. Let's start with the heart of the technology ecosystem introduced in Chapter 1—the hardware, software, and content.

For this discussion, let's use the same definitions used in Chapter 1, with *hardware* referring to the physical equipment used, *software* referring to instructions used by the hardware to manipulate content, and *content* referring to the messages transmitted using these technologies. Some companies produce both hardware and software, ensuring compatibility between the equipment and programming, but few companies produce both equipment and content. (A notable exception is Sony, which operates a movie and television production and distribution company in addition to its hardware manufacturing.)

The next distinction has to be made between production and distribution of both equipment and content. As these names imply, companies involved in production engage in the manufacture of equipment or content, and companies involved in distribution are the intermediaries between production and consumers. It is a common practice for some companies to be involved in both production and distribution, but, as discussed below, a large number of companies choose to focus on one or the other.

These two dimensions interact, resulting in separate functions of equipment production, equipment distribution, content production, and content distribution. As discussed below, distribution can be further broken down into national and local distribution. The following section introduces these dimensions, which then help us identify the role played by specific companies in communication industries.

One other note: These functions are hierarchical, with production coming before distribution in all cases. Let's say you are interested in creating a new type of telephone, perhaps a "high-definition telephone." You know that there is a market, and you want to be the person who sells it to consumers. But you cannot do so until someone first makes the device. Production always comes before distribution, but you cannot have successful production unless you also have distribution—hence the hierarchy in the model. Figure 4.1 illustrates the general pattern, using the U.S. television industry as an example.

Hardware

When you think of hardware, you typically envision the equipment you handle to use a communication technology. But it is also important to note that there is a second type of hardware for most communication industries—the equipment used to make the content. Although most consumers do not deal with this equipment, it plays a critical role in the system.

Content Production Hardware

Production hardware is usually more expensive and specialized than other types. Examples in the television industry include TV cameras, microphones, and editing equipment. A successful piece of production equipment might sell only a few hundred or a few thousand units, compared with tens of thousands to millions of units for consumer equipment. The profit margin on production equipment is usually much higher than on consumer equipment, making it a lucrative market for manufacturing companies.

Consumer Hardware

Consumer hardware is the easiest to identify. It includes anything from a television to a mobile phone or DirecTV satellite dish. A common term used to identify consumer hardware in consumer electronics industries is CPE, which stands for customer premises equipment. An interesting side note is that many companies do not actually make their own products, but instead hire manufacturing facilities to make products they design, shipping them directly to distributors. For example, Microsoft does not manufacture the Xbox One; Flextronics and Foxconn do. As you consider communication technology hardware, consider the lesson from Chapter 1—people are not usually motivated to buy equipment because of the equipment itself, but because of the content it enables, from the conversations (voice and text!) on a wireless phone to the information and entertainment provided by a high-definition television (HDTV) receiver.

Figure 4.1

Structure of the Traditional Broadcast TV Industry

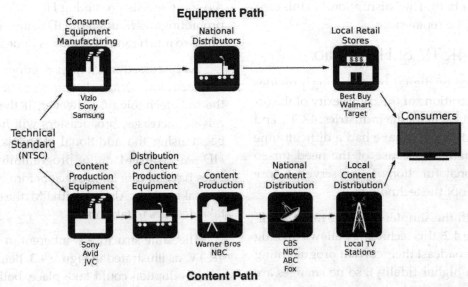

Source: R. Grant & G. Martin

Distribution

After a product is manufactured, it has to get to consumers. In the simplest case, the manufacturer sells directly to the consumer, perhaps through a company-owned store or a website. In most cases, however, a product will go through multiple organizations, most often with a wholesaler buying it from the manufacturer and selling it, with a mark-up, to a retail store, which also marks up the price before selling it to a consumer. The key point is that few manufacturers control their own distribution channels, instead relying on other companies to get their products to consumers.

Content Path: Production and Distribution

The process that media content goes through to get to consumers is a little more complicated than the process for hardware. The first step is the production of the content itself. Whether the product is movies, music, news, images, etc., some type of equipment must be manufactured and distributed to the individuals or companies who are going to create the content. (That hardware production and distribution goes through a similar process to the one discussed above.)

The content must then be created, duplicated, and distributed to consumers or other end users.

The distribution process for media content/software follows the same pattern for hardware. Usually there will be multiple layers of distribution, a national wholesaler that sells the content to a local retailer, which in turn sells it to a consumer.

Disintermediation

Although many products go through multiple layers of distribution to get to consumers, information technologies have also been applied to reduce the complexity of distribution. The process of eliminating layers of distribution is called disintermediation (Kottler & Keller, 2005); examples abound of companies that use the internet to get around traditional distribution systems to sell directly to consumers.

Netflix is a great example. Traditionally, digital videodiscs (DVDs) of a movie were sold by the studio to a national distributor, which then delivered them to thousands of individual movie rental stores, which, in turn rented or sold them to consumers.

Netflix cut one step out of the distribution process, directly bridging the movie studio and the consumer. (As discussed below, the Apple Store serves the same function for the music industry, simplifying music distribution.) The result of getting rid of one middleman is greater profit for the companies involved, lower costs to the consumer, or both. The "disruption," in this case, was the demise of the rental store.

Illustrations: 4K TV & HD Radio

The emergence of digital broadcasting provides two excellent illustrations of the complexity of the organizational structure of media industries. 4K TV and its distant cousin HD radio have had a difficult time penetrating the market because of the need for so many organizational functions to be served before consumers can adopt the technology.

Let's start with the simpler one: HD radio. As illustrated in Figure 4.2, this technology allows existing radio stations to broadcast their current programming (albeit with much higher fidelity), so no changes are needed in the software production area of the model. The only change needed in the software path is that radio stations simply need to add a digital transmitter. The complexity is related to the consumer hardware needed to receive HD radio signals. One set of companies makes the radios, another distributes the radios to retail stores and other distribution channels, then stores and distributors have to agree to sell them.

The radio industry has therefore taken an active role in pushing diffusion of HD radios throughout the hardware path. In addition to airing thousands of radio commercials promoting HD radio, the industry is promoting distribution of HD radios in new cars (because so much radio listening is done in automobiles).

As discussed in Chapter 8, adoption of HD radio has been slow because listeners see little advantage in the new technology. However, if the number of receivers increases, broadcasters will have a reason to begin using the additional channels available with HD. As with FM radio, programming and receiver sales have to *both* be in place before consumer adoption takes place. Also, as with FM, the technology may take decades to take off.

The same structure is inherent in the adoption of 4K TV, as illustrated in Figure 4.3. Before the first consumer adoption could take place, both programming and receivers (consumer hardware) had to be available. Because a high percentage of primetime TV programming was recorded in digital formats at the time 4K TV receivers first went on sale in the United States, that programming could easily be transmitted in 4K, providing a nucleus of available programming. On the other hand, local news and TV programs shot on video required entirely new production and editing equipment before they could be distributed to consumers in 4K.

Figure 4.2
Structure of the HD Radio Industry

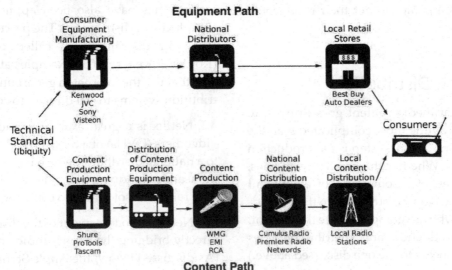

Source: R. Grant & G. Martin

Figure 4.3

Structure of the 4K TV Industry

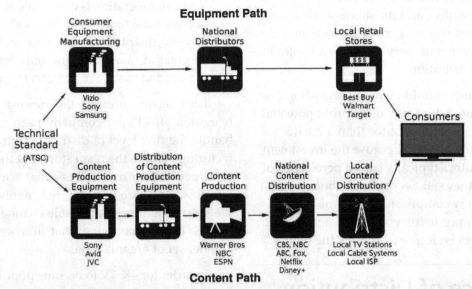

Source: R. Grant & G. Martin

As discussed in Chapter 6, the big force behind the diffusion of 4K TV and other digital TV formats was a set of regulations issued by the Federal Communications Commission (FCC) that enabled television stations to broadcast 4K signals.

The success of this format will require modifications of virtually all areas of the television system diagrammed in Figure 4.3, with new production equipment manufactured and then used to produce content in the new format, and then distributed to consumers—who are not likely to buy until all of these functions are served.

From Target to iTunes

One of the best examples of the importance of distribution comes from an analysis of the popular music industry. In the last century, music was recorded on physical media such as CDs and audio tapes and then shipped to retail stores for sale directly to consumers. At one time, the top three U.S. retailers of music were Target, Walmart, and Best Buy.

Once digital music formats that could be distributed over the internet were introduced in the late 1990s, dozens of start-up companies created online

stores to sell music directly to consumers. The problem was that few of these online stores offered the top-selling music. Record companies were leery of the lack of control they had over digital distribution, leaving most of these companies to offer a marginal assortment of music.

The situation changed in 2003 when Apple introduced the iTunes store to provide content for its iPods, which had sold slowly since appearing on the market in 2001. Apple obtained contracts with major record companies that allowed them to provide most of the music that was in high demand.

Initially, record companies resisted the iTunes distribution model that allowed a consumer to buy a single song for $0.99; they preferred that a person have to buy an entire album of music for $13 to $20 to get the one or two songs they wanted.

Record company delays spurred consumers to create and use file-sharing services that allowed listeners to get the music for free—and the record companies ended up losing billions of dollars. Soon, the $0.99 iTunes model began to look very attractive to the record companies, and they trusted Apple's digital rights management system to protect their music.

Today, as discussed in Chapter 8, digital distribution dominates the music industry. The music is similar, but the distribution process is dramatically different from what it was in 2001. The change took years of experimentation, and the successful business model that emerged required cooperation from dozens of separate companies serving different roles in production and distribution.

Two more points should be made regarding distribution. First, there is typically more profit potential and less risk in being a distributor than a creator (of either hardware or software) because the investment is less and distributors typically earn a percentage of the value of what they sell. Second, distribution channels can become very complicated when multiple layers of distribution are involved—the easiest way to unravel these layers is simply to "follow the money."

Importance of Distribution

As the above discussion indicates, distributors are just as important to new technologies as manufacturers and service providers. When studying these technologies, and the reasons for success or failure, the distribution process (including economics) must be examined as thoroughly as the product itself.

Diffusion Threshold

Analysis of the elements in Figure 4.1 reveals an interesting conundrum—there cannot be any consumer adoption of a new technology until all of the production and distribution functions are served, along both the hardware and software paths.

This observation adds a new dimension to Rogers' (2003) diffusion theory, discussed in the previous chapter. The point at which all functions are served has been identified as the "diffusion threshold," the point at which diffusion of the technology can begin (Grant, 1990).

It is easier for a technology to "take off" and begin diffusing if a single company provides a number of different functions, perhaps combining production and distribution, or providing both national and local distribution. The technical term for owning multiple functions in an industry is "vertical integra-

tion," and a vertically integrated company has a disproportionate degree of power and control in the marketplace.

Vertical integration is easier said than done, however, because the "core competencies" needed for production and distribution are so different. A company that is great at manufacturing may not have the resources needed to sell the product to end consumers.

Let's again consider the newest generation of television, 4K TV. A company such as Vizio might handle the first level of distribution, from the manufacturing plant to the retail store, but they do not own and operate their own stores—that is a very different business. They are certainly not involved in owning the streaming services or cable channels that distribute 4K TV programming; that function is served by another set of organizations.

In order for 4K TV to become popular, one organization (or set of organizations) has to make the television receivers, another has to get those televisions into stores, a third has to operate the stores, a fourth has to make 4K cameras and technical equipment to produce content, and a fifth has to package and distribute the content to viewers, through the internet using internet protocol television (IPTV), cable television systems, or over-the-air.

Most companies that would like to grow are more interested in applying their core competencies by buying up competitors and commanding a greater market share, a process known as "horizontal integration." For example, it makes more sense for a company that makes televisions to grow by making other electronics rather than by buying television stations. Similarly, a company that already owns television stations will probably choose to grow by buying more television stations or cable networks rather than by starting to make and sell television receivers.

The complexity of the structure of most communication industries prevents any one company from serving every needed role. Because so many organizations have to be involved in providing a new technology, many new technologies end up failing.

The lesson is that understanding how a new communication technology makes it to market requires comparatively little understanding of the technology itself compared with the understanding needed of the industry in general, especially the distribution processes.

A "Blue" Lesson

One of the best examples of the need to understand (and perhaps exploit) all of the paths illustrated in Figure 4.1 comes from the earliest days of the personal computer. When the PC was invented in the 1970s, most manufacturers used their own operating systems, so that applications and content could not easily be transferred from one type of computer to other types. Many of these manufacturers realized that they needed to find a standard operating system that would allow the same programs and content to be used on computers from different manufacturers, and they agreed on an operating system called CP/M.

Before CP/M could become a standard, however, IBM, the largest U.S. computer manufacturer—mainframe computers, that is—decided to enter the personal computer market. "Big Blue," as IBM was known (for its blue logo and its dominance in mainframe computers, typewriters, and other business equipment) determined that its core competency was making hardware, and they looked for a company to provide them an operating system that would work on their computers. They chose a then-little-known operating system called MS-DOS, from a small start-up company called Microsoft.

IBM's open architecture allowed other companies to make compatible computers, and dozens of companies entered the market to compete with Big Blue. For a time, IBM dominated the personal computer market, but, over time, competitors steadily made inroads on the market. (Ultimately, IBM sold its personal computer manufacturing business in 2006 to Lenovo, a Chinese company.)

The one thing that most of these competitors had in common was that they used Microsoft's operating systems. Microsoft grew... and grew... and grew. (It is also interesting to note that, although Microsoft has dominated the market for software with its operating systems and productivity software such as Office,

it has had less success in most areas of hardware manufacturing. Notable failures include its routers and home networking hardware, keyboards and mice, and WebTV hardware. The only major successes Microsoft has had in manufacturing hardware are with its Surface computers and Xbox video game systems.)

The lesson is that there is opportunity in all areas of production and distribution of communication technologies. All aspects of production and distribution must be studied in order to understand communication technologies. Companies have to know their own core competencies, but a company can often improve its ability to introduce a new technology by controlling more than one function in the adoption path.

What are the Industries?

We need to begin our study of communication technologies by defining the industries involved in providing communication-related services in one form or another. Broadly speaking, these can be divided into:

- *Mass media*—including books, newspapers, periodicals, movies, radio, and television.

- *Telecommunications*—including networking and all kinds of telephony (landlines, long distance, wireless, and voice over internet protocol).

- *Computers*—including hardware (desktops, laptops, tablets, etc.) and software.

- *Consumer electronics*—including audio and video electronics, video games, and cameras.

- *Internet*—including enabling equipment, network providers, content providers, and services.

These industries were introduced in Chapter 2 and are discussed in more detail in the individual chapters that follow.

At one point, these industries were distinct, with companies focusing on one or two industries. The opportunity provided by digital media and convergence enables companies to operate in numerous industries, and many companies are looking for synergies across industries.

Table 4.1

Examples of Major Communication Company Industries, 2020

	TV/Film/Video Production	TV/Film/Video Distribution	Print	Landline Telephone	Mobile Telephony	Internet
AT&T	●●●	●●●		●●●	●●●	●●●
Charter Cable		●●●		●●	●●	●●●
Comcast	●●●	●●●		●●	●●	●●●
Disney	●●●	●●●	●			●●
Google		●			●	●●●
Sony	●●●	●				●
Verizon	●	●		●●●	●●●	●●●
ViacomCBS	●●●	●●				●

The number of dots is proportional to the importance of this business to each company.   *Source:* A. Grant

Table 4.1 lists examples of well-known companies in the communication industries, some of which work across many industries, and some of which are (as of this writing) focused on a single industry. Part of the fun in reading this chart is seeing how much has changed since the book was printed in mid-2020.

There is a risk in discussing specific organizations in a book such as this one; in the time between when the book is written and when it is published, there are certain to be changes in the organizational structure of the industries. For example, as this chapter was being written in early 2020, Netflix and DirecTV were distributing 4K television through online services. By the time you read this, some broadcasters are sure to have adopted the set of technical standards known as ATSC 3.0, allowing them to deliver 4K HD signals directly to viewers.

Fortunately, mergers and takeovers that revolutionize an industry do not happen that often—only a couple a year! The major players are more likely to acquire other companies than to be acquired, so it is fairly safe (but not completely safe) to identify the major players and then analyze the industries in which they are doing business.

As in the AT&T story earlier in this chapter, the specific businesses a company is in can change dramatically over the course of a few years.

Future Trends

The focus of this book is on changing technologies. It should be clear that some of the most important changes to track are changes in the organizational structure of media industries. The remainder of this chapter projects organizational trends to watch to help you predict the trajectory of existing and future technologies.

Disappearing Newspapers

For decades, newspapers were the dominant mass medium, commanding revenues, consumer attention, and significant political and economic power. Since the dramatic drop in newspaper revenues and subscriptions began in 2005, newspaper publishers have been reconsidering their core business. Some prognosticators have even predicted the demise of the printed newspaper completely, forcing newspaper companies to plan for digital distribution of their news and advertisements.

Before starting the countdown clock, it is necessary to define what we mean by a "newspaper publisher." If a newspaper publisher is defined as an organization that communicates and obtains revenue by smearing ink on dead trees, then we can easily predict a steady decline in that business. If, however, a newspaper publisher is defined as an organization that gathers news and advertising messages, distributing them via a wide range of available media, then newspaper publishers should be quite healthy through the century.

The current problem is that there is no comparable revenue model for delivery of news and advertising through new media that approaches the revenues available from smearing ink on dead trees. It is a bad news/good news situation.

The bad news is that traditional newspaper readership and revenues are both declining. Readership is suffering because of competition from the web and other new media, with younger cohorts increasingly ignoring print in favor of other sources.

Advertising revenues are suffering for two reasons. The decline in readership and competition from new media are impacting revenues from display advertising. More significant is the loss in revenues from classified advertising, which at one point comprised up to one-third of newspaper revenues.

The good news is that newspapers remain profitable—at least on a cash flow basis. But many newspaper companies borrowed extensively to expand their reach, with interest payments often exceeding these gross profits. Stockholders in newspaper publishers have been used to much higher profit margins, and the stock prices of newspaper companies have been punished for the decline in profits.

Some companies reacted by divesting themselves of their newspapers in favor of TV and new media investments. Others used the opportunity to buy up other newspapers; in an attempt to wring out the last bit of profit from a dying industry.

Advertiser-Supported Media

For advertiser-supported media organizations, the primary concern is the impact of the internet and other new media on revenues. As discussed above, some of the loss in revenues is due to loss of advertising dollars (including classified advertising), but that loss is not experienced equally by all advertiser-supported media.

The internet is especially attractive to advertisers because online advertising systems have the most comprehensive reporting of any advertising medium. For example, advertisers using Google AdWords (discussed in Chapter 1) gets comprehensive reports on the effectiveness of every message—but "effectiveness" is defined by these advertisers as an immediate response such as a click-through.

As Grant & Wilkinson (2007) discuss, not all advertising is "call-to-action" advertising. There is another type of equally important advertising—image advertising, which does not demand immediate results, but rather works over time to build brand identity, increasing the likelihood of a purchase.

Any medium can carry any type of advertising, but image advertising is more common on television (especially national television) and magazines, and call-to-action advertising is more common online and in newspapers. As a result, newspaper advertising revenue has been more impacted by the increase in internet advertising.

Interestingly, local advertising is more likely to be call-to-action advertising, but local advertisers have been slower than national advertisers to move to the internet, most likely because of the global reach of the internet. This paradox could be seen as an opportunity for an entrepreneur wishing to earn a million or two by exploiting a new advertising market (Wilkinson, Grant & Fisher, 2012).

The "Mobile Revolution"

Another trend that can help you analyze media organizations is the shift toward mobile communication technologies. Companies that are positioned to produce and distribute content and technology that further enable the "mobile revolution" are likely to have increased prospects for growth.

Areas to watch include mobile internet access (involving new hardware and software, provided by a mixture of existing and new organizations), mobile advertising, new applications of GPS technology, and new applications designed to take advantage of internet access available anytime, anywhere.

Consumers—Time Spent Using Media

Another piece of good news for media organizations is the fact that the amount of time consumers are spending with media is increasing, with much of that increase coming from simultaneous media use (Papper, et al., 2009). Advertiser-supported media thus have more "audience" to sell, and subscription-based media have more prospects for revenue.

Furthermore, new technologies are increasingly targeting specific messages at specific consumers, increasing the efficiency of message delivery for advertisers and potentially reducing the clutter of irrelevant advertising for consumers. Already, advertising services such as Google's Double-Click and Google's AdWords provide ads that are targeted to a specific person or the specific content on a web page, greatly increasing their effectiveness.

Imagine a future where every commercial on TV that you see is targeted—and is interesting—to you! Technically, it is possible, but the lessons of previous technologies suggest that the road to customized advertising will be a meandering one.

Principle of Relative Constancy

The potential revenue from consumers is limited because they devote a fixed proportion of their disposable income to media, the phenomenon discussed in Chapter 3 as the "Principle of Relative Constancy." The implication is emerging companies and technologies have to wrest market share and revenue from established companies. To do that, they can't be just as good as the incumbents. Rather, they have to be faster, smaller, less expensive, or in some way better so that consumers will have the motivation to shift spending from existing media.

Conclusions

The structure of the media system may be the most dynamic area in the study of new communication technologies, with new industries and organizations constantly emerging and merging. In the following chapters, organizational developments are therefore given significant attention. Be warned, however; between the time these chapters are written and published, there is likely to be some change in the organizational structure of each technology discussed. To keep up with these developments, visit the *Communication Technology Update and Fundamentals* home page at www.tfi.com/ctu.

Bibliography

AT&T. (2008). *Milestones in AT&T history*. Retrieved from http://www.corp.att.com/history/ milestones.html.

Dizard, W. (1989). *The coming information age: An overview of technology, economics, and politics, 2nd ed*. New York: Longman.

Grant, A. E. (1990, April). The "pre-diffusion of HDTV: Organizational factors and the "diffusion threshold. Paper presented to the Annual Convention of the Broadcast Education Association, Atlanta.

Grant, A. E. & Wilkinson, J. S. (2007, February). Lessons for communication technologies from Web advertising. Paper presented to the Mid-Winter Conference of the Association of Educators in Journalism and Mass Communication, Reno.

Hush-A-Phone Corp. v. AT&T, et al. (1955). FCC Docket No. 9189. Decision and order (1955). 20 FCC 391.

Hush-A-Phone Corp. v. United States. (1956). 238 F. 2d 266 (D.C. Cir.). Decision and order on remand (1957). 22 FCC 112.

Kottler, P. & Keller, K. L. (2005). *Marketing management, 12th ed*. Englewood Cliffs, NJ: Prentice-Hall.

Microwave Communications, Inc. (1969). FCC Docket No. 16509. Decision, 18 FCC 2d 953.

Papper, R. E., Holmes, M. A. & Popovich, M. N. (2009). Middletown media studies II: Observing consumer interactions with media. In A. E. Grant & J. S. Wilkinson (Eds.) *Understanding media convergence: The state of the field*. New York: Oxford.

Rogers, E. M. (2003). *Diffusion of innovations, 5th ed*. New York: Free Press.

Wilkinson, J.S., Grant, A. E. & Fisher, D. J. (2012). *Principles of convergent journalism (2nd ed.)*. New York: Oxford University Press.

5

Regulation and Communication Technologies

Eric P. Robinson, J.D., Ph.D.*

Introduction
The Race Between Technology & Law

Do you use TikTok, YouTube, Instagram, or Snapchat? Do you post pictures or videos to these sites? Do you forward posts to others? In other words, do you use social media? A majority of American adults do, with particularly high rates among those ages 18 to 24 (Perrin & Anderson, 2019)

If you do, you are—whether you know it or not—involved in a race with the law, in which the law is inevitably doomed to always be behind. As has been

true with the development of what we now call "legacy media," the courts and the legal system are struggling to keep up with new developments and uses of the internet and social media. As these uses change and evolve, so do the legal issues and dilemmas. For example, privacy may mean a different thing today than it did when cameras used film as cameras on mobile phones allow instant, worldwide distribution. The emergence of social media has raised privacy issues—for example, access to and use of social media posts for advertising and other purposes—that were never previously contemplated. As the pace of innovation has increased, so have the legal dilemmas and the need for rules on these activities.

---

* Assistant Professor, School of Journalism and Mass Communications, University of South Carolina (Columbia, South Carolina).

The race between law and technology—its past, its present, and its future—is the focus of this chapter.

Always Playing Catch-up

Oliver Wendell Holmes, Jr., later a U.S. Supreme Court Justice, said in 1881 that "The life of the law has not been logic; it has been experience" (Holmes, 2013). Surveying regulation of media in the United States—which varies wildly by medium, from almost complete *lassis-faire* approach towards print media to content restrictions imposed on broadcast media—shows the wisdom and accuracy of this statement. American media regulation as a whole, lacks cohesive logic, it is based on the learned experience of each particular medium's historic, political, and social role in American culture.

The *Telecommunications Act of 1996* was a deviation from this pattern. Anticipating the convergence of previously separate media technologies, this statute established the groundwork for media entities to stray from their traditional businesses into competition with other entities in different telecommunications sectors. It also allowed for government regulation of such activities. But while the *Act* was in many ways anticipated developments in media, there were inevitably unexpected developments that led to new issues that the law did not foresee (Hendricks, 1999; Furchtgott-Roth & Roth, 2016).

Media regulation is both reactive and anticipatory, in that it reflects not only the real issues that have arisen with each form of media, but also the perceived—and, oftentimes, feared—concerns that each type of medium engenders. It is also constrained—to a greater or lesser extent that also depends on the particular medium—by the free speech and press provisions of the First Amendment.

Background

The First Amendment & Media Regulation

Any regulation of the media in the United States is limited by the First Amendment of the United States Constitution, which provides that "Congress shall make no law … abridging freedom of speech, or of the press" (U.S. Const., amend. I).

The seeming absolutism of the phrase "no law" in the First Amendment is deceptive. Indeed, within a decade of ratification of the First Amendment (and the other first ten amendments to the Constitution, known as the Bill of Rights), a Congress dominated by the Federalist party passed and Federalist president John Adams signed a Sedition Act in 1798 which criminalized all speech disparaging the president, Congress, or the federal government as a whole (Slack, 2015; Stone, 2004). Arguments to trial courts that the statute violated the First Amendment were unavailing, and 26 cases under the statute were brought against 32 individuals, resulting in 13 convictions (Bird, 2016).

While by its own terms the *Sedition Act* expired with the end of the Adams presidency—evidence that it was intended to target his political opponents—the attitude embodied by that statute predominated in the first 125 years after ratification of the First Amendment: that the Amendment barred the federal government from prohibiting speech before it was uttered or published, but did not place any restrictions on subsequent penalties that could be imposed (*Patterson v. Colorado*, 1907; Rabban, 1997; Rosenberg, 1986). This allowed for the continuation of English common law concepts such as civil and criminal punishment for defamatory speech, although early cases and later statues in the states during the early years of American independence established truth as a defense to a defamation case, and later recognized that public officials, in particular, should be subject to legitimate criticism for their public actions (Rabban, 1997). The doctrine that obscene speech—however defined—could be prohibited and punished also survived (Rabban, 1997).

This understanding of the First Amendment began to change in the early 20th Century in a series of prosecutions of dissent from U.S. participation in World War I. For example, in *Schenk v. United States* several individuals were prosecuted for distributing flyers alleging that the military draft constituted slavery and violated the Thirteenth Amendment banning most involuntary servitude. The U.S. Supreme Court affirmed their conviction, holding that the defendants' speech could be punished because it constituted a "clear and present danger" (*Schenk v. United States*, 1919).

Fifty years later, the U.S. Supreme Court set the current standard, which allows for punishment of speech only when it carries the threat of causing "imminent lawless action" (*Brandenburg v. Ohio*, 1969). And in 1971, when the federal government sought to stop publication of the *Pentagon Papers* history of United States' involvement in southeast Asia leading up to the Vietnam War, the Supreme Court held that speech could be prohibited only when the restriction serves a compelling government interest and is the least restrictive means of protecting that interest (*New York Times v. United States*, 1971).

There have also been questions over whether First Amendment protection varies by medium. As new media have been invented and moved to mainstream use, policy-makers and courts have had to determine what, if any, government control and regulation would be imposed on each new medium.

The United States inherited its initial laws from England, which had already confronted this issue regarding print media. Initially, when the printing press ended the religious and political elites' control of and access to manuscripts, the religious and government authorities instituted licensing schemes for the new medium. But by the late 17th century England replaced direct control through licensing with indirect forms of control, such as imposition of taxes on printed material. While the British colonies in North American were not included in the licensing, the taxes on printed matter became a cause of the American Revolution (Starr, 2004).

The introduction of the telegraph in the 1840s and the telephone in the 1870s raised questions of how these new media would be regulated. While European countries generally favored government ownership and control of these electronic media, in the United States these media eventually ended up being privately-owned and operated but subject to extensive government regulation (Starr, 2004). As these industries matured, they came to dominated by single, nationwide firms—the telegraph by Western Union, the telephone by American Telephone & Telegraph (AT&T)—through the machinations of their owners with an assist from government (Lloyd, 2006). Classified legally as "common carriers," these companies were required to accept all customers able to pay their rates and were immune from liability for the contents of the communications that they transmitted. (Lloyd, 2006; Nonnenmacher, 2001).

The 20th Century saw the emergence of motion pictures as a mass medium, followed by local governments' efforts to control it. In 1915, the U.S. Supreme Court upheld the ability of state and local governments to screen and censor movies based on their content, holding that "the exhibition of moving pictures is a business, pure and simple, originated and conducted for profit like other spectacles, and not to be regarded as part of the press of the country or as organs of public opinion within the meaning of freedom of speech and publication guaranteed by the Constitution of Ohio. … We cannot regard [the censorship of movies] as beyond the power of government." (*Mutual Film Corp. v. Industrial Comm'n of Ohio*, 1915). This led the motion picture industry to adopt its own limitations on content, known as the Motion Picture Production Code or Hays Code (Starr, 2004). But the influence of the code waned after independent producers and distributors began to challenge the limitations, and in 1952 the U.S. Supreme Court reversed course and held that "expression by means of motion pictures is included within the free speech and free press guaranty of the First and Fourteenth Amendments," overruling its prior decision (*Burstyn v. Wilson*, 1952).

The initial interest in "wireless" radio communications was for military purposes during World War I. But eventually it came into civilian use, with commercial radio beginning in the 1920s in the United States. This new medium was based on a structure of licensure of private entities to use publicly-owned airwaves in the "public convenience, interest or necessity." Thus, broadcasting in America, unlike print and other prior media, has always been a "licensed press." (Powe, Jr., 1987).

But while U.S. broadcasting regulators are specifically barred from censoring content or restricting free speech, the government does use its licensing power to exert some control over content. In 1930 the Federal

Radio Commission denied license renewal from one station whose programming consisted mainly of promotions for dubious medical remedies sold by the station's owner, a doctor, and from another that featured diatribes against local government officials, Catholics, and Jews (Starr, 2004; Powe, Jr., 1987; *KFKB Broadcasting Ass'n, Inc. v. FRC*, 1932; *Trinity Methodist Church, South v. FRC*, 1932).

Commercial television emerged after World War II under a similar regulatory scheme as radio, based on individual station licensing, resulting in a similar structure of oligopoly networks (Lloyd, 2006). And as it did with radio, the government has continued to use its licensing power to exert some control over content on television. In 1943 the U.S. Supreme Court upheld Federal Communications Commission efforts to rein in network control over affiliates (Powe, Jr., 1987; *Nat'l Broad. Co. v. United States*, 1943). Then, in 1966 the Commission was forced by the federal courts to accept public input into broadcast license renewal decisions, and later the court overturned a license renewal decision that was contrary to such input (Lloyd, 2006; *Office of Communication of United Church of Christ v. FCC*, 1966; *Office of Communication of United Church of Christ v. FCC*, 1969).

One area of regulation is political content. As early as 1927 Congress required radio broadcasters to offer equal opportunities to political candidates to buy advertising time (Starr, 2004). In 1949, the Federal Communications Commission formalized a requirement that both radio and television broadcasters allow the airing of differing opinions on public issues: the so-called "fairness doctrine" (Arbuckle, 2017; Lloyd, 2006). This policy, and government regulation of broadcast radio and television content more generally in the "public interest," was upheld by the U.S. Supreme Court in 1969 (Lloyd, 2006; *Red Lion Broadcasting Co. v. FCC*, 1969). The FCC later revoked the fairness doctrine under the theory that it violated the First Amendment, and the repeal was upheld by the courts (Arbuckle, 2017; Lloyd, 2006; *Syracuse Peace Council v. FCC*, 1989).

Another type of regulation limited broadcast of "indecent" material to hours when children would not be expected to accessing broadcast media. The U.S. Supreme Court upheld these restrictions in 1978,

on the grounds that broadcasting is uniquely ubiquitous and readily accessible to children (*FCC v. Pacifica Found.*, 1978). These regulations persist, although enforcement has waxed and waned over the years as successive presidential administrations vowed crackdowns while others eased back on application of these rules (O'Connor, 2019).

Scandals in the 1950s led to more regulation (Lloyd, 2006). The discovery that several television game shows were fixed led to Congressional hearings and laws making any contest or game intended to deceive the audience illegal (Anderson, 1978), while the revelation that radio and television music programs were receiving payment for playing particular songs ("payola") led to laws and FCC regulations banning such payments unless they are disclosed (Coase, 1979). Because these restrictions are FCC rules, they apply to broadcasters but not to other forms of media. The FCC also got involved in some regulation of cable television when it emerged in the 1960s, with authority derived from the government's power over broadcast stations whose signals cable was originally established to convey to difficult-to-reach locations (Lloyd, 2004; *U.S. v. Southwestern Cable Corp.*, 1968).

The emergence of the internet, and then social media sites and apps, presented an entirely new medium which raised several new social, legal and regulatory questions. A fundamental question was which model of government regulation would be imposed online: the relative free-for-all of print media, or the content and other restrictions of broadcast media. The U.S. Supreme Court answered this question in 1997 when rejecting Congress's attempt to impose restrictions on obscene, indecent and other material harmful to children online, rejecting the government's argument that the internet is like broadcasting in that it is ubiquitous and readily accessible to children. "...our cases," the Court majority ruled, "provide no basis for qualifying the level of First Amendment scrutiny that should be applied to [the internet]." (*Reno v. ACLU*, 1997).

The explosive growth of the internet and social media since the *Reno* decision has created new policy issues regarding government regulation of both these new forms of media, as well as older, legacy media. Some of these are new permutations of issues that arose with older media, while others are unique to the current digital age and the changes in technology, the economy, and society that it has brought.

Current Issues

Privacy

In the United States, the legal right to privacy is limited and tenuous. But the rise and dominance of social media and its collection and sale of personal information has led to calls for a major revamp and strengthening of privacy rights.

The "right of privacy," as a legal concept, did not exist in American law until the 1910s. In 1890, future U.S. Supreme Court Justice Louis Brandeis and his law partner Samuel Warren argued for creation of privacy rights, writing in the *Harvard Law Review* that "…recent inventions and business methods call attention to the next step which must be taken for the protection of the person, and for securing to the individual … the right 'to be let alone'" (Warren & Brandeis, 1890). States then began passing statutes barring unauthorized use of an individual's image in commercial advertisements, an action was known variously as "appropriation" or "misappropriation," or as a violation of the individual's "right of publicity." By the 1960s, this was one of four generally-recognized privacy torts, the others being intrusion, publication of private information, and "false light," which is presentation of true information in a manner that conveys a false impression (Prosser, 1960). In 1965 the U.S. Supreme Court recognized that a privacy right was implicit in the "penumbras" of the constitutional guarantees such as free speech and against unreasonable search and seizure (*Griswold v. Connecticut*, 1965).

Congress and the states have passed various privacy provisions since then. But these are mostly limited in scope, with federal laws covering individual information held by the federal government (federal Privacy Act); health information (provisions of the federal Health Insurance Portability Protection Act); and information collected from minors (federal Children's Online Privacy Protection Act, COPPA). Other laws protect information in specific modalities, such as federal and state laws prohibiting wire-tapping, accessing aural and digital communications as they are transmitted.

The Federal Trade Commission has taken some action against misuse or breaches of personal data, but due to the lack of federal privacy provisions these actions are based on violations of the companies' stated privacy policies or terms of service, or violations of prior FTC consent decrees. Cases include a $22.5 million fine against Google in 2012 and a record $5 billion fine paid by Facebook in 2019 over its failure to follow its own privacy policy (Feiner & Rodriguez, 2019; Yarrow, 2012).

There is significantly more protection for privacy in the European Union. A right of privacy is included in the EU's *Charter of Fundamental Rights of the European Union*, and has been implemented through the EU's *General Data Protection Regulation* (GDPR). The regulation went into effect on May 25, 2018, replacing the *1995 Data Protection Directive*. As this is written, the application of the rule to information transfers between Great Britain and the remaining EU countries after Brexit is subject to negotiation (Telegraph Reporters, 2019; Dibene, 2017).

GDPR and its predecessor require entities that collect and retain personal information to clearly explain the data collected, its purpose, and whether it is shared with other entities. Data collectors must also give individuals access to data about them, control over how and whether such data is retained, and the right to have the data deleted. In May 2014, the European Court of Justice held that these rights include a "right to be forgotten," which is the right for EU residents to request removal from search engine results links to data that "appear to be inadequate, irrelevant or no longer relevant or excessive in the light of the time that had elapsed" (Court of Justice of the European Union, 2014). This removal applies only to sites directed towards EU residents: an effort to apply the doctrine to sites not intended for EU residents was rejected by the court (Amaro & Taylor, 2019).

The EU regulations, and their implementation and enforcement by EU member states, has resulted in hefty fines against Google and Facebook (Osborne, 2018; Stoller & Merken 2019). In 2018 Facebook pledged to follow GDPR for users worldwide (Tarantola, 2018).

California has adopted its own statute modeled after the EU law. *The California Consumer Privacy Act* (CCPA), effective in January 2020, generally provides state residents with the right to control what individual data is collected about them by large corporations, including the rights to access such data, to request its deletion, and to control how is it is used, including whether it is sold to other entities. Other states, including Maine and Nevada, have passed their own data privacy laws, and more states are considering such laws (Burt, 2018; Determann & Engfeldt, 2019). But because of its prominence, it is likely that California's law will become a *de facto* national standard in the United States, unless the federal government takes action (*Brennan, et al., 2019*; Brill, 2019).

As this is written in early 2020, there are discussion in Congress about legislation to strengthen individual privacy rights for United States residents, but partisan differences have kept any bills from passing (Birnbaum, 2019; Laslo, 2019). Americans largely support the creation of a legal right of privacy (Bode & Jones, 2017), although any such statute would have to overcome several First Amendment obstacles (Antani, 2015).

Online Provider Liability

As part of their status as common carriers, telegraph and telephone companies are usually not held liable for the legal harms caused by the material they transmit: a rule applied to other entities that convey content created entirely by others, such as bookstores (Branscomb, 1995; *Smith v. California*, 1959).

In the early days of the internet, courts differed on whether online publishers—website operators—were liable for material posted on their sites by others. In order to resolve this confusion, section 230 of the *Communications Decency Act* provides that "…no provider or user of an interactive computer service shall be treated as the publisher or speaker of any information provided by another information content provider." (section 230(c)(1)). Under this provision, operators of online bulletin boards, social media platforms, and other sites and platforms that allow users to post material, are not legally liable for such posts unless the site or service is actively involved in the creation of the content. (Frieden, 1997; Ou, 2013). In *Fair Housing Council of San Fernando Valley v. Roommates.Com, LLC* the court found curating and deleting problematic posts is not enough involvement is to lead to liability,

but forcing users to select from unlawful preferences on a roommates matching site may make the site liable under housing anti-discrimination laws (Ou, 2013; Fair Hous. Council of San Fernando Valley v. Roommates.Com, LLC, 2008). Courts have applied the immunity conferred by section 230 broadly, finding web site operators and others such as internet service providers immune from claims including defamation, privacy, other tort claims, and civil rights claims (Ou, 2013). But providers are not immune from copyright claims and criminal offenses.

Some legislators and scholars have expressed concern over the courts' generally broad application of section 230 immunity (Ou, 2013). In 2018, President Trump signed the first legislative scaleback of section 230 since the statute's creation: the *Allow States and Victims to Fight Online Sex Trafficking Act* ("FOSTA"), which excluded sex trafficking from section 230 immunity (Goldman, 2019). This may presage further attempts to limit section 230 for other sympathetic plaintiffs, or for certain types of offensive speech (Goldman, 2019; Wakabayashi, 2019).

Ownership/Anti-Trust

Federal and state anti-trust laws were first adopted in the late 19th century in an effort to rein in large companies that emerged and dominated various industries such as oil, steel and railroads. *The Sherman Anti-Trust Act*, enacted in 1890, outlaws unreasonable contracts, combinations, or conspiracies in restraint of trade and any monopolization, attempted monopolization, or conspiracy or combination to monopolize, Agreements to fix prices, divide markets, or rig bids are *"per se"* violations of the Act. The U.S. Department of Justice is the primary enforcer of this law, although private parties may also bring lawsuits under the statute.

Congress passed two additional antitrust laws in 1914: the *Federal Trade Commission Act* and the *Clayton Act*. The first created the Federal Trade Commission to create and enforce regulations banning "unfair methods of competition," "unfair or deceptive acts or practices," and violations of the *Sherman Act*. The *Clayton Act*, including subsequent amendments, bars interlocking directorates (a single person making business decisions for competing companies); discriminatory prices, services, and allowances in dealings between merchants; and mergers and acquisitions where the effect "may be substantially to lessen competition, or to tend to create a monopoly." Additionally, the *Hart-Scott-Rodino Antitrust Improvements Act* (1976) requires companies to give advance notice to the government of plans for large mergers or acquisitions, for antitrust review. The FTC and private entities may file suits under these laws.

While the anti-trust laws allow for substantial monetary fines and even imprisonment, in many cases the government avoids litigation by reaching a consent decree in which the offensive conduct is curbed, limited or prohibited. Any such decree must be approved the court, based on the public interest.

While these anti-trust laws sought to limit market power in manufacturing and other industries, in broadcasting the government's allocation of frequencies — primarily meant to avoid interference — combined with commercial imperatives to foster development of local stations affiliated with an oligopoly of a few national radio and television networks (Starr, 2004; Lloyd, 2006). An anti-trust case against the networks brought in 1972, alleging that they monopolized prime time programming, ended with the networks agreeing to scale back their ownership of such programs (Shanahan, 1976; Pincus & Lardner, Jr., 1997).

Notable anti-trust cases involving communications include actions against the major movie studios, IBM, AT&T, and Microsoft. The case against the movie studios went through several iterations until it ended with a consent decree which forced the studios to sell the movie theaters that they owned and prohibited "block booking," the practice of offering films for exhibition only in collections, rather than individually (United States v. Paramount Pictures, Inc., 1948; Maddaus, 2019). (In 2019, the U.S. Department of Justice moved to vacate the consent decrees from this case. (Maddaus, 2019)). The IBM case alleged that the company unfairly dominated mainframe business computing. But after 13 years, the government dropped the case, in part because of the shift from to personal computing. The AT&T case challenged the company's nationwide dominance of telephone service, including individual connections and local and long-distance calling, and after nine years resulted in a settlement which required the separation of the company from regional subsidiaries that provided connections and local calling, while creating a competitive market for long-distance calling. (Lloyd, 2006; *U.S. v. AT&T*, 1982). The Microsoft case, which alleged that the company had leveraged the dominance of its Windows computer operating system to promote its web browsing software and discourage use of competitors' products, ended with the company surviving but with restrictions on its activities. (*Massachusetts v. Microsoft Corp.*, 2004).

There have also been instances in which anti-trust enforcement — or the threat of it — has been used to manipulate the media. President Lyndon Johnson demanded a "letter of fidelity" from the owner of the *Houston Chronicle* before approving the merger of a bank of which the *Chronicle* publisher was the president (Caro, 2012, pp. 523-27). And President Nixon explicitly discussed using an anti-trust suit against the major television networks to leverage favorable news coverage (Pincus and Lardner, Jr., 1997).

There has been little use of antitrust laws against media entities since the Microsoft case. During the Clinton Administration, the *Telecommunications Act of 1996* paved the way to unification of various telecommunications and media services, and also towards more corporate consolidation across media (Lloyd, 2006). Since then, the federal government has approved the combinations of several media entities, leading to a relative handful of major media companies controlling most prominent online and offline media in the United States.

In recent years, several members of the U.S. Congress and President Trump have alleged that Facebook and other social media services are biased towards "liberal" news sources, and downgrade "conservative" sources and voices (Nunez, 2016). In response, Facebook replaced human content selectors with algorithms, but this resulted in a multitude of verifiably false news and claims appearing in the section (Ohlheiser, 2016).

The claims of bias led to calls for investigation of whether Facebook and other social media sites are illegal monopolies. A House of Representative subcommittee began an inquiry into the platforms' dominance, and may also consider changing anti-trust law to reflect modern technology and business practices.

Other concerns regarding antitrust involve user privacy and security of users' data and platforms' dominance in online advertising, The Justice Department has warned that misuse of user data could be the basis of antitrust charges, besides privacy claims (Romm, 2019a).

Some concerns are specific to individual companies. Concerns about Google/Alphabet include its control of the dominant Android cell phone operating system; its dominance in online search; and its dominant web browser Chrome. Regarding Amazon, a major concern is its practice of favoring its own products over competitors' on its retail website, and the company's dominance in cloud storage and processing services, including for several competing online marketplaces. Facebook, meanwhile, faces skepticism because of its dominance in social media, including its ownership of Instagram and WhatsApp (Brandom, 2018; Hirsch & Feiner, 2019).

In 2019, the federal Department of Justice and Federal Trade Commission announced that they would divide responsibilities for antitrust investigations against tech companies (Bartz & Wolfe, 2019). Meanwhile, state attorneys general from 47 states also began their own effort against the companies (Romm, 2019b).

Other concerns have arisen over foreign ownership or control of prominent technology companies. In 2019, the Trump administration barred import and use of Huawei brand smartphones in the United States over concerns that they allowed the Chinese government to access information stored on the phones. In 2019, similar privacy concerns arose with the popular TikTok social media app, which is owned by the Chinese company ByteDance.

Meanwhile, the European Union has vigorously pursued antitrust cases against several new media organizations, resulting in fines against Google totaling €8.25 billion (≈ US$9 billion); and a total of €2.2 billion (≈ US$2.45 billion) against Microsoft.

Net Neutrality

"Net neutrality" is the concept that all traffic and content on the internet should be treated equally, without any particular content either receiving priority or being hampered in its transmission to end users. Proponents of the concept assert that it is necessary to protect minority voices, and allow new entrants into online businesses. Opponents counter that the government should not dictate how private companies allow use of their proprietary internet infrastructure, and that the internet has robustly developed without any such rules in place.

Legally, net neutrality is based on the "common carrier" requirements applied to earlier forms of media, such a telegraph and telephone, which required carriers in these industries to accept all customers who were able to pay their rates and barred carriers from blocking (and liability for) most content transmitted by users (Wu, 2003).

The Federal Communications Commission has made three attempts to adopt net neutrality rules, but each time the rules have been thwarted by either court rulings or political decisions.

The Telecommunications Act of 1996 divides telecommunications services into two types for regulatory purposes: "telecommunications" services, covered by title 2 of the *Act*, are those that simply transmit information, while title 1 covers "information" services, in which data is processed or altered before or as it is transmitted. The Act allows for intense regulation of title 2 "common carrier" services, while title 1 "information" services are subject to little regulation. FCC rules determine into which category particular services fall, and the FCC had previously determined that telephone and telegraph services fell under title 2 while online and computer processing services fell under title 1.

The FCC's first attempt to adopt net neutrality rules in 2008, and the second in 2010, were both struck down in the courts on the grounds that the regulations exceeded those permitted for Title 1 services. In response, the FCC reclassified internet services as Title 2 services, subject to more regulation. It then adopted a new set of net neutrality rules in 2015 based on this new classification.

The 2015 regulations were upheld in court, based on the reclassification (*U.S. Telecomm. Ass'n v. FCC*, 2016). But Donald Trump was elected president a few months after the court decision, and in February 2017 the reconstituted FCC repealed the 2015 rules. That repeal has been upheld in court (*Mozilla Corporation v. FCC*, 2019).

The repeal of the FCC net neutrality rules is unlikely to be the end of the issue. California has adopted its own net neutrality rules, and the issue is likely to be a political and electoral issue for several years.

Frequency Reallocation & New Technology in Broadcasting

The government's role in regulating radio and then television broadcasting was initially limited to allocation of frequencies to avoid interference between stations. In practice, however, this also allowed the government to shape the form and content of the broadcast media. For example, government allocation policies combined with commercial imperatives to foster the development of local stations affiliated with a few national radio networks (Starr, 2004; Lloyd, 2006).

In the past decade, the Federal Communications Commission has, at the direction of Congress, undertaken a process of reallocating frequencies in the electromagnetic spectrum to make space for increasing demand for faster wireless internet services. Some of this reallocation has been accomplished by an auction process, in which existing broadcasters volunteered to relocate to another frequency to make way for other uses, in return for a share of the bid of the winning new user (Moore, 2013). But exponential increases in demand may require other innovative solutions, such as facilitation and subsidization of shared frequencies among government and commercial uses, and even sharing among competing commercial providers (Merwaday, et al., 2018; Rebato & Zorzi, 2018).

Television broadcasters and television set manufacturers are also introducing ATSC 3.0 or "NextGen" broadcasting standards and equipment, which allows for broadcast signals to be received not only by televisions but also by mobile and other devices. The new technology will also allow for tighter integration of broadcast and online sources, and for precise monitoring of viewership to allow for targeted advertising (Morrison, 2020).

Intellectual Property

The ease of reproduction and distribution of data with modern technologies presents the industries based on selling access to content—such as the book, music, film, and television industries—with both new opportunities and novel challenges. As physical musical and video recordings gave way to digitized media streams, established methods of distribution, payment, and compensation for the creators and distributors of such content have been fundamentally challenged.

Most of these works are covered by copyright, which protects works of expression for a set period of time. In the United States, this period lasts until 75 years after the death of the author; for works created by or for a corporate entity, the period is either 95 years from first publication, or 120 years from creation, whichever expires first. Words and logos that are used to identify specific products or services can be protected as trademarks, for 10-year renewable terms at the federal level; there is also less robust protection at the state level. In contrast to content, inventions and other new concepts or processes are eligible for protection by patents, which last for one, non-renewable 20-year term beginning when the patent application it is filed with the government.

The ease of duplicating copyrighted work with modern technology challenges the legal concept of "derivative works." Under traditional copyright law, the right to make derivative works—such as a film version of a novel—exclusively belonged to the author of the original work; rights which the author could use themselves or sell to someone else. On the other hand, the law allows for some, limited use of a copyrighted work without authorization or compensation to the author, under the doctrine of "fair use." But easy duplication allows for more discrete usage, making the difference between a fair use and an infringing work less clear.

Just as in other areas of the law, the copyright laws underlying these content industries, regarding ownership and distribution, have not kept up. Most nations around the world now comport their domestic copyright laws with the *International Berne Convention for the Protection of Literary and Artistic Works,* which was negotiated in 1886. (The United States did not ratify the convention until 1986.) But while there have been some more recent international agreements on some copyright issues, most intellectual property issues are now often negotiated in bilateral and regional trade agreements, known as *Trade-Related Aspects of Intellectual Property Rights* (abbreviated TRIPS). Another issue is that ownership of intellectual property normally vests in its (human) creator, ignoring the possibility of creation by mechanical or artificial intelligence (Tarantola, 2017).

Different Media, Different Rules

While the *Telecommunications Act of 1996* tried to anticipate media convergence and to facilitate coherent governance of such combined services, many areas of law and regulation remain separated by medium. These differences remain as a result of various factors, including the history of a particular medium, the government's initial approach towards regulation of the medium, the social and political role that the particular medium took in society and the technical requirements and limitations of the technology.

For example, the law provides that a song that is protected by copyright may be used for commercial purposes only with permission, usually including payment of a licensing fee to both the songwriter of a work and the group/artist that performs it. The specific rate is customarily set and collected by organizations such as the American Society of Composers and Publishers (ASCAP), but different according to medium and the specifics of the use. Restaurants, bars, most retail stores, and even live performance venues such as stadia and auditoriums must pay a "performance rate" for their use of music. Producers and distributors of video programs pay a different, "synchronization" rate for music to be used in their productions. There is yet another rate for music distributed via streaming services. Radio and television broadcasters pay royalties to songwriters, but not to artists. And there are some uses of music, such as background or bumper music on radio, that are royalty-free (BMI, n.d.; Music Royalty Exchange, 2014).

These rules are complex and confusing and have led to some calls for more equal treatment of different media (Branigan, 2018).

Conclusion

The rules regarding copyright royalties are partially result of the fact that government in the United States has taken a variety of approaches towards media regulation, with rules for a specific medium based on the historical, social and political context in which the media emerged and now operates. And all regulation of the media in the United States is shaped and limited by the First Amendment.

The emergence of problems with new online media has recently led to a rethinking of the general *lassis-faire* approach towards regulation of internet media, which may change the paradigm of media regulation—both online and offline—in the coming years.

Bibliography

Arbuckle, M.R. (2017). How the FCC Killed the Fairness Doctrine: A Critical Evaluation of the 1985 Fairness Report Thirty Years After Syracuse Peace Council, *First Amendment Law Review 15*, 331-379.

Amaro, S. & Taylor, C. (2019, Sept. 24), *Google does not have to apply 'right to be forgotten' globally, EU court rules*, CNBC, https://www.cnbc.com/2019/09/24/eu-rules-on-google-right-to-be-forgotten-case.html.

Anderson, J. K. (1979). *Television Fraud: The History and Implications of the Quiz Show Scandals*. Greenwood Press.

Antani, R. (2015). The Resistance of Memory: Could The European Union's Right To Be Forgotten Exist In The United States? *Berkeley Technology Law Journal 30*, 1173-1210.

BMI [Broadcast Music International] (n.d.). *How We Pay Royalties*. https://www.bmi.com/creators/royalty/general_information.

Bartz, D. & Wolfe, J. (2019, June 3). *U.S. moving toward major antitrust probe of tech giants*. Reuters. https://www.reuters.com/article/us-usa-technology-antitrust-idUSKCN1T42JH.

Bird, W. (2016). *Press and Speech Under Assault: The Early Supreme Court Justices, the Sedition Act of 1798, and the Campaign Against Dissent*. Oxford University Press.

Birnbaum, E. (2019, Nov. 18). *Senate Democrats unveil priorities for federal privacy bill.* The Hill. https://thehill.com/policy/technology/470963-senate-democrats-unveil-priorities-for-federal-privacy-bill.

Bode, L. & Jones, M.L. (2017). Ready to forget: American attitudes toward the right to be forgotten. *The Information Society 33(2),* 76-85. http://dx.doi.org/10.1080/01972243.2016.1271071.

Brandenburg v. Ohio, 395 U.S. 444 (1969).

Branigan, D. (2018, Nov. 28). Open Music Initiative: Seeking To Drive The Beat On Global Standards, Rights Attribution. *Intellectual Property Watch.* https://www.ip-watch.org/2018/11/28/open-music-initiative-seeking-drive-beat-global-standards-rights-attribution/.

Brandom, R. (2018, Sept. 5). *The monopoly-busting case against Google, Amazon, Uber, and Facebook.* The Verge. https://www.theverge.com/2018/9/5/17805162/monopoly-antitrust-regulation-google-amazon-uber-facebook.

Branscomb, A. (1995). Anonymity, autonomy, and accountability: Challenges to the first amendment in cyberspaces. *Yale Law Journal, 104,* 1639-1680.

Brennan, T., Schneider, S.D. & Shukla, R. (2019). California Sets De Facto National Data Privacy Standard. *Corporate Counsel Business Journal.* https://ccbjournal.com/articles/california-sets-de-facto-national-data-privacy-standard.

Brill, J. (2019, Nov. 11). *Microsoft will honor California's new privacy rights throughout the United States.* Microsoft on the Issues. https://blogs.microsoft.com/on-the-issues/2019/11/11/microsoft-california-privacy-rights/.

Burstyn v. Wilson, 343 U.S. 495 (1952).

Burt. A. (2018, Aug. 21). *States are leading the way on data privacy.* The Hill. https://thehill.com/opinion/technology/402775-states-are-leading-the-way-on-data-privacy.

Caro, R.A. (2012), *The Passage of Power: The Years of Lyndon Johnson.* Alfred A. Knopf, Inc.

Coase, R. (1979). Payola in Radio and Television Broadcasting. *The Journal of Law & Economics, 22(2),* 269-328. www.jstor.org/stable/725120.

Communications Decency Act, section 230(c)(1), 47 U.S.C. § 230(c)(1) (adopted 1996).

Court of Justice of the European Union (2014, May 13), C-131/12, *Google Spain SL, Google Inc. v. Agencia Espanola de Proteccion de Datos, Mario Costeja Gonzalez.* http://curia.europa.eu/jcms/upload/docs/application/pdf/2014-05/cp140070en.pdf.

Determann, L. & Engfeldt, H.J. (2019, June 30). *Maine and Nevada's New Data Privacy Laws and the California Consumer Privacy Act Compared,* Baker McKenzie. https://www.bakermckenzie.com/en/insight/publications/2019/06/maine-and-nevada-new-data-privacy-laws.

Dibene, C. (2017). Sir, the Radar Sir, It Appears to Be . . . Jammed: The Future of "The Right to Be Forgotten" in a Post-Brexit United Kingdom. *San Diego International Law Journal, 19(1),* 161-191.

Fair Housing Council of San Fernando Valley v. Roommates.Com, LLC, 521 F.3d 1157 (9th Cir. 2008), *later proceeding at* 666 F.3d 1216 (9th Cir. 2012).

Federal Communications Commission v. Pacifica Foundation, 438 U.S. 726 (1978).

Feiner, L. & Rodriguez, S. (2019, July 24). *FTC slaps Facebook with record $5 billion fine, orders privacy oversight.* CNBC. https://www.cnbc.com/2019/07/24/facebook-to-pay-5-billion-for-privacy-lapses-ftc-announces.html.

Frieden, R. (1997). Schizophrenia Among Carriers: How Common and Private Carriers Trade Places. *Michigan Telecommunications & Technology Law Review 3,* 19-44. http://www.mttlr.org/volthree/frieden.pdf.

Furchtgott-Roth, H. and Roth, A. (2016). Answering four questions on the anniversary of the telecommunications act of 1996. *Federal Communications Law Journal, 68(1),* 83-94.

Goldman, E. (2019). The Complicated Story of FOSTA and Section 230. *First Amendment Law Review 17,* 279-293. https://ssrn.com/abstract=3362975 .

Griswold v. Connecticut, 381 U.S. 479 (1965).

Hart-Scott-Rodino Antitrust Improvements Act, Pub. L. No. 94-435, 90 Stat. 1390 (1976).

Hendricks, J.A. (1999) The Telecommunications Act of 1996: Its Impact on the Electronic Media of the 21st Century. *Communications and the Law 21(2),* 39-53.

Hirsch, L. & Feiner, L. (2019, Nov. 14). *States' massive Google antitrust probe will expand into search and Android businesses.* CNBC. https://www.cnbc.com/2019/11/14/states-google-antitrust-probe-to-expand-into-search-android-businesses.html.

Holmes, O.W. (2013). *The Common Law.* Project Guttenberg. (Originally published 1881). https://www.gutenberg.org/files/2449/2449-h/2449-h.htm.

KFKB Broadcasting Association, Inc. v. Federal Radio Commission, 47 F.2d 670 (D.C. Cir. 1932).

Laslo, M. (2019, Nov. 29). *Hey Congress, How's That Privacy Bill Coming Along?* Wired. https://www.wired.com/story/congress-privacy-bill-copra/.

Lloyd, M. (2006). *Prologue to a Farce: Communication and Democracy in America*. University of Illinois Press.

Maddaus, G. (2019, Nov. 22). *Justice Department Goes to Court to Lift Paramount Consent Decrees*. Variety. https://variety.com/2019/biz/news/paramount-doj-consent-decrees-court-filing-1203413811/.

Massachusetts v. Microsoft Corp., 373 F.3d 1199 (D.C. Cir. 2004).

Merwaday, A., Yuksel, M., Quint, T., Güvenç, I., Saad, W. & Kapucu, N. (2018). Incentivizing spectrum sharing via subsidy regulations for future wireless networks. *Computer Networks 135*, 132-146, https://doi.org/10.1016/j.comnet.2018.02.011.

Moore, L.K. (2013). *Spectrum Policy in the Age of Broadband: Issues for Congress*. Congressional Research Service, Library of Congress. https://fas.org/sgp/crs/misc/R40674.pdf.

Morrison, G. (2020, Jan. 15). Next Gen TV is free 4K TV with an antenna, and it's coming to TVs this year, *Cnet*. https://www.cnet.com/news/next-gen-tv-is-free-4k-tv-with-an-antenna-and-its-coming-this-year/.*Mozilla Corporation v. Federal Communications Commission*, 940 F.3d 1 (D.C. Cir. Oct. 1, 2019), *pet. for en banc review filed*, No. 18-1051 (D.C. Cir. Dec. 13, 2019).

Mozilla Corporation v. FCC, 940 F.3d 1 (D.C. Cir. 2019).

Music Royalty Exchange (2014, June 16). *Music Royalties Guide*. https://www.royaltyexchange.com/blog/music-royalties

Mutual Film Corp. v. Industrial Commission of Ohio, 236 U.S. 230 (1915).

National Broadcasting Co. v. United States, 319 U.S. 190 (1943).

New York Times v. United States, 403 U.S. 713 (1971).

Nonnenmacher, T. (2001). State Promotion and Regulation of the Telegraph Industry, 1845-1860. *The Journal of Economic History, 61*(1), 19-36. www.jstor.org/stable/2697853.

Nunez, M. (2016, May 9). *Former Facebook Workers: We Routinely Suppressed Conservative News*. Gizmodo. https://gizmodo.com/former-facebook-workers-we-routinely-suppressed-conser-1775461006.

O'Connor, C.E. (2019). American Horror Story: The FCC's Chilling Indecency Policy, *Notre Dame Journal of Law, Ethics and Public Policy 33*, 531-545.

Office of Communication of United Church of Christ v. Federal Communications Commission, 359 F.2d 994 (D.C. Cir. 1966).

Office of Communication of United Church of Christ v. Federal Communications Commission, 425 F.2d 543 (D.C. Cir. 1966).

Ohlheiser, A. (2016, Aug. 26). *Three days after removing human editors, Facebook is already trending fake news*. Washington Post, https://www.washingtonpost.com/news/the-intersect/wp/2016/08/29/a-fake-headline-about-megyn-kelly-was-trending-on-facebook/.

Osborne, C. (2018, July 11). *Facebook faces £500,000 fine in UK over Cambridge Analytica scandal*. ZDNet. https://www.zdnet.com/article/uk-watchdog-to-give-facebook-500000-fine-over-data-scandal/.

Ou, J. (2013). The Overexpansion of the Communications Decency Act Safe Harbor, *Hastings Communications & Entertainment Law Journal 35*, 455-470.

Pardau, S.L. (2018). The California Consumer Privacy Act: Towards a European-Style.

Privacy Regime in the United States, *Journal of Technology Law & Policy, 23*, 68-114.

Patterson v. Colorado, 205 U.S. 454 (1907).

Perrin, A. and Anderson, M. (2019, Apr. 10). Share of U.S. adults using social media, including Facebook, is mostly unchanged since 2018. Pew Research Center. https://www.pewresearch.org/fact-tank/2019/04/10/share-of-u-s-adults-using-social-media-including-facebook-is-mostly-unchanged-since-2018/.

Pincus, W. and Lardner, Jr., G. (1997, Dec. 1). *Nixon Hoped Antitrust Threat Would Sway Network Coverage*. Washington Post. https://perma.cc/C42R-HKN8.

Porter, W.E. *Assault on the Media: The Nixon Years*. University of Michigan Press.

Powe, Jr., L.A. (1987). *American Broadcasting and the First Amendment*. University of California Press.

Prosser, W.L. (1960), Privacy, *California Law Review 48*, 383-423. https://doi.org/10.15779/Z383J3C.

Rabban, D.M. (1997). *Free Speech in its Forgotten Years*. Cambridge University Press.

Rebato, M. & Zorzi, M. (2018). A Spectrum Sharing Solution for the Efficient Use of mmWave Bands in 5G Cellular Scenarios. *2018 IEEE International Symposium on Dynamic Spectrum Access Networks (DySPAN)*, Seoul, 2018, pp. 1-5. https://ieeexplore.ieee.org/document/8610481.

Red Lion Broadcasting Co. v. Federal Communications Commission, 395 U.S. 367 (1969).

Reno v. American Civil Liberties Union, 521 U.S. 844 (1997).

Romm, T. (2019a, Nov. 8). *DOJ issues new warning to big tech: Data and privacy could be competition concerns*. Washington Post, https://www.washingtonpost.com/technology/2019/11/08/doj-issues-latest-warning-big-tech-data-privacy-could-be-competition-concerns/.

Romm, T. (2019b, Oct. 22). *Forty-six attorneys general have joined a New York-led antitrust investigation of Facebook.* Washington Post, https://www.washingtonpost.com/technology/2019/10/22/forty-six-attorneys-general-have-joined-new-york-led-antitrust-investigation-into-facebook/.

Rosenberg, N.L. (1986). *Protecting the Best Men: An Interpretive History of the Law of Libel.* University of North Carolina Press.

Schenk v. United States, 249 U.S. 47 (1919).

Shanahan, E. (1976, Nov. 18). *NC Agrees to Settle Suit by U.S.; Will Curb Financial Stake in Shows.* The New York Times. https://www.nytimes.com/1976/11/18/archives/nbc-agrees-to-settle-suit-by-us-will-curb-financial-stake-in-shows.html.

Sherman Anti-Trust Act, 15 U.S.C. § 1, et seq. (enacted 1890).

Slack, C. (2015). *Liberty's First Crisis: Adams, Jefferson, and the Misfits Who Saved Free Speech.* Grove Press.

Smith v. California, 361 U.S. 147 (1959).

Starr, P. (2004). *The Creation of the Media: Political Origins of Modern Communications.* Basic Books.

Stoller, D.R. & Merken, S. (2019, Jan. 24). *Big Google Privacy Fine May Set Bar for EU Privacy Penalties.* Bloomberg Law. https://news.bloomberglaw.com/privacy-and-data-security/big-google-privacy-fine-may-set-bar-for-eu-privacy-penalties.

Stone, G.R. (2004). *Perilous Times: Free Speech in Wartime from the Sedition Act of 1798 to the War on Terrorism.* W.W. Norton & Co.

Syracuse Peace Council v. Federal Communications Commission, 867 F.2d 654 (D.C. Cir. 1989), *certiorari denied*, 493 U.S. 1019 (1990).

Telegraph Reporters (2019, Dec. 26). *Britain is running out of time to agree a data deal with Brussels.* Telegraph. https://news.yahoo.com/britain-running-time-agree-data-131521687.html.

Tarantola, A. (2017, Dec. 13). *Modern copyright law can't keep pace with thinking machines.* Engaget. https://www.engadget.com/2017/12/13/copyright-law-AI-robot-thinking-machines/. Tarantola, A. (2018, Apr. 12). *Can Facebook really apply the EU's data-privacy rules worldwide? Yes, yes it can.* Engaget. https://www.engadget.com/2018/04/12/Facebook-EU-GDPR/.

Trinity Methodist Church, South v. Federal Radio Commission, 62 F.2d 850 (D.C. Cir. 1932), *certiorari denied*, 288 U.S. 599 (1933).

U.S. Const., amend. I.

U.S. Telecommunications Association v. Federal Communications Commission, 825 F.3d 674 (D.C. Cir. 2016), *en banc review denied*, 855 F.3d 381 (D.C. Cir. 2017), *certiorari denied*, 139 S.Ct. 454, 202 L.Ed.2d 361 (U.S. Nov. 5, 2018).

United States v. American Telephone & Telegraph Co., 552 F. Supp. 131, 177 (D.D.C. 1982), *affirmed as* Maryland v. United States, 460 U.S. 1001 (1983).

United States v. Paramount Pictures, Inc., 334 U.S. 131 (1948).

United States v. Southwestern Cable Corp., 392 U.S. 157 (1968).

Wakabayashi, D. (2019, Aug. 6). *Legal Shield for Websites Rattles Under Onslaught of Hate Speech.* New York Times. https://www.nytimes.com/2019/08/06/technology/section-230-hate-speech.html.

Warren, S.D. & Brandeis, L.D. (1890). The Right to Privacy. *Harvard Law Review*, 4, 193-220. https://www.jstor.org/stable/1321160.

Wu, T. (2003). Network Neutrality, Broadband Discrimination, *Journal of Telecommunications & High Technology Law*, 2, 141-176. http://www.jthtl.org/content/articles/V2I1/JTHTLv2i1_Wu.PDF.

Yarrow, J. (2012, Aug. 9). *Google Violates Apple Users' Privacy, Gets Hit With A $22.5 Million Fine.* Business Insider. https://www.businessinsider.com/google-ftc-fine-2012-8.

Electronic Mass Media

Section II

Electronic
Mass Media

6

Digital Television & Video

Peter B. Seel, Ph.D.*

Abstract

There are four significant trends in digital television (DTV) and video technologies: home screens are getting larger, they have higher resolution with 4K displays as the norm and massive 8K screens as an emerging technology, and streaming video is fast supplanting broadcast content as the sheer amount of online content expands at dramatic rate. The fourth trend is that 20% of U.S. DTV viewers do not own a television set—they are watching on their mobile devices.

In the United States, the voluntary broadcast transition to the new ATSC 3.0 digital television standard is progressing slowly due to multiple factors outlined below. U.S. television stations have either sold their digital spectrum for other wireless applications or are sharing it with partner stations in their markets. The broadcaster's share of the viewing audience continues to shrink as online streaming services, such as Netflix, Hulu, and Amazon are joined by new digital programming providers Disney+, Apple TV Plus, the Peacock network, and HBO Max.

The global transition to digital television is now largely complete, with the widespread diffusion of 4K programming via broadcast and streaming video providers—and even the start of 8K program transmission in Japan in 2019. The inclusion of high-quality 4K video cameras in the next generation of 5G mobile phones will improve their image quality and fuel the explosion of online videos posted by every connected global citizen. The posting of digital videos is a driving force for social media and legacy media sites worldwide.

* Professor Emeritus in the Department of Journalism and Media Communication, Colorado State University (Fort Collins, Colorado).

Introduction

Digital television and video displays are simultaneously getting larger and smaller (in home OLED screens that now exceed seven feet in diagonal size, while head-mounted digital displays such as Microsoft's Hololens2 have screens that are measured in centimeters). There is also improved image resolution at both ends of this visual spectrum. The improvement in video camera quality in mobile phones has led to a global explosion in the number of videos posted online. Digital program delivery is also rapidly shifting from an over-the-air (OTA) broadcast model where viewers have content "pushed" to them in real-time to an expanding online environment where they "pull" what they want to watch when they want to see it. This transformation in delivery is affecting every aspect of motion media viewing in countries worldwide.

In the United States, this online-delivery process is known as *Over-The-Top* television—with OTT as the obligatory acronym now used in promoting these services. It should more accurately be called over-the-top **video**, as the term refers to non-broadcast, internet-streamed video content viewed on a digital display, but the confusion reflects the blurring of the line between broadcast/cable/DBS-delivered "linear" television and that of on-demand streamed content (Seel, 2012). The term "Over-The-Top" is unique to the U.S.—in other regions of the world it is called "Connected" video or television.

Over-The-Top or *Connected* television indicates that there is an increasingly popular means of accessing television and video content that is independent of traditional linear over-the-air television broadcasters and multichannel video programming distributors (MVPDs, which include satellite, cable, and telco content providers). OTT pioneers Netflix, Hulu, and Amazon Prime were joined by new programming providers Disney Plus and Apple TV Plus in late 2019, and by NBC Universal's Peacock network and Warner Media's HBO Max in the spring of 2020 (Littleton & Lowe, 2020). The sheer amount of new streamed video programming is daunting to home viewers and is overwhelming to media critics who need to review it (Lawler, 2020).

In the United States, a massive retooling is occurring in broadcast and video production operations as the nation undergoes a second digital television transition. The first ended on June 12, 2009 with the cessation of analog television transmissions and the completion of the *first* national transition to digital broadcasting using the Advanced Television Systems Committee (ATSC 1.0) DTV standard (*DTV Delay Act*, 2009). Development work on improved DTV technology for the U.S. has been underway for the past decade and the Federal Communication Commission voted in November of 2017 for a national *voluntary* adoption of a new ATSC 3.0 standard (FCC, 2017; Eggerton, 2017). An update on the U.S. conversion to the U.S. ATSC 3.0 standard is included below in the Recent Developments section.

Most flat-screen displays larger than 42-inches sold globally in 2020 and beyond will be "smart" televisions that can easily display these motion media sources and websites such as Facebook and Pinterest. An example is Sony's smart TV interface in its "Bravia" brand displays that offers voice control in accessing a panoply of internet-delivered and MVPD-provided content. Remote controls for these displays can easily toggle between broadcast/cable/DBS content and internet-streamed OTT videos. (See Figure 6.1)

Figure 6.1

Smart Video TV

Apple TV, YouTube TV, Prime Video, and other OTT viewing options seen on the interface screen for a "smart" digital television.

Source: P. Seel

Digital television displays and high-definition video recording are increasingly common features in mobile devices used by more than five billion of the world's population of 7.7 billion people (Bankmycell, 2020). Digital video cameras in mobile phones have become more powerful and have new features. An example is the $1,000 Apple iPhone 11 (see Figure 6.3) which records 4K-resolution video at 24, 30, and 60 frames/second and has dual lenses for capturing standard and wide-angle 12 megapixel images.

Figures 6.2 and 6.3

Two Contemporary High-Definition Digital Cameras.

Left: lightweight $49,000 Arriflex Alexa Mini LF camera was used to capture the remarkable action scenes in the feature film "1917" with 4K image resolution (2160 X 4096 pixels).
Right: the $1,000 Apple iPhone 11 records 4K-resolution video at 24, 30, and 60 frames/second with dual camera lenses, normal and wide-angle. It can also multitask as a mobile phone.

Sources: Arriflex and Apple

The democratization of "television" production generated by the explosion in the number of devices that can record digital video has created a world where over 1.9 billion users watch more than one billion videos each day at the YouTube.com site. Local YouTube sites are available in more than 100 countries, with content available in 80 languages, and 70% of all viewing globally is on mobile devices (YouTube for Press, 2020). The online distribution of digital video and television programming is an increasingly disruptive force to established broadcasters and program producers. They have responded by making their content available online (for free or via subscription) as increasing numbers of viewers seek to "pull" digital television content on request rather than watch at times when it is "pushed" as broadcast programming. The increasing

ubiquity of digital video recording capability also bodes well for the global free expression and exchange of ideas via the internet. The expanding "universe" of digital television and video is being driven by improvements in high-definition video cameras for professional production and the simultaneous inclusion of higher-quality video capture capability in mobile phones.

Another key trend is the recently completed global conversion from analog to digital television (DTV) technology. The United States completed its national conversion to digital broadcasting in 2009, Japan completed its transition in 2012, India in 2014, and most European nations did so in 2015. At the outset of high-definition television (HDTV) development in the 1980s, there was hope that one global television standard might emerge, easing the need to perform format conversions for international program distribution. There are now multiple competing DTV standards based on regional affiliations and national political orientation. In many respects, global television has reverted to a "Babel" of competing digital formats reminiscent of the advent of analog color television. However, DTV programming in the widescreen 16:9 aspect ratio is now a commonplace sight in all nations that have made the conversion.

Background

The global conversion from analog to digital television technology is the most significant change in television broadcast standards since color images were added in the 1960s. Digital television combines higher-resolution image quality with improved multichannel audio, and new "smart" models include the ability to seamlessly integrate internet-delivered OTT "television" programming into these displays. In the United States, the Federal Communications Commission (FCC, 1998) defines DTV as "any technology that uses digital techniques to provide advanced television services such as high definition TV (HDTV), multiple standard definition TV (SDTV) and other advanced features and services" (p. 7420).

Digital television programming can be accessed via linear over-the-air (OTA) fixed and mobile transmissions, through cable/telco/satellite multichannel video program distributors MVPDs), and through

Over-the-Top (OTT) internet-delivered sites. OTT is a "pull" technology in that viewers seek out a certain program and watch it in a video stream or as a downloaded file. OTA linear broadcasting is a "push" technology that transmits a digital program to millions of viewers at once. Connected or OTT TV, like other forms of digital television, is a scalable technology that can be viewed as lower quality, highly compressed content or programs can be accessed in high-definition quality on sites such as Vimeo.com.

In the 1970s and 1980s, Japanese researchers at NHK (Japan Broadcasting Corporation) developed two related analog HDTV systems: an analog "Hi-Vision" *production* standard with 1125 scanning lines and 60 fields (30 frames) per second; and an analog "MUSE" *transmission* system with an original bandwidth of 9 MHz designed for satellite distribution throughout Japan. The decade after 1996 was a significant era in the diffusion of HDTV technology in Japan, Europe, and the United States. In April 1997, the FCC defined how the United States would make the transition to DTV broadcasting and set December 31, 2006 as the target date for the phase-out of NTSC broadcasting (FCC, 1997). In 2005, after it became clear that this deadline was unrealistic due to the slow consumer adoption of DTV sets, it was reset at February 17, 2009 for the cessation of analog full-power television broadcasting (*Deficit Reduction Act*, 2005).

However, as the February 17, 2009 analog shutdown deadline approached, it was apparent that millions of over-the-air households with analog televisions had not purchased the converter boxes needed to continue watching broadcast programs. Cable and satellite customers were not affected, as provisions were made for the digital conversion at the cable headend or with a satellite set-top box. Neither the newly inaugurated Obama administration nor members of Congress wanted to invite the wrath of millions of disenfranchised analog television viewers, so the shut-off deadline was delayed by an act of Congress 116 days to June 12, 2009 (*DTV Delay Act*, 2009). The number of U.S. television households using digital HDTV displays climbed from 24% to over 90% by July 2016, which was a significant adoption take-up over seven years (DTV Household Penetration, 2016).

There are two primary areas affecting the diffusion of digital television and video in the United States: an ongoing battle over the digital television spectrum between broadcasters, regulators, and mobile telecommunication providers and the diffusion of new digital display technologies. The latter include now ubiquitous 4K-resolution Ultra-High-Definition (UHD) sets with display enhancements such as improved High-Dynamic Range (HDR) and the evolution of quantum dot technology. The spectrum battle is significant in the context of what is known as the *Negroponte Switch*, which describes the conversion of "broadcasting" from a predominantly over-the-air service to one that is now wired for many American households—and the simultaneous transition of telephony from a traditional wired service to a wireless one for an increasing number of users (Negroponte, 1995). The growth of wireless broadband services has placed increasing demands on the available radio spectrum—and broadcast television is a significant user of that spectrum. The advent of digital television in the United States made this conflict possible in that the assignment of new DTV channels demonstrated that spectrum assigned to television could be "repacked" at will without the adjacent-channel interference problems presented by analog transmission (Seel, 2011).

The DTV Spectrum Auction in the United States

Some context is necessary to understand why the federal government was seeking the return of digital spectrum that it had allocated in the 2009 U.S. DTV switchover. As a key element of the transition, television spectrum between former channels 52-70 was auctioned off for $20 billion to telecommunication service providers, and the existing 1,500 broadcast stations were "repacked" into lower terrestrial broadcast frequencies (Seel & Dupagne, 2010). The substantial auction revenue generated in the transition was not overlooked by federal officials and, as the U.S. suffered massive budget deficits in the global recession of 2008-2010, they proposed auctioning more of the television spectrum as a way to increase revenue without raising taxes in a recession. The U.S. Congress passed the *Middle Class Tax Relief and Job Creation Act of 2012* (*Middle Class*, 2012) and it was signed into law by President Barack Obama. The *Act* authorized the

Federal Communication Commission to reclaim DTV spectrum assigned to television broadcasters and auction it off to the highest bidder (primarily wireless broadband providers).

The success of this unique auction depended on several key factors. Television spectrum holders had to weigh the substantial value of their surrendered spectrum against three possible options: taking their OTA signal off the air and returning their assigned 6 MHz TV channel, moving to a lower VHF (Very High Frequency) channel in a local repack, or moving to a higher VHF channel in their market. Stations also had the option to share an OTA channel with another broadcaster, which DTV technology permits. The auction value to the broadcaster declined with each option. Broadcasters seeking to participate in the auction had to file an application with the FCC by January 12, 2016 and specify which of the three options they selected (Federal Communications Commission, 2016).

The auction concluded in April 2017 with bids of nearly $20 billion for the TV spectrum sold by 175 broadcast licensees to telecommunication service providers (Pressman, 2017). This amount is very close to that generated in the repacking of the television spectrum in the original U.S. transition to digital television. The daunting task for 957 television stations was that they only had 36 months (to a deadline of July 5, 2020) to shift their transmission frequencies into lower VHF bands to clear the former UHF spectrum for the telecommunication companies such as AT&T and T-Mobile which bid on it. Of the 175 stations that sold their former UHF spectrum, 12 went off the air, 30 shifted their channel assignments to lower VHF frequencies, and the balance shared another station's spectrum (Pressman, 2017).

How will local viewers be affected? In most of the U.S., a majority of television viewers watch local channels via a cable company or from a satellite provider, so the number of OTA viewers are often comparatively small. Cable or satellite viewers would likely *not* notice a change after the channel repack, as their providers assign their own channels on their respective networks. OTA viewers in each market will have to rescan their digital television channels after the repack, as they will be on different frequencies. Viewers will not need to purchase a new digital television set

as they did by 2009, as only the digital channel assignments will have changed (Federal Communications Commission, 2016). The U.S. incentive auction is a win-win for most participants, as it opens more of the former TV spectrum for an expansion of wireless services (including watching OTT videos) and compensates 175 television stations with half of the $20 billion proceeds from the auction for giving up all or part of their spectrum.

The ATSC 3.0 DTV Standard

As noted above, the ATSC 3.0 transmission standard was adopted by the FCC on November 16, 2017 as a *voluntary* digital television option for the United States (FCC, 2017). Broadcast television companies at the local and national network level wanted to be more competitive with internet-based video content providers. These companies lobbied Congress and the Commission to create a new system that would combine OTA broadcasting with related over-the-top programming, which would allow broadcasters to more precisely target advertising to consumers in ways that resemble internet-based marketing. It is a convergent technology where broadcast HEVC-compressed programming meets web-delivered content for simulcast delivery to millions of viewers (McAdams, 2015). Table 6.1 compares the technical attributes of the first ATSC 1.0 digital transmission standard with the ATSC 3.0 technology, which will supplant the 1.0 standard over time as broadcasters switch over to the more spectrum-efficient technology at their new channel assignments.

There are multiple benefits of the transition to the ATSC 3.0 standard for broadcasters:

- Over-the-Air digital data throughput increases from 19.4 Mbps to 25 Mbps or better in a 6 MHz

- OTA broadcast channel. This increased data rate would facilitate the OTA transmission of 4K programming, especially if improved High-Efficiency Video Coding (HEVC) is used in this process, and broadcast programming could contain 22.2 channel audio (Baumgartner, 2016).

- Consumers could use flash drives (much like those used for Chromecast reception) to upgrade older DTV sets for the display of program content that might be a blend of broadcast programs with related content delivered via an internet connection to a smart display (see Figure 6.4, which illustrates the future of television).

Table 6.1

U.S. ATSC 1.0 & ATSC 3.0 Digital Television Transmission Standards

Feature	ATSC 1.0	ATSC 3.0
Resolution	480-1,080 Lines	2,160 Lines x 3,840 Pixels Wide —"4K"
Scanning	Interlaced and Progressive	Progressive
Aspect Ratio	4:3 and 16:9	16:9
Audio	Up to 5.1 Channels	Up to 22.2 Channels
Video Encoding	MPEG-2	HEVC
Signal Modulation	8-VSB	OFDM
Data Payload	19.4 Mbps	25-30 Mbps

Source: M. Dupagne

Recent Developments

It is unlikely that all U.S. broadcasters will be on the air at full power with their new channel assignments in the spectrum repack by the FCC's deadline of July 5, 2020, complicated by the 2020 COVID-19 economic disruption. The planned 36-month conversion schedule made a number of assumptions that stations could share transmission towers and that the needed hardware could be provided by transmitter and other manufacturers. Tower modifications were delayed by bad weather, a shortage of helicopter crews to make these installations, and extended timelines to acquire new transmitters (despite working double shifts by some manufacturers). By November 2019, only 73% of stations (388 of 528 nationwide) had completed their conversion to the new ATSC 3.0 transmission technology (O'Neal, 2019). Many of the stations who found their conversion plans stymied by

construction or equipment delivery delays applied to the FCC for special temporary authorizations (STAs) to use transmitters operating at reduced power on their new channel assignments. Some stations will not be broadcasting at full power until 2021—well past the July 2020 deadline set by the FCC (O'Neal, 2019). Television viewers watching broadcast channels via cable, satellite, or OTT likely will not notice any loss of image or sound quality due to these delays. However, some viewers using OTA antennas far from station transmitters may not be able to receive some of the channels that are broadcasting with reduced power under FCC special temporary authorizations.

Figure 6.4

The Future of Television

The future of television is a blended display of broadcast content—in this case, a severe weather alert—combined with an internet-delivered run-down of the local newscast for random access and other streamed media content around the broadcast image.

Source: P.B. Seel.

Another complication is that most, older, digital televisions in the U.S. cannot decode the new ATSC 3.0 signals without some type of conversion—a major hurdle for OTA viewers who have dropped their cable service or removed their satellite dishes. At the *2020 Consumer Electronics Show* in Las Vegas, a number of manufacturers introduced their first digital televisions that include ATSC 3.0 chips, which can decode the over-the-air transmission of the new DTV signals (Morrison, 2020). Cable "cord cutters" and satellite TV "dish ditchers" will be pleased to see that they can watch 4K OTA broadcast programming in 2020 and beyond with a new display with the 3.0 chipset (or a signal conversion dongle) and a simple antenna on their homes.

The other significant recent development has been the rapid global adoption of 4K display technology. A market survey of U.S. consumers in mid-2019 found that 48% were interested in 4K televisions and were planning to purchase one in the following 12 months. A related finding was that, despite the proliferation of mobile devices in the home such as phones, tablets, and notebook computers, the mean number of televisions per household has remained constant at 2.6 sets over the five-year period from 2014-2019 (Leichtman Research Group, 2019). As Table 6.2 shows, the dramatic increase in the sales of 4K displays has been driven by a reduction in set prices of more than 50% since 2017. Consumers are able to purchase larger screens with higher resolutions and more features for lower equivalent prices than in 2017.

Current Status
4K Technology

Display options. There are a wide variety of 4K displays on the market worldwide in screen sizes that range from 42 to 98 diagonal inches. The key features of 4K display technologies are:

More pixels. These Ultra-High-Definition (UHD) televisions have 8,294,400 pixels (3840 wide x 2160 high)—four times that of a conventional HDTV display. One problem with 4K display technologies is that the data processing and storage requirements are four times greater than with 2K HDTV. Just as this appeared to be a significant problem for 4K program production and distribution, a new compression scheme emerged to deal with this issue: High-Efficiency Video Coding (HEVC). It uses advanced compression algorithms in a new global H.265 DTV standard adopted in January 2013 by the International Telecommunications Union. It superseded their H.264 AVC standard and uses half the bit rate of MPEG-4 coding, making it ideal for 4K production and transmission (Butts, 2013).

Organic Light Emitting Diode (OLED). The Sony Corporation introduced remarkably bright and sharp OLED televisions in 2008 that had a display depth of 3mm–about the thickness of three credit cards. Mobile phone manufacturers such as Apple and Samsung have adopted AMOLED versions of these displays, as they are very sharp, colorful, and bright. Korean manufacturer Samsung Electronics recently introduced a 98-inch 8K QLED display that features a remarkable pixel resolution of 7,680 wide X 4,320 high (also referenced as 4320p) (see Figure 6.5).

High Dynamic Range (HDR) Imaging. Another recent development in television technology has been the inclusion of HDR enhancements to HD and UHD displays. Improving the dynamic range between the light and dark areas of an image adds to its visual impact. Greater detail can be seen in the shadow areas of a displayed scene and also in the highlights. The technology was developed by Dolby Laboratories in its plan for the future of television known as Dolby Vision. While also increasing the range of contrast in televised images, HDR-enabled displays have much higher screen brightness than those without it (Waniata, 2015).

Figure 6.5

Samsung Electronics introduced a new 98-inch 8K display at the 2020 CES Show.

Source: Samsung Electronics.

Quantum Dots. A new term for consumers considering purchasing a UHD display is the added feature of *quantum dots*. A familiarity with theoretical physics is not required to understand that this term refers to the addition of a thin layer of nano-crystals between the LCD backlight and the display screen. This unique application of nanotechnology increases image color depth by 30% without adding extra pixels to the display (Pino, 2016). Displays that feature this technology are called "QLED" by manufacturers such as Samsung and "Nanocell" by LG, and they typically cost more than sets without quantum dots.

MicroLED and MiniLED Displays. These are emerging flat-panel technologies that utilize arrays of microscopic light emitting diodes (LEDs) to form the pixel elements. Dividing the pixels into smaller LED elements improves image contrast, refresh rates, color saturation, and energy consumption. Unlike OLED screens, microLED and miniLED displays are made of gallium nitride (GaN), which offer higher screen brightness and lower power consumption (Triggs, 2017). Originally used in smart-watch displays and mobile phone screens, these technologies are now being used for television displays as large as 292 inches (24.33 feet), as they can scale larger than OLEDs. However, the manufacture of large MicroLED screens is a complicated process and the early models are more expensive than OLED displays (Hruska, 2019). From an environmental perspective, the lower energy consumption of microLED and miniLED displays is an improvement over older conventional OLED displays.

Both Sony and Samsung have developed large flat-screen displays that use microLED technologies. Sony created a 55-inch "Crystal LED" display in 2012 as a demonstration product. In 2019, they began marketing the "CLEDIS" (Crystal LED Integrated Structure) brand of microLED displays that ranged from 146 to 790-inches in diagonal screen size (Wilkinson, 2019). Not to be outdone, Samsung introduced a 146-inch microLED display at CES in 2018 branded as "The Wall." They followed with a new line in 2019 of large flat-screen displays labeled "The Wall Luxury MicroLED" that ranged in size and resolution from 73 inches in 2K displays to 292 inches in 8K screens (Hruska, 2019). Apple has also been involved in MicroLED research for iPhone, iPad, and Apple Watch screens that are more durable, brighter, and less power hungry than their present OLED displays (Gurman, 2018).

4K Sales

Display sales. Sales of 4K UHD models have exploded worldwide since 2018 and set prices have plummeted, which is good news for consumers. Note in Figure 6.2 that a 55-inch 4K UHD television with HDR image enhancement that cost an average price of $800 in 2017 can now be purchased for less than half that amount, averaging $290 in 2019. Most higher-end 4K displays are "smart" that can easily display any desired programming source, whether they are broadcast, cable or satellite, or internet-delivered content.

Programming. With the advent of 4K UHD televisions, consumers are now demanding more 4K content to watch. Netflix has obliged by producing and distributing new series in 4K, beginning with the Emmy-award-winning *House of Cards* series in 2013. The video streaming service now offers multiple original series, documentaries, and motion pictures in 4K resolution, but subscribers have to sign up for the premium tier to see these. Amazon has emerged as another provider of 4K content and the company is starting to produce their own content in 4K, much like Netflix (Pino, 2016). MVPD networks such as ESPN are now routinely telecasting major sporting events in 4K. Both Sony and Samsung offer Blu-ray players that can up-convert HD programs for display on 4K televisions.

Factors to Watch

The global diffusion of DTV technology will evolve over the third decade of the 21st century as prices continue to drop for UHD 4K displays. While 8K displays are presently very expensive—at $70,000 for the largest 98-inch models sold by Sony and Samsung—as their prices decline, viewers can expect to see these in select venues such as sports bars. In the United States, the future inclusion of UHD-quality video recording capability in most mobile phones will enhance this diffusion on a personal and familial level. In the process, digital television and video will influence what people watch, especially as smart televisions, tablet computers, and the diffusion of 5G mobile phones between 2020 and 2022 will offer easy access to internet-delivered content that is more interactive than traditional television.

Table 6.2

Average U.S. Retail Prices of 2K LED, OLED 4K, QLED 4K & QLED 8K Television Models from 2015–2019

Display Sizes (diagonal)	Average Retail Price in 2015	Average Retail Price in 2017	Average Retail Price in 2019
40-42-inch LED HDTV	$ 330 (1080p)	$ 265 (1080p)	$ 175 (1080p)
55-inch 4K UHD TV with HDR	$ 3,000 (2160p)	$ 800 (2160p)	$ 290 (2160p)
55-inch 4K QLED TV with HDR	N.A.	N.A.	$ 700 (2160p)
65-inch 4K UHD TV with HDR	$ 5,000 (2160p)	$ 1,400 (2160p)	$ 554 (2160p)
65-inch 4K QLED TV with HDR	N.A.	N.A.	$ 1,000 (2160p)
75-inch 4K UHD TV with HDR	N.A.	N.A.	$900
75-inch 4K QLED TV with HDR	N.A.	N.A.	$1,750
75-inch 8K UHD TV with HDR	N.A.	N.A.	$ 5,000 (4320p)
86-inch 4K Super-UHD TV	$ 10,000 (2160p)	$ 4,500 (2160p)	$ 2,600 (2160p)
86-inch 4K QLED TV with HDR	N.A.	$ 18,000 (2160p)	$ 3,050 (2160p)
88-inch 8K OLED TV	N.A.	$ 20,000 (4320p)	$ 13,000 (4320p)
98-inch 8K QLED TV	N.A.	N.A.	$ 70,000 (4320p)

n.a. = set not available in these years. "QLED" indicates that these sets utilize quantum dot technology for enhanced image quality and are typically priced higher than UHD models.

Sources: U.S. retail surveys by P.B. Seel

Key Trends in 2020–2022

• **Evolving 8K UHD technologies**—The continued development of higher-resolution video production and display technologies will enhance the viewing experience of television audiences around the world. While there are substantial manufacturing, data processing/storage, and transmission issues with 8K Ultra High-Definition technologies, these will be resolved in the coming decade. The roll-out of fiber-to-the-home (FTTH) in many nations will enhance the online delivery of these high-resolution programs. When combined with ever-larger, high-resolution displays in homes, audiences can have a more immersive viewing experience (with a greater sense of telepresence) by sitting closer to their large screens.

• **Cord-Shavers and Cord-Nevers**—As noted above, the number of younger television viewers seeking out alternative viewing options on the internet is steadily increasing. A recent consumer survey found that 31% of adult respondents watched video on a connected (OTT) device daily. (Leichtman Research Group, 2019). In 2019, OTT/Connected video viewing surpassed that of digital video reorder (DVR) viewing for the first time (Frankel, 2019). Expect to see more mainstream television advertisers placing ads on popular internet multimedia sites such as YouTube to reach this key demographic. With a smart television, an internet connection, and a small antenna for receiving over-the-air digital signals, viewers can see significant amounts of streamed and broadcast content without paying monthly cable or satellite fees. MVPDs such as cable and satellite companies are concerned about this trend as it enables these viewers to "cut the cord" or "ditch the dish," so these companies are seeking unique programming such as premium sports events (e.g., the Super Bowl and the NCAA Basketball Tournament) to hold on to their customer base.

• **Two-Screen Viewing**—This is a growing trend where viewers watch television programs while accessing related content (via Tweets, texts, and related social media sites) on a mobile device, or via picture-in-picture on their DTV. Smart TV technology would facilitate this type of multitasking while watching television. Producers of all types of television programing are planning to reach out to dual-screen viewers with streams of texts and supplementary content while shows are on the air

- **U.S. Television Station Adoption of the ATSC 3.0 Standard**—The ATSC 3.0 Next Gen TV standard will facilitate the merger of television broadcasting and the delivery of streamed on-demand programing in a hybrid model. For example, a large-screen smart television could display conventional broadcast programming surrounded by additional windows of related internet-delivered content around it. Stations face a July 5, 2020 FCC deadline to begin transmitting HDTV programming on their new channel assignments and ATSC 3.0 broadcasts are planned for the top 40 U.S. markets in 2020 (Dickson, 2019). The first television displays with new chipsets that will decode ATSC 3.0 telecasts were exhibited at the *2020 Consumer Electronics Show* and it is expected that all new televisions sold in the U.S. will include these in the future. However, station group experts, such as Anne Schelle of Pearl TV, state that it may take five years (2019-2024) for enough new TVs with the 3.0 chipsets to be sold to consumers and enough stations transmitting the 3.0 signal for the critical mass required for viewers to benefit from the merger of TV and internet programming—and for

broadcasters to see the return on their investment in 3.0 technology (Dickson, 2019).

The era of the blended television-computer or "tele-computer" has arrived. Most large-screen televisions are now "smart" models capable of displaying multimedia content from multiple sources: broadcast, cablecast, DBS, and all internet-delivered motion media. As the typical home screen has evolved toward larger (and smarter) models, it is also getting thinner with the arrival of new OLED and MicroLED models. The image quality of ultra-high-definition displays is startling and will improve further with the delivery of 4K UHD sets enhanced by quantum dot technologies and the arrival of new lower-cost 8K displays in the near future. Television, which has been called a "window on the world," will literally resemble a large window in the home for viewing live sports, IMAX-quality motion pictures, and any video content that can be steamed over the internet. On the other end of the viewing scale, mobile customers with 5G phones will have access to this same diverse content on their high-definition MicroLED screens, thanks to new wireless technology. Human beings are inherently visual creatures and this is an exciting era for those who like to watch motion media programs of all types in ultra-high-definition at home and away.

Gen Z

The GenZ cohort of young adults have been immersed in digital technology for their entire lives and are avid media technology users in most areas of the internet-connected world. New developments in digital television and video will affect their lives at both ends of the viewing spectrum. On mobile phones and similar access devices such as notebooks, tablets, and lightweight laptops, improvements in digital cameras and sharper, brighter displays will make them more useful than prior models. The development of brighter, ultra-sharp screens using MicroLED and similar display technologies, combined with multiple optical lenses on recent model phones, will encourage more video capture for live streaming and recording. The inclusion of these improved display technologies in AR and VR headsets, will make game-play and augmented viewing experiences even more immersive and compelling.

At the other end of the viewing spectrum, the rapid expansion of flat-screen sizes thanks to OLED and MicroLED technologies, will bring IMAX-type viewing experiences to homes, condos, and apartments of all sizes. It will be common after 2022 to see six and seven-foot flat-screen displays in homes as prices drop by more than 50% every two years (see Table 6.2). More GenZ viewers will access "televised" programming as "à la carte" streamed content over high-speed Gigabit connections in communities rewired with greater internet bandwidth. They can routinely teleconference using these high-resolution digital displays with friends and family in a world that will appear to shrink using fiber optics and users with camera and screens in every hand.

Global Concerns & Sustainability

When the world's first large electronic computer, Eniac, was constructed in a large room at the University of Pennsylvania in Philadelphia in 1945, a large air conditioning system was installed to make the room temperature bearable for the operators. In this early electronic era before the invention of the transistor, the massive 30-ton computer used 20,000 vacuum tubes as processing circuits and memory storage. It required 150 kilowatts of electricity to operate, enough to power a small city. It was rumored that when the computer was powered up at night, the lights in the city of Philadelphia dimmed.

The power requirements of a contemporary mobile phone (with computing power thousands of times more powerful than the Eniac) are miniscule in comparison. Early television sets in the 1950s also used hot vacuum tubes. The cool, bright, and sharp OLED and MicroLED screens on a phone today enable the vivid display of recorded and live-streamed video and photos. As communities around the world turn to renewable energy sources such as solar and wind, it is incumbent on power users to find ways to reduce demand at peak periods. The good news is that our mobile devices are typically recharged while we sleep (at periods of the lowest electrical loads), but the sun is not shining and the wind often dies in the middle of the night. The development of low-cost household batteries will be required to power the multiple digital screens and recharge mobile devices in a home at night. The development of new display technologies such as MicroLED will reduce power demands on screens as small as a digital watch and as large as an entire wall. The reduction in heat produced by these innovative displays will also reduce air conditioning loads in hot season months around the world. As the planet warms, reducing power demands and air conditioning loads will be vital for the survival of all nations.

BIBLIOGRAPHY

Bankmycell (2020). How Many Smart Phones are in the World? https://www.bankmycell.com/blog/how-many-phones-are-in-the-world.

Baumgartner, F. (2016, January 4). April SBE Ennis workshop to focus on ATSC 3.0. *TV Technology*. http://www.tvtechnology.com/broadcast-engineering/0029/april-sbe-ennes-workshop-to-focus-on-atsc-30/277680.

Butts, T. (2013, January 25). Next-gen HEVC video standard approved. *TV Technology*. http://www.tvtechnology.com/article/nexgen-hevc-video-standard-approved/217438.

Deficit Reduction Act of 2005. Pub. L. No. 109-171, § 3001-§ 3013, 120 Stat. 4, 21 (2006).

Dickson, G. (2019, October 24). Broadcasters take long view on 3.0 payoff. *TV News Check*. https://tvnewscheck.com/article/240408/broadcasters-take-long-view-on-3-0-payoff/.

DTV Delay Act of 2009. Pub. L. No. 111-4, 123 Stat. 112 (2009).

DTV Household Penetration in the United States from 2012-2016. (2016). Statista.com. https://www.statista.com/statistics/736150/dtv-us-household-penetration/.

Eggerton, J. (2017, November 16). FCC launches next-gen broadcast TV standard. *Broadcasting & Cable*. http://www.broadcastingcable.com/news/washington/fcc-launches-next-gen-broadcast-tv-standard/170165.

Federal Communications Commission. (1997). Advanced Television Systems and Their Impact Upon the Existing Television Broadcast Service (*Fifth Report and Order*), 12 FCC Rcd. 12809.

Federal Communications Commission. (1998). Advanced Television Systems and Their Impact Upon the Existing Television Broadcast Service (*Memorandum Opinion and Order on Reconsideration of the Sixth Report and Order*), 13 FCC Rcd. 7418.

Federal Communications Commission. (2016, March 23). *Broadcast Incentive Auction*. https://www.fcc.gov/about-fcc/fcc-initiatives/incentive-auctions.

Federal Communications Commission. (2017, November 16). *Authorizing Permissive Use of the "Next Generation" Broadcast Television Standard*. https://apps.fcc.gov/edocs_public/attachmatch/FCC-17-158A1.docx.

Frankel, D. (2019, July 2). OTT has surpassed DVR usage in U.S., now 'mainstream,' Comscore declares. *Multichannel News*. https://www.multichannel.com/news/ comscore-declares-ott-mainstream-in-us.

Gurman, M. (2018, March 18). Apple is said to develop gadget displays in secret facility. *Bloomberg*. https://www.bloomberg.com/news/articles/2018-03-19/apple-is-said-to-develop-displays-to-replace-samsung-screens.

Hruska, J. (2019, October 4). Samsung's massive 292-inch MicroLED TV wall now shipping. *ExtremeTech*. https://www.extremetech.com/electronics/299639-samsungs-massive-292-inch-microled-tv-wall-now-shipping.

Lawler, K. (2019, November 15). Too much: Why the streaming wars between Apple, Disney, HBO and more are ruining TV. *USA Today*. https://www.usatoday.com/story/entertainment/tv/2019/11/14/disney-plus-apple-tv-plus-streaming-wars-ruining-tv/2516655001/.

Littleton, C. & Low, E. (2020). Adapt or die: Why 2020 will be all about entertainment's new streaming battleground. *Variety*. https://variety.com/2019/biz/features/streaming-2020-disney-plus-netflix-hbo-max-apple-tv-amazon-1203439700/.

Leichtman Research Group. (2019). *4K TV Entering the Growth Stage*. (Research report). https://www.leichtmanresearch.com/wp-content/uploads/2019/07/LRG-Research-Notes-2Q-2019.pdf.

McAdams, D. (2015, February 12). HPA 2015: ATSC 3.0 prototypes expected in 2016. *TV Technology*. http://www.tvtechnology.com/exhibitions-&-events/0109/hpa—atsc—prototypes-expected-in-/274513.

Middle Class Tax Relief and Job Creation Act of 2012. Pub. L. No. 112-96.

Morrison, G. (2020, January 4). Next Gen TV is free 4K TV with an antenna, and it's coming this year. *CNet*. https://www.cnet.com/news/next-gen-tv-is-free-4k-tv-with-an-antenna-and-its-coming-this-year/.

Negroponte, N. (1995). *Being Digital*. New York: Alfred A. Knopf.

O'Neal, J. E. (2019, November 4). FCC says repack progressing nicely; Others disagree.*TV Technology*. https://www.tvtechnology.com/repack/fcc-says-repack-progressing-nicely-others-disagree.

Pino, N. (2016, February 16). 4K TV and UHD: Everything you need to know about Ultra HD. *TechRadar*. http://www.techradar.com/us/news/television/ultra-hd-everything-you-need-to-know-about-4k-tv-1048954.

Pressman, A. (2017, April 19). Why almost 1,000 TV stations are about to shift channels. *Fortune*, http://fortune.com/2017/04/19/tv-stations-channels-faq/.

Seel. P. B. (2011). Report from NAB 2011: Future DTV spectrum battles and new 3D, mobile, and IDTV technology. *International Journal of Digital Television*. 2, 3. 371-377.

Seel. P. B. (2012). The 2012 NAB Show: Digital television goes over the top. *International Journal of Digital Television*, 3, 3. 357-360.

Seel, P. B., & Dupagne, M. (2010). Advanced television and video. In A. E. Grant & J. H. Meadows (Eds.), *Communication Technology Update* (12th ed., pp. 82-100). Boston: Focal Press.

Triggs, R. (2017, October 6). MicroLED explained: The next-gen display technology. *Android Authority*. https://www.androidauthority.com/micro-led-display-explained-805148/.

Waniata, R. (2015, May 3). What are HDR TVs and why is every manufacturer selling them now? *DigitalTrends*. http://www.digitaltrends.com/home-theater/hdr-for-tvs-explained.

Wilkinson, S. (2019, August 2). Eyes on Sony's CLED (Crystal LED) display technology: Samsung isn't the only player in the Micro LED game. *TechHive*. https://www.techhive.com/article/3429636/sony-crystal-led-display-technology.html.

YouTube for Press. (2020). Global Reach. https://www.youtube.com/intl/en-GB/about/press/.

7

Multichannel Television Services

Paul Driscoll, Ph.D. & Michel Dupagne, Ph.D.*

Abstract

Consumers are using four methods to receive packages of television channels and/or programming, including cable, satellite, telephone lines, and internet delivery (OTT—Over-the-Top Television). This multichannel video programming distributor (MVPD) market is more competitive than ever, with consumers actively switching from one service to another, as well as subscribing to multiple services at the same time. Overall, the MVPD industry continued to face substantial subscriber losses in 2019 due to cord-cutting trends. In fact, the net traditional MVPD subscriber losses accelerated between 2013 and 2019, with an estimated total loss of more than 15 million subscribers (or 15% of the estimated 101 million multichannel video subscribers in 2013). In 2019, traditional MVPDs lost nearly 6 million subscribers while the industry only gained about 1 million virtual MVPD customers. Cable operators still dominated the MVPD market with a 58% market share, followed by direct broadcast satellite (DBS) providers with a 30% share and telephone companies with a 12% share. The top four U.S. pay-TV providers (Comcast, AT&T, Charter, and DISH) served 80% of the mainstream video subscribers in 2019.

Introduction

In 2020, the U.S. MVPD industry—which includes cable operators, direct broadcast satellite (DBS) providers, telephone companies, and other delivery entities that offer multichannel television services— is standing at the crossroads of its very existence. Indeed, after 40 years of steady growth and prosperity, the traditional MVPD business model is facing an unprecedented challenge—possibly an existential threat— with a multi-year subscribership decline due to growing

* Driscoll is Associate Professor and Vice Dean for Academic Affairs, Department of Journalism and Media Management, University of Miami (Coral Gables, Florida) and Dupagne is Professor in the same department.

cord-cutting habits and rising streaming platform popularity. From a peak of 102 million subscribers in 2012, multichannel video operators shed about 15 million customers by the end of 2019 (see Table 7.1). Not only is there is no sign of abatement, but there is evidence the cord-cutting phenomenon (i.e., a traditional MVPD subscriber's decision to cancel video service) has quickened the pace. As of early 2020, it is unknown when and how cord-cutting will begin to stabilize.

In retrospect, the former Chief Executive Officer of Cablevision (a large multiple system operator that is now part of Altice USA), James Dolan, could not have been more prescient in 2013 when he acknowledged the possibility that his company would one day stop offering multichannel video service and would instead concentrate on the high-speed data (broadband) segment (Ramachandran & Peers, 2013). According to Leichtman Research Group, the top U.S. cable operators served 67% of the broadband internet subscribers by the end of 2019 (Frankel, 2020b).

While Dolan's prediction has not materialized yet, there is a growing realization that "video customers are worth substantially less than their broadband counterparts as viewership and focus continues to shift away from providing pay TV service and more towards broadband connectivity" (Farrell, 2019a). Industry analyst Craig Moffett went even further by stating that

"[v]ideo just doesn't matter much anymore," pointing to the low cash flow multiple per video customer and the growing importance of programmers' direct-to-consumer (DTC) video offerings, such as HBO Now and ESPN+ (Farrell, 2019a; see also Nator, 2020).

Although the current negative performance results are real and should not be minimized, one must also remember that 68% of the 125.6 million U.S. television households continued to subscribe to an MVPD service in 2019 (see Table 7.2). As such, MVPD companies still represent a formidable economic force in the TV industry, delivering popular video content to a large majority of American households and often benefiting from program creation assets through vertical integration.

"One reason why consumers [might be] reluctant to leave their cable and satellite provider is that they can't get the same viewing experience from streaming" (Berr, 2017). While the MVPD business will likely continue its transformation in the years to come and become smaller and more integrated with the programming side, due to changing video consumption patterns, it has developed some strategies (e.g., "skinny" bundles and virtual MVPD offerings), discussed below, to confront the cord-cutting downturn. In sum, to paraphrase Mark Twain, the reports of the MVPD industry's death are greatly exaggerated, given its resilience in earlier decades.

Table 7.1
Net Subscriber Adds & Losses for the Top Pay-TV Providers/MVPDs, 2009-2019[*]

Year	All Top MVPDs	Top Traditional MVPDs	Top Virtual MVPDs[**]
2019	(4,915,000)	(5,925,000)	1,010,000
2018	(1,585,000)	(3,525,000)	1,940,000
2017	(1,125,000)	(3,125,000)	2,000,000
2016	(775,000)	(1,920,000)	1,145,000
2015	(385,000)	(910,000)	525,000
2014	(145,000)	(185,000)	40,[†]
2013	(65,000)	(105,000)	40,000[†]
2012	155,000	145,000	10,000[†]
2011	380,000	380,000	NA
2010	550,000	550,000	NA
2009	1,950,000	1,950,000	NA

Note. NA=not available. *The top pay-TV Providers or MVPDs represent about 95% of the market. **Subscriber data for the top virtual MVPDs only include publicly available counts for Hulu + Live TV, Sling TV, and AT&T TV NOW. According to Leichtman Research Group, there were more than three million additional, but not publicly reported, vMVPD subscribers by the end of the 2019. †Subscriber data for the top virtual MVPDs from 2012 to 2014 originated from DISH's Sling TV International service.

Source: Leichtman Research Group

Table 7.2
U.S. Multichannel Video Industry Benchmarks (in millions), 2016-2020

Category	2016	2017	2018	2019	2020	CAGR (%) 2016-2020
Basic cable subscribers	52.8	51.7	50.5	49.0	47.5	(2.6)
Digital service cable subscribers	50.5	50.1	49.9	48.7	47.3	(1.6)
High-speed data cable subscribers	63.5	66.4	69.5	72.3	74.1	3.9
Voice service cable subscribers	28.8	28.5	28.0	27.0	26.0	(2.5)
DBS subscribers	33.2	31.5	29.1	25.9	24.2	(7.6)
Telco video subscribers	11.6	10.7	10.2	9.9	9.2	(5.5)
All multichannel video subscribers	97.7	93.9	89.8	84.8	80.9	(4.6)
TV households	121.7	123.2	124.5	125.6	126.7	1.0
Occupied households	123.0	124.5	125.9	127.0	128.0	1.0
Housing units	136.0	137.8	138.9	140.1	141.2	1.0

Note. All figures are estimates. These counts exclude virtual MVPD subscribers.

Source: Kagan (2020a). Reprinted with permission

This chapter will first provide background information about multichannel television services to situate the historical, regulatory, and technological context of the industry. Special emphasis will be given to the cable industry. The other sections of the chapter will describe and discuss major issues and trends affecting the MVPD industry.

Definitions

The Cable Television Consumer Protection and Competition Act of 1992 defines a multichannel video programming distributor as "a person such as, but not limited to, a cable operator, a multichannel multipoint distribution service, a direct broadcast satellite service, or a television receive-only satellite program distributor, who makes available for purchase, by subscribers or customers, multiple channels of video programming" (p. 244). For the most part, MVPD service refers to multichannel video service offered by cable operators (e.g., Comcast, Charter), DBS providers (DirecTV, DISH), and telephone companies (e.g., AT&T's U-verse, Verizon's Fios). This chapter focuses on these distributors.

Pay television designates a category of TV services that offers programs uninterrupted by commercials for an additional fee on top of the basic MVPD service (Federal Communications Commission [FCC], 2013a). These services primarily consist of premium channels, pay-per-view (PPV), and video on demand (VOD). Subscribers pay a monthly fee to receive such premium channels as HBO and Starz. In the case of PPV, they pay a per-unit or transactional charge for ordering a movie or another program that is scheduled at a *specified time*. VOD, on the other hand, allows viewers to order a program from a video library at *any given time* in return for an individual charge or a monthly subscription fee. The latter type of VOD service is called subscription VOD or SVOD. We should note that the terms "pay television" and "MVPD" are often used interchangeably to denote a TV service for which consumers pay a fee, as opposed to "free over-the-air television," which is available to viewers at no cost.

As of early 2020, in spite of a *Notice of Proposed Rulemaking* (FCC, 2014a) now in limbo, the Federal Communications Commission (FCC) considered all over-the-top (OTT) providers (linear or not), such as Netflix, YouTube, and Sling TV, as online video distributors (OVD) for the sake of regulatory classification. An OVD "offers video content by means of the internet or other Internet Protocol (IP)-based transmission path provided by a person or entity other than the OVD" (FCC, 2015, p. 3255).

Background
The Early Years

While a thorough review of the history, regulation, and technology of the multichannel video industry is beyond this chapter (see Baldwin & McVoy [1988],

Parsons [2008], Parsons & Frieden [1998] for more information), a brief overview is necessary to understand the broad context of these technologies.

As TV broadcasting grew into a new industry in the late 1940s and early 1950s, many households were unable to access programming because they lived too far from local stations' transmitter sites or because geographic obstacles blocked reception of terrestrial electromagnetic signals. Without access to programming, consumers were not going to purchase TV receivers—a problem for local stations seeking viewers and appliance stores eager to profit from set sales.

The solution was to erect a central antenna capable of capturing the signals of local market stations, then amplify and distribute them through wires to prospective viewers for a fee. Thus, cable extended local stations' reach and provided an incentive to purchase a set. The first non-commercial Community Antenna TV (CATV) service was established in 1949 in Astoria, Oregon, but multiple communities claim to have pioneered this retransmission technology, including a commercial system launched in Lansford, Pennsylvania, in 1950 (Besen & Crandall, 1981).

Local TV broadcasters initially valued cable's ability to extend their household reach, but tensions arose when cable operators began using terrestrial microwave links to import programming from stations located in "distant markets," increasing the competition for audiences that local stations rely on to sell advertising. Once regarded as a welcome extension to their over-the-air TV signals, broadcasters increasingly viewed cable as a threat and sought regulatory protection from the government.

Evolution of Federal Communications Commission Regulations

At first, the FCC showed little interest in regulating cable TV. But given cable's growth in the mid-1960s, the FCC, sensitive to TV broadcasters' statutory responsibilities to serve their local communities and promote local self-expression, adopted rules designed to protect over-the-air TV broadcasting (FCC, 1965). Classifying cable as only a supplementary service to over-the-air broadcasting, regulators mandated that cable systems carry local TV station signals and placed limits on the duplication of local programming by distant station imports (FCC, 1966). The Commission saw such rules as critical to protecting TV broadcasters, especially struggling UHF-TV stations, although actual evidence of such harm was largely nonexistent. Additional cable regulations followed in subsequent years, including a program origination requirement and restrictions on pay channels' carriage of movies, sporting events, and series programming (FCC, 1969, 1970).

The FCC's robust protection of over-the-air broadcasting began a minor thaw in 1972, allowing an increased number of distant station imports depending on market size, with fewer imports allowed in smaller TV markets (FCC, 1972a). The 1970 pay cable rules were also partially relaxed. Throughout the rest of the decade, the FCC continued its deregulation of cable, sometimes in response to court decisions.

Although the U.S. Supreme Court held in 1968 that the FCC had the power to regulate cable television as reasonably ancillary to its responsibility to regulate broadcast television, the scope of the FCC's jurisdiction under the *Communications Act of 1934* remained murky (*U.S. v. Southwestern Cable Co.*, 1968). FCC regulations came under attack in a number of legal challenges. For instance, in *Home Box Office v. FCC* (1977), the Court of Appeals for the D.C. Circuit questioned the scope of the FCC's jurisdiction over cable and rejected the Commission's 1975 pay cable rules, designed to protect broadcasters against possible siphoning of movies and sporting events by cable operators.

In the mid-1970s, cable began to move beyond its community antenna roots with the use of domestic satellites to provide additional programming in the form of superstations, cable networks, and pay channels. These innovations were made possible by the FCC's 1972 *Open Skies Policy*, which allowed qualified companies to launch and operate domestic satellites (FCC, 1972b). Not only did the advent of satellite television increase the amount of programming available to households well beyond the coverage area of local TV broadcast channels, but it also provided cable operators with an opportunity to fill their mostly 12-channel cable systems.

Early satellite-delivered programming in 1976 included Ted Turner's "Superstation" WTGC (later WTBS) and the Christian Broadcasting Network, both initially operating from local UHF-TV channels. In 1975, Home Box Office (HBO), which had begun as a pay programming service delivered to cable systems via terrestrial microwave links, kicked off its satellite distribution offering the "Thrilla in Manila" heavyweight title fight featuring Mohammad Ali against Joe Frasier.

Demand for additional cable programming, offering many more choices than local television stations and a limited number of distant imports, stimulated cable's growth, especially in larger cities. According to the National Cable & Telecommunications Association (2014), there were 28 national cable networks by 1980 and 79 by 1990. This growth started an alternating cycle where the increase in the number of channels led local cable systems to increase their capacity, which in turn stimulated creation of new channels, leading to even higher channel capacity.

Formal congressional involvement in cable TV regulation began with the *Cable Communications Policy Act of 1984*. Among its provisions, the *Act* set national standards for franchising and franchise renewals, clarified the roles of Federal, State, and local governments, and freed cable rates except for systems that operated without any "effective competition," defined as TV markets having fewer than three over-the-air local TV stations. Cable rates soared, triggering an outcry by subscribers.

The quickly expanding cable industry was increasingly characterized by a growing concentration of ownership. Congress reacted to this situation in 1992 by subjecting more cable systems to rate regulation of their basic and expanded tiers, and instituting retransmission consent options (requiring cable systems to give something of value to local TV stations not content with simple must-carry status).

The Cable Television Consumer Protection and *Competition Act of 1992* also required that MVPDs make their programming available at comparable terms to satellite and other services. The passage of the *Telecommunications Act of 1996* also reflected a clear preference for competition over regulation. It rolled back much of the rate regulation put in place in 1992 and opened the door for telephone companies to enter the video distribution business.

Distribution of TV Programming by Satellite

By the late 1970s and early 1980s, satellites were being used to distribute TV programming to cable systems, expanding the line-up of channels available to subscribers. But in 1981, about 11 million people lived in rural areas where it was not economical to provide cable; five million residents had no TV at all, and the rest had access to only a few over-the-air stations (FCC, 1982).

One solution for rural dwellers was to install a Television Receive Only (TVRO) satellite dish. Sometimes called "BUGS" (Big Ugly Dishes) because of the 8-to-12-foot diameter dishes needed to capture signals from low-power satellites, TVROs were initially the purview of hobbyists and engineers. In 1976, H. Taylor Howard, a professor of electrical engineering at Stanford University, built the first homemade TVRO and was able to view HBO programming for free (his letter of notice to HBO having gone unanswered, Owen, 1985).

Interest in accessing free satellite TV grew in the first half of the 1980s, and numerous companies began to market satellite dish kits directly to consumers. From 1985 to 1995, two to three million dishes were purchased (Dulac & Godwin, 2006). However, consumer interest in home satellite TV stalled in the late 1980s when TV programmers began to scramble their satellite feeds.

In 1980, over the strenuous objections of over-the-air broadcasters, the FCC began to plan for direct broadcast satellite (DBS) TV service (FCC, 1980). In September 1982, the Commission authorized the first commercial operation. At the 1983 Regional Administrative Radio Conference, the International Telecommunication Union (ITU), the organization that coordinates spectrum use internationally, awarded the United States eight orbital positions of 32 channels each (Duverney, 1985). The first commercial attempt at DBS service began in Indianapolis in 1983, offering five channels of programming, but ultimately all of the early efforts failed, given the high cost of operations, technical challenges, and the limited number of desirable channels available to subscribers. (Cable operators also used their leverage to dissuade programmers from licensing their products to DBS services.) Still, the early effort demonstrated the technical feasibility of using medium-power Ku-band satellites (12.2 to 12.7 GHz downlink) and foreshadowed a viable DBS service model.

In 1994, the first successful DBS providers were DirecTV, a subsidiary of Hughes Corporation, and U.S. Satellite Broadcasting (USSB) offered by satellite TV pioneer Stanley Hubbard. Technically competitors, the two services offered largely complementary programming and together launched the first high-power digital DBS satellite capable of delivering over 200 channels of programming to a much smaller receiving dish (Crowley, 2013).

DirecTV bought USSB in 1998. EchoStar (now the DISH Network) was established in the United States in 1996 and became DirecTV's primary competitor. In 2005, the FCC allowed EchoStar to take over some of the satellite channel assignments of Rainbow DBS, a failed 2003 attempt by Cablevision Systems to offer its Voom package of 21 high-definition (HD) TV channels via satellite. In 2003, the U.S. Department of Justice blocked an attempted merger between DirecTV and EchoStar. DirecTV was later acquired by AT&T in 2015.

Direct Broadcast Satellite Expansion

DBS systems proved popular with the public but faced a number of obstacles that allowed cable to remain the dominant provider of multichannel television service. One competitive disadvantage of early DBS operators over their cable counterparts was that subscribers were unable to view their local-market TV channels without either disconnecting the TV set from the satellite receiver or installing a so-called A/B switch that made it possible to flip between over-the-air and satellite signals.

In 1988, Congress passed the *Satellite Home Viewer Act* that allowed the importation of distant network programming (usually network stations in New York or Los Angeles) to subscribers unable to receive the networks' signals from their over-the-air local network affiliates. Congress followed up in 1999 with the *Satellite Home Viewer Improvement Act* (SHVIA) that afforded satellite companies the opportunity (although

not a requirement) to carry the signals of local market TV stations to all subscribers living in that market. SHIVA also mandated that DBS operators carrying one local station carry all other local TV stations requesting carriage in a local market, a requirement known as "carry one, carry all."

Today, almost all of U.S. TV households subscribing to DBS service can receive their local stations via satellite. DISH Network supplies local broadcast channels in HD to all 210 U.S. TV markets and DirecTV covers 198 markets (Jacobson, 2019). Local station signals are delivered from the satellite using spot beam technology that transmits signals into specific local TV markets. Satellites have dozens of feedhorns used to send their signals to earth, from those that cover the entire continental United States (CONUS) to individual spot beams serving local markets.

Cable System Architecture

Early cable TV operators built their systems based on a "tree and branch" design using coaxial cable, a type of copper wire suitable for transmission of radio frequencies (RF). As shown in Figure 7.1, a variety of video signals are received at the cable system's "headend," including satellite transmissions of cable networks, local TV station signals (delivered either by wire or from over-the-air signals received by headend antennas), and feeds from any public, educational, and governmental (PEG) access channels the operator may provide. Equipment at the headend processes these incoming signals and translates them to frequencies used by the cable system.

Most cable companies use a modulation process called quadrature amplitude modulation (QAM) to encode and deliver TV channels. The signals are sent to subscribers through a trunk cable that, in turn, branches out to thinner feeder cables, which terminate at "taps" located in subscribers' neighborhoods. Drop cables run from the taps to individual subscriber homes. Because electromagnetic waves attenuate (lose power) as they travel, the cable operator deploys amplifiers along the path of the coaxial cable to boost signal strength. Most modern TV receivers sold in North America contain QAM tuners, although set-top boxes provided by the cable system are usually used because of the need to descramble encrypted signals.

Figure 7.1

Tree & Branch Design

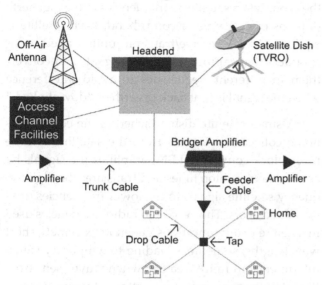

Source: P. Driscoll and M. Dupagne

Over the years, the channel capacity of cable systems has increased dramatically, from 12-22 channel systems through the mid-1970s to today's systems capable of carrying more than 300 TV channels and bi-directional communications. Modern cable systems have 750 MHz or more of radio frequency capacity, up to 1GHz in the most advanced systems (Afflerbach, DeHaven, Schulhuf, & Wirth, 2015). This capacity allows not only plentiful television channel choices, but also enables cable companies to offer a rich selection of pay-per-view (PPV) and video-on-demand (VOD) offerings. It also makes possible high-speed internet and voice over internet protocol (VoIP) telephone services to businesses and consumers. Most cable operators have moved to all-digital transmission, although a few still provide some analog channels, usually on the basic service tier.

As shown in Figure 7.2, modern cable systems are hybrid fiber-coaxial (HFC) networks. Transmitting signals using modulated light over glass or plastic, fiber optic cables can carry information over long distances with much less attenuation than copper coaxial, greatly reducing the need for expensive amplifiers and offering almost unlimited bandwidth with high reliability. A ring of fiber optic cables is built out from the headend, and fiber optic strands are dropped off at each distribution hub. From the hubs, the fiber optic

strands are connected to optical nodes, a design known as fiber-to-the-node or fiber-to-the-neighborhood (FTTN). From the nodes, the optical signals are transduced back to electronic signals and delivered to subscriber homes using coaxial cable or thin twisted-pair copper wires originally installed for telephone service. The typical fiber node today serves about 500 homes, but additional nodes can be added as data-intensive services like high speed broadband, video on demand, and Internet Protocol (IP) video are added. Depending on the number of homes served and distance from the optical node, a coaxial trunk, feeder, and drop arrangement may be deployed.

Rollouts of HFC cable networks and advances in compression have added enormous information capacity compared to systems constructed with only coaxial copper cable, boosting the number of TV channels available for live viewing, and allowing expansive VOD offerings and digital video recorder (DVR) service. Additionally, given the HFC architecture's bi-directional capabilities, cable companies can offer consumers both telephone and broadband data transmission services, generating additional revenues.

Figure 7.2

Hybrid Fiber-Coaxial Design

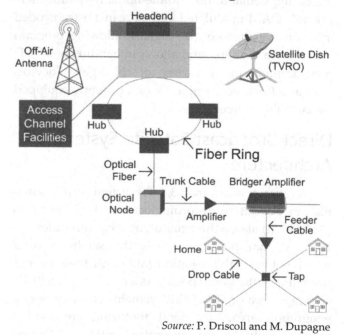

Source: P. Driscoll and M. Dupagne

Some MVPDs, such as Verizon's Fios service and AT&T Fiber, extend fiber optic cable directly to the consumer's residence, a design known as fiber-to-the-home (FTTH) or fiber-to-the-premises (FTTP). Fios uses a passive optical network (PON), a point-to-multipoint technology that optically splits a single fiber optic beam sent to the node to multiple fibers that run to individual homes, ending at an optical network terminal (ONT) attached to the exterior or interior of the residence (Corning, 2005). The ONT is a transceiver that converts light waves from the fiber optic strand to electrical signals carried through the home using appropriate cabling for video, broadband (high-speed data), and telephony services.

Although the actual design of FTTH systems may vary, fiber has the ability to carry data at higher speeds compared to copper networks. Fiber broadband services can deliver up to a blazing one gigabit (one billion bits) of data per second. Fiber-to-the-home may be the ultimate in digital abundance, but it is expensive to deploy and may actually provide more speed than most consumers currently need. In addition, technological advances applied to existing HFC systems have significantly increased available bandwidth, making a transition to all-fiber networks a lower priority for existing cable systems in the United States. However, with ever-increasing demand, fiber-to-the-home may ultimately prevail. QAM-modulated signals can be transcoded into Internet Protocol (IP) signals, allowing program distribution via apps, and allows integration of MVPD programming on a variety of third-party devices such as a Roku box, Apple TV or a properly equipped "smart" TV receiver.

Direct Broadcast Satellite System Architecture

As shown in Figure 7.3, DBS systems utilize satellites in geostationary orbit, located at 22,236 miles (35,785 km) above the equator, as communication relays. At that point in space, the satellite's orbit matches the earth's rotation (although they are not travelling at the same speed). As a result, the satellite's coverage area or "footprint" remains constant over a geographic area. In general, footprints are divided into those that cover the entire continental United States, certain portions of the United States (because

of the satellite's orbital position), or more highly focused spot beams that send signals into individual TV markets. DBS operators send or "uplink" data from the ground to a satellite at frequencies in the gigahertz (billions of cycles per second) band. Each satellite is equipped with receive/transmit units called transponders that capture the uplink signals and convert them to different frequencies to avoid interference when the signal is sent back to earth or "downlinked."

A small satellite dish attached to the consumer's home collects the signals from the satellite. A low-noise block converter (LBN) amplifies the signals by as much as a million times and translates the high frequency satellite signals to the lower frequencies used by TV receivers. The very high radio frequencies used in satellite communications generate extremely short wavelengths, sometimes leading to temporary signal interference in rainy weather when waves sent from the satellite are absorbed or scattered by raindrops.

Figure 7.3

Basic Direct Broadcast Satellite Technology

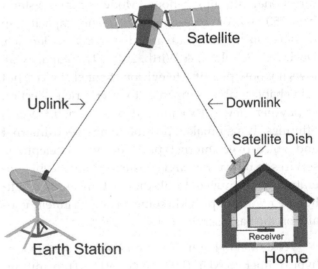

Source: P. Driscoll and M. Dupagne

Unlike cable service that features bi-directional data flows, commercial DBS is a one-way service that relies on broadband connections to offer most of its VOD services. But, with its nationwide footprint, DBS service is available in rural areas where the infrastructure cost of building a cable system would be prohibitive. Even though DBS TV services may be falling out of favor, swarms of low earth orbiting (LEO) satellites may someday provide a breakthrough in closing the broadband divide and bringing broadband low-latency video services to neglected areas (Ritchie, 2019).

Video Compression

The bounty of available TV programming from cable, satellite and online services is possible due to digital video compression. Developed by the ITU and the MPEG group, video compression dramatically reduces the number of bits needed to store or transmit a video signal by removing redundant or irrelevant information from the data stream. Video compression yields increased capacity and permits MVPD distributors to carry 4K Ultra High Definition (UHD) channels, albeit in limited number.

Impressive advances in compression efficiency resulted in the adoption by DBS and some cable operators of MPEG-4 Advanced Video Coding (AVC) that reduced by half the bit rate needed compared with MPEG-2 (Crowley, 2013). High Efficiency Video Coding (HEVC), approved in 2013 by the ITU and MPEG, doubled the program carrying capacity compared with MPEG-4 compression. (Baumgartner, 2017a). Although HEVC/H.264/265 and VP9 are two primary video compression standards today, increasingly powerful codecs will be needed to drive the bits necessary for 8K displays (containing 16 times the pixels of 1080P HD) with associated features that add information to the data stream: high dynamic range (HDR), wide color gamut (WCG), and higher frame rates (Burnichon, 2019). New codecs expected in early 2020 include Essential Video Coding (EVC), Versatile Video Coding (VVC) and AV2. Initial test results suggest 60 to 80 Mbps may be needed to send an 8K television signal using HEVC compression and about half that rate using VVC coding (Burnichon, 2019).

Recent Developments

Statewide Video Franchising

Prior to 2005, the required process of cable franchising (i.e., authorizing an entity to construct and operate a cable television system in an area) was generally handled by local governments. The *Cable Act of 1984* did not mandate that the franchising authority be a local body, but subjecting cable operators to local jurisdiction made sense since they used public rights-of-way to deploy their infrastructure.

On the other hand, cable operators have long argued that local franchising requirements are onerous and complex, which often involve negotiations and agreements with multiple local governments (e.g., county and city).

In 2005, Texas became the first state to enact a law shifting the responsibilities of cable/video franchise requirements from the local to the state government (see Parker, 2011). More than 20 other states followed suit. In those states, it is typically the department of state or the public utilities commission that issues cable franchises (sometimes called certificates) to wireline MVPDs (National Conference of State Legislatures, 2014). Other state agencies may share other cable television service duties (e.g., quality complaints) that previously fell within the purview of local governments. As of May 2019, 26 states awarded cable/video franchising agreements (Morton, 2019).

A la Carte Pay-TV Pricing

A la carte MVPD service would allow subscribers to order individual channels as a customized programming package instead of relying on standard bundled packages or tiers. This issue has attracted considerable policy interest, especially in the mid-2000s, because consumers have often complained to their elected officials or policymakers about the high cost of cable service in relation to other services. Indeed, the FCC has documented for decades that cable rate increases outpace the general inflation rate (see FCC, 2013b). In its latest *Report on Cable Industry Prices* (FCC, 2018), the Commission reported that the average price of expanded basic cable service (the most popular cable tier or package) grew by a compound annual growth rate (CAGR) of 4.7% from 2006 to 2016. In contrast, the consumer price index, all urban consumer (CPI-U), for all items rose by only 1.8% per year during these 10 years.

Using a different methodology (i.e., a telephone survey to estimate mean consumer spending for all types of pay-TV service), Leichtman Research Group

(2019) reported in press releases that surveyed MVPD subscribers spent $103.10 in 2016 and $109.60 in 2019. The CAGR from 2016 to 2019 was then 2.1%, about the same as the CPI-U for all items per year for this period. With the proper methodological caveats (i.e., FCC and LRG data are not directly comparable), it seems that the amount of the expected increases in monthly MPVD bills has not risen as much in recent years and has paralleled annual inflation rate changes, presumably due to cord-cutting pressures.

The Commission also found that the number of channels available in the expanded cable package soared from 71 in 2006 to 181 in 2016, or by a CAGR of 7.0% (FCC, 2018). But the per-channel price declined by 1.6% per year, which bolsters the case of the cable industry for value-related price hikes.

On paper, a bundling strategy makes more economic sense than a pure components strategy (a la carte) when the marginal cost of channel delivery is low, which is often the case with popular channels, and when consumers price channel packages similarly (Hoskins, McFadyen, & Finn, 2004). But studies have shown that consumers would be receptive to a la carte selection if afforded the opportunity. For instance, TiVo (2020) reported that about 68% of the respondents favored the option of customizing their own MVPD package in late 2019, down from 81% in late 2017 (TiVo, 2018). They ranked ABC, CBS, Discovery Channel, A&E, and NBC as the top five most desirable channels to have in a self-selected package. Respondents also indicated that they would spend an average of $33.30 per month for their a la carte pay-TV package and would select an average of 22 channels (TiVo, 2020).

Bundling is hardly an insignificant pricing matter for MVPDs. Not only does bundling play a key role in monetizing MVPDs' video content, but it can also influence the entire revenue stream of cable operators that offer triple-play packages. Triple play, which refers to the combination of video, voice, and data products, is beneficial to both customers who can save money and some MVPDs, particularly cable operators, who can secure a competitive advantage over other MVPDs and counter the effect of cord-cutting or slow growth in the video segment. Media research group Kagan estimated that, as of December 2018, 63% of the customers served by the top five public multiple system operators (MSO) subscribed to more than one service, and 35% opted for the triple-play bundle (Lenoir, 2019a).

In May 2019, Maine became the first U.S. State to enact a law requiring cable operators to "offer subscribers the option of purchasing access to cable channels, or programs on cable channels, individually" (State of Maine, 2019). In so doing, the Maine legislature argued that there was no federal law preventing them from requiring a la carte pricing. But the cable industry viewed this state statute quite differently, contending that it violated its First Amendment rights to provide licensed programming to the general public the way it saw fit. Furthermore, it claimed that L.D. 832 contravened several provisions of the 1984 Cable Act, including a section (544(f)) that prohibits local franchising authorities from imposing provision and content obligations on cable services (Brodkin, 2019). In September 2019, Comcast and several cable programmers filed a complaint for a preliminary injunction against the enforcement of the law in federal court. In December 2019, the U.S. District Court ruled in favor of the plaintiffs, concluding that, even though L.D. 832 was deemed content-neutral, the State of Maine had not presented sufficient evidence ("an important or substantial governmental interest") as to how the a la carte pricing law would have reduced rising cable bills (*Comcast v. Janet Mills*, 2019). In what is likely to give more headaches to cable operators, Judge Torresen also stated that "LD 832 is likely not preempted by §544(f) of the Cable Act" (p. 20). In January 2020, the State of Maine decided to appeal the judgment.

Adapting to Changes in the Linear Viewing Model

MVPDs continue to adapt their technology and content offerings to accommodate subscribers' shifting patterns of TV consumption. As of 2020, the average adult American watches 5h:46m of video content daily, mostly on live plus time shifted TV (4h:27m), and 0h:54m using TV connected devices (Nielsen, 2019). Yet disruption continues, especially with younger consumers.

Almost three-quarters of U.S. TV households have at least one internet-connected TV device (including smart TVs), and 31% say they watch streamed video on TV daily (Leichtman, 2019). The home digital video recorder (DVR), first introduced in 1999, is now present in 58% of television households (Frankel, 2019a). Although available as stand-alone units, DVRs are usually integrated into powerful set-top boxes provided by cable and DBS operators. Increasingly, these set-top boxes also provide access to subscription over-the-top programming like Netflix. Many subscribers to virtual MVPD services (and some cable systems) can access a "cloud DVR" service that relies on remote servers to store customer program content; most charge a monthly fee for any beyond minimal storage. Smart TVs and a host of streaming media devices will continue to churn the MVPD marketplace.

Retransmission Consent

Few other issues in the MVPD business have created more tensions between distributors and broadcasters (and angered more subscribers) than retransmission consent negotiations going awry.

As noted above, the *Cable Act of 1992* allowed local broadcasters to seek financial compensation for program carriage from MVPDs, a process known as retransmission consent. While most of these agreements are concluded with little fanfare (Lenoir, 2014), some negotiations between parties can degenerate into protracted, mercurial, "who-will-blink-first" disputes, which sometimes lead to high-profile programming blackouts when the broadcaster forces the removal of its signal from the MVPD's line-up. According to the American Television Alliance, the number of retransmission-related blackouts surged to 278 by the end of 2019, up from the 165 station blackouts recorded in 2018 (Farrell, 2019b). In a separate analysis, Kagan found that 63 of the 113 signal disruptions reported from publicly available sources between 2014 and 2019 occurred in the last three years (Zubair, 2020).

The economic stakes of retransmission negotiations are high. Broadcasters claim that, like cable networks, they deserve fair rates for their popular programs and increasingly consider retransmission fees as a second revenue stream. Kagan estimated that these fees could account for close to 50% of local station owners' revenues in 2019 (Farrell, 2019b). Television stations expect to collect more than $12 billion in broadcast retransmission fees in 2020 (see Table 7.3). Not surprisingly, the amount of broadcast retransmission fees varies by market, with broadcasters in large markets commanding higher carriage fees (Barbour, 2019). Few industry observers would disagree that "[r]etrans saved TV broadcasting, but it cannot save it forever" (Jessell, 2017).

Table 7.3

Annual Broadcast Retransmission Fee Projections by Medium, 2018-2020

Medium	2018	2019	2020
Cable ($ millions)	5,452,1	6,083.3	6,670.8
Average cable fee/sub/month ($)	8.89	10.19	11.52
DBS ($ millions)	3,249.0	3,430.8	3,623.8
Average DBS fee/sub/month ($)	8.93	10.24	11.56
Telco ($ millions)	1,534.6	1,684.8	1,802.9
Average telco fee/sub/month ($)	12.17	13.89	15,71
Total retransmission fees ($ millions)	10,235.7	11,198.9	12,097.5
Average fee/sub/month ($)	9.48	10.90	12.25

Note. All figures are estimates as of June 2019. The average fee per month per subscriber for each video medium is obtained by dividing the amount of annual retransmission fee for the medium by the average number of subscribers for that medium. The average number of subscribers, which is calculated by taking the average of the subscriber count from the previous year and the subscriber count of the current year, is meant to estimate the number of subscribers at any given time during the year and reflect better the retransmission fee charged during the year instead of at the year-end. The average estimates refer to the fees paid to television stations on behalf of each subscriber, not to the payment per station.

Source: Kagan (2020b). Reprinted with permission

On the other hand, the MVPDs, fully aware that excessive customer bills could intensify cord-cutting, have attempted to rein in broadcasters' demands for higher retransmission fees, generally with little success, reflecting consumers' robust appetite for broadcast network programming (Baumgartner, 2016). Generally, larger MVPDs are more capable of negotiating volume discounts than smaller ones (Barbour, 2019). With both sides having so much at stake, it is likely that major retransmission consent negotiations will continue to become contentious in 2020 and 2021.

The Communications Act (47 U.S.C. § 325(b)(3) (C)) imposes a statutory duty on both MVPDs and broadcasters to exercise "good faith" in retransmission consent negotiations, and regulators have increasingly taken note of clashes that too often leave consumers in the dark. In 2014, the FCC barred the "top-4" broadcast stations in a TV market (that are not commonly owned) from banding together to leverage their bargaining power in retransmission consent negotiations (FCC, 2014b). In 2019, AT&T filed a complaint with the FCC against nine TV station licensees for allegedly pulling 20 stations from its systems... "with no end in sight" (Eggerton, 2019).

Current Status

TV Everywhere

According to the FCC (2015), TV Everywhere (TVE) "allows MVPD subscribers to access both linear and video-on-demand ("VOD") programming on a variety of in-home and mobile internet-connected devices" (p. 3256). Ideally, as its name indicates, TVE should provide these subscribers the ability to watch programs, live or not, in any location as long as they have access to a high-speed internet connection.

Launched in 2009, TVE has grown into a major marketing strategy by multichannel video operators to retain current subscribers, undercut cord-cutting trends, and attract Millennials to pay-TV subscription by offering channel line-ups on mobile devices (Winslow, 2014). According to a February 2019 Kagan survey, 56% of the respondents reported using TVE at least once in the last three months (Bacon, 2019). They were more likely to use TVE for viewing live content inside the home (49%) than live content outside the

home (43%), on-demand content inside the home (42%), or on-demand content outside the home (38%). Thus, TVE was more frequently used as a home-based viewing activity than outside the home. Not surprisingly, Generation Z respondents (53%) relied more often on TVE than Millennial (50%), Generation X (43%), or Baby Boomer respondents (30%; Bacon, 2019).

Decline of MVPD Subscription

As shown earlier in the chapter in Table 7.2, the MVPD penetration (MVPD subscribers/TV households) was 68% in 2019, down from 76% in 2017. The number of basic cable subscribers decreased by 7% between 2016 (52.8 million) and 2019 (49.0 million). For the DBS and telco (telephone company) providers, the subscriber decline was even worse (22%% and 15%, respectively). Kagan projected that this gradual decline in MVPD subscription would continue in 2020. It expected that the (traditional) MVPD penetration would fall to 64% in 2020.

From a high revenue of $116.4 billion in 2016, the MVPD industry lost an estimated $11 billion by the end of 2019 ($105 billion). Kagan also predicted that multichannel video revenues would decrease by another $4 billion (or 4%) in 2020 (Lenoir, 2019b).

The ranking of the top MVPDs changed dramatically in 2015 with AT&T's acquisition of DirecTV for $49 billion. As indicated in Table 7.4, the top four providers served 80% of all MVPD subscribers by September 2019, generating a high degree of concentration in that industry (see Hoskins et al., 2004). But there was little concentration difference between December 2017 (79%) and September 2019. All ten MVPDs listed in the table accounted for a 95% share of the estimated total MVPD market.

In March 2020, AT&T President and Chief Operating Officer, John Stankey, announced that while the company would "continue to offer satellite and DirecTV where it has a rightful place in the market, places where cable broadband is not prevalent," it would also de-emphasize marketing DirecTV service and promote instead a software-driven approach to pay-TV subscription in the future (i.e., the new AT&T TV service, Brodkin, 2020). In 2019, AT&T lost more than 4 million subscribers across its satellite, wireline (U-verse), and streaming (AT&T TV Now) platforms (Frankel, 2020a).

Table 7.4

Top 10 MVPDs in the United States, as of September (Q3) 2019 (in thousands)

Company	Subscribers
Comcast	21,403
AT&T (DirecTV/U-verse)	20,384
Charter Communications	16,245
DISH Network	9,494
Verizon Communications (Fios)	4,280
Cox Communications	3,459
Altice USA*	3,179
Mediacom Communications	729
Frontier Communications	698
WideOpenWest	380
Total estimated U.S. MVPD market (2019)	84,826

Note. These counts exclude virtual MVPD subscribers. *Altice USA includes Optimum (Cablevision) and Suddenlink, but the company did not report separate counts for the two brands in its 2019 annual report.

Source: Kagan (2020c). Reprinted with permission

Factors to Watch

MVPD Consolidation

Scale is possibly the most important economic strategy in the global media business because it allows a larger media firm to reduce input costs (e.g., programming costs) and derive competitive advantages against rivals (e.g., channel revenues and advertising opportunities). For example, AT&T's John Stankey pointed out that "[w]e didn't buy DirecTV because we love satellite exclusively as a distribution medium; we bought it because it gave us scale in entertainment and scale in distribution of entertainment" (*Broadcasting & Cable*, 2016, p. 19). The consolidation of distribution and programming sectors continues against a backdrop of substantial existing horizontal ownership concentrations.

For example, in 2018, AT&T purchased Time-Warner for $85 billion, adding Time Warner's holdings including Turner Networks, HBO, CNN, music interests, and more to the AT&T conglomerate. In March 2019, Disney, the world's biggest content creator, paid over $52 billion to acquire key programming assets from 20th Century Fox.

Government has several tools to protect consumers from potentially anticompetitive consequences from completed mergers, including divestiture, imposition of behavioral conditions, or both, before granting approval. For example, in April 2016, the U.S. Department of Justice approved Charter Communications' $65.5 billion acquisition of two major MSOs, Time Warner Cable and Bright House Networks, subject to pro-competition and consumer-friendly restrictions. Charter agreed to comply with the following key conditions (Eggerton, 2016; Kang & Steel, 2016):

- Not to strike any agreement with video programmers that would limit programming availability on online video distributors (OVDs) like Netflix and Amazon Prime Video for seven years
- Not to use usage-based pricing and enforce data caps on broadband users for seven years
- Not to charge OVDs interconnection fees for seven years
- To expand availability of high-speed internet service to another two million homes
- To offer a low-cost broadband service option to low-income households

The seismic consolidations are likely to continue, but companies ripe for takeover are shrinking. Some companies are exploring more expansion abroad, with Comcast acquiring satellite service Sky for $31 billion in 2018 (Farrell, 2018a). Normal consolidation will likely continue between smaller-size operators, but only once the giants sort it out.

Cord-Cutting

The continued decline in subscriptions to traditional MVPDs from so-called cord-cutting has been called dreadful, brutal, and "freaking ugly" (Munson, 2019a, 2019b). Consumers cancelling their MVPD service sometimes subscribe to a vMVPD service as a substitute, usually at a lower cost for a more limited selection of linear channels adversely affecting revenues. Of course, many consumers choosing to remain with an MVPD service also watch OTT services, but about 19% of OTT households are cord-cutters and 16% are cord-nevers (Engleson, 2019). Those in the cordless category have significantly lower incomes compared with MVPD subscribers and spend much more time watching OTT TV programming (Engleson, 2019).

From Table 7.1. it is clear that the annual subscriber losses due to cord-cutting are substantial for traditional MVPDs. These operators shed more than an estimated 15 million subscribers since 2013, and there are few signs of abatement for 2020. One writer even compared the evolution of cord-cutting to the continuous gradual decline of broadcast network audiences: "Each year the networks argued to advertisers, that at some point, the ratings decline would begin to stabilize" (Adgate, 2017).

The main reasons for cutting the cord have remained consistently the same over time: cost of MVPD service (factor mentioned by 87% of respondents without pay-TV service in late 2017); use of streaming services (40%), and use of an over-the-air antenna for reception of basic channels (23%; TiVo, 2018).

The open question remains whether cord-cutting will eventually stop, even heralding a return to traditional MVPD subscription, or whether MVPD service will follow the path of the music or newspaper industries. For pay-TV providers, the options range from bad to catastrophic and may require them to rethink their video revenue model. On the very negative side, the Diffusion Group predicted that 30 million homes—about one third of the 94 million MVPD subscribers in 2017—would go without pay-TV service by 2030 (Baumgartner, 2017b).

It is also possible that viewers decide to re-subscribe to traditional MVPD service after missing the breadth of channels available in the MVPD line-up or being disappointed by the technical quality and content diversity of virtual MVPD service (see Baumgartner, 2018; Lafayette, 2018). For the long term, Nielsen has argued that the decline in pay-TV subscription may reverse itself by the changing viewing habits of older Millennials who have children; these cord-never Millennials may decide to become subscribers once they start families (Lafayette, 2016).

"Skinny" Bundles & Virtual MVPD Service

"Skinny" bundles—or packages—refer to "smaller programming packages with fewer channels at potentially lower costs to consumers (and, less altruistically, also allow providers to save on the cost of programming)" (Leichtman Research Group, 2015, p. 1). Even though the basic service tier can be technically considered as a slimmed-down bundle (Hagey & Ramachandran, 2014), the term really applies to selected groups of channels from the (expanded) programming service tier. Perhaps the best example of an early "skinny" bundle is DISH Network's Sling TV streaming service. Introduced in 2015, it initially featured a dozen channels for $20 a month (Fowler, 2015). As of March 2020, the Sling TV brand was available in three "flavors:" orange (32 channels for $30), blue (47 channels for $30), and orange & blue (53 channels for $45). By the end of 2019, more than 2.5 million viewers subscribed to Sling TV (Munson, 2020).

As more companies followed DISH Network's example and launched streaming-based "skinny" packages, such as DirecTV Now (now AT&T TV Now) and YouTube TV, the term "virtual multichannel video distributor" (vMVPD) began to appear in the industry vernacular. Today, the "vMVPD" acronym is more commonplace than "skinny" bundles, even though both terms signify the same thing because all primary "skinny" bundles are linear streaming options. vMVPDs can also be defined as over-the-top (OTT) providers that normally offer lower-cost and limited-channel packages via internet streaming technology. The big four vMVPD services—delivering linear TV content that may include subscription video on demand (SVOD) and "cloud" DVR service—include: Hulu + Live TV, Sling TV, AT&T TV Now, and YouTube TV. In 2019, there were approximately 8.5 million traditional vMVPD subscribers (Olgeirson, Lenoir, & Pabst, 2019). Combined with so-called "pure-play" vMVPD services like Netflix and Disney+, there were about 93.6 million vMVPD subscribers (Olgeirson et al., 2019). Also in 2019, it was estimated that 63.7 million U.S. households streamed OTT services where about 225 OTT programming services competed for attention (Kurz, 2019). Also joining the marketplace of OTT offerings are advertising-supported video-on-demand services (AVOD), such as Pluto TV and Tubi TV, featuring multiple free linear channels of programming and/or VOD programming in return for watching ads.

Normally, vMVPD services provide consumers with less pricey alternatives and may mitigate some revenue losses resulting from cord-cutting—while enticing cord-nevers—in an increasingly consolidated ecosystem. The market faces ongoing innovation and disruption while competitive pressures continue to build. For example, citing high programming costs, Sony dropped its Play Station Vue vMVPD service in 2020 while Comcast/NBCUniversal plans to launch its Peacock vMVPD service in summer 2020 (Coster, 2019). The service will have a number of tiers including a limited free program base followed by an ad-supported tier that can also be purchased as a premium ad-free tier. Traditional MVPD subscribers to Comcast and Cox (and maybe more) cable services will be able to access the ad-supported tier for free, but will have to pay for the premium service (Alexander, 2020). Additionally, traditional MVPDs like Comcast and AT&T are also rolling out free or low-cost devices that provide a proprietary interface to facilitate access to streaming services for broadband-only customers (Takiff, 2019).

Reacting to business imperatives, many services sharply increased prices in 2019 and decreased promotional offers (Engebreston, 2019; Frankel, 2019b). Some critics, however, have argued that "[b]eing a vMVPD will remain a truly lousy business" (Baumgartner, 2018) because its average revenue per user (ARPU) is lower and its churn is higher than that of traditional MVPD services (see Farrell, 2018b). Charles Ergen, CEO of DISH, recognized that the churn for low-cost Sling TV was higher than for other services, but he also added that subscribers "tend to come back over a period of time" (Williams, 2016). In the tumult of today's pay TV market, vMVPD services face a variety of headwinds to become financially sustainable and ultimately, perhaps, to prosper (Ball, 2018).

Not to be overlooked are plans by wireless cellular companies to provide wireless broadband services over fixed 5G cellular networks, eventually increasing competition in the broadband television business. For example, Verizon plans to roll out its nascent 5G Home service to 30 cities (Dano, 2019).

Sustainability

Sustainability efforts play a critical role in the MVPD and vMVPD marketplace. Samsung's 2019 sustainability report (141 pages) reflects a growing belief that business evaluation goes beyond profit and must include stewardship and vigorous commitment to practices assisting future generations (Samsung Electronics, 2019). Profits cannot come at the expense of ethics. Failure to address sustainability will eventually wreak havoc; consumers will punish companies for failure to consider our planet. For example, increasing public and corporate demand for green financial instruments could reshape finance, leading to a punishing tipping point for companies that fail to implement environmental and other stewardship efforts (Connaker & Madsbjerg, 2019; Sorkin, 2018).

The major MVPD/vMVPD players have extensive sustainability plans in place. For example, Comcast/NBCUniversal has reduced truck rolls to customers, cutting emissions; it continues to install energy-saving products for its deployed MVPD TV equipment, and saved 51.7 million pounds of cable from landfills since 2014 (Comcast/NBCUniversal, 2019). Among its environmental initiatives, the Consumer Electronics Association promotes the eco-design of products, packaging and recycling, and green facilities (Consumer Technology Association, 2018). Huge media conglomerates like Disney and AT&T have also made numerous commitments to environmental and community improvements (AT&T, 2018; Walt Disney Co., 2018).

Companies must also extend their sustainability principles through oversight of their supply chains, purchasing materials and equipment from only environmentally responsible and socially conscious businesses.

Bibliography

Adgate, B. (2017, December 7). Cord cutting is not stopping any time soon. *Forbes*. https://www.forbes.com.

Afflerbach, A, DeHaven, M., Schulhuf, M., & Wirth, E. (2015, January/February). Comparing cable and fiber networks. *Broadband Communities*. http://www.bbcmag.com.

Alexander, J. (2020, January 16). NBC's Peacock streaming service will launch on July 15th with three different price tiers. *The Verge*. https://www.theverge.com.

AT&T. (2018). *Corporate social responsibility summary*. https://about.att.com/ecms/dam/csr/sustainability-report-ing/PDF/2019/ATT-Corporate-Responsibility-Summary.pdf.

Bacon, B. (2019, October 15). TV Everywhere or just at home? *S&P Global Market Intelligence*. https://platform.mi.spglobal.com

Baldwin, T. F., & McVoy, D. S. (1988). *Cable communication* (2nd ed.). Englewood Cliffs, NJ: Prentice Hall.

Ball, M. (2018, August 25). The artificial fantasy of virtual pay-TV. *Redef Media Original*. https://redef.com.

Barbour, N. (2019, February 7). Broadcast fees in step with estimated retrans costs at national level. *S&P Global Market Intelligence*. https://platform.mi.spglobal.com.

Baumgartner, J. (2016, April 27). TiVo: Big four broadcasters top 'must keep' list. *Multichannel News*. http://www.multichannel.com.

Baumgartner, J. (2017a, August 28). Direct-to-consumer models lead multiscreen push. *Broadcasting & Cable*. http://www.broadcastingcable.com.

Baumgartner, J. (2017b, November 29). 'Legacy' pay TV market to fall to 60% by 2030: Forecast. *Multichannel News*. http://www.multichannel.com.

Baumgartner, J. (2018, February 12).Virtual MVPDs ended 2017 with 5.3M subs: Study. *Multichannel News*. http://www.multichannel.com.

Berr, J. (2017, March 20). Why reports of pay TV's demise are exaggerated. *CBS News*. https://www.cbsnews.com

Besen, S. M., & Crandall, R.W. (1981). The deregulation of cable television. *Journal of Law and Contemporary Problems, 44*, 77-124.

Broadcasting & Cable. (2016, March 7). *146*(9), 19.

Brodkin, J. (2019, September 10). Comcast sues Maine to stop law requiring sale of individual TV channels. *Ars Technica*. https://arstechnica.com.

Brodkin, J. (2020, March 4). Struggling AT&T plans "tens of billions" in cost cuts, more layoffs. *Ars Technica*. https://arstechnica.com.

Burnichon, T. (2019, April 10). 8K video compression [PowerPoint slides]. *ATEME*. https://8kassociation.com/wp-content/uploads/2019/04/ATEME-8K-Video-Compression.pdf.

Cable Communications Policy Act of 1984, 47 U.S.C. §551 (2011).

Cable Television Consumer Protection and Competition Act of 1992, 47 U.S.C. §§521-522 (2011).

Comcast v. Janet Mills. (2019). Case 1:19-cv-00410-NT. https://www.govinfo.gov.

Comcast/NBCUniversal. (2019). *Sustainability*. https://corporate.comcast.com/values/sustainability.

Connaker, A., & Madsbjerg, S. (2019, January 19). The state of socially responsible investing. *Harvard Business Review*. https://hbr.org/2019/01/the-state-of-socially-responsible-investing.

Consumer Technology Association. (2019). Sustainability report. *CTA(CES)*. https://www.cta.tech/sustainability-report/index.html .

Corning, Inc. (2005). *Broadband technology overview white paper: Optical fiber*. http://www.corning.com/ docs/opticalfiber/wp6321.pdf.

Coster, H. (2019, January 16). Comcast, late to the streaming party, to give details of NBC's Peacock service. *Reuters*. https://www.reuters.com.

Crowley, S. J. (2013, October). *Capacity trends in direct broadcast satellite and cable television services*. Paper prepared for the National Association of Broadcasters. http://www.nab.org.

Dano, M. (2019, October 21). Verizon relaunches fixed wireless service with 5G NR, DIY installs. *Light Reading*. https://www.lightreading.com.

Dulac, S., & Godwin, J. (2006). Satellite direct-to-home. *Proceedings of the IEEE, 94*, 158-172. doi: 10.1109/JPROC.2006.861026.

Duverney, D. D. (1985). Implications of the 1983 regional administrative radio conference on direct broadcast satellite services: A building block for WARC-85. *Maryland Journal of International Law & Trade, 9*, 117-134.

Eggerton, J. (2016, April 25). FCC proposes Charter-Time Warner Cable merger conditions. *Broadcasting & Cable*. http://www.broadcastingcable.com.

Eggerton, J. (2019, June 19). AT&T files 'bad faith' retrans complaint against nine broadcasters. *Broadcasting & Cable*. http://www.broadcastingcable.com.

Engebreston, J. (2019, October 18). Report: AT&T TV Now basic price to increase to $65 in November, highlighting rising streaming TV pricing. *Telecompetitor*. https://www.telecompetitor.com.

Engleson, S. (2019, June 11). The state of OTT [PowerPoint slides]. *Comscore*. https://www.comscore.com/Insights/Presentations-and-Whitepapers/2019/State-of-OTT.

Farrell, M. (2018a, March 1). Burke: Comcast has long had its eye on Sky. *Broadcasting & Cable*. http://www.broadcastingcable.com

Farrell, M. (2018b, March 5). Small dish, deep decline. *Multichannel News*. http://www.multichannel.com.

Farrell, M. (2019a, February 25). Moffett: Video just doesn't matter. *Multichannel News*. http://www.multichannel.com.

Farrell, M. (2019b, October 7). Is the retrans cash flow cow running low? *Multichannel News*. http://www.multichannel.com.

Federal Communications Commission. (1965). Rules re microwave-served CATV (*First Report and Order*), 38 FCC 683.

Federal Communications Commission. (1966). CATV (*Second Report and Order*), 2 FCC2d 725.

Federal Communications Commission. (1969). Commission's rules and regulations relative to community antenna television systems (*First Report and Order*), 20 FCC2d 201.

Federal Communications Commission. (1970). CATV (*Memorandum Opinion and Order*), 23 FCC2d 825.

Federal Communications Commission. (1972a). Commission's rules and regulations relative to community antenna television systems (*Cable Television Report and Order*), 36 FCC2d 143.

Federal Communications Commission. (1972b). Establishment of domestic communications- satellite facilities by non-governmental entities (*Second Report and Order*), 35 FCC2d 844.

Federal Communications Commission. (1980). *Notice of Inquiry*, 45 F.R. 72719.

Federal Communications Commission. (1982). Inquiry into the development of regulatory policy in regard to direct broadcast satellites for the period following the 1983 Regional Administrative Radio Conference (*Report and Order*), 90 FCC2d 676.

Federal Communications Commission. (2013a). *Definitions*, 47 CFR 76.1902.

Federal Communications Commission. (2013b). *Report on Cable Industry Prices*, 28 FCCR 9857.

Federal Communications Commission. (2014a). Promoting innovation and competition in the provision of multichannel video programming distribution services (*Notice of Proposed Rulemaking*), 29 FCCR 15995.

Federal Communications Commission. (2014b). Amendment of the Commission rules related to retransmission consent (*Report and Order and Further Notice of Proposed Rulemaking*), 29 FCCR 3351.

Federal Communications Commission. (2015). Annual assessment of the status of competition in the market for the delivery of video programming (*Sixteenth Report*), 30 FCCR 3253.

Federal Communications Commission. (2018). *Report on Cable Industry Prices*, 33 FCCR 1268.

Fowler, G. A. (2015, January 26). Sling TV: A giant step from cable. *The Wall Street Journal*. http://www.wsj.com

Frankel, D. (2019a, July 2). OTT has surpassed DVR usage in U.S., now 'mainstream,' Comscore declares. *Multichannel News*. http://www.multichannel.com.

Frankel, D. (2019b, November 15). Hulu+ Live TV jumps to No. 1 among vMVPDs, raises prices. *Multichannel News*. https://www.multichannel.com.

Frankel, D. (2020a, March 5). AT&T's Stankey: DirecTV now sold only 'in places where cable broadband is not prevalent.' *Next TV*. https://www.nexttv.com.

Frankel, D. (2020b, March 6). Comcast and other cable operators control 67% of U.S. broadband market. *Multichannel News*. http://www.multichannel. Com.

Hagey, K., & Ramachandran, S. (2014, October 9). Pay TV's new worry: 'Shaving' the cord. *The Wall Street Journal*. http://www.wsj.com.

Home Box Office v. FCC, 567 F.2d 9 (D.C. Cir. 1977).

Hoskins, C., McFadyen, S., & Finn, A. (2004). *Media economics: Applying economics to new and traditional media*. Thousand Oaks, CA: Sage.

Jacobson, A. (2019, March 15). Congress to DirecTV: Every DMA needs their local TV. *Radio+Television Business Report*. https://www.rbr.com.

Jessell, H. A. (2017, September 29). Retrans saved local TV, now what? *TVNewsCheck*. http://www.tvnewscheck.com.

Kagan. (2020a). U.S. multichannel industry benchmarks. *S&P Global Market Intelligence*. https://platform.mi.spglobal.com

Kagan. (2020b). Total broadcast retransmission & virtual sub carriage fees projections, 2006-2024. *S&P Global Market Intelligence*. https://platform.mi.spglobal.com

Kagan. (2020c). Global top operators. *S&P Global Market Intelligence*. https://platform.mi.spglobal.com

Kang, C., & Steel, E. (2016, April 25). Regulators approve Charter Communications deal for Time Warner Cable. *The New York Times*. http://www.nytimes.com.

Kurz, P. (2019, February 1). Report: OTT competition intense, providers look beyond borders for revenue growth. *TV Technology*. https://www.tvtechnology.com.

Lafayette, J. (2016, March 28). Nielsen: Pay TV subs could stabilize. *Broadcasting and Cable, 146*(12), 4.

Lafayette, J. (2018, March 14). Cord cutters not returning to pay TV, TiVo Q4 study finds. *Broadcasting and Cable*. http://www.broadcastingcable.com.

Leichtman Research Group. (2015, Q3). 83% of U.S. households subscribe to a pay-TV service. *Research Notes*. http://www.leichtmanresearch.com.

Leichtman Research Group. (2019, Q2). 31% of adults watch video via a connected TV device daily. *Research Notes*. https://www.leichtmanresearch.com.

Lenoir, T. (2014, January 14). High retrans stakes for multichannel operators in 2014. *S&P Global Market Intelligence*. https://platform.mi.spglobal.com.

Lenoir, T. (2019a, March 11). Q4'18 multiproduct analysis sheds more light on video's fall from grace. *S&P Global Market Intelligence*. https://platform.mi.spglobal.com.

Lenoir, T. (2019b, December 13). US multichannel video revenues reel from deep sub erosion in 10-year outlook. *S&P Global Market Intelligence*. https://platform.mi.spglobal.com.

Morton, H. (2019, July 13). Personal communication.

Munson, B. (2019a, July 29). Cord cutting is getting 'freaking ugly,' analyst says. *FierceVideo*. https://www.fiercevideo.com.

Munson, B. (2019b, October 31). Cord cutting even worse than 'freaking ugly' in Q3. *FierceVideo*. https://www.fiercevideo.com.

Munson, B. (2020, February 2020). Sling TV loses subscribers for the first time in Q4. *FierceVideo*. https://www.fiercevideo.com.

National Cable and Telecommunications Association. (2014). *Cable's story*. https://www.ncta.com.

National Conference of State Legislatures. (2014). *Statewide video franchising statutes*. http://www.ncsl.org/ research/ telecommunications-and-information-technology/statewide-video-franchising-statutes.aspx.

Nator, K. (2020, January 7). Cable programmers focus on streaming-heavy future as revenue growth slows. *S&P Global Market Intelligence*. https://platform.mi.spglobal.com.

Nielsen. (2019, Q1). The Nielsen total audience report. New York: Author.

Olgeirson, T., Lenoir, T., & Pabst, E. (2019, November 13). Q3 traditional video decline hits nearly 1.9 million amid modest virtual gains. *S & P Global Market Intelligence*. https://platform.marketintelligence.spglobal.com.

Owen, D. (1985, June). Satellite television. *The Atlantic Monthly*, 45-62.

Parker, J. G. (2011). Statewide cable franchising: Expand nationwide or cut the cord? *Federal Communications Law Journal, 64*, 199-222.

Parsons, P. (2008). *Blue skies: A history of cable television*. Philadelphia: Temple University Press.

Parsons, P. R., & Frieden, R. M. (1998). *The cable and satellite television industries*. Boston: Allyn and Bacon.

Ramachandran, S., & Peers, M. (2013, August 4). Future of cable might not include TV. *The Wall Street Journal*. http://www.wsj. com

Ritchie, G. (2019, August 9). Why low-earth orbit satellites are the new space race. *Bloomberg Businessweek*. https://www.bloomberg.com.

Samsung Electronics. (2019). A fifty year journey towards a sustainable future: Samsung electronics sustainability report 2019. Samsung. https://images.samsung.com/is/content/samsung/p5/global/ir/docs/sustainability_report_2019_en_new.pdf.

Satellite Home Viewer Act. (1988). Pub. L. No. 100-667, 102 Stat. 3949 (codified at scattered sections of 17 U.S.C.).

Satellite Home Viewer Improvement Act. (1999). Pub. L. No. 106-113, 113 Stat. 1501 (codified at scattered sections in 17 and 47 U.S.C.).

Sorkin, A. (2018, January 15). BlackRock's message: Contribute to society, or risk losing our support. *The New York Times*. https://www.nytimes.com.

State of Maine. (2019). *An act to expand options for consumers of cable television in purchasing individual channels and programs* (L.D. 832). https://legislature.maine.gov.

Takiff, J. (2019, October 9). Xfinity Flex review: Comcast's "free" streaming hardware/service combo is a work in progress. *TechHive*. https://www.techhive.com.

Telecommunications Act. (1996). Pub. L. No. 104-104, 110 Stat. 56 (codified at scattered sections in 15 and 47 U.S.C.).

TiVo. (2018). *Q4 2017 Video trends report*. http://blog.tivo.com.

TiVo. (2020). *Q4 2019 Video trends report*. http://blog.tivo.com.

U.S. v. AT&T, Inc. (2017). Case 1:17-cv-02511. http://www.justice.gov.

U.S. v. Southwestern Cable Co., 392 U.S. 157 (1968).

Walt Disney Company. (2018). *Corporate social responsibility update 2018*. https://www.thewaltdisneycompany.com/ wp-content/uploads/2019/03/2018-CSR-Report.pdf.

Williams, J. (2016, March 1). Feeling the churn: A new model for OTT. *S&P Global Market Intelligence*. https://platform. mi.spglobal.com.

Winslow, G. (2014, December 8). Operators look for TV Everywhere to live up to its name. *Broadcasting & Cable, 144*(44), 8-9.

Zubair, A. (2020, March 3). History of retrans deals: Negotiations getting more contentious. *S&P Global Market Intelligence*. https://platform.mi.spglobal.com.

Radio & Digital Audio

Heidi D. Blossom, Ph.D.*

Abstract

This is a momentous time for radio and digital audio. 2020 marked the 100th anniversary of commercial radio broadcasting that originally brought the world to the listener. Now listeners have the world in the palms of their hands with countless ways of listening to music, news, and other forms of audio entertainment. Now digital audio streaming services such as Spotify, Pandora, and YouTube are threatening the existence of traditional radio. For the first time in history, time spent listening to digital audio streaming has surpassed time spent listening to radio. Radio will have to reinvent itself at a time when artificial intelligence is making music selections for an eager audience, and new forms of audio hardware and software are pulling users away from the centenarian audio broadcast medium.

Introduction

Radio is one of the oldest forms of mass media. It was on November 2, 1920 that KDKA, the first commercial radio station in the United States, first went in the air in Pittsburgh broadcasting the returns of the Harding-Cox presidential election. The first radio news program also debuted that same year in Detroit at WWJ, which remains an all-news radio station to this day (Douglas, 2004). Radio has experienced a remarkable evolution over the past 100 years surviving significant innovations in communication technology.

Despite radios' longevity, many critics claim that radio is dead, and they wouldn't be entirely wrong. The fact is that the radio broadcasting of the past *is* dead, and what we called radio in those early days of broadcasting has little resemblance to the radio broadcasting of today. In an attempt to survive the times and stay relevant to audiences, the radio industry has had to adapt to technological advances in both hardware and content.

Radio's relationship with its audience has evolved over the past century as it reinvents itself every couple of decades to be in line with its listeners' needs. First, it was the introduction of sound in film, the Great

*Educator and Digital Media Consultant, RealVision, LC (Greenville, South Carolina).

Depression, then television came on the scene and naysayers proclaimed the death of radio. Developments in digital audio created new challenges, and now we are in the midst of a new transformation as radio fights to stay relevant in a socially connected, digital world. Newer forms of digital media delivery, such as smartphones, streaming audio services, satellite radio, and podcasting are all threatening traditional radio.

In the following pages we will explore how radio evolved and learn how newer digital alternatives to radio will advance in the years to come.

Figure 8.1

KDKA Studios circa 1922

Source: Public Domain

Background

"Mary had a little lamb, its fleece was white as snow"

These were the first words ever recorded, launching a global audio recording revolution. Little did Thomas Alva Edison realize what he had invented when he discovered that sound could be recorded and played back. While the audio quality on the first phonograph was barely recognizable, the phonograph was a true media innovation. Recording technology brought music, theater, politics, and education into the living rooms of a nation, but by the 1920s the recorded music industry was in decline, mostly due to WWI and the invention of radio, which had hit its stride by the end of the 1920s (Lule, 2014).

Radio

No one expected radio to revolutionize the world, but more than 100 years ago that is exactly what happened. On March 1, 1893 Nikola Tesla, the father of modern communication technology, gave his first public demonstration of radio at the National Electric Light Association in St. Louis, paving the way to further innovations and uses of his new technology (Cheney, 2011). Tesla held more than 700 patents, including wireless communications, cellular communications, robotics, remote control, radar, and many other communication technologies that we still use today. Although many have credited Guglielmo Marconi with the invention of radio, Marconi combined the inventions of Tesla and others to advance already established radio technologies. In 1943 the U.S. Patent office restored credit for the original radio patents back to Nikola Tesla (Coe, 1996).

Figure 8.2

Nikola Tesla

Source: Public Domain

The first radio transmission of actual sound came on December 24, 1906 from Ocean Bluff-Brant Rock, Massachusetts when Reginald Fessenden played *O Holy Night* on his violin and then read a passage from the Bible to ship wireless operators at sea. The ships had been informed to be listening at 9pm on that Christmas Eve and they were amazed to hear music and voices—something that had never been heard before over a wireless connection (O'Neal, 2006). That first live broadcast laid the foundation of the live music programming format. After the first commercial radio station went on the air, other stations emerged in major urban areas and thus began the Golden Age of Radio known for live drama, comedy, and music shows that challenged the imagination. Americans gathered around their radios tuning in to hear programming that was inaccessible to the masses before. It was a magical time in the history of media and the public was hooked.

After a ban on radio during World War I, department stores, universities, churches, and entrepreneurs started their own radio stations in the early 1920s. In just four years' time, the number of commercial radio stations in the U.S. rose from 34 to 378 stations (Scott, 2008). By the 1930s, more than 80% of Americans owned a radio and the medium exerted its influence and marking its prominence in mass media. Radio connected individuals to the world and allowed listeners to have a front-row seat to the boxing match, hear first-hand the news from around the world, and experience politics and culture like never before—live. It was a powerful mass medium and the only communication technology that was accessible to most, which continues today with the vast majority of Americans listening to the radio each week (Nielsen, 2020).

One of the main reasons radio was so successful through the Great Depression of the 1930s was because there was no competition that compared to radio until the introduction of television. By 1949, television's popularity had risen to a level that caused many to predict the demise of radio. Television eventually replaced radio's prime-time dominance as many popular radio shows moved to television; regardless, radio penetrated places of isolation that television could not go, namely, the automobile. Nevertheless, radio was able to maintain a captive audience as the industry adjusted by shifting its programming to what it still does best—music (Cox, 2002). By the 1950s, the radio DJ was born, and America's Top 40 format ruled the airwaves (Brewster, 2014).

During the counterculture of the 1960s, young people looked to radio to listen to popular musical artists such as The Beatles and The Rolling Stones who spoke to the culture of the day. Although FM stereo was invented in the 1930s, it wasn't until the late 1970s that FM radio manufacturers caught up to the demand as FM radio stations surpassed the reach of AM stations in the United States (Lule, 2014). Consumers wanted quality-sounding music, and FM radio was their free, go-to source. It was that shift of purpose for the radio industry that redefined it and saved it.

At the same time that FM radio was surging, digital audio technology was being developed. By the late 1980s and 1990s, Compact Discs (CD) came into mainstream use, ending the long-standing analog era of music recordings. In the meantime, radio technology languished, with no real innovations until Hybrid Digital (HD) radio nearly two decades later. To understand this part of radio's history, it is important to know the difference between analog and digital signals.

Radio Technology

The technology behind radio is remarkable. The idea that sound could be transmitted from across a city or even across the ocean is beautifully complex and yet so simple. Radio technology works by using invisible radio waves to send music, news, or sports though the air to a box that receives those otherwise undetectable radio waves and converts the waves into sound.

A radio wave is created by modulating an electric current and sending it through an antenna. These electrical waves are then detected by receivers (radios), which convert the electrical waves into music, speech or other sounds. By tuning a radio receiver to a specific wave frequency, you can listen to a radio station. In the U.S., the Federal Communication Commission, or FCC, allocates the frequency a radio station uses and issues a license that allows that radio station to use that frequency within a designated broadcast area. When a radio station announcer says, "You are listening to Magic 98.9 FM, WSPA in Greenville, South Carolina" they mean you are listening to an FM radio station that broadcasts on frequency 98.9 with the FCC allocated call letters of WSPA in the designated broadcast area.

All FM radio stations transmit their signals in a frequency band between 88 and 108 megahertz. AM radio is similar but is confined to a band from 535 kilohertz to 1,700 kilohertz. It is important to know that every wireless technology has its own frequency band. For example, some cell phones transmit at 824 to 849-megahertz, global positioning systems (GPS) transmit between 1,227 and 1,575 megahertz, and a baby monitor transmits at 49 megahertz (Brain, 2010).

At the time that FM radio was surging, digital audio technologies were being innovated. There were several issues the radio industry had to overcome in the transition from analog to digital. In the 1980s the cost of storing and broadcasting digital signals was exorbitant. In 1983, 1 Mb of storage cost $319 and would hold about one quarter of one song (Komorowski, 2014). The speed at which audio could be broadcast was so slow that it was inconceivable to move to a digital broadcast system. In order for digital technology to be practical, the size of the files had to be decreased, and the bandwidth had to be increased.

Figure 8.3

Radio Sound Waves

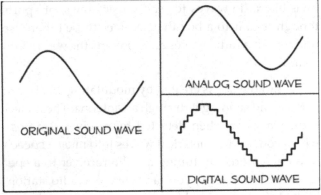

Source: H. Blossom

The difference between analog and digital audio is in how the audio is reproduced. Digital recordings are made by taking thousands of "samples" of an analog signal, with a number for each sample representing the size or strength of the signal at that moment. The numbers created for digital audio are created through binary code, or a series of 0's and 1's which creates a code that instructs the digital player on what to play (Hass, 2013). Figure 8.3 represents the stair step form of a digital audio wave. The more samples you take, the smoother the wave—and the more accurate the sound reproduction.

By the early 1980s, digital audio hit mainstream in the form of the compact disc (CD), which was marketed as small ("compact") and virtually indestructible with near perfect sound as opposed to its predecessors, the vinyl record that could scratch, break, or warp and the cassette tape that stretched and produced a hissing sound as the tape aged (Sterling & Kittross, 2001). Consumers quickly embraced CD technology as prices dropped and the music industry transitioned to a digital era in music production. The public was hooked, and the radio industry had to adjust to higher expectations from consumers who demanded clearer sound; however, converting analog radio to digital radio would be a complex and costly process that would take decades to accomplish. Digital audio recording technology would advance with the introduction of codecs that shrunk digital audio into sizes that were more manageable and enabled faster transmission of the audio data. The compression of audio is important for digital radio broadcasting because it affects the quality of the audio delivered on the digital signal. The codec that emerged as a standard was the MPEG-1 Audio Layer 3, more commonly known as MP3. MP3 compressed the audio to about one-tenth the size without perceptibly altering the quality of the sound for FM listeners (Sellars, 2000).

Figure 8.4

Multimedia Radio Studio

Source: HISRadio

Chapter 8 ✦ Radio & Digital Audio

The Shift to Digital Audio Broadcasting

European broadcasters beat American broadcasters in adopting digital transmission with their development and adoption of the Eureka 147 DAB codec, otherwise known as MP2 (O'Neill & Shaw, 2010). The National Association of Broadcasters (NAB) was very interested in Eureka 147 DAB in 1991; however, broadcasters eventually lobbied for the proprietary iBiquity IBOC codec designed to deliver a higher-quality audio with lower bit rates (or bandwidth) than the Eureka system.

iBiquity's proprietary In-band On-channel (IBOC) digital radio technology layers the digital and analog signals, allowing radio broadcasters to keep their assigned analog frequencies and simply add the digital conversion hardware to upgrade their current broadcast systems (FCC, 2004). Known as HD Radio (Hybrid Digital), this technology also allows for metadata transmission including station identification, song titles and artist information, as well as advanced broadcasting of services such as breaking news, sports, weather, traffic and emergency alerts. The IBOC technology also allows for tagging services allowing listeners to tag music to purchase and download on mobile devices. Ultimately, this technology gives more control to the consumer and engages them in a way that traditional analog radio could not do (Hoeg & Lauterbach, 2004). The FCC approved iBiquity's HD Radio system in 2002, and radio stations began broadcasting the HD signals in 2004, delivering CD quality audio on both AM and FM frequencies (HDRadio, 2004).

HD radio opens up the amount of spectrum available to broadcasters, allowing single radio stations to split their signal into multiple digital channels offering more options in programming, otherwise known as multicasting. For example, a radio station may primarily be a country music formatted station (95.5 FM) but offer a news-talk station on the split frequency (95.5-2) and a Spanish language music station on a third channel (95.5-3).

HD radio is free to listeners but requires users to purchase an HD ready radio. This requirement caused a delay in adoption as the HD radio sets were costly and people were not willing to invest in a costly radio until there were stations who were broadcasting in HD. Alternative music delivery systems and the lack of consumer knowledge about HD Radio were also barriers to this technology's success. The other barrier is found in the automotive industry. Since most people listen to radio in their vehicle, it is imperative that vehicles come equipped with HD Radio. As of mid-2020, HD Radio is not standard in all vehicles being shipped in the U.S. and until that happens, HD adoption will continue its slow growth.

Radio Automation

In the early 1990's the United States entered a recession that caused the media industries to search for ways to economize. In an effort to better allocate resources, radio stations and networks across the globe began using automated operations. Radio automation is the use of broadcast programming software to automate delivery of pre-recorded programming elements such as music, jingles, commercials, announcer segments, and other programming elements.

Radio automation allows for better management of personnel allowing announcers to pre-record their segments in a way that sounds like they are live on the air when in reality they may have recorded the segment hours or even days before. Even radio personalities Ryan Seacrest and John Tesh, whose nationally syndicated radio shows draw in millions of listeners, pre-record part or all of their programs.

Radio automation continues to be used in both traditional radio broadcasting as well as streaming radio stations on the web.

Recent Developments
Radio's New Threat

Digital audio streaming represents a threat to traditional radio broadcasting. For the first time in history, users are turning to digital audio streaming an average of 4.4 minutes *more* per day compared to traditional radio (Nielsen, 2019).

Given the 94% market penetration of smartphones among adult cellphone users, it is no wonder that radio listeners are using their mobile devices to expand their audio choices beyond traditional terrestrial radio (Pew, 2019). More than 75% of American adults 18+ listen to streaming radio via their smartphones (Shearer, 2019). Music streaming services and podcasting have also grown exponentially as consumers subscribe to streaming services such as Spotify, Pandora, or Apple Music. Listeners are seeking less-complex ways of accessing their entertainment, and smartphones break down the complexity barriers with easy to use apps. These apps connect the user to literally millions of streaming options.

Smartphone use has influenced the way people listen to all forms of digital audio. Users spend an average of 17.2 hours a week with their favorite broadcast radio stations, streaming music service, or podcasts, and 68% of Americans listen to digital audio online (Edison, 2019). These options are forcing the radio industry to consider how to keep the online listeners they have now and how to keep them listening.

Streaming services have immense brand awareness and usage among a younger audience, with 85% of 13-15-year-olds streaming music on mobile devices (IFPI, 2017). The radio industry is taking notice as they fight to maintain audience control of the lucrative younger demographics.

Figure 8.5

Music Streaming Reaches One-Trillion Milestone in 2019

Number of on-demand music streams (audio and video) in the United States

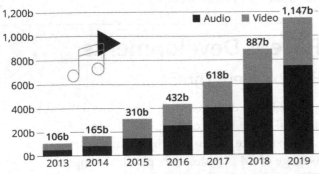

Source: F. Richter/Nielsen Music, MRC

The Streaming Music Obsession

Americans are obsessed with their music evidenced by a record 1.14 trillion music streams in 2019. To put it into perspective, that is nearly 1 billion hours of music or an average of 3,500 streams per person in the United States (Richter, 2020). These numbers do not include internet radio or video streaming services. Music streaming increased by 23% in 2019 mostly due to the continued growth of mobile smart devices (Nielsen, 2019). Users find that the convenience of having music in the palm of their hands, along with instant access to a massive selection of music, has driven them towards digital music streaming service (IFPI, 2019).

Figure 8.6

Ways to Stream Music

Source: Spotify

Music Streaming services are digital audio subscription services that offer unlimited access to up to 40 million songs; the widest diversity of music offering in history (IFPI, 2017). Most audio streaming services offer a free subscription that include commercial breaks; however, paid subscriptions offer commercial free listening and direct access to particular songs, artists, and albums. Paid monthly subscriptions make up more than 90% of music streaming services' revenue. The other 10% comes from other sources including advertising (RIAA, 2019).

The Record Music Association of America (RMAA) reported that "overall revenues in the U.S. recorded music industry grew 12% to $9.8 billion" in 2019, and streaming services accounted for 80% of the overall revenue for the industry that same year.

Audio streaming services are a lucrative prospect for advertisers. In 2019, advertising on music streaming services such as Pandora, Spotify, Apple, and Amazon jumped 26%, surpassing the growth rates of search and digital video advertising. The sale of physical albums and downloaded music is plummeting. Fifty-eight percent of adults said they never buy music albums and 63% said the same about singles. As more cellular companies offer unlimited data plans and more people adopt 5G technologies, revenues are expected to grow even more as consumers tune to digital audio streaming services on their mobile devices (RIAA, 2019).

Podcasting

Podcasting, a term that comes from combining the term "iPod" and "Broadcasting," continues to grow in popularity. More than 50% of Americans claimed to have listened to a podcast in the last month and that number has grown year after year (Edison, 2019). Podcasts are digital audio recordings of music or talk, available for download onto a mobile device or computer.

Figure 8.7
Podcasting Studio

Source: Photo by A. Distel

Podcasts are easily accessible. They can be accessed on a mobile device or computer and listened to on demand when an individual has the time, making podcasting a convenience medium. In 2019, there were more than 750,000 active podcasts sharing knowledge and entertainment to more than 150 million Americans who regularly listen to podcasts (Edison & NPR, 2020).

Podcasts are typically produced as a series and available as individual downloads or automatic subscription downloads. Radio broadcasts converted to podcasts are some of the most popular, with radio programs such as *This American Life, TED Radio Hour,* and *Fresh Air,* all popular NPR radio shows, leading the charge of weekly shows converted to podcasts. *This American Life,* an hour long themed journalistic, informational, and sometimes comedic radio program, has more than five million downloads every month of their popular podcast (Podtrac, 2019).

In 2020, it was reported that more than 50 billion podcasts had been downloaded from Apple in 2019 propelling podcasting into mainstream. (Shearer, 2019).

Figure 8.8
Vinyl Records Rebound

Source: Photo by D. Bayer

A Step Back to Vinyl

While physical CD album sales and paid downloads continue their slide into irrelevance, a phenomenon has occurred in the U.S. with the resurrection of a decade's old technology—vinyl records. The first vinyl record was sold in 1930, starting a long run of popularity that lasted through the late 1980's. In an odd turn of events, vinyl album sales have experienced a revival over the past several years and are set to surpass the sales of CDs for the first time since 1986. In 2019, nearly 19 million vinyl albums were sold in the

U.S., a 14.5% increase over the year before (Nielsen, 2020). The vinyl resurgence is powered mostly by decades old musical greats such as The Beatles, The Rolling Stones, Bob Marley, and Michael Jackson who collectively sold millions of their albums on vinyl LPs in 2019. Newer musicians are also experiencing the vinyl renaissance. In less than eight months, award winning American singer and songwriter, Billie Eilish sold more than 176,000 vinyl copies of her debut album, *When We All Fall Asleep, Where Do We Go?* (Nielsen, 2020).

Copyright Royalties

Copyright infringement remains the biggest challenge for the music industry. A recent study found that 27% of people used unlicensed methods to listen to music in the past month and 23% used illegal streaming services (IFPI, 2019). While music piracy is not a new issue, it is a major concern to those who are entitled to be paid for the use of their music.

Copyright issues have plagued the music industry since its inception. The antiquated laws set to protect music creators do not apply to newer digital technologies creating challenges for those who have the right to be paid royalties for their copyrighted music. Copyright gives the creator(s) of an artistic work control over how their work is used and gives them the right to be paid royalties for the use of their music. If someone uses that work without permission, the copyright has been violated and the owner of the copyright could sue to protect their work.

In order for a radio station or streaming service to be able to play music, they have to pay for the licensing rights, or royalties, to use that piece of music. Before 2018, radio stations and streaming services obtained rights to play music on a song-by-song basis. Music performing rights organizations like the Broadcast Music Inc. (BMI) and the American Society of Composers, Authors, and Publishers (ASCAP) negotiate the rights to play individual pieces of music and serve as the middleman between the radio station or streaming service and the musicians who own the copyright. BMI and ASCAP charged for the royalties or rights to play the copyrighted recording and would, in turn, pay out the royalties to the copyright owner. Streaming music service escalated copyright issues because many streaming services believed they were not under the same copyright restrictions. This created a new fight with artists struggling to get paid royalties for music being streamed or downloaded.

The biggest update to music copyright laws in the past 40 years was unanimously passed by both houses of Congress on October 11, 2018. *The Orrin G. Hatch–Bob Goodlatte Music Modernization Act* or MMA unites requirements from across the music community under one legislative umbrella to ensure protections for hundreds of thousands of music creators nationwide who have the right to be paid royalties for their work. The MMA creates a system for paying royalties to music creators including musicians, performers, sound engineers, and others. The new legislation establishes a legal blanket license for digital music services to distribute sound recordings as long as the required royalties are paid and other rules are followed.

Before the MMA was passed, recordings created before 1972 were not protected under federal copyright laws. The MMA corrects that by giving sound recordings a 95-year protection after the year of first publication, meaning that a recording that was first published in 1970 will go into public domain after the year 2065 (MMA, 2018).

The MMA promises more streamlined licensing procedures for streaming music services, a new centralized registration database, and an updated system for creators and rights owners to be paid royalties. The MMA is set to take effect in 2021. For the next several years, creators and music rights owners wishing to be paid royalties will be busy registering their works with the new collective database.

Satellite Radio

SiriusXM satellite radio is a subscription-based radio service that uses geosynchronous satellite technology to deliver its programming, meaning that no matter where subscribers are in the U.S., they can hear their favorite SiriusXM station (Keith, 2009). Satellite radio services have been offered since 2001 and have seen slow but steady growth. Satellite radio comes standard in most new and many pre-owned cars and trucks, but requires a subscription activation. In 2010, there were 27.5 million satellite radio subscribers, and that figure has steadily grown to more than 34 million subscribers in 2019, making satellite radio a steady growth industry (SiriusXM, 2020).

In early 2019, SiriusXM Holdings, Inc. purchased Pandora for $3.5 billion (SiriusXM, 2019). Soon after, SiriusXM announced a new interactive station that would create a new musical experience for listeners. The Pandora NOW music station allows users on both platforms to interact with the song selections by skipping tracks they don't like or hitting the thumbs up or down buttons to give feedback, which influences what gets played next. Pandora NOW will also be available as a continually updated playlist for Pandora's Premium subscribers. SiriusXM and Pandora together reach more than 100 million users each month with their audio and streaming products (SiriusXM, 2020).

Artificial Intelligence

Pandora was founded in 2004 as an alternative to traditional radio. It is a streaming music service that creates a personalized experience for each user utilizing machine learning and artificial intelligence. Pandora's AI can create "stations" that are based off of songs from a favorite genre of music or a favorite artist. Pandora accomplished this by incorporating the Music Genome Project, the most comprehensive analysis of music ever undertaken (Westergren, 2020).

Named after the Human Genome Project that analyzed DNA sequence of millions of human genomes, the Music Genome Project analyzes the core characteristics of each song using 450 unique attributes to analyze everything from style of music, cord progression, BPM (beats per minute), instrumentation, combination of notes, qualities of performers voices, and more. Musicologists work with data scientists to test the accuracy of machine learning and teach artificial intelligence programs to detect the DNA of the music and the nuances that make each song unique. The Music Genome Project has taken Pandora's song library of more than 2 million songs and compared the song DNA with user feedback, such as a thumbs up or down, to yield customized programming (Forbes, 2020).

Factors to Watch

Smart Speakers

Smart speakers are taking the audio and radio industries by storm creating new opportunities for these industries to connect with their targeted audiences. More than 24% of U.S. adults—or 60 million people—own at least one smart speaker device (Edison & NPR, 2020).

A smart speaker is a web-connected wireless virtual voice assistant. Whether you want to hear the local weather forecast, listen to your favorite band, get news headlines, or connect with other smart devices in your home, Amazon Echo, Google Home, or Apple HomePod are waiting for your voice command to assist you. U.S. households have been quick to purchase these speakers. The number of smart speakers in households in the U.S. grew dramatically between 2017 and 2019 with more than 90 million households using a smart speaker (Edison, 2020 & NPR). (For more on smart speakers, see Chapter 22.)

When a consumer chooses to listen to audio on a smart speaker, streaming audio dominates the listening time split between streaming services, satellite radio, and podcasts. Nearly 70% of users say they listen to streaming music on their smart speaker and more than 40% of users say they listen to AM/FM radio (Voicebot.ai, 2019).

The use of smart speakers to connect to radio and music streaming services is creating a shift in hardware usage from the use of mobile phones and computers to the use of a small speaker to play their favorite music and podcasts. This change is creating a vacuum of people dropping one hardware for another more-innovative hardware making an entire industry scramble to adapt to a new consumer behavior.

5G Technology

5G wireless technology, launched in the United States in 2019, is the fifth generation of wireless technology for digital cellular companies. 5G is expected to have a transformative effect on both digital audio and video streaming, eventually making streamed content universally available without dropout and delivered with superior quality. The speed of 5G is also faster and far exceeds the speed of 3G and 4G/LTE with download speeds on both AT&T and Verizon networks averaging more than 1 gigabyte per second (Dolcourt, 2019). To put that into perspective, with 5G running at 1 gbps, a person could download approximately 250 high quality songs each second.

At the time this text was written, 5G capabilities had only been introduced in a few urban areas, and less than 1% of cell phone users in the U.S. had adopted this new technology (Ball, 2020). Forecasters predict that by the year 2025, the total number of 5G connections will reach 1.5 billion globally (Juniper, 2020).

This innovation of wireless has great implications for both radio broadcasting and digital audio delivery; however, the full technical evolution of 5G is still years away, giving the radio and digital audio industries a little time to reveal how it will take advantage of the next-generation of cellular networks (Stine, 2020).

AM Radio Set for a Makeover

A makeover for AM radio is well overdue. In late 2019 the Federal Communications Commission (FCC) voted unanimously to adopt a proposal for rulemaking that would allow AM radio stations to convert to fully digital broadcasting using the all-digital mode of HD Radio. AM radio has struggled for decades with a steady decline in listenership caused by interference and reception issues and new audio technologies pulling audiences away. But AM radio has an opportunity to undergo a big change in the coming years if this proposal is passed by the FCC (FCC, 2019). At the time this text was written in mid-2020, AM stations were broadcasting using either a traditional analog or a hybrid HD signal. An all-digital broadcast would improve the quality of AM sound. Digital broadcasting would also improve the reliability of the AM broadcast signal that has constrained many AM stations to a voice-only format. On the other hand, a move to all-digital would no longer be receivable by the traditional radio receivers and would require users to have a special HD radio that can accept both analog and digital signals. The majority of people who listen to AM radio listen in their vehicles, but only about 50%

of vehicles on the road in the U.S. are equipped with HD radios, which could impede listenership (Rilsmandel, 2019).

With fewer people having access to all-digital AM radio, this change could pose an issue during emergency events when many turn to radio for vital information.

If this ruling passes, it will give AM Radio stations the option of adopting HD Radio technology. If enough AM stations adopt this broadcast technology, the FCC could mandate a change to all-digital much like it did for HD television.

Advances in AI Technology

There is no question that the radio and digital audio industries are undergoing dramatic change as technology creates demand for personalized music experiences across multiple sectors. The industries have been adapting and searching for innovative solutions to waning audiences that are choosing more on-demand options. Some of the top radio and digital audio companies are already turning to artificial intelligence as their savior. iHeartMedia, Universal Music Group, Napster, Dischord Records, and others are jumping on the AI bandwagon.

Artificial intelligence capabilities will allow radio networks and music streaming services the ability to create individualized listening experiences that mimic the production of a live radio broadcast. For example, traditional radio DJs not only voice announcements and continuity, but they also create an energy, character, and sound that is unique to that radio station through seamless transitions, segueing from talk to music, and transitioning between songs. At the time this text was written, Super Hi-Fi was the only company dedicated to the integration of AI and radio and digital audio. Their AI system replaces human touch with trillions of data points that allow a supercomputer to make decisions in real time executing transitions between songs by layering in music, sound effects, voice-over ads, and delivering the style of smooth, seamless playback to create a live sounding broadcast just as a seasoned human DJ would do.

The transition to this technology is already disrupting the industry. In an effort to adapt to changes in the industry, recapture listeners, and fend off bankruptcy, iHeartMedia, the dominant player in radio and music streaming in the U.S. with 850 local radio stations across the country and one of the top music streaming services, made a bold move by laying off an estimated 1,000 employees in early 2020 (Harwell, 2020). The layoff included hundreds of local talk show hosts, news anchors, producers, program directors, reporters and drive-time DJs. Executives for iHeartMedia stated that AI was the muscle it needs to fend off rivals, recapture listeners, and emerge from bankruptcy, and "employee dislocation was necessary as it pivoted to AI" (Harwell, 2020).

What we do know is that AI is creating a new way for the audience to interact with radio and digital audio technology. In the past people would have to look at a screen, click or turn a knob, and make adjustments tactilely. There was direct interaction with the technology and the user was still in control of making decisions about what to do next. The way we use technology now is shifting, and for the first time in history, technology is speaking back and making decisions for users based on voice commands and AI technologies.

Brain-Computer Interface (BCI)

Smart speakers and artificial intelligence are both creating a new way of interacting with media, but there are innovations in technology that will completely change the way we interact with all forms of technology using our brains and nothing else. Brain-Computer Interfaces (BCIs) are systems that translate brain activity into actual messages or commands for interactive applications enabling users to control machines by using their own brain activity.

Brain-Computer Interface works by reading brain activity. As a user thinks about a physical command, BCI acquires that thought brain signal, decodes the signal, and then applies the command to an external device. This technology is still in the research stage, but has already had astounding results. Researchers have used BCI to help restore feeling and movement for people with motor disabilities and have helped people who are completely paralyzed and cannot speak to communicate through a computer (Shih, et.al., 2012).

BCI has huge implications for media as consumers search for entertainment and information that will help them in their lives. Youth 16-24 listen to music to relieve stress. Researchers are working on a system that could analyze a person's emotional state using their brain activity, and then automatically develop an appropriate piece of music that either matches the emotion or helps correct emotions that are considered harmful (Best, 2019).

The research and testing for BCI in a media application have not yet been conducted; however, with investors like Elon Musk pouring millions of U.S. dollars into BCI research, this technology could become a reality sooner than we realize. For now, researchers have to explore the ethical issues that could rise out of the application of BCI, and then use those ethics as a guide for determining which methods of brain interaction could be incorporated into real-world applications.

The future of radio and digital audio in the next five years will be determined by improvements in bandwidth including 5G, advancements in AI, and hardware connected to the Internet of Things (IoT) These technologies will propel digital audio streaming forward and will, once again, force radio to re-invent itself to survive yet another day.

Gen Z

Gen Z is the first generation to be born in a completely digital world. They are the first smartphone generation and they live in a digital world where their cell phones keep them connected to what they love. More than 90% of this generation over the age of 16 uses a smart phone, and 83% say that music is the number one activity they do on their phone (IFPI, 2019).

Music plays a key role in the lives of Gen Z and is their chosen activity to relieve stress, they are listening to their curated music the only way they know how—digital audio streaming services. Music on demand streaming service, YouTube, reached 72% of respondents aged 16 to 19 years old. They like to listen to what they want, when they want it, and streaming services offer accessibility and convenience (MusicBiz, 2019).

Gen Z still listens to traditional radio in the car; however, their time spent with audio media is mostly spent with digital audio content they can engage with. They want to share with their friends, connect with the talent, and be an integral part of the media experience they choose, which is why traditional radio is not as appealing. They don't like to be talked to—they like the feeling of a community where their voice is just as important as the content they are listening to.

Sustainability

In order to begin to understand sustainability related to radio and digital media you have to consider the life cycle—cradle to grave, from extracting raw materials to designing and manufacturing products and services that have value. It includes using components/products that don't pollute or damage the environment either in use or at end of use, and it considers the energy, materials and water used to make the products, as well as waste management. In addition, the human factors can't be ignored, such as whether child labor was used in production and whether any landfill waste has contaminated water supplies that may in turn affect humans and our food chain. From a broadcast business point of view, sustainability must consider the factors above along the entire value chain from content and product creation to post-production and formatting, to transmission and distribution, through to the energy used by the consumer. Few users appreciate the energy required to use their device which goes orders of magnitude beyond the energy to charge the device. A simple click on a search engine or uploading a video starts servers spinning around not to mention the antivirus running in the background. This energy is hidden from users.

Janet West (2020)

Environmental
Scientist & Strategist

Bibliography

Ball, M. (2020, January 13) Data suggests China has outpaced US in 5G adoption. M Science Insights. https://mscience.com/insights/will-2020-be-the-year-of-5g.html.

Best, J. (2019, October 10) Mind-reading systems: Seven ways brain computer interfaces are already changing the world. *ZDNet: Building the Bionic Brain.* https://www.zdnet.com/article/mind-reading-systems-seven-ways-brain-computer-interfaces-are-already-changing-the-world/.

Brain, M. (2010, April 1) How the Radio Spectrum Works. How Stuff Works. https://electronics.howstuffworks.com/radio-spectrum.htm.

Brewster, B. (2014). *Last night a DJ saved my life: The history of the disc jockey.* Grove/Atlantic, Inc.

Cheney, M. (2011). *Tesla: Man out of time.* Simon and Schuster.

Coe, L. (1996). *Wireless radio: A brief history.* McFarland.

Cox, J. (2002). *Say goodnight, Gracie: The last years of network radio.* McFarland.

Dolcourt, J. (2019, July 23) We ran 5G speed tests on Verizon, AT&T, EE and more: Here's what we found. *CNET*. https://www.cnet.com/features/we-ran-5g-speed-tests-on-verizon-at-t-ee-and-more-heres-what-we-found/.

Douglas, S. J. (2004). *Listening in: Radio and the American imagination*. U of Minnesota Press.

Edison (2019, March). The infinite dial 2019. http://www.edisonresearch.com.

Edison & NPR (2020, January) The Smart Audio Report. https://www.nationalpublicmedia.com/insights/reports/smart-audio-report/.

FCC (2004) Digital Audio Broadcasting Systems and Their Impact on the Terrestrial Radio Broadcast Service: Further Notice of Proposed Rulemaking and Notice of Inquiry. MM Docket No. 99-325. Washington, DC.

FCC (2019) Proposal for rulemaking: All-Digital AM Radio Broadcasting: Revitalization of the AM Radio Service. https://docs.fcc.gov/public/attachments/DOC-360519A1.pdf.

Forbes (2019, October 1) How Pandora Knows What You Want to Hear. Forbes Insights. https://www.forbes.com/sites/insights-teradata/2019/10/01/how-pandora-knows-what-you-want-to-hear-next/#6faeeff63902.

Harwell, D. (2020, January 30). iHeartMedia laid off hundreds of radio DJs. Executives blame AI. DJs blame the executives. *The Washington Post*. https://www.washingtonpost.com/technology/2020/01/31/iheartmedia-radio-artificial-intelligence-djs/.

Hass, J. (2013) *Introduction to Computer Music: Volume One*. Indiana University.

HD Radio (2004) http://hdradio.com/us-regulatory/fcc_approval_process.

Hoeg, W., & Lauterbach, T. (Eds.). (2004). *Digital audio broadcasting: principles and applications of digital radio*. John Wiley & Sons.

IFPI (2017, September) *IFPI Connecting with Music: Music Consumers Insight Report*. http://www.ifpi.org/downloads/Music-Consumer-Insight-Report-2017.pdf.

IFPI (2019, September) Music Listening 2019: A Look at How Music is Enjoyed Around the World. *Insight Report*. https://www.ifpi.org/downloads/Music-Listening-2019.pdf.

Juniper R (2020, January) 5g: The 5-Year Roadmap. *Juniper Research*. https://Www.Juniperresearch.Com/Document-Library/White-Papers/5g-The-5-Year-Roadmap.

Keith, M. C. (2009). *The Radio Station: broadcast, satellite & Internet*. Taylor & Francis.

Komorowski, M. (2014) A history of storage cost. http://www.mkomo.com/cost-per-gigabyte.

Lule, J. (2014). *Understanding Media and Culture: An Introduction to Mass Communication New York*: Flat World Education.

MusicBiz (2019). Audio Monitor US: The Overall Music Landscape. *MusicBiz*. https://musicbiz.org/wp-content/uploads/2018/09/AM_US_2018_V5.pdf.

Music Modernization Act (2018). U.S.C. 115–264 § 1551.

NextRadio (2018) What is NextRadio? http://nextradioapp.com/.

Nielsen (2019, December). *Audio Today 2019: How America Listens*. https://www.nielsen.com/us/en/insights/report/2019/audio-today-2019/.

Nielsen (2020). *Nielsen Music U.S. Music Year-End Report*, 2019.

O'Neal, J. E., (2006) Fessenden: World's First Broadcaster? *Radio World*. New York, NY http://www.radioworld.com/article/fessenden-world39s-first-broadcaster/15157.

O'Neill, B., & Shaw, H. (2010). *Digital Radio in Europe: Technologies, Industries and Cultures*. Bristol, U.K.: Intellect.

Pew Research Center (2019, June 12) Mobile Fact Sheet.

Podtrac (2019). Podcast Industry Audience Rankings. http://analytics.podtrac.com/industry-rankings/.

Pluskota, J. P. (2015). The perfect technology: Radio and mobility. *Journal of Radio & Audio Media*, 22(2), 325-336.

RIAA (2019). Mid-Year 2019 RIAA Music Revenues Report. https://www.riaa.com/wp-content/uploads/2019/09/Mid-Year-2019-RIAA-Music-Revenues-Report.pdf.

Rilsmandell, P. (2019, November 22) FCC Opens Proceeding for All-Digital AM Radio. RadioSurvivor.com.

Richter, F. (January 13, 2020). Music Streaming Hits One-Trillion Milestone in 2019 [Digital image]. https://www-statista-com.ezproxy.regent.edu/chart/14647/music-streams-in-the-united-states/.

Scott, C. (2008). "History of the Radio Industry in the United States to 1940". *EH.Net*, edited by Robert Whaples. March 26, 2008. http://eh.net/encyclopedia/thehistoryoftheradioindustryintheunitedstatesto1940/.

Schafer, D. (2019, December 23) Bandwidth vs. Latency: What is the Difference? HighspeedInternet.com https://www.highspeedinternet.com/resources/bandwidth-vs-latency-what-is-the-difference.

Sellars, P. (2000). Perceptual coding: How MP3 compression works. http://www.soundonsound.com/sos/may00/articles/mp3.htm.

Shearer, E. (2019). Audio and Podcasting: fact sheet. Pew Research Center, 15. https://www.journalism.org/fact-sheet/audio-and-podcasting/.

Shih, J. J., Krusienski, D. J., & Wolpaw, J. R. (2012, March 8). Brain-computer interfaces in medicine. In Mayo Clinic Proceedings (Vol. 87, No. 3, pp. 268-279). Elsevier.

SiriusXM (2019, February 1) SiriusXM Completes Acquisition of Pandora. SiriusXM Press Release. http://investor.siriusxm.com/investor-overview/press-releases/press-release-details/2019/SiriusXM-Completes-Acquisition-of-Pandora/default.aspx.

SiriusXM (2020, January 7) SiriusXM Beats 2019 Subscriber Guidance and Issues 2020 Guidance, http://investor.siriusxm.com/investor-overview/press-releases/press-release-details/2020/SiriusXM-Beats-2019-Subscriber-Guidance-and-Issues-2020-Guidance/default.aspx.

Statista (2019) Share of monthly online radio listeners in the United States from 2000 to 2019. https://www-statista-com.ezproxy.regent.edu/statistics/475950/online-radio-reach-age-usa/.

Sterling, C. H., & Kittross, J. M. (2001). *Stay tuned: A history of American broadcasting*. Routledge.

Stine, R. (2020, January 21) For Radio, It's a Wait and See About 5G. *Radio World*. https://www.radioworld.com/news-and-business/for-radio-its-wait-and-see-about-5g.

This American Life (2016) About Us. http://www.thisamericanlife.org/about.

Voicebot.ai (2019, March) Smart Speaker Consumer Adoption Report. https://voicebot.ai/wp-content/uploads/2019/03/smart_speaker_consumer_adoption_report_2019.pdf.

West, J. A. (2020, February 11) Sustainability in Broadcast and Digital Media.

Westergren, T. (2020) The Music Genome Project. Pandora. https://www.pandora.com/about/mgp.

Zarecki, T. (2017, March 8). Is Facebook Live Upstaging Your Station? https://radioink.com/2017/03/08/facebook-live-upstaging-station/.

9

Digital Signage

Jennifer H. Meadows, Ph.D.[*]

Abstract

Digital signage is a ubiquitous technology that we seldom notice in our everyday lives. Usually in the form of LCD digital displays with network connectivity, digital signage is used in many market segments including retail, education, banking, corporate, and healthcare. Digital signage has many advantages over traditional print signs including the ability to update quickly, use dayparting, and incorporate multiple media including video, social media, and interactivity. Interactive technologies are increasingly being incorporated into digital signage, allowing the signs to deliver customized messages. New technologies including artificial intelligence are being used to communicate with users individually, and digital signs can now collect all sorts of information on viewers.

Introduction

When thinking about the wide assortment of communication technologies, digital signage probably didn't make your top-10 list. But consider that digital signage is fast becoming omnipresent, although many of us go through the day without ever really noticing it. When you hear the term digital signage perhaps the huge outdoor signs that line the Las Vegas Strip or Times Square in New York City come to mind. Digital signage is actually deployed in most fast food restaurants, airports, and on highways. Increasingly it can be found in schools, hospitals, and even vending machines.

The major markets for digital signage are retail, entertainment, transportation, education, hospitality, corporate, and health care. Digital signage is usually used to lower costs, increase sales, get information or wayfinding, merchandising, and to enhance the customer experience (Intel, 2016). Throughout this chapter, examples of these markets and uses will be provided.

The Digital Screenmedia Association defines digital signage as "the use of electronic displays or screens (such as LCD, LED, OLED, plasma or projection, discussed in Chapter 6) to deliver entertainment, information and/or advertisement in public or private spaces, outside of home" (Digital Screenmedia, n.d.).

[*] Professor, Department of Media Arts, Design, and Technology, California State University, Chico (Chico, California).

Digital signage is frequently referred to as DOOH or digital outside of the home.

The Digital Signage Federation expands the definition a bit to include the way digital signage is networked and defines digital signage as "a network of digital displays that is centrally managed and addressable for targeted information, entertainment, merchandising and advertising" (Digital Signage Federation, n.d.). Either way, the key components to a digital signage system include the display/screen, the content for the display, and some kind of content management system.

The displays or screens for digital signage can take many forms. LCD, including those that use LED technology, is the most common type of screen, but systems also use plasma, OLED, and projection systems. Screens can be flat or curved, rigid or flexible, and can include touch and gesture interactivity and cameras. Screens can range in size from a tiny digital price tag to a huge stadium scoreboard.

Content for digital signage can also take many forms. Similar to a traditional paper sign, digital signs can contain text and images. This is where the similarity ends, though. Digital signage content can also include video (live and stored, standard, 3D, high definition, 4K, and 8K), animation, augmented reality (AR), RSS feeds, social networking, and interactive features.

Unlike a traditional sign, digital signage content can be changed quickly. If the hours at a university library were changed for spring break, a new traditional sign would have to be created and posted while the old sign was taken down. With a digital sign, the hours can be updated almost instantly.

Another advantage of digital signage is that content can also be delivered in multiple ways besides a static image. The sign can be interactive with a touch screen or gesture control. The interaction can be with a simple touch or multi-touch where the users can swipe, and pinch. For example, restaurants can have digital signage with the menu outside. Users can then scroll through the menu on the large screen just like they would on a smartphone. Digital signs can also include facial recognition technologies that allow the system to recognize the viewer's age and sex, then deliver a custom message based upon who is looking at the sign.

All of these technologies allow digital signage to deliver more targeted messages in a more efficient way, as well as collect data for the entity that deploys the sign. For example, a traditional sign must be reprinted when the message changes, and it can only deliver one message at a time. Digital signage can deliver multiple messages over time, and those messages can be tailored to the audience using techniques such as dayparting.

Dayparting is the practice of dividing the day into segments so a targeted message can be delivered to a target audience at the right time (Dell, 2013). For example, a digital billboard may deliver advertising for coffee during the morning rush hour and messages about the 6:00 news in the evening. Those ads could be longer because drivers are more likely to be stopped or slowed because of traffic. An advertisement for dog food during mid-day would likely be very short because traffic will be flowing.

Digital signage also allows viewers to interact with signs in new ways such as the touch and gesture examples above and through mobile devices. In some cases, viewers can move content from a sign to their mobile device using near field communication (NFC) technology or Wi-Fi. For example, visitors to a large hospital could interact with a digital wayfinding sign that can send directions via NFC to their phones.

Another advantage with digital signage is social networking integration. For example, retailers can deploy a social media wall within a store that shows feeds from services such as Instagram, Snapchat, and Twitter.

New communication technologies are bringing change to digital signage. For example, artificial intelligence (AI) technologies might enable a virtual hotel concierge who can answer questions and provide local information. Gamification helps engage customers with brands and IoT (Internet of Things) technology, especially sensors, can be incorporated into digital signage.

All this interactivity brings great benefits to those using digital signs. Audience metrics can be captured such as dwell time (how long a person looks at the sign), where the person looks on the screen (eye tracking), age and sex of the user, what messages were used interactively, and more.

Table 9.1

Paper versus Digital Signage

Traditional/Paper Signs	Digital Signage
• Displays a single message over several weeks	• Displays multiple messages for desired time period
• No audience response available	• Allows audience interactivity
• Changing content requires new sign creation	• Content changed quickly and easily
• Two-dimensional presentation	• Mixed media = text, graphics, video, pictures
• Lower initial costs	• High upfront technology investment costs

Source: Janet Kolodzy

Considering all the advantages of digital signage, why are people still using traditional signs? Although digital signage has many advantages such as being able to deliver multiple forms of content; allowing interactivity, multiple messages, and dayparting; easily changeable content; and viewer metrics; the big disadvantage is cost. Upfront costs for digital signage are much higher than for traditional signs. Over time, the ROI (return on investment) is generally high for digital signage, but for many businesses and organizations this upfront cost is prohibitive. See Table 9.1 for a comparison of traditional and digital signs.

Cost is often a factor in choosing a digital signage system. There are two basic types of systems: premise based and software as a service (SaaS). A premise-based system means that the entire digital signage system is in-house. Content creation, delivery, and management are hosted on location. Advantages of premise-based systems include control, customization, and security. In addition, there isn't an ongoing cost for service. Although a doctor's office with one digital sign with a flash drive providing content is an inexpensive option, multiple screen deployments with a premise-based system are generally expensive and are increasingly less popular.

SaaS systems use the internet to manage and deliver content. Customers subscribe to the service, and the service usually provides a content management system that can be accessed over the internet. Templates provide customers an easy interface to create sign content, or customers can upload their own content in the many forms described earlier. Users then pay the SaaS provider a regular fee to continue using

the service. Both premise and SaaS systems have advantages and disadvantages and many organizations deploy a mix of both types.

So, to review the technology that makes digital signage work, first there needs to be a screen or display. Next, there should be some kind of media player which connects to the screen. Content can be added to the media player using devices such as flash drives and DVDs, but more commonly, the media player resides on a server and content is created, managed, and delivered using a server-on-site or accessed over the internet.

Digital signage, then, is a growing force in out-of-home signage. Kelsen (2010) describes three major types of digital signage: point of sale, point of wait, and point of transit. These categories are not mutually exclusive or exhaustive but they do provide a good framework for understanding general categories of digital signage use.

Point of Sale digital signage is just what it sounds like—digital signage that sells. The menu board at McDonald's is an example of a point of sale digital sign. At a sports venue point of sale signs can be found at ticketing windows as well as concession stands. Those same signs can quickly be changed to show big replays and other customized messages (Hastings, 2019). Micro-digital-signage (MDS) allows what is called "intelligent shelving." These small digital signs are placed on shelves in retail stores to give price, promotion, and product information using eye catching features such as HDTV (Retail Solution, 2016).

Point of Wait digital signage is placed where people are waiting and there is dwell time, such as a bank, elevator, or doctor's office. One example is the digital signage deployed inside taxis. Point of wait signage can convey information such a weather, news, or advertising.

Point of Transit digital signs target people on the move. This includes digital signs in transit hubs such as airports and train stations, signage that captures the attention of drivers, like digital billboards, and walkers, such as street facing retail digital signs. In downtown Atlanta's Arts & Entertainment District, street facing digital signage is being embraced for advertising and public art (Green, 2019) Figure 9.1 gives examples of where you might find the three different types of digital signage.

Figure 9.1

Types of Digital Signage

Point of Sale
Restaurant Menu Board
Price Tags
Point of Payment
Mall

Point of Wait
Taxi
Doctor's Office
Elevator
Bank Line

Point of Transit
Airport
Billboard
Bus Shelter
Store Window

Source: J. Meadows

Arguably there is a fourth category that I'll call "point of mind/place." This is digital signage employed to create an environment or state of mind. Another term often used is "techorating." Hotel lobbies, restaurants, and offices use digital signage to create an environment for guest/clients. For example, the Salesforce Tower in San Francisco features a huge digital wall that displays images and video. As shown in Figure 9.2, the 4 mm LED wall stands 12'7" high and is 106'5" wide,

containing 7 million pixels (Salesforce: 50 Freemont, n.d.) The designers and artists at Obscura Digital worked with Salesforce designers to transform the plain lobby into an environment that would inspire awe and wonder. The images and video are also day-parted with specific imagery such as the redwood forests of Northern California in the morning and a falling water wall during the afternoon.

Figure 9.2

Point of Mind/Place Digital Signage at the Salesforce Building in San Francisco

Source: A. Draper

Background

The obvious place to begin tracing the development of digital signage is traditional signage. The first illuminated signs used gas, and the P.T. Barnum Museum was the first to use illuminated signs in 1840 (National Park Service, n.d.). The development of electricity and the light bulb allowed signs to be more easily illuminated, and then neon signs emerged in the 1920s (A Brief History, 1976). Electronic scoreboards such as the Houston Astros scoreboard in the new Astrodome in 1965 used 50,000 lights to create large messages and scores (Brannon, 2009).

Billboards have been used in the United States since the early 1800s. Lighting was later added to allow the signs to be visible at night. In the 1900s billboard structure and sizes were standardized to the sizes and types we see today (History of OOH, n.d.).

The move from illuminated signs to digital signs developed along the same path as the important technologies that make up digital signage discussed earlier: screen, content, and networks/content management. The development of the VCR in the 1970s contributed to digital signage as the technology provided a means to deliver a custom flow of content to television screens. These would usually be found indoors because of weather concerns.

A traditional CRT television screen was limited by size and weight. The development of projection and flat screen displays advanced digital signage. James Fergason developed LCD display technology in the early 1970's that led to the first LCD watch screen (Bellis, n.d.).

Plasma display monitors were developed in the mid 1960's but, like the LCD, were also limited in size, resolution, and color. It wasn't until the mid-1990s that high definition flat screen monitors were developed and used in digital signage. Image quality and resolution are particularly important for viewers to be able to read text, which is an important capability for most signage.

As screen technologies developed, so did the content delivery systems. DVDs replaced VCRs, eventually followed by media servers and flash memory devices. Compressed digital content forms such as jpeg, gif, flv, mov, and MP3 allowed signage to be quickly updated over a network.

Interactivity with screens became popular on ATM machines in the 1980s. Touch screen technology was developed in the 1960s but wasn't developed into widely used products until the 1990s when the technology was incorporated in personal digital assistant devices (PDAs) like the Palm Pilot.

Digital billboards first appeared in 2005 and immediately made an impact, but not necessarily a good one. Billboards have always been somewhat contentious. Federal legislation, beginning with the *Bonus Act* in 1958 and the *Highway Beautification Act* in 1965, tie Federal highway funding to the placement and regulation of billboards (Federal Highway Administration, n.d.). States also have their own regulations regarding outdoor advertising including billboards.

The development of digital billboards that provide multiple messages and moving images provoked attention because the billboards could be driver distractors. Although Federal Transportation Agency rules state billboards that have flashing, intermittent or moving light or lights are prohibited, the agency determined that digital billboards didn't violate that rule as long as the images are on for at least 4 seconds and are not too bright (Richtel, 2010).

States and municipalities also have their own regulations. For example, a law passed in Michigan in early 2014 limits the number of digital billboards and their brightness, how often the messages can change, and how far apart they can be erected. Companies can erect a digital billboard if they give up three traditional billboards or billboard permits (Eggert, 2014).

While distraction is an issue with billboards, privacy is an overall concern with digital signage, especially as new technologies allow digital signs to collect data from viewers. The Digital Signage Federation adopted its *Digital Signage Privacy Standards* in 2011. These voluntary standards were developed to "help preserve public trust in digital signage and set the stage for a new era of consumer-friendly interactive marketing" (Digital Signage Federation, 2011).

The standards cover both directly identifiable data such as name, address, date of birth, and images of individuals, as well as pseudonymous data. This type of data refers to that which can be reasonably associated with a particular person or his or her property. Examples include IP addresses, internet usernames, social networking friend lists, and posts in discussion forums. The standards also state that data should not be knowingly collected on minors under 13.

The standards also recommend that digital signage companies use other standards for complementary technologies, such as the *Mobile Marketing Association's Global Code of Conduct* and the *Code of Conduct* from the Point of Purchase Association International.

Other aspects of the standard include fair information practices such as transparency and consent. Companies should have privacy policies and should give viewers notice.

Audience measurement is divided into three levels:

- *Level I*—Audience counting such as technologies that record dwell time but no facial recognition.

- *Level II*—Audience targeting. This is information that is collected and aggregated and used to tailor advertisements in real time.

- *Level III*—Audience identification and/or profiling. This is information collected on an individual basis. The information is retained and has links to individual identity such as a credit card number.

The standards recommend that Level I and II measurement should have opt-out consent while Level III should have opt-in consent (Digital Signage Federation, 2011).

As the use of digital signage has grown, so too has the number of companies offering digital signage services and technologies. Some of the long-standing companies in digital signage include Scala, BrightSign, Dynamax, Broadsign, and Four Winds Interactive. Other more traditional computer and networking companies including IBM, AT&T, Intel, and Cisco have also begun to offer full service digital signage solutions.

Many digital signage solutions include ways for signs to deliver customized content using short wave wireless technologies such as NFC (near field communication), RFID (radio-frequency identification), and BLE (Bluetooth low energy) beacons.

For example, using NFC, a shopper could use their phone to tap a sign to get more information on a product or even pay for a product. With RFID, chips placed in objects can then communicate with digital signs. For example, Scala's connected wall allows users to pick up an object such as a shoe with embedded RFID and then see information about that shoe on a digital sign.

BLE Beacon technology works a little differently. Small low-power beacons are placed within a space such as a store. Then shoppers can receive customized messages from the beacons. Users will have to have an app for the store for the beacons to work. So, if a user has an app for a restaurant, as they are walking by the restaurant, the beacon can send a pop-up message with specials or a coupon to entice the user to enter the restaurant.

One fun trend in digital signage is augmented reality. "Augmented reality (AR) is a live, direct or indirect, view of a physical, real-world environment whose elements are augmented by computer-generated sensory input such as sound, video, graphics or GPS data" (Augmented Reality, n.d.). So, imagine a store where you can virtually try on any item—even ones not in stock at the time.

AR and digital signage has been used previously in some creative ways. In one innovative application Pepsi used AR technology to create all kinds of unusual surprises on a street in London. People in a bus shelter could look through a seemingly transparent window at the street, and then monsters, wild animals, and aliens would appear, interacting with people and objects on the street. The transparent sign was actually a live video feed of the street (Pepsi Max Excites, 2014). A video of the deployment can be seen at www.youtube.com/watch?v=Go9rf9GmYpM.

While that AR installation was pretty specialized, nowadays, AR can be found in retail stores across the country. One of the most common ways that AR is being used is with beauty products. Instead of having to try out makeup with samples that everyone else has touched, AR technology allows customers to "try on" different makeup with an AR mirror. Expect AR applications to increase for retail to allow users to try on clothes, makeup, glasses, and even hair styles and colors. Both VR and AR are discussed in more depth in Chapter 13.

Recent Developments

Innovations in digital signage continue to emerge at a fast pace making it a must-have technology rather than a nice-to-have technology. Some of the major developments in digital signage are connection with the internet of things, artificial intelligence, E Ink, emergency alert systems, better analytics, 5G, and community deployment.

"Alexa, what's on sale?" Virtual assistants like Amazon Echo and Google Home are now popular home technologies, and this technology is now going beyond just home devices to be included in a wide range of environments from cars to digital signage. Imagine going into a store with questions; there's no need to find a salesperson when an Alexa enabled sign can answer your questions.

Alexa isn't the only technology; digital signs can now respond to users' voices or other factors using AI technology. Biometric identification such as finger print and retinal scans, as well as facial recognition is being incorporated into digital signage. These features can bring an unprecedented level of customization to signage messages. For example, fast casual restaurant chain, BurgerFi, has deployed touch screen kiosks in some locations with facial recognition to allow customers to order quickly using saved payment information and previous orders. Customers must opt in on the service (Rallo, 2018). CaliBurger in Pasadena, CA and Malibu Poke in Dallas, TX have similar systems. As facial recognition technologies improve, expect to see more deployment across the digital signage market (Maras, 2019). This technology does have challenges, though, including privacy and accuracy. Some facial recognition technologies have poor accuracy with people of color and women, especially black women (Martin, 2019).

Another application of AI in digital signage is on the processing of information collected. AI technologies can deliver better analytics by collecting huge amounts of data. For example, signs can have proximity sensors and cameras with facial recognition to capture sex, mood, age, and touch locations.

One major part of a digital sign is the actual screen and media player. Advances in LED, Mini-LED, QLED, OLED, 4K, Ultra HD, and 8K screens (discussed in Chapter 6) and displays have made digital signs clearer even at very large sizes.

OLED screens are almost as thin as the old paper posters. LG is the top manufacturer of OLED screens in the world. The company has OLED signage in many formats include curved, flat and transparent (OLED Signage, 2020). As of early 2020 the super thin screens aren't used much in digital signage because of cost and fragility, but expect their deployment to

rise—especially for interior installations. The sturdy LCD remains king of outdoor digital signage.

Most digital signs use traditional LCD display technologies. A related technology, E Ink was thought to be limited to smaller e-readers such as the Amazon Kindle. Easy to read in sunlight and using less energy, E Ink has better readability, and now it is available for digital signage in the form of e-paper. E-paper is often used with solar energy sources because power is only needed when the display changes. Look for more e-paper display installations in the future (E-Ink, 2018).

One company is using E Ink signage deployments to improve and enhance community engagement and communication. Soofa's signage is being used in Atlanta, Georgia to do just that. The network of large, E Ink signs are in five neighborhoods, City Hall, and downtown. There are 30 signs altogether that provide real time information on public transportation, upcoming local events, community posts, and local business advertising. The signage, shown in Figure 9.3, runs on solar power and is designed for people on foot. Soofa signs can also be found in Boston and Miami.

Figure 9.3:
Soofa E Ink Sign

Source: Soofa

How does information get to networked digital signs? They can be networked with wires or wirelessly through a WiFi or cellular network. The deployment of 5G wireless technologies will have a tremendous impact on digital signage as more robust (and data heavy) content can be delivered to signs with great speed. This can allow digital signs to deliver more content faster than ever.

Ubiquitous internet connections will allow connection between digital signs and all of devices connected through the internet of things (IoT). So, for example, a digital sign could have a sensor that can tell whether someone is standing by it, deliver a customized message and/or content, and accept payment. A digital sign in Norway used augmented reality technology to deliver an anti-hate message to people who stood near the sign. To viewers it looked as though the street was taken over by a white nationalist hate group demonstration. A message then urged viewers to donate to an anti-hate charity using an NFC enabled device on the sign itself (Cooper, 2018).

While advertising is one major use for digital signage, emergency alerting is a relatively new application. In 2019, emergency alert signs were integrated with IPAWS, the integrated public alert and warning systems for the private sector. Part of the Federal Emergency Management Administration (FEMA), IPAWS allows quick alerts in case of emergency (FEMA, 2019). This feature allows digital signage to instantly deliver necessary information such as AMBER alerts and weather notifications. For example, Daktronics integrated IPAWS into its Venus Control Suite software for digital billboards and outdoor displays (Daktronics, 2019).

Current Status

The digital signage industry is growing steadily. MarketsandMarkets reported that the digital signage market is expected by grow from $20.8 billion in 2019 to $29.6 billion by 2024 (MarketsandMarkets, 2019). The same report noted that hardware and indoor applications will dominate the market through 2024.

The top markets in the U.S. for digital signage are retail, corporate, banking, healthcare, education, and transportation. Transportation and public displays are expected to have the largest growth through 2024 (MarketsandMarkets, 2019).

Digital billboards, despite their relative unpopularity with some states and municipalities, continue to proliferate. According to Guttmann (2019), in 2019 there were 8,800 digital billboards across the USA—up from 6400 in 2016. Guttmann also notes that $3.5 billion was spend on digital out of home advertising in 2019.

According to the Nielsen OOH Advertising Study (2019), 90% of U.S. residents over 16 noticed out-of-home advertising in the past month, and 81% of billboard viewers look at the advertising message all or most of the time. The study found that 55% of respondents noticed a digital billboard in the past month. Billboards are the most noticed out of home advertising with 83% of viewers making a point to see the message. The same study found that 66% of smartphone users take action after viewing the signs.

Factors to Watch

The digital signage industry continues to mature and develop. Trends worth noting include increasing deployment in public spaces as local governments loosen regulations. The downtown Atlanta example noted earlier in this chapter is an example of that deregulation.

Another trend to look for is increased use of artificial intelligence technologies within digital signs offering customers better service while also eliminating workers. The increasing use of order kiosks in restaurants is an example of this trend.

Display technologies will continue to develop offering larger screen and better resolution. Look for more custom screen sizes from the tiny to the gigantic. Another display technology which might appear in the future? Holograms. At the 2020 *Consumer Electronics Show*, Looking Glass showed off an 8K immersive holographic display (Looking Glass, 2020).

With 5G networks, digital signs will be able to offer seamless and robust content in more places—even those considered off limits in the past.

Gen Z

While other technologies discussed in this book may be on their way out, digital signage is just getting started. Gen Z have always been closely associated with their screens, and digital signage takes the idea of a personal screen like a smartphone or tablet and extends those functions to other screens. While some of these screens remain for the passive viewer like a billboard, increasingly these screens require interactivity. Gen Z users are often more comfortable interacting with a screen than an actual person. Take for example, the restaurant kiosks mentioned earlier. Digital signage's future is strong considering the demands of this generation and what this signage can deliver.

Global Concerns & Sustainability

When it comes to digital signage, the most obvious concern is the screens themselves. Digital screens generate a large amount of waste in production and are difficult to dispose. They also consume power, but the industry is making increasing use of solar energy to power digital signs. In addition, E Ink displays use much less power than traditional digital displays.

However, when considering the replacement of paper signs, digital signs offers a green alternative. Replacing many paper signs with one digital sign also saves water, oil and electricity. Digital signage also greatly reduces paper use. Dewitt (2019) found that one digital sign costs about the same as 4 reams of paper—about 20% of the paper a worker uses in a year. Digital signs can also be placed in low energy mode when not in use.

Bibliography

A Brief History of the Sign Industry. (1976). American Sign Museum. http://www.signmuseum.org/a-brief-history -of-the-sign-industry/.

Augmented Reality. (n.d.). *Mashable*. http://mashable.com/category/augmented-reality/.

Bellis, M. (n.d.). Liquid Crystal Display — LCD. http://inventors.about.com/od/lstartinventions/a/LCD.htm.

Brannon, M. (2009). The Astrodome Scoreboard. *Houston Examiner*. http://www.examiner.com/article/the-astrodome-scoreboard.

Cooper, B (2018). AR digital sign battles racism with contactless payments. https://www.digitalsignagetoday.com/articles/augmented-reality-digital-signage-battles-racism-with-contactless-payments/.

Daktronics. (2019). Daktronics integrates control solution with IPAWS for automatic emergency alerts. https://www.daktronics.com/news/Pages/Daktronics-Integrates-Control-Solution-with-IPAWS-for-Automatic-Emergency-Alerts.aspx.

Dell, L. (2013). Why Dayparting Must Be Part of Your Mobile Strategy. *Mashable*. http://mashable.com /2013/08/14/dayparting-mobile/.

Dewitt, D. (2019). Using digital signage for greener communications. Sixteen Nine. https://www.sixteen-nine.net/2019/09/19/using-digital-signage-for-greener-communications/.

Digital Screenmedia Association. (n.d.). Frequently Asked Questions. http://www.digitalscreenmedia.org/faqs#faq15.

Digital Signage Federation. (n.d.). Digital Signage Glossary of Terms. http://www.digitalsignagefederation .org/glossary#D.

Digital Signage Federation. (2011). Digital Signage Privacy Standards. http://www.digitalsignagefederation. org/. Resources/Documents/Articles%20and%20Whitepapers/DSF%20Digital%20Signage%20Privacy%20Standards %2002-2011%20(3).pdf.

Eggert, D. (2014, February 1). Mich law may limit growth of digital billboards. *Washington Post*. http://www.washington-times.com/news/2014/feb/1/mich-law-may-limit-growth-of-digital-billboards/?page=all.

E-Ink delivers e-paper display to Japan (2018). Digital Signage Today. https://www.digitalsignagetoday. com/news/e-ink-delivers-e-paper-display-to-japan.

Federal Highway Administration (n.d.) Outdoor Advertising Control. http://www.fhwa.dot.gov/real_estate/practitioners/oac/oacprog.cfm.

FEMA (2019). Integrated public alert & warning system for the private sector. https://www.fema.gov/integrated-public-alert-warning-system-private-sector.

Green, J. (2019). Downtown's push for a digital lights, arts district is starting to yield results. Curbed Atlanta. https://atlanta.curbed.com/2019/12/16/21024205/downtown-atlanta-signs-digital-lights-tourism.

Guttman, A. Number of digital billboards in the U.S. 2016-2019. Statista. https://www.statista.com/statistics/659381/number-billboards-digital-usa/.

Hastings, J. (2019). 3 digital signage trends that will shape the sporting fan experience. Digital Signage Today. https://www.digitalsignagetoday.com/blogs/3-digital-signage-trends-that-will-shape-the-sporting-fan-experience/.

History of OOH. (n.d.). Outdoor Advertising Association of America. http://www.oaaa.org/outofhomeadvertising/historyofooh.aspx.

Intel. (2016). Digital Signage At-a—Glance. http://www.intel.com/content/dam/www/public/us/en/documents/guides/digital-signage-guide.pdf.

Kelsen, K. (2010). *Unleashing the Power of Digital Signage*. Focal Press.

Looking Glass (2020). Looking Glass 8K immersive display. https://lookingglassfactory.com/product/8k.

MarketsandMarkets (2019). Digital signage market. https://www.marketsandmarkets.com/Market-Reports/digital-signage-market-513.html?gclid=EAIaIQobChMI_8HIw-6X5gIVjMhkCh0UrAVUEAAYASAAEgKt6PD_BwE.

Martin, N. (2019). The major concerns around facial recognition technology. Forbes. https://www.forbes.com/sites/nicolemartin1/2019/09/25/the-major-concerns-around-facial-recognition-technology/#16856e364fe3.

Maras, E. (2019) Facial recognition continues to advance. Digital Signage Today. https://www.digitalsignagetoday.com/articles/facial-recognition-continues-to-advance/.

National Park Service. (n.d.). Preservation Briefs 24-34. Technical Preservation Services. U.S. Department of the Interior.

Nielsen (2019). Out of home advertising study. Nielsen. https://www.driveads.com/wp-content/uploads/2019/07/Nielsen-Out-of-Home-Advertising-Study-compressed.pdf.

OLED Signage (2020). LG. https://www.lg.com/global/business/information-display/oled-signage

Pepsi Max Excites Londoners with Augmented Reality DOOH First. (2014). Grand Visual. http://www.grandvisual.com/pepsi-max-excites-londoners-augmented-reality-digital-home-first/.

Rallo, N. (2018). Meet BurgerFi, where facial recognition software remembers your burger. Dallas Observer. https://www.dallasobserver.com/restaurants/burgerfi-brings-branded-burgers-and-facial-recognition-software-to-dallas-10845067.

Retail Solution: Digital Self Displays (2016). http://www.intel.com/content/www/us/en/retail/solutions/digital-shelf-display.html.

Richtel, M. (2010, March 2). Driven to Distraction. *New York Times*. http://www.ntimes.com/2010/03/02/technology/02billboard.html?pagewanted=all&_r=0.

Salesforce: 50 Freemont. (n.d.). SNA Displays. https://snadisplays.com/projects/salesforce-headquarters/.

10

Cinema Technologies

Michael R. Ogden, Ph.D. *

Abstract

Film's history is notably marked by the collaborative dynamic between the art of visual storytelling and the technology that makes it possible. Filmmakers have embraced each innovation—be it sound, color, special effects, or widescreen imagery—and pushed the art form in new directions. Digital technologies have accelerated this relationship. As today's filmmakers come to terms with the inexorable transition from film to digital acquisition, they also find new, creative possibilities. Among the enabling technologies are cameras equipped with larger image sensors, expanded color gamut and high dynamic range, 2K, 4K, ultra-high definition (8K plus) image resolutions, and high frame rates for smoother motion. Emerging standards in 2D and 3D acquisition and display are competing with the rise of VR and immersive cinema technologies. Satellites, fiber optics, and hard drives are now the standards for theater distribution, while advances in laser projection deliver more vibrant colors and sharper contrast. These developments promise moviegoers a more immersive cinematic experience. It is also hoped they will re-energize moviegoers, lure them back into the theaters, and increase profits.

Introduction

Storytelling is a universally human endeavor. In his book, *Tell Me A Story: Narrative and Intelligence* (1995), computer scientist and cognitive psychologist, Roger Schank, conjectures that not only do humans think in terms of stories—our very understanding of the world is in terms of stories we already understand. "We tell stories to describe ourselves, not only so others can understand who we are, but also so we can understand ourselves.... We interpret reality through our stories and open our realities up to others when we tell our stories" (Schank, 1995, p. 44). Stories touch all of us, reaching across cultures and generations.

* Professor, Media Production & Storytelling, College of Communication & Media Sciences, Zayed University (Dubai, United Arab Emirates).

Robert Fulford, in his book, *The Triumph of Narrative: Storytelling in the Age of Mass Culture* (1999), argues that storytelling formed the core of civilized life and was as important to preliterate peoples as it is to us living in the information age. However, with the advent of mass media—and, in particular, modern cinema—the role of storyteller in popular culture shifted from the individual to the "Cultural Industry" (c.f. Horkheimer & Adorno, 1969; Adorno, 1975; Andrae 1979), or what Fulford calls the "industrial narrative" (1999)—an apparatus for the production of meanings and pleasures involving aesthetic strategies and psychological processes (Neale 1985) and bound by its own set of economic and political determinants made possible by contemporary technical capabilities.

In other words, cinema functions as a social institution providing a form of social contact desired by citizens immersed in a world of "mass culture." It also exists as a psychological institution whose purpose is to encourage the movie-going habit (Belton, 2013).

Simultaneously, cinema evolved as a technological institution, its existence premised upon the development of specific technologies, most of which originated during the course of industrialization in Europe and America in the late 19th and early 20th centuries (e.g., film, the camera, the projector, and sound recording). The whole concept and practice of filmmaking evolved into an art form dependent upon the mechanical reproduction and mass distribution of "the story"—refracted through the aesthetics of framing, light and shade, color, texture, sounds, movement, the shot/countershot, and the *mise en scène* of cinema.

However, the name most associated with cinematic storytelling, Kodak, after 131 years in business as one of the largest producers of film stock in the world, filed for Chapter 11 bankruptcy in 2012 (De La Merced, 2012). The company was hit hard by the recession and the advent of the RED Cinema cameras, the ARRI Alexa, and other high-end digital cinema cameras. In the digital world of bits and bytes, Kodak became just one more 20th century giant to falter in the face of advancing technology.

Perhaps it was inevitable. The digitization of cinema began in the 1980s in the realm of special visual effects. By the early 1990s, digital sound was widely propagated in most theaters, and digital nonlinear editing began to supplant linear editing systems for post-production.

Background

Until recently "no matter how often the face of the cinema… changed, the underlying structure of the cinematic experience… remained more or less the same" (Belton, 2013, p. 6). Yet, even that which is most closely associated with cinema's identity—sitting with others in a darkened theater watching "larger-than-life" images projected on a big screen—was not always the norm.

From Novelty to Narrative

The origins of cinema as an independent medium lie in the development of mass communication technologies evolved for other purposes (Cook, 2004). Specifically, photography (1826-1839), roll film (1880), the Kodak camera (1888), George Eastman's motion picture film (1889), the motion picture camera (1891-1893), and the motion picture projector (1895-1896) each had to be invented in succession for cinema to be born.

Figure 10.1

Eadweard Muybridge's Zoopraxiscope, 1893

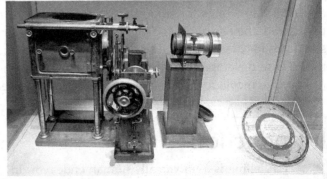

Source: Wikimedia Commons

Early experiments in series photography for capturing motion were an important precursor to cinema's emergence. In 1878, Eadweard Muybridge set up a battery of cameras triggered by a horse moving through a set of trip wires. Adapting a Zoëtrope (a parlor novelty of the era) for projecting the photographs, Muybridge arranged his photograph plates around the perimeter of a disc that was manually rotated. Light from a "Magic Lantern" projector was shown through each slide as it passed in front of a lens. The image produced was then viewed on a large screen (Neale, 1985). If rotated rapidly enough, a phenomenon known as *persistence of vision* (an image appearing in front of the eye lingers a split second in the retina after removal of the image), allowed the viewer to experience smooth, realistic motion.

Perhaps the first movie projector, Muybridge called his apparatus the Zoopraxiscope, which was used to project photographic images in motion for the first time to the San Francisco Art Association in 1880 (Neale, 1985).

In 1882, French physiologist and specialist in animal locomotion, Étienne-Jules Marey, invented the Chronophotographic Gun in order to take series photographs of birds in flight (Cook, 2004). Shaped like a rifle, Marey's camera took 12 instantaneous photographs of movement per second, imprinting them on a rotating glass plate coated with a light-sensitive emulsion. A year later, Marey switched from glass plates to paper roll film. But like Muybridge, "Marey was not interested in cinematography…. In his view, he had invented a machine for dissection of motion similar to Muybridge's apparatus but more flexible, and he never intended to project his results" (Cook, 2004, p. 4).

In 1887, Hannibal Goodwin, an Episcopalian minister from New Jersey, first used celluloid roll film as a base for light-sensitive emulsions. George Eastman later appropriated Goodwin's idea and in 1889, began to mass-produce and market celluloid roll film on what would eventually become a global scale (Cook, 2004). Neither Goodwin nor Eastman were initially interested in motion pictures. However, it was the introduction of this durable and flexible celluloid film, coupled with the technical breakthroughs of Muybridge and Marey, that inspired Thomas Edison to attempt to produce recorded moving images to accompany the recorded sounds of his newly-invented phonograph (Neale, 1985).

Capitalizing on these innovations, W.K.L. Dickson, a research assistant at the Edison Laboratories, invented the first authentic motion picture camera, the Kinetograph—first constructed in 1890 with a patent granted in 1894. The basic technology of modern film cameras is still nearly identical to this early device. All film cameras, therefore, have the same five basic functions: a "light tight" body that holds the mechanism which advances the film and exposes it to light; a motor; a magazine containing the film; a lens that collects and focuses light on to the film; and a viewfinder that allows the cinematographer to properly see and frame what they are photographing (Freeman, 1998).

Thus, using Eastman's new roll film, the Kinetograph advanced each frame at a precise rate through the camera, thanks to sprocket holes that allowed metal teeth to grab the film, advance it, and hold the frame motionless in front of the camera's aperture at split-second intervals. A shutter opened, exposing the frame to light, then closed until the next frame was in place. The Kinetograph repeated this process 40 times per second. Throughout the silent era, other cameras operated at 16 frames per second; it wasn't until sound was introduced that 24 frames per second became standard, in order to improve the quality of voices and music (Freeman, 1998). When the processed film is projected at the same frame rate, realistic movement is presented to the viewer.

However, for reasons of profitability alone, Edison was initially opposed to projecting films to groups of people. He reasoned (correctly, as it turned out) that if he made and sold projectors, exhibitors would purchase only one machine from him—a projector—instead of several Kinetoscopes (Belton, 2013) that allowed individual viewers to look at the films through a magnifying eyepiece. By 1894, Kinetographs were producing commercially viable films. Initially the first motion pictures (which cost between $10 and $15 each to make) were viewed individually through Edison's Kinetoscope "peep-shows" for a nickel apiece in arcades (called Nickelodeons) modeled on the phonographic parlors that had earlier proven so successful for Edison (Belton, 2013).

It was after viewing the Kinetoscope in Paris that the Lumière brothers, Auguste and Louis, began thinking about the possibilities of projecting films onto a screen for an audience of paying customers. In 1894, they began working on their own apparatus, the Cinématograph. This machine differed from Edison's machines by combining both photography and projection into one device at the much lower (and thus, more economical) film rate of 16 frames per second. It was also much lighter and more portable (Neale, 1985).

In 1895, the Lumière brothers demonstrated their Cinématograph to the Société d'Encouragement pour l'Industries Nationale (Society for the Encouragement of National Industries) in Paris. The first film screened was a short actuality film of workers leaving the Lumière factory in Lyons (Cook, 2004). The actual engineering contributions of the Lumière brothers were quite modest when compared to that of W.K.L. Dickson—they merely synchronized the shutter movement of the camera with the movement of the photographic film strip. Their real contribution is in the establishment of cinema as an industry (Neale, 1985).

Figure 10.2

First Publicly Projected Film: Sortie des Usines Lumière à Lyon, 46 seconds, 1895

Source: Screen capture courtesy Lumière Institute

The early years of cinema were ones of invention and exploration. The tools of visual storytelling, though crude by today's standards, were in hand, and the early films of Edison and the Lumière brothers were fascinating audiences with actuality scenes—either live or staged—of everyday life. However, an important pioneer in developing narrative fiction film was Alice Guy

Blaché. Remarkable for her time, Guy Blaché was arguably the first director of either sex to bring a story-film to the screen with the 1896 release of her one-minute film, *La Fée aux Choux* (The Cabbage Fairy) that preceded the story-films of Georges Méliès by several months.

Figure 10.3

Alice Guy Blaché Portrait, 1913

Source: Collection Solax, Public Domain

One could argue that Guy Blaché was cinema's first story "designer." From 1896 to 1920 she wrote and directed hundreds of short films including over 100 synchronized sound films and 22 feature films and produced hundreds more (McMahan, 2003). In the first half of her career, as head of film production for the Gaumont Company (where she was first employed as a secretary), Guy Blaché almost single-handedly developed the art of cinematic narrative (McMahan, 2003) with an emphasis on storytelling to create meaning.

By the turn of the century, film producers were beginning to assume greater editorial control over the narrative, making multi-shot films and allowing for greater specificity in the story line (Cook, 2004). Such developments are most clearly apparent in the work of Georges Méliès. A professional magician who owned and operated his own theater in Paris, Méliès was an important early filmmaker, developing cinematic narrative which demonstrated a created cause-and-effect reality. Méliès invented and employed a number of important narrative devices, such as the fade-in and fade-out, "lap" (overlapping) dissolve as well as impressive visual effects such as stop-motion photography (*Parce qu'on est des geeks!* 2013). Though he didn't employ much editing within individual scenes, the scenes were connected in a way that supported a linear, narrative reality.

By 1902, with the premiere of his one-reel film *Le Voyage Dans La Lune* (A Trip to the Moon), Méliès was fully committed to narrative filmmaking.

Figure 10.4

Méliès, Le Voyage Dans La Lune, 1902, 13 minutes

Source: Screen capture, M.R. Ogden.

Middle-class American audiences, who grew up with complicated plots and fascinating characters from such authors as Charles Dickens and Charlotte Brontë, began to demand more sophisticated film narratives. Directors like Edwin S. Porter and D.W. Griffith began crafting innovative films in order to provide their more discerning audiences with the kinds of stories to which theatre and literature had made them accustomed (Belton, 2013).

Influenced by Méliès, American filmmaker Edwin S. Porter is credited with developing the "invisible technique" of continuity editing. By cutting to different angles of a simultaneous event in successive shots, the illusion of continuous action was maintained. Porter's *Life of an American Fireman* and *The Great Train Robbery*, both released in 1903, are the foremost examples of this new style of storytelling (Cook, 2004).

Figure 10.5

Porter, The Great Train Robbery, 1903, 11 minutes

Source: Screen capture, M.R. Ogden.

Taking this a step further, D.W. Griffith, a former actor in some of Porter's films, went on to become one of the most important filmmakers of all time, and truly the "father" of modern narrative form. He altered camera angles, employed close-ups, and actively narrated events, thus shaping audience perceptions of them. Additionally, he employed "parallel editing" (or cross-cutting)—cutting back and forth from two or more simultaneous events taking place in separate locations—to create suspense (Belton, 2013). In D.W. Griffith's 1915 film *The Birth of a Nation*, parallel editing was employed to establish relationships and drive the narrative. Griffith's understanding of the importance of editing in establishing such relationships on screen lead to the production of more complex and dynamic cinematic stories.

Even though Edison's Kinetograph camera had produced more than 5,000 films (Freeman, 1998), by 1910, other camera manufacturers such as Bell and Howell, and Pathé (which acquired the Lumière patents in 1902) had invented simpler, lighter, more compact cameras that soon eclipsed the Kinetograph.

Nearly all of the cameras of the silent era were hand-cranked, yet, camera operators were amazingly accurate in maintaining proper film speed (16 fps) and could easily change speeds to suit the story. Cinematographers could crank a little faster (over-crank) to produce slow, lyrical motion, or they could crank a little slower (under-crank) and when projected back at normal speed, they displayed the frenetic, sped-up motion apparent in the silent slapstick comedies of the Keystone Film Company (Cook, 2004).

In the United States, the early years of commercial cinema were tumultuous as Edison sued individuals and enterprises over patent disputes in an attempt to protect his monopoly and his profits (Neale, 1985). However, by 1908, the film industry was becoming more stabilized as the major film producers "banded together to form the Motion Picture Patents Company (MPPC) which sought to control all aspects of motion picture production, distribution and exhibition" (Belton, 2013, p. 12) through its control of basic motion picture patents.

In an attempt to become more respectable, and to court middle-class customers, the MPPC began a campaign to improve the content of motion pictures by engaging in self-censorship to control potentially offensive content (Belton, 2013). The group also provided half-price matinees for women and children and improved the physical conditions of theaters. Distribution licenses were granted to 116 exchanges that could distribute films only to licensed exhibitors who paid a projection license of two dollars per week.

Unlicensed producers and exchanges continued to be a problem, so in 1910 the MPPC created the General Film Company to distribute their films. This development proved to be highly profitable and "was… the first stage in the organized film industry where production, distribution, and exhibition were all integrated, and in the hands of a few large companies" (Jowett, 1976, p. 34) presaging the emergence of the studio system 10 years later.

The Studio System

For the first two decades of cinema, nearly all films were photographed outdoors. Many production facilities were like that of George Méliès, who constructed a glass-enclosed studio on the grounds of his home in a Paris suburb (Cook, 2004). However, Edison's laboratory in West Orange, New Jersey, dubbed the "Black Maria," was probably the most famous film studio of its time. It is important to note that the "film technologies created in [such] laboratories and ateliers

would come to offer a powerful system of world building, with the studio as their spatial locus" (Jacobson, 2011, p. 233). Eventually, the industry outgrew these small, improvised facilities and moved to California, where the weather was more conducive to outdoor productions. Large soundstages were also built in order to provide controlled staging and lighting.

Figure 10.6

Edison's Black Maria, World's First Film Studio, circa 1890s

Source: Wikimedia Commons

By the second decade of the 20th Century, dozens of movie studios were operating in the U.S. and across the world. A highly specialized industry grew in southern California, honing sophisticated techniques of cinematography, lighting, and editing. Hollywood studios divided these activities into preproduction, production, and post-production. During preproduction, a film was written and planned. The production phase was technology intensive, involving the choreography of actors, cameras and lighting equipment. Post-production consisted of editing the films into coherent narratives and adding titles—in fact, film editing is the only art that is unique to cinema.

The heart of American cinema was now beating in Hollywood, and the institutional machinery of filmmaking evolved into a three-phase business structure of production, distribution, and exhibition to get their films from studios to theater audiences. Although the MPPC was formally dissolved in 1918 as a result of an antitrust suit initiated in 1912 (Cook, 2004), powerful new film companies, flush with capital, were emerging. With them came the advent of vertical integration.

Through a series of mergers and acquisitions, formerly independent production, distribution, and exhibition companies congealed into five major studios; Paramount, Metro-Goldwin-Mayer (MGM), Warner Bros, RKO (Radio-Keith-Orpheum), and Fox Pictures. "All of the major studios owned theater chains; the minors—Universal, Columbia, and United Artists—did not" (Belton, 2013, p. 68), but distributed their pictures by special arrangement to the theaters owned by the majors. The resulting economic system was quite efficient. "The major studios produced from 40 to 60 pictures a year… [but in 1945 only] owned 3,000 of the 18,000 theaters around the country… [yet] these theaters generated over 70% of all box-office receipts" (Belton, 2013, p. 69).

As films and their stars increased in popularity, and movies became more expensive to produce, studios began to consolidate their power, seeking to control each phase of a film's life. However, since the earliest days of the Nickelodeons, moralists and reformers had agitated against the corrupting nature of the movies and their effects on American youth (Cook, 2004). A series of scandals involving popular movie stars in the late 1910s and early 1920s resulted in ministers, priests, women's clubs, and reform groups across the nation encouraging their membership to boycott the movies.

In 1922, frightened Hollywood producers formed a self-regulatory trade organization—the Motion Picture Producers and Distributors of America (MPPDA). By 1930, the MPPDA adopted the rather draconian Hayes Production Code. This "voluntary" code, intended to suppress immorality in film, proved mandatory if the film was to be screened in America (Mondello, 2008). Although the code aimed to establish high standards of performance for motion-picture producers, it "merely provided whitewash for overly enthusiastic manifestations of the 'new morality' and helped producers subvert the careers of stars whose personal lives might make them too controversial" (Cook, 2004, p. 186).

Sound, Color & Spectacle

Since the advent of cinema, filmmakers hoped for the chance to bring both pictures and sound to the screen. Although the period until the mid-1920s is considered the silent era, few films in major theaters actually were screened completely silent. Pianists or organists—sometimes full orchestras—performed musical accompaniment to the projected images. At times, actors would speak the lines of the characters and machines and performers created sound effects. "What these performances lacked was fully synchronized sound contained within the soundtrack on the film itself" (Freeman, 1998, p. 408).

By the late 1920s, experiments had demonstrated the viability of synchronizing sound with projected film. When Warner Bros Studios released The *Jazz Singer* in 1927, featuring synchronized music and dialog using their Vitaphone process, the first "talkie" was born. Vitaphone was a sound-on-disc process that issued the audio on a separate 16-inch phonographic disc. While the film was projected, the disc played on a turntable indirectly coupled to the projector motor

---

Figure 10.7

Vitaphone Projection Setup, 1926 Demonstration

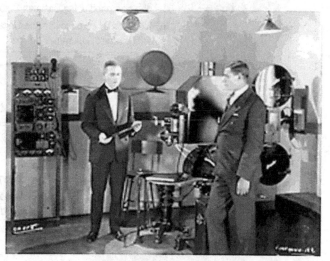

Source: Wikimedia Commons.

(Bradly, 2005). Other systems were also under development during this time and Warner Bros Vitaphone process had competition from Movietone, DeForest Phonofilm, and RCA's Photophone.

Figure 10.8

Cinema History Highlights

Late 1800s	• Early experiments between 1870s & 1890s in motion photography by Muybridge & Marey inspire Thomas Edison to invent the **Kinetograph** camera in 1894. **Kinetoscope** "peep-shows" become popular entertainment. • Eastman markets first commercially available transparent roll film in 1889. • In 1895, the Lumière brothers demonstrated their **Cinématograph**.
1900–1920s	• Alice Guy Balché, Georges Mèliés, Edwin S. Porter & D.W. Griffith develop and perfect narrative film techniques enhancing storytelling. • **Pathé camera** (based on Lumière patents) becomes the dominant camera. • In 1927, Warner Brothers released *The Jazz Singer*, with synchronized dialog & music using their **Vitaphone** process .
1930s–1940s	• Invented in the 1920s, the **Moviola** becomes the dominant film editing device in the 1930 through the 1970s (with successive improvements). • Disney uses the **Technicolor** three-color process in his *Silly Symphonies* cartoon series. • "Deep focus" cinematography demonstrated by Orson Welles & Gregg Toland in the 1941 film *Citizen Kane*.
1950s–1960s	• **Cinerama** (1952) launches a widescreen revolution. • **Eastman Color** used on *The Robe* (1953), by 1955, most widescreen films photographed using Eastman Color film. • Network TV introduced in the United States in 1949, film box office receipts plummet.
1970s–1980s	• 1976 *Futureworld* features 1st use of 3-D Computer Generated Imagery (CGI). • 1970s, Dolby Labs introduced Dolby noise reduction & Dolby Stereo, by 1984 **Dolby 5.1** surround sound technology • 1980s saw studio consolidations & rise of VHS rental market.
1990s–2000s	• 1st all-CGI animated film, *Toy Story* (1995), released by Pixar. • *Star Wars Episode II: Attack of the Clones* (2002) is 1st **completely digital film** (acquired, distributed & projected). • 2009 James Cameron releases the **digital 3-D** blockbuster *Avatar* which grosses nearly $3 billion worldwide.
2010s–Beyond	• Declining DVD revenue force Hollywood to adopt new business model—switch to on-demand services & **Internet streaming**. • 2015, 95% of world cinema screens are digital. • 2017, *Guardians of the Galaxy 2* becomes first commercial film **shot in 8K**. • **AI algorithms** increasingly used to automatically render advanced VFX shots.

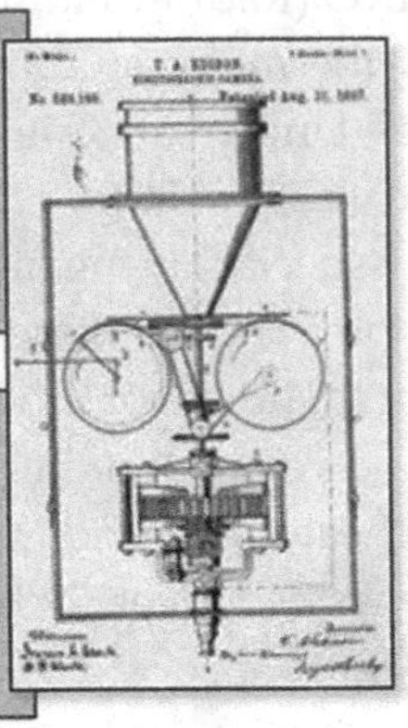

Source: M.R. Ogden

Though audiences were excited by this new novelty, from an aesthetic standpoint, the advent of sound actually took the visual production value of films backward. Film cameras were loud and had to be housed in refrigerator-sized vaults to minimize the noise; as a result, the mobility of the camera was suddenly limited. Microphones had to be placed very near the actors, resulting in restricted blocking and the odd phenomenon of actors leaning close to a bouquet of flowers as they spoke their lines; the flowers, of course, hid the microphone. No question about it, though, sound was here to stay.

Once sound made its appearance, the established major film companies acted cautiously, signing an agreement to only act together. After sound had proved a commercial success, the signatories adopted Movietone as the standard system—a sound-on-film method that recorded sound as a variable-density optical track on the same strip of film that recorded the pictures (Neale, 1985). The advent of the "talkies" launched another round of mergers and expansions in the studio system. By the end of the 1920s, more than 40% of theaters were equipped for sound (Kindersley, 2006), and by 1931, "...virtually all films produced in the United States contained synchronized soundtracks" (Freeman, 1998, p. 408).

Movies were gradually moving closer to depicting "real life." But life isn't black and white, and experiments with color filmmaking had been conducted since the dawn of the art form. Finally, however, in 1932, Technicolor introduced a practical three-color dye-transfer process that slowly revolutionized moviemaking and dominated color film production in Hollywood until the 1950s (Higgins, 2000). Though aesthetically beautiful, the Technicolor process was extremely expensive, requiring three times the film stock, complicated lab processes, and strict oversight by the Technicolor Company, who insisted on strict secrecy during every phase of production. As a result, most movies were still produced in black and white well into the 1950s; that is, until single-strip Eastman Color Negative Safety Film (5247) and Color Print Film (5281) were released, "revolutionizing the movie industry and forcing Technicolor strictly into the laboratory business" (Bankston, 2005, p. 5).

By the late 1940s, the rising popularity of television drove the early impetus for widescreen cinema technology because film studios were losing money at the box office. In the early years of film the "Academy Standard" 4:3 aspect ratio set by Edison (4 units wide by 3 units high, also represented as 1.33:1) was assumed to be more aesthetically pleasing than a square box and was the most common aspect ratio for most films until the 1960s (Freeman, 1998). This was also the aspect ratio adopted for broadcast television until the switch to HDTV in 1998. The Hollywood studios' initial response to television was characteristically cautious, initially choosing to release fewer but more expensive films (still in the standard Academy aspect ratio) hoping to lure audiences back to theaters with more spectacle and higher quality product (Belton, 2013).

It was through the efforts of independent filmmakers during the 1950s and early 1960s, that the most pervasive technological innovations in Hollywood since the introduction of sound were realized. "A series of processes changed the size of the screen, the shape of the image, the dimensions of the films, and the recording and reproduction of sound" (Bordwell, Staiger & Thompson, 1985, p. 358).

Cinerama (1952) launched a widescreen revolution that would permanently alter the shape of the motion picture screen. Cinerama was a widescreen process that required filming with a three-lens camera and projecting with synchronized projectors onto a deeply curved screen extending the full width of most movie theaters. This viewing (yielding a 146° by 55° angle of view) was meant to approximate that of human vision (160° by 60°) and fill a viewer's entire peripheral vision. Mostly used in travelogue-adventures, such as *This is Cinerama* (1952) and *Seven Wonders of the World* (1956), the first two Cinerama fiction films—*The Wonderful World of the Brothers Grimm* and *How the West Was Won*—were released in 1962, with much fanfare and critical acclaim. However, three-lens camera productions and three-projector system theaters like Cinerama and CineMiracle (1957) were extremely expensive technologies and quickly fell into disuse.

Figure 10.9

Cinerama's 3-Camera Projection

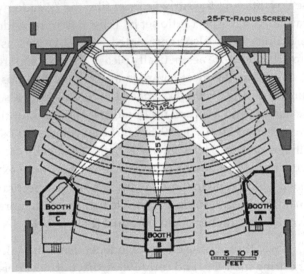

Source: Wikimedia Commons.

Anamorphic processes used special lenses to shoot or print squeezed images onto the film as a wide field of view. In projection, the images were un-squeezed using the same lenses, to produce an aspect ratio of 2.55:1—almost twice as wide as the Academy Standard aspect ratio (Freeman, 1998). When Twentieth Century-Fox released *The Robe* in 1953 using the CinemaScope anamorphic system, it was a spectacular success and just the boost Hollywood needed. Soon, other companies began producing widescreen films using similar anamorphic processes such as Panascope and Superscope. Nearly all these widescreen systems—including CinemaScope—incorporated stereophonic sound reproduction.

Figure 10.10

Film Aspect Ratios

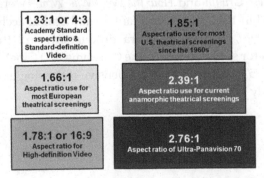

Source: M.R. Ogden

If widescreen films were meant to engulf audiences, pulling them into the action, "3D assaulted audiences—hurling spears, shooting arrows, firing guns, and throwing knives at spectators sitting peacefully in their theatre seats" (Belton, 2013, p. 328).

The technology of 3D is rooted in the basic principles of binocular vision. Early attempts at reproducing monochromatic 3D used an anaglyphic system: two strips of film, one tinted red, the other cyan, were projected simultaneously for an audience wearing glasses with one red and one cyan filtered lens (Cook, 2004). When presented with slightly different angles for each eye, the brain processed the two images as a single 3D image. The earliest 3D film using the anaglyphic process was *The Power of Love* in 1922.

In the late 1930s, MGM released a series of anaglyphic shorts, but the development of polarized filters and lenses around the same time permitted the production of full-color 3D images. Experiments in anaglyphic films ceased in favor of the new method. In 1953, Milton Gunzberg released *Bwana Devil*, a "dreadful" film shot using a polarized 3D process called Natural Vision. It drew in audiences and surprisingly broke box office records, grossing over $5 million by the end of its run (Jowett, 1976). Natural Vision employed two interlocked cameras whose lenses were positioned to approximate the distance between the human eyes and record the scene on two separate negatives. In the theater, when projected simultaneously onto the screen, spectators wearing disposable glasses with polarized lenses perceived a single three-dimensional image (Cook, 2004). Warner Bros released the second Natural Vision feature, *House of Wax* (1953), which featured six-track stereophonic sound and was a critical and popular success, returning $5.5 million on an investment of $680,000 (Cook, 2004).

Although Hollywood produced 69 features in 3D between 1953 and 1954, most were cheaply made exploitation films. By late 1953, the stereoscopic 3D craze had peaked. Although 3D movies were still made decades later for special short films at Disney theme parks, 3D was no longer part of the feature-film production process (Freeman, 1998). One reason for 3D's demise was that producers found it difficult to make serious narrative films in such a gimmicky process (Cook, 2004). Another problem was the fact that audiences disliked wearing the polarized glasses; many also complained of eyestrain, headaches and nausea.

But, perhaps, the biggest single factor in 3D's rapid fall from grace was cinematographers' and directors' alternative use of deep-focus widescreen photography—especially anamorphic processes that exploited depth through peripheral vision—and compositional techniques that contributed to the feeling of depth without relying on costly, artificial means. Attempts to revive 3D, until most recently, met with varying degrees of success, seeing short runs of popularity in the 1980s with films like *Friday the 13th, Part III* and *Jaws 3D* (both 1983).

In 1995, with the release of the IMAX 3D film, *Wings of Courage*—and later, *Space Station 3D* in 2002—the use of active display LCD glasses synchronized with the shutters of dual-filmstrip projectors using infrared signals presaged the eventual rise of digital 3D films 10 years later.

Hollywood Becomes Independent

"The dismantling of the studio system began just before World War II when the U.S. Department of Justice's Antitrust Division filed suit against the [five] major [and three minor] studios, accusing them of monopolistic practices in their use of block booking, blind bidding, and runs, zones, and clearances" (Belton, 2013, p. 82).

In 1948, in the case of *U.S. vs. Paramount*, the U.S. Supreme Court ruled against the block booking system and recommended the breakup of studio–theater monopolies. The major studios were forced to divorce their operations from one another, separate production and distribution from exhibition, and divest themselves of their theater chains (Belton, 2013). "RKO and other studios sold their film libraries to television stations to offset the losses from the *Paramount* case. The studios also released actors from contracts who became the new stars of the television world" (Bomboy, 2015). Other factors also contributed to the demise of the studio system, most notably changes in leisure-time entertainment, the aforementioned competition with television, and the rise of independent production (Cook, 2004). Combined with the extreme form of censorship Hollywood imposed upon itself

through the Hayes Production Codes—and "after World War II, with competition from TV on the family front, and from foreign films with nudity on the racy front" (Mondello, 2008)—movie studios were unable (or unwilling) to rein in independent filmmakers who chafed under the antiquated Code.

In another landmark ruling (known as the "Miracle decision"), the U.S. Supreme Court decided in 1952 that films constitute "a significant medium of communication of ideas" and were therefore protected by both the First and Fourteenth Amendments (Cook, 2004, p. 428). By the early 1960s, supported by subsequent court rulings, films were "guaranteed full freedom of expression" (Cook, 2004, p. 428). The influence of the Hayes Production Code had all but disappeared by the end of the 1960s, replaced by the MPAA ratings system (MPAA, 2020) instituted in 1968, revised in 1972, and now in its latest incarnation since 1984.

Though the 1960s still featured big-budget, lavish movie spectacles, a parallel movement reflected the younger, more rebellious aesthetic of the "baby boomers." Actors and directors went into business for themselves, forming their own production companies, and taking as payment lump-sum percentages of the film's profits (Belton, 2013). The rise and success of independent filmmakers like Arthur Penn (*Bonnie and Clyde*, 1967), Stanley Kubrick (*2001: A Space Odyssey*, 1968), Sam Peckinpah (*The Wild Bunch*, 1969), Dennis Hopper (*Easy Rider*, 1969), and John Schlesinger (*Midnight Cowboy*, 1969), demonstrated that filmmakers outside the studio system were freer to experiment with style and content. For some, the writing was on the wall, the major studios would no longer dominate popular filmmaking as they had in the past.

In the early 1960s, an architectural innovation changed the way most people see movies—the move from single-screen theaters (and drive-ins) to multi-screen cineplexes. Although the first multi-screen house with two theaters was built in the 1930s, it was not until the late 1960s that film venues were built with four to six theaters. What these theaters lacked was the atmosphere of the early "movie palaces." While some owners put effort into the appearance of the lobby and concessions area, in most cases the "actual theater was merely functional" (Haines, 2003, p. 91). The number of screens in one location continued to grow; 1984 marked the opening of the first 18-plex theater in Toronto (Haines, 2003).

The next step in this evolution was the addition of stadium seating—offering moviegoers a better experience by affording more comfortable seating with unobstructed views of the screen (EPS Geofoam, n.d.). Although the number of screens in a location seems to have reached the point of diminishing returns, many theaters are now working on improving the atmosphere they create for their patrons. From bars and restaurants to luxury theaters with a reserved $29 movie ticket, many theater owners are once again working to make the movie-going experience something different from what you can get at home (Gelt & Verrier, 2009).

Recent Developments

Since the first cinematic "trickery" of Georges Méliès thrilled audiences, much has changed in the past century. Most films occupying a list of the top 100 highest grossing films of all time—not adjusted for inflation—feature heavy reliance on visual effects in their storytelling. According to *The Numbers* (2020a), the top five all-time worldwide box office revenue earners are: *Avengers: End Game* (2019, $2.79 billion), *Avatar* (2009, $2.78 billion), *Titanic* (1997, $2.2 billion), *Star Wars: The Force Awakens* (2015, $2.06 billion), and *Avengers: Infinity War* (2018, $2.04 billion). When examining the inflation adjusted highest grossing films (pattap-21567, 2018), the list is topped by *Gone with the Wind* (1939, originally $402 million, inflation adjusted to $3.8 billion), a film that featured, for its day, innovative matte shots. The next two films on the list are James Cameron's *Titanic* (1997) and *Avatar* (2009), followed by George Lucas' *Star Wars: A New Hope* (1977), arguably the most iconic visual effects movie ever made (Bredow, 2014). Disney's capstone film to the Marvel Avengers saga rounds out the top five.

Such "trickery"—composite special effects, more commonly known as "visual effects" (VFX)—are divided into mechanical, optical, and computer-generated imagery (CGI). "Mechanical effects include those devices used to make rain, wind, cobwebs, fog, snow, and explosions. Optical effects allow images to be combined... through creation of traveling mattes run through an optical printer" (Freeman, 1998, p. 409).

Early optical effects, like matte composites, required the camera to remain stationary and any foreground movement restricted to one side of the "matte line" dividing live action from a mostly static background. In 1918, cinematographer Frank Williams patented a double matte technique that photographed foreground subjects against a black background—blue or green screens were introduced with the advent of color film (Hess, 2017). This footage was used to create a high-contrast silhouette of actor movement. Originally called the "Williams Process," it became known as a traveling matte, and consisted of a "holdout matte" contained a black silhouette of actor movement against a clear background and the inverse "cover matte" with a clear silhouette of actors against a black background (Edwards, 2014). The final composite was created using previously shot background footage "bi-packed" (two strips of film together in the same projector) with the "holdout matte" to combine the image on to a third, unexposed reel of film leaving an unexposed area on the film corresponding to actor movement. The exposed film would then be rewound and the actor footage bi-packed with the "cover matte" to fill the unexposed area and create the final effect (Edwards, 2014). Fast-forward to today's highly realistic, CGI effects and the basic principles (and terminology) are the same. The major difference is the digital tools have greatly improved the workflow, and AI algorithms afford extraordinary levels of finesse over the photochemical processes of the past.

Figure 10.11

Holdout & Cover Mattes Used In *Return of the Jedi* (1983)

Source: Lucasfilm Ltd.

Inspired by the traveling matte effects used in *King Kong* (1933), Ray Harryhausen began to experiment with stop-motion animation and split-screen action in *The 7ᵗʰ Voyage of Sinbad* (1958) and the invention of "Dynamation" (Martin, 2014)—a technique of rear and front projected footage one frame at a time—most famously employed in *Jason and the Argonauts* (1963), as well as in two additional Sinbad films (1973 & 1977) and the 1981 film, *Clash of the Titans*.

Four years in the making, the 1968 Stanley Kubrick film, *2001: A Space Odyssey*, created a new standard for VFX credibility (Martin, 2014). Kubrick used sophisticated traveling mattes combined with "hero" miniatures of spacecraft (ranging from four to 60 feet in length) and live-action to stunning effect (Cook, 2004). The film's "star gate" sequence dazzled audiences with controlled streak photography, macrophotography of liquids, and deliberate misuse of color-records. Also, throughout the opening "Dawn of Man" sequence, audiences witnessed the first major application of the front projection technique (Martin, 2014).

Arguably the first movie ever to use computers to create a visual effect—a two-dimensional rotating structure on one level of the underground lab—was *The Andromeda Strain* in 1971. This work was considered extremely advanced for its time.

In 1976, American International Pictures released *Futureworld*, which featured the first use of 3D CGI—a brief view of a computer-generated face and hand. In 1994, this groundbreaking effect was awarded a Scientific and Engineering Academy Award. Since then, CGI technology has progressed rapidly.

"In the history of VFX, there is a before-and-after point demarcated by the year 1977—when *Star Wars* revolutionized the industry" (Martin, 2014, p. 71-72). VFX supervisor John Dykstra invented an electronic motion-controlled camera capable of repeating its movements (later called the "Dykstraflex") and developed methodologies for zero-gravity explosions. Likewise, George Lucas' visual effects company, Industrial Light & Magic (ILM), took a big step forward for CGI with the rendering of a 3D wire-frame view of the Death Star trench depicted as a training aid for rebel pilots in *Star Wars* (1977).

Star Trek: The Wrath of Kahn (1982) incorporated a one-minute sequence created by Pixar (a LucasFilm spin-off), that simulated the "Genesis Effect" (the birth and greening of a planet) and is cinema's first totally computer-generated VFX shot. It also introduced a fractal-generated landscape and a particle-rendering system to achieve a fiery effect (Dirks, 2009).

Tron (1982) was the first live-action movie to use CGI for a noteworthy length of time (approximately 20 minutes) in the most innovative sequence of its 3D graphics world inside a video game and "showed studios that digitally created images were a viable option for motion pictures" (Bankston, 2005, p. 1).

In *Young Sherlock Holmes* (1985), LucasFilm/Pixar created perhaps the first fully photorealistic CGI character in a full-length feature film with the sword-wielding medieval "stained-glass" knight who came to life when jumping out of a window frame.

Visual impresario James Cameron has always relied on VFX in his storytelling dating back to the impressive low-budget miniature work on *The Terminator* (1984), later expanded on for *Aliens* (1986). Cameron's blockbuster action film, *Terminator 2: Judgment Day* (1991) received a Best Visual Effects Oscar thanks to its depiction of Hollywood's first CGI main character, the villainous liquid metal T-1000 cyborg (Martin, 2014).

Toy Story (1995), was the first successful animated feature film from Pixar, and was also the first all-CGI animated feature film (Vreeswijk, 2012).

In *The Lord of the Rings* trilogy (2001, 2002 and 2003), a combination of motion-capture performance and key-frame techniques brought to life the main digital character Gollum (Dirks, 2009) by using a motion-capture suit (with reflective sensors) and recording the movements of actor Andy Serkis.

CGI use has grown exponentially and hand-in-hand with the increasing size of the film's budget it occupies. *Sky Captain and the World of Tomorrow* (2004) was the first big-budget feature to use only "virtual" CGI back lot sets. Actors Jude Law, Gwyneth Paltrow, and Angelina Jolie were filmed in front of blue screens; everything else was added in post-production (Dirks, 2009).

More an "event" than a movie, James Cameron's *Avatar* (2009) ushered in a new era of CGI. Many believe that *Avatar*, a largely computer-generated, 3D film—and the first film in history to earn nearly $3 billion worldwide—changed the movie-going experience (Muñoz, 2010). New technologies used in the film included CGI Performance Capture techniques for facial expressions, the Fusion Camera System for 3D shooting, and the Simul-Cam for blending real-time shoots with CGI characters and environments (Jones, 2012).

One of the greatest obstacles to CGI has been effectively capturing facial expressions. In order to overcome this hurdle, Cameron built a technology he dreamed up in 1995, a tiny camera on the front of a helmet that was able to "track every facial movement, from darting eyes and twitching noses to furrowing eyebrows and the tricky interaction of jaw, lips, teeth and tongue" (Thompson, 2010).

Whereas, the original 1977 *Star Wars: A New Hope* had an unprecedented (for its time) number of S/VFX shots (360) conjured up by ILM, the 2020 Oscar nominations for Best Visual Effects continue the trend in VFX-heavy Hollywood feature films—each a masterwork in computer generated imagery. Disney's three entrants into the award category: *Avengers: Endgame* has nearly 2,500 VFX shots; *The Lion King*, a CGI animation is comprised of 1,490 VFX shots; while *Star Wars: The Rise of Skywalker* had the CGI recreation of Carrie Fisher's character (following the actor's death) from previously shot footage adding to the 1,900 total VFX shots in the film (Hogg, 2020). The other two VFX Oscar contenders include Paramount's *The Irishman* with 1,750 VFX shots, mostly for de-aging the main characters, and Universal Picture's *1917* which was edited to look like a single continuous shot with the VFX teams having a hand in shaping an estimated 91% of the film (Giardia, 2020). With the VFX team at MPC Films so heavily involved, little surprise then that *1917* won the 2020 Oscars for both Best Cinematography (Rodger Deakins) and Best Visual Effects (Guillaume Rocheron, Greg Butler and Dominic Tuohy).

Outside of the Hollywood film industry, the Hindi-language film, *Ra One* (2011), about a gaming programmer who creates an indestructible character who escapes the game and kills its creator, amassed 3,500 VFX shots, while the 2012 Bollywood action-comedy film, *Son of Saadar* had 3,800 VFX shots, and the Indian epic action film, *Baahubali: The Beginning* (2015), currently holds the record with over 4,500 VFX shots (VFXMovies, n.d.). These films illustrate the appeal and global trend in producing VFX enhanced storytelling as well as advances in the digital tools used to create them.

Continued advances in motion-capture and the ability to provide realistic facial expressions to CGI characters captured from the actor's performance were demonstrated to great emotional effect by actor Andy Serkis who did the motion-capture performances for Caesar in the *Planet of the Apes* trilogy (2011, 2014 and 2017). In the last film of the trilogy, *War for the Planet of the Apes* (2017), VFX supervisor Dan Lemmon from Weta Digital, stated that "[every] time we reviewed shots with Caesar, we had Andy Serkis' performance side-by-side with Caesar to get the emotion right" (Verhoeven, 2017). Perhaps the most anticipated film of 2019, *Avengers: Endgame*, presented challenges for the VFX teams working to deploying new technology for *Endgame's* 2,500 VFX shots. To accomplish this, WETA Digital employed AI algorithms and machine learning to create "Deep Shape," a new tool, according to VFX Supervisor Matt Aitkens, allowing WETA to add more fine details to facial performance of such characters as Thanos (Josh Brolin) and Professor Hulk (Mark Ruffalo) among others (WETA Digital, 2019). Deep Shape is also one of the reasons that Ang Lee's *Gemini Man* (2019) possibly has the industry's most believable digital human in Will Smith's cinematic clone. "[it's] 50-year-old Will Smith (Henry) fighting 23-year-old Will Smith (Junior): aging assassin vs. his younger clone assassin.

Of course, the unstated goal of the VFX industry has always been to make a photorealistic, CGI character so realistic that the audience can't tell the difference between the CGI character and a real one (Media Insider, 2018).

Multichannel Sound

Sound plays a crucial role in making any movie experience memorable. Perhaps none more so than the 2018 Oscar nominated *Dunkirk* (2017), in which sound editor Richard King combines the abstract and experimental score of Hans Zimmer with real-world soundscapes to create a rich, dense, and immersive experience (Andersen, 2017).

Developed by Lucasfilm and named after *THX1138* (George Lucas' 1971 debut feature film), THX is a standardization system that strives to "reproduce the acoustics and ambience of the movie studio, allowing audiences to enjoy a movie's sound effects, score, dialogue, and visual presentation with the clarity and detail of the final mastering session" (THX, 2020).

At the time of THX's initial development in the early 1980s, most of the cinemas in the U.S. had not been updated since World War II. Projected images looked shoddy, and the sound was crackly and flat. "All the work and money that Hollywood poured into making movies look and sound amazing was being lost in these dilapidated theaters" (Denison, 2013).

Even if movies were not being screened in THX-certified theaters, the technical standards set by THX illustrated just how good the movie-going experience could be and drove up the quality of projected images and sound in all movie theaters. THX was introduced in 1983 with *Star Wars Episode VI: Return of the Jedi* and quickly spread across the industry. To be a THX Certified Cinema, movie theaters must meet the standards of best practices for architectural design, acoustics, sound isolation, and audio-visual equipment performance (THX, 2020). By 2017, there were over 5,000 THX certified theaters worldwide (Dcinema Today, 2017). In 2016, it was acquired by Razer, the gaming peripherals and laptop maker dual-headquartered in Irving, CA and Singapore. According to Razer CEO Min-Liang Tan, keen to establish the company as "an entertainment powerhouse [across] music, movies, and games" (Savov, 2016). Since its founding, THX certification efforts were broadened to include consumer electronics, automotive sound systems, and live performances.

While THX set the standards, Dolby Digital 5.1 Surround Sound is one of the leading audio delivery technologies in the cinema industry. In the 1970s, Dolby Laboratories introduced Dolby noise reduction (removing hiss from magnetic and optical tracks) and Dolby Stereo—a highly practical 35mm stereo optical release print format that fit the new multichannel soundtrack into the same space on the print occupied by the traditional mono track (Hull, 1999).

Dolby's unique quadraphonic matrixed audio technique allows for the encoding of four channels of information (left, center, right and surround) on just two physical tracks on movie prints (Karagosian & Lochen, 2003). The Dolby stereo optical format proved so practical that today there are tens of thousands of cinemas worldwide equipped with Dolby processors (Hull, 1999).

By the late 1980s, Dolby 5.1 was introduced as the cinematic audio configuration documented by various film industry groups as best satisfying the requirements for theatrical film presentation (Hull, 1999). Dolby 5.1 uses "five discrete full-range channels—left, center, right, left surround, and right surround—plus a… low-frequency [effects] channel" (Dolby, 2010a). Because this low-frequency effects (LFE) channel is felt more than heard, and because it needs only one-tenth the bandwidth of the other five channels, it is refered to as a ".1" channel (Hull, 1999).

Dolby also offers Dolby Digital Surround EX, a technology developed in partnership with Lucasfilm's THX that places a speaker behind the audience to allow for a "fuller, more realistic sound for increased dramatic effect in the theatre" (Dolby, 2010b).

Dolby Surround 7.1 is the newest cinema audio format developed to provide more depth and realism to the cinema experience. By resurrecting the full range left extra, right extra speakers of the earlier Todd-AO 70mm magnetic format, but now calling them left center and right center, Dolby Surround 7.1 improves the spatial dimension of soundtracks and enhances audio definition thereby providing full-featured audio that better matches the visual impact on the screen.

Film's Slow Fade to Digital

With the rise of CGI-intensive storylines and a desire to cut costs, celluloid film is quickly becoming an endangered medium for making movies as more filmmakers use digital cinema cameras capable of creating high-quality images. "While the debate has raged over whether or not film is dead, ARRI, Panavision, and Aaton quietly ceased production of film cameras in 2011 to focus exclusively on design and manufacture of digital cameras" (Kaufman, 2011).

But film is not dead—at least, not yet. Although there were only 39 movies released domestically in 2014 that were shot on 35mm film, there was an uptick to 64 in 2015. But since 2016, the number of movies originating on 35mm film (in whole or in part) has not risen above 31 titles (Rizov, 2020). After the banner year for film in 2015, Steve Bellamy, the president of Motion Picture and Entertainment for Kodak, wrote in an open letter, "There have been so many people instrumental in keeping film healthy and vibrant. Steven Spielberg, JJ Abrams and… [generations] of filmmakers will owe a debt of gratitude to Christopher Nolan and Quentin Tarantino for what they did with the studios. Again, just a banner year for film!" (cited in Fleming Jr., 2016). It is evident from the more or less constant number of movies originating in film over the past few years that these films are in "an increasingly limited number of categories: auteur films by directors too old or stubborn to change and with the clout to follow through on that; period pieces; and enormous blockbusters" with big budgets (Rizov, 2018).

Digital cinema only came on the scene in 2002 with *Star Wars Episode II: Attack of the Clones* being the first full-on high definition digital release. In 2013, only four of the nine films nominated for Best Picture at the Academy Awards were shot on film, and *The Wolf of Wall Street* (2014) was the first to be distributed exclusively digitally (Stray Angel Films, 2014). At the 2018 Oscars, of the ten nominations for Best Picture, seven were shot digitally (ARRI Alexa XT and Alexa Mini), including *The Shape of Water* which won Best Picture. For the 2020 Best Picture Oscars, four films were shot digitally and five on film (35mm or 70mm). *The Irishman* was at least partially shot using the Alexa Mini and the RED Helium, thus giving film a 52% to 48% lead in the "film vs. digital" comparison (Mendelovich, 2020). However, the surprise winner of the 2020 Best Picture Oscar (and three other Oscar nods for Directing, Original Screenplay, and International Feature Film) was writer, director and producer Bong Joon-Ho's *Parasite*, shot entirely digital using the ARRI Alexa 65 camera.

Although visual aesthetics are important in cinematic storytelling, so too is budget. Digital presents a significant savings for low-budget and independent filmmakers. Production costs using digital cameras and non-linear editing are a fraction of the costs of film production.

Maverick filmmaker, Steven Soderbergh, has a habit of being an iconoclast. His latest film, *High Flying Bird* (2019; production budget $2 million) was released on Netflix and is Soderbergh's second smartphone film shot entirely with an iPhone. In 2018, Soderbergh shot *Unsane* in only two weeks using an iPhone 7+ with a $1.5 million production budget generating box office receipts of $14.3 million. "[Just] prior to the release of *Unsane*, Soderbergh declared phones 'the future' of filmmaking… 'You really don't need much more than what's in your pocket and some software, and off you go'" (Lindbergh, 2019).

It is obvious that digital acquisition "offers many economic, environmental, and practical benefits" (Maltz, 2014). For the most part, this transition has been a boon for filmmakers, but as "born digital" productions proliferate, a huge new headache emerges: preservation (Maltz, 2014). This is why the Academy of Motion Picture Arts and Sciences sounded a clarion call in 2007 over the issue of digital motion picture data longevity in the major Hollywood studios. In their report, titled *The Digital Dilemma*, the Academy concluded that, although digital technologies provide tremendous benefits, they do not guarantee long-term access to digital data compared to traditional filmmaking using motion picture film stock. Digital technologies make it easier to create motion pictures, but the resulting digital data is much harder to preserve (Science & Technology Council, 2007).

The Digital Dilemma 2, the Academy's 2012 update to this initial examination of digital media archiving, focused on the new challenge of maintaining long-term archives of digitally originated features created by the burgeoning numbers of independent and documentary filmmakers (Science & Technology Council, 2012). Digital preservation issues notwithstanding, film preservation has always been of concern. According to the Library of Congress Film Preservation Study, less than half of the feature films made in the United States before 1950 and less than 20% from the 1920s are still around. Even films made after 1950 face danger from threats such as color-fading, vinegar syndrome, shrinkage, and soundtrack deterioration (Melville & Simmon,

1993). However, the "ephemeral" nature of digital material means that "digital decay" is as much a threat to films that were "born digital" as time and the elements are to nitrate and celluloid films of the past. The good news is that there is a lot of work being done to raise awareness of the risk of digital decay generally and to try to reduce its occurrence (Maltz, 2014).

Along with digital acquisition technology, new digital distribution platforms have emerged making it easier for independent filmmakers to connect their films with target audiences and revenue streams (through video-on-demand, pay-per-view, and online distribution). However, these platforms have not yet proven themselves when it comes to archiving and preservation (Science & Technology Council, 2012).

Digital 3D

The first digital 3D film released was Disney's *Chicken Little* (2005), shown on Disney Digital 3D (PR Newswire, 2005). Dolby Laboratories outfitted about 100 theaters in the 25 top markets with Dolby Digital Cinema systems in order to screen the film. The idea of actually shooting live-action movies in digital 3D did not become a reality until the creation of the Fusion Camera, a collaborative invention by director James Cameron and Vince Pace (*Hollywood Reporter*, 2005). The camera "fuses" two Sony HDC-F950 HD cameras "2½ inches apart to mimic the stereoscopic separation of human eyes" (Thompson, 2010). The camera was used to film 2008's *Journey to the Center of the Earth* and 2010's *Tron Legacy*.

Cameron used a modified version of the Fusion Camera to shoot 2009's blockbuster *Avatar*. The altered Fusion allows the "director to view actors within a computer-generated virtual environment, even as they are working on a 'performance-capture' set that may have little apparent relationship to what appears on the screen" (Cieply, 2010).

Another breakthrough technology born from *Avatar* is the swing camera. For a largely animated world such as the one portrayed in the film, the actors must perform through a process called motion capture which records 360 degrees of a performance, but with the added disadvantage that the actors do not know where the camera will be (Thompson, 2010). Likewise, in the past, the director had to choose the shots desired

once the filming was completed. Cameron tasked virtual-production supervisor, Glenn Derry, with creating the swing camera, which "has no lens at all, only an LCD screen and markers that record its position and orientation within the volume [the physical set space] relative to the actors" (Thompson, 2010). An effects switcher feeds back low-resolution CG images of the virtual world to the swing camera allowing the director to move around shooting the actors photographically or even capturing other camera angles on the empty stage as the footage plays back (Thompson, 2010).

Figure 10.12

Vince Pace & James Cameron with Fusion 3D Digital Cinema Camera

Source: CAMERON | PACE Group
Photo credit: Marissa Roth

When Hollywood studio executives saw the $2.8 billion worldwide gross receipts for *Avatar* (2009), they took notice of the film's ground-breaking use of 3D technology; and when Tim Burton's 2010 *Alice in Wonderland*—originally shot in 2D—had box office receipts reach $1 billion thanks to a meticulous 3D conversion and audiences willing to pay an extra $3.50 "premium" up-charge for a 3D experience, the film industry decided that 3D was going to be their savior. Fox, Paramount, Disney, and Universal collectively shelled out $700 million to help equip domestic theaters with new projectors, and the number of 3D releases jumped from 20 in 2009 to 45 in 2011. However, by 2013 only 33 films were released domestically in 3D (Renner, 2013), and by 2015, domestic 3D film releases fell to only 15 (Renner, 2015). 2016 saw a slight up-tick of 37 films released in 3D (Renner, 2016), but unfortunately, the downward trend continued in subsequent years with 2018 reporting merely eight 3D film releases and 2019 marking a slight increase of 12 domestic 3D films (Renner, 2019). According to *The Numbers* (2020b), most of the over 450 3D films released since *Avatar* (2009) were post-production conversions while 206 were shot or rendered (animation) in 3D. Projecting into the mid-2020's, only a few studios have announced plans to release 3D films domestically, with three on the books for 2020, James Cameron's much anticipated *Avatar 2* slated for a 2021 release with *Avatar 3*, *Avatar 4*, and *Avatar 5* to follow in 2023, 2025, and 2027 respectively (Renner, 2020).

What accounts for this initially up, but mostly downward trend in North American 3D film releases? Does the success of landmark 3D films like *Avatar* (2009), *Hugo* (2011), *Gravity* (2013), *The Martian* (2015), and Disney's 3D release of *Star Wars: The Last Jedi* (2017), represent true innovations in cinematic storytelling, or has 3D been subsumed by moneymaking gimmickry? Perhaps, the answer is a bit of both. "Some movies use the technology to create more wonderfully immersive scenes, but others do little with the technology or even make a movie worse" (Lubin, 2014). In part, it could also be that the novelty has worn off with film audiences tired of having their expectations go unfulfilled. Citing the analysis of 1,000 random moviegoers' Tweets, Phillip Agnew, an analyst with Brandwatch (a social media monitoring and analytics company), observed that, "Before taking their seats, the majority of cinema goers [600 out of 1,000] were positive about 3D films" (Agnew, 2015). However, comments made after screening the 3D movie showed that viewers were mostly disappointed. "Consumers expecting to see the 'future of cinema' were instead paying more for a blurrier viewing experience, and in many cases, headaches" (Agnew, 2015).

Global 3D box office revenues of $6.7 billion in 2018 represent a decrease of 20% compared to 2017, and comprise only 16% of total global box office revenues in 2018 (MPAA, 2019). Why then does Hollywood continue to make 3D movies if domestic revenues and audience attendance has collapsed and globally 3D are trending downward? The answer is quite simple, 3D cinema screens in the Asia Pacific region, Europe, the Middle East, and Africa represent nearly 23 times the number of 3D screens in North and South America combined (MPAA, 2019). Despite the drop in global revenue, the Asia Pacific region saw the lowest decline (-14%) even as the Asia Pacific region continues to have the highest number of 3D digital screens versus 2D digital screens at a ratio of 3:4 (MPAA, 2019). Hollywood studios have taken notice and now release 3D versions of some big films exclusively in China (Epstein, 2017).

Digital Theater Conversions

In 1999, *Star Wars I: The Phantom Menace* was the first feature film projected digitally in a commercial cinema—although there were very few theaters capable of digital projection at the time.

January 2012, marked the film industry's crossover point when digital theater projection surpassed 35mm. Nearly 90,000 movie theaters were converted to digital by the end of 2012 (Hancock, 2014).

The Motion Picture Association of America's 2018 *THEME Report* (2019), observed that, in 2018, the number of digital screens in the United States now account for 99% of all U.S. cinema screens with the majority of the screens located at venues with five or more screens. Reflecting the broader downward trend in 3D film production, the number of 3D screens remained roughly the same as in 2017 at approximately 15,500 screens reflecting approximately 39% of all screens in the U.S. (MPAA, 2019). Global cinema screens increased 7% reaching nearly 190,000 due in large measure to continued double digit growth in the Asia Pacific region (13%). By the end of 2018, approximately 97% of the world's cinema screens were digital, however, approximately 5,000 analog screens in the Asia Pacific have yet to be converted (MPAA, 2019). The number of 3D digital screens increased globally by 13% and now account for 58% of all digital screens with all regions except U.S./Canada seeing gains (MPAA, 2019).

With 97% of global screens digitally converted, and the U.S. market at near 100% digital, the global digital theater conversion is in its endgame. Exhibitors and technology manufacturers have turned their attention to developing other technologies to enhance the audience experience, drive innovation, and, hopefully, new revenues. The most visible (and audible) technologies being explored and developed include laser projection and immersive audio (Hancock, 2015).

Current Status

As part of the second phase of the digital cinema technology rollout, cinema operators turned their attention to the electronic distribution of movies. In 2007, hard drive delivery prices were in the $150–$250 range and represented a significant savings over the four-figure print costs that had come before the push for digital projection (Pahle, 2019). Still, studio executives thought there had to be a cheaper and more efficient way. In 2013, the Digital Cinema Distribution Coalition (DCDC) was formed by "AMC Theatres, Cinemark Theatres, Regal Entertainment Group, Universal Pictures and Warner Bros. to provide the industry with theatrical digital delivery services across North America. It [would be] capable of supporting feature, promotional, pre-show and live content distribution into theaters" (DCDC, 2016). Randy Blotky, CEO of DCDC, described the movie distribution network as a "... 'smart pipe' made up of sophisticated electronics, software and hardware, including satellites, high-speed terrestrial links, with hard drives used as backup" (Blotky, 2014). "DCDC pays for all of the equipment that goes in the theatres [sic], we maintain all of that equipment, and we install all of that equipment at no cost to exhibition" (Blotky, cited it Fuchs, 2016). To recover their costs, DCDC charges fees for delivery to both the theater and content providers that is "way less expensive than for delivering hard drives and physical media to the theatres [sic]" (Blotky, cited it Fuchs, 2016). In 2019, the DCDC served more than 32,000 theaters representing 80% of U.S. movie screens (Pahle, 2019). Through 2025, DCDC's priority is to optimize its service in the U.S. and extended their service to Canada. Also, DCDC would "...love to find ways to create the same kind of partnerships that gave birth to DCDC elsewhere" (cited in Pahle, 2019) with hopes of finding partners in Europe, Latin America, Australia, and Asia.

Hopeful signs of the expansion of digital cinema technology are also evident in India as "the old, celluloid prints that where physically ferried to cinema halls have disappeared from India's 12,000 screens.... Instead, the digital file of the film is downloaded by satellite or other means to those cinema theatres [sic] that have paid for it.... Also, [electronic] distribution of d-cinema is simpler, faster, cheaper, and piracy can be better controlled" (BW Online Bureau, 2015).

Global box office revenues for all films in 2018 reached $41.1 billion, only a 1% increase from 2017 (MPAA, 2019). Domestic U.S. and Canadian box office revenues grew 7% over 2017's intake, but international box office receipts were down 2% and accounted for 71% of total global box office revenue (MPAA, 2019). In 2018, U.S. gross domestic revenues were $11.9 billion, up 7% over 2017 and 4% from 2016, while ticket sales were up 5% at just over 1.3 billion with an average ticket price of $9.11 (and has remained the same into 2020) representing only a 2% increase over 2017 and less than the 3% cost of inflation (MPAA, 2019). Global box office revenues—excluding U.S. and Canadian—across all regions increased by 5% compared to 2017 (MPAA, 2019). The strong growth in the Asia Pacific was, again, driven by China with a 12% increase in box office revenue. However, Europe, the Middle Ease & Africa box office receipts decreased by 3% and was mostly driven by a 14% drop in revenues in Germany and Russia. Latin America also dropped 22% in box office revenues with Brazil reporting a 22% drop while the depreciation of currencies in Argentina (-41%) and Mexico (-2%) also contributed to the overall decline in regional box office receipts (MPAA, 2019).

The MPAA's Classification and Rating Administration (CARA) issue ratings (G, PG, PG-13, R, NC-17) on the basis of graphic sex or violence, drug use, and dark or adult themes prior to a movie's release in order to provide parents with guidelines for deciding what films they would allow their children to watch (CARA, 2013). Reviewing the market share for each MPAA rating for films released in 2019, The Numbers (2020c) reports 111 PG-13 rated films topped the market with $4.8 billion in gross revenues and 530.6 million tickets sold. If combined with PG (66) and G

(13) rated films, these films represent nearly 75% of all domestic releases and $8.46 billion total gross revenues while R rated films (167) at 24% and unrated films (341) making up less than 1% of total films released and combined gross box office revenues of $3.26 billion (The Numbers 2020c). When examining the top grossing domestic films for the last 10 years, eight were rated PG-13, and one each for PG and G (The Numbers, 2020d).

Not surprisingly, nine of the top 10 earning films for the 2010-2019 decade were produced by Disney; only *Harry Potter and the Deathly Hallows: Part II* (2011), produced by Warner Brothers, kept Disney from owning all 10 of the top earning films of the 2010s (The Numbers 2020d). Nine of the top 10 grossing films of 2019 were in the Action/Adventure genre while one (*Joker*, Warner Brothers) was a Thriller/Suspense genre. In fact, the Action/Adventure genre dominated box office receipts in 2019 with 104 films accruing gross revenues of $6.8 billion, while Thriller/Suspense genre (55 films) accounted for $1.2 billion, and Drama (228 films) brought in $1.4 billion in gross box office revenues (The Numbers, 2020e). Horror films (39) rounded out the top five genres with $748 million while Westerns (4 films) occupied last place with a little over $2 million in gross revenue (The Numbers 2020e).

When it comes to a film's profitability, nothing is looked forward to by film studios as much as getting an "Oscar Bounce," the ticket sales boost an Academy Award nomination brings to the film's bottom line. For films like Danny Boyle's *Slumdog Millionaire* (2008), the 10 Oscar nominations received was the difference between making $30 million (with a $15 million production budget) before the nominations, and $377 million—well over two-thirds after Academy recognition (Mondello, 2020). *Parasite* (2019), 2020's Best Picture Award winner, saw its box office receipt jump by 18% after the nominations and nearly doubling gross box office receipts since its release in 2019 (Boxoffice Pro, 2020). This accomplishment is even more amazing since *Parasite* also became available on Amazon and iTunes the day after it won 4 Oscars—quickly becoming the number one rental on both streaming platforms. With the advent of Netflix, Amazon, and iTunes (among others), the decade of the 2010s brought on a much shorter pipeline from a film's theater run to on-demand streaming that has made the Oscar Bounce much less impactful. The 2019 films, *The Irishman* and *Marriage Story*—both produced by Netflix—were released in early November to limited theatrical release and began streaming on Netflix the following month, even before garnering 10 and 6 Oscar nominations, respectively, in 2020. In fact, movies released by Netflix have garnered a total of 54 Oscar nominations since 2014 with 24 (two more than Disney) in 2020 alone (Clark, 2020). Hollywood took a dim view of Netflix's decision to release its films to subscribers online the same day as their theatrical release, nor were they happy with the small number of theaters that Netflix does allow to screen its movies. However, "[as] Netflix's impact on the world of cinema becomes increasingly undeniable, the younger and more diverse film academy is no longer shunning the streaming service" (Epstein, 2020). "Though [Netflix films] might not be in movie theaters [very long], more people are seeing these films than if they were given a traditional theatrical release" (Epstein, 2020).

Factors to Watch

Of all of the current trends in digital cinema camera designs, perhaps one of the most interesting is the shift toward larger sensors. Full-frame sensors (36mm x 24mm) have been used to shoot video in DSLR cameras since the Canon 5D Mark II was introduced in September 2008. Large format sensors, however, are a completely different matter. As Rich Lackey, writing for *Digital Cinema Demystified* observed, "It is clear that more is more when it comes to the latest high-end cinema camera technology. More pixels, more dynamic range, larger sensors, and more data. Enough will never be enough… however there are points at which the technology changes so significantly that a very definite generational evolution takes place" (Lackey, 2015b). Since the introduction of the full-frame sensor, the quest for higher resolution, higher dynamic range and better low light performance in digital cinema cameras inevitably resulted in a "devil's choice" in sensor design trade-offs. "High dynamic range and sensitivity require larger photosites, and simple mathematics dictate that this places limits on overall resolution for a given size sensor… unless, of course, you increase the size of the sensor" (Lackey, 2015a). Answering the call for larger sensors that can provide more pixels as well as greater dynamic range, the Phantom 65 was introduced with a 4K (4096 x 2440) sensor (52.1mm x 30.5mm), while ARRI Alexa 65, used to shoot 2015's Best Cinematography Oscar winner, *The Revenant*, and 2020's Best Picture Oscar award, *Parasite*, has a 54.12mm x 25.59mm sensor yielding 6.5K (6560 x 3100). The first feature film shot in 8K (8192 x 4320) was 2017's *Guardians of the Galaxy Vol. 2* using the RED Weapon with a 40.96mm x 21.6mm sensor. Likewise, five of the 42 initial award nominees for Best Cinematography in 2020 were shot in full format 8K using the RED Monstro camera. Of course, large frame sensors require optics that can cover the sensor, and for this, the lens manufacturers are a critical component.

If 4K and 8K-plus image resolutions seem inevitable, then apparently, so are high frame rates (HFR) for digital acquisition and screening. Since the introduction of synchronous sound in movies, the standard frame rate has been 24 frames per second (fps). However, this was primarily a financial decision. Using more frames meant more costs for film and processing, and studio executives found that 24-fps was the cheapest, minimally acceptable frame rate they could use for showing sound-synchronous movies with relatively smooth motion. Nearly a century later, it is still the standard. However, since the late 1920s, projectors have been using shutter systems that show the same frame two or three times to boost the overall frame rate and reduce the "flicker" an audience would otherwise experience. But even this is not enough to keep up with the fast motion of action movies and sweeping camera movements or fast panning. The visual artifacts and motion-blur that's become part of conventional cinema is visually accentuated in 3D because our eyes are working extra hard to focus on moving objects.

For most of the digital 3D movies already released, the current generation of digital projectors can operate at higher frame rates with a mere software update, but theaters are still showing 3D movies at 24-fps. To compensate for any distractions that can result from barely perceivable flashing due to the progression of frames (causing eyestrain for many viewers), each frame image is shown three times per eye. Called "triple flashing," the actual frame rate is tripled, resulting in viewers receive 72 frames per second per eye for a total frame rate of 144-fps. Audiences watching a 3D film produced at 48-fps would see the same frame flashed twice per second (called, "double flashing"). This frame rate results in each eye seeing 96-fps and 192-fps overall, nearly eliminating all flicker and increasing image clarity. Any 3D films produced and screened at 60-fps, and double-flashed for each eye, would result in movie-goers seeing a 3D film at an ultra-smooth 240-fps.

As a leading proponent of HFR, director Peter Jackson justified the release of his *The Hobbit* film trilogy (2012, 2013, and 2014) in a 2014 interview, stating, "science tells us that the human eye stops seeing individual pictures at about 55 fps. Therefore, shooting at 48 fps gives you much more of an illusion of real life. The reduced motion blur on each frame increases sharpness and gives the movie the look of having been shot in 65mm or IMAX. One of the biggest advantages is the fact that your eye is seeing twice the number of images each second, giving the movie a wonderful immersive quality. It makes the 3D experience much more gentle and hugely reduces eyestrain. Much of what makes 3D viewing uncomfortable for some people is the fact that each eye is processing a lot of strobing, blur and flicker. This all but disappears in HFR 3D" (Jackson, 2014).

However, HFR film has its detractors who opine that 24-fps films deliver a depth, grain and tone that is unique to the aesthetic experience and not possible to recreate with digital video—this lack of "graininess" is often jarring and uncomfortable to first-time viewers of HFR 3D visual images. Such adjectives as "blurry," or "Hyper-real," "plastic-y," "weirdly sped-up," or the more derisive "Soap Opera Effect" are thrown around a lot. This appears to be a red herring argument since 99% of all cinema screens in the U.S. are now projected digitally and the film grain look has become a post-production aesthetic. Confronted with such criticisms leveled at *The Hobbit: An Unexpected Journey* (2012), Peter Jackson continued to use the 48-fps 3D digital format for his two other *Hobbit* films, but added lens filters and post-production work to relax and blur the imagery, "hoping it wouldn't look so painfully precise" (Engber, 2016).

For cinematic futurist like Peter Jackson, James Cameron (who is using HFR 3D for his anxiously anticipated *Avatar* sequels), and Ang Lee, the rejection of HFR digital technology seems bizarre. As James Cameron himself said, "I think [HFR] is a tool, not a format… I think it's something you want to weave in and out and use when it soothes the eyes, especially in 3D during panning, movements that [create] artifacts that I find very bothersome. I want to get rid of that stuff, and you can do it through high frame rates" (cited in Giardina, 2016). Perhaps then, Ang Lee's *Billy Lynn's Long Halftime Walk* —shot in 3D with 4K resolution at 120-fps—breaks as much ground in the storytelling front as in the technical. As Daniel Engber, writing for *Slate Magazine*, noted in his extensive review of Lee's film, "Most exciting is the way Lee modulates the frame rate from one scene to the next. At certain points he revs the footage up to 120 fps, while at [other times] the movie slides toward more familiar speeds…. If there's any future for HFR in Hollywood, this must be it—not as a hardware upgrade on the endless path to total cinema, but as a tool that can be torqued to fill a need" (Engber, 2016).

An innovation that has been lurking on the fringes of cinematography could soon become a potential game-changer. Light-field technology—also referred to as computational or plenoptic imaging and better understood as volumetric capture—is expected to be a "disruptive" technology for cinematic virtual reality (VR) filmmakers. Whereas, ordinary cameras are capable of receiving 3D light and focusing this on an image sensor to create a 2D image, a plenoptic camera samples the 4D light field on its sensor by inserting a microlens array between the sensor and main lens (Ng, et al., 2005). Not only does this effectively yield 3D images with a single lens, but the creative opportunities of light-field technology enable such typical production tasks as refocusing, virtual view rendering, shifting focal plains, and dolly zoom effects all capable of being accomplished during post-production. Companies including Raytrix and Lytro are currently at the forefront of light field photography and videography. In late 2015, "Lytro announced… the *Lytro Immerge*… a futuristic-looking sphere with five rings of light field cameras and sensors to capture the entire light field volume of a scene. The resulting video [was intended to] be compatible with major virtual reality platforms and headsets such as *Oculus,* and allow viewers to look around anywhere from the *Immerge's* fixed position, providing an immersive, 360 degree live-action experience" (Light-Field Forum, 2015). By February 2017, having raised an additional $60 million to continue developing the light-field technology, the company decided to switch to a flat (planar) capture design (Lang, 2017). With this approach, capturing a 360-degree view requires the camera to be rotated to individually shoot each side of an eventual pentagonal capture volume and then stitch the image together in post-production. The advantage of this is that the director and the crew can remain behind the camera and out of the shot throughout the production process.

In 2016, the cinematic world was introduced to the prototype Lytro Cinema Camera at the National Association of Broadcasters Convention in Las Vegas, NV. The Lytro Cinema Camera is large "…and unwieldy enough to remind DPs of the days when cameras and their operators were encased in refrigerator-sized sound blimps. But proponents insist the Lytro has the potential to change cinematography as we know it… It produces vast amounts of data, allowing the generation of thousands of synthetic points of view. With the resulting information, filmmakers can manipulate a range of image characteristics, including frame rate, aperture, focal length, and focus" (Heuring, 2016). The Lytro Cinema Camera has 755 RAW megapixels of resolution per frame utilizing a high resolution active scanning system, up to 16 stops of dynamic range and can shoot at up to 300-fps and generates a data stream of 300Gb per second while processing takes place in the cloud where Google spools up thousands of CPUs to compute each thing you do, while you work with real-time proxies (Sanyal, 2016). But independent filmmakers will probably not be using a Lytro Cinema Camera any time soon. Rental packages for the Lytro Cinema Camera are reported to start at $125K.

Only a few years ago terms such as "immersive", "experiential," "volumetric," and "virtual reality" (VR) were on the outer fringe of filmmaking. Today, however, predictions are that Cinematic VR, consisting of high-quality 360° 3D video experiences, ambisonic audio, and possibly with interactive elements (Ogden, 2019), will change the face of cinema in the next decade—but only if content keeps up with the advances in technology. It is projected that, by 2021, augmented reality (AR) and VR spending will increase from 2017's $11.4 billion to $215 billion world-wide (International Data Corporation, 2017). VR is already being heavily promoted by the tech giants, with Facebook's Oculus and Microsoft launching new headsets they hope will ensure the format goes mainstream. As if to emphasize this point, in early 2017, IMAX opened its first VR cinema in Los Angeles (Borruto, 2017), while the leading film festivals—including Sundance, Cannes, Venice, and Tribeca—now have sections dedicated to recognizing ground-breaking work in VR. Patric Palm, the CEO and co-founder of *Favro*, a collaboration application used by VR and AR studios, sees Cinematic VR technology pushing the boundaries of filmmaking, stating that "[when] you're making a film, as a filmmaker, you are in the driver's seat of what's going to happen. You control everything. With a 360 movie, you have a little bit less of that because where the person's going to look is going to be a little different. More interactive experiences, where the consumer is in the driver's seat, are the single most interesting thing. Right now, the conversation is all about technology. What will happen when brilliant creative minds start to get interested in this space and they're trying to explore what kind of story we can do in this medium? There's a lot of money stepping into the VR and AR space, but a lot of creative talent is moving in this direction, too" (Volpe, 2017). (For more on VR and AR, see Chapter 15.)

In September 2016, 20th Century Fox announced that it used the artificial intelligence (AI) capabilities of IBM's Watson to make a trailer for its 2016 sci-fi thriller *Morgan*. Not surprisingly, everyone was eager to see how well Watson could complete such a creative and specialized task (Bludov, 2017). IBM research scientists "taught" Watson about horror movie trailers by feeding it 100 horror trailers, cut into scenes. Watson analyzed the data and "learned" what makes a horror trailer scary. The scientists then uploaded the entire 90-minute film and Watson "instantly zeroed in on 10 scenes totaling six minutes of footage" (Kaufman, 2017). To everyone's surprise, Watson did remarkably well. However, it's important to note that a real person did the actual trailer editing using the scenes that were selected by Watson, so AI did not actually "cut" the trailer—a human was still needed to do that.

Figure 10.13

Morgan (2016), First Movie Trailer Created by IBM's Watson AI System

Source: 20th Century FOX, Official Trailer

For the 2016 Sci-Fi London 48 Hour Film Challenge, director Oscar Sharp submitted, *Sunspring*, a short film written entirely by an AI system that named itself "Benjamin." Sharp and his longtime collaborator, Ross Goodwin, an AI researcher at New York University, wanted to see if they could get a computer to generate an original script. Benjamin is a Long Short-Term Memory recurrent neural network, a type of AI used for text recognition. To train Benjamin in screenplay writing, Goodwin fed the AI dozens of sci-fi scripts he found online. Benjamin learned to imitate the genre's structure and produced stage directions and well-formatted character lines (Newitz, 2016). Unfortunately, the script had an incoherent plot, quirky and enigmatic dialogue and almost impossible stage directions.

Jack Zhang, through his company Greenlight Essentials, has taken a different approach to AI script generation. Using what Zhang calls "augmented intelligence software" to analyze audience response data to help writers craft plot points and twists that connect with viewer demand (Busch, 2016). The patent-pending predictive analytics software helped write, *Impossible Things*, a feature-length screenplay Zhang describes as "the scariest and creepiest horror film out there" (Nolfi, 2016). Additionally, the AI software suggested a specific type of trailer including key scenes from the script to increase the likelihood that the target audience would like it. Greenlight Essentials produced the trailer to use in a Kickstarter campaign to fund the film.

Presently, there are more "misses" then there are "hits" with the use of AI. The very human talent of creativity, a specialty in the entertainment industry, appears safe for the foreseeable future. (Kaufman, 2017).

There is little doubt that emerging technologies will continue to have a profound impact on cinema's future. As Steven Poster, ASC and President of the International Cinematographers Guild stated, "Frankly I'm getting a little tired of saying we're in transition. I think we've done the transition and we're arriving at the place we're going to want to be for a while. We're finding out that software, hardware, and computing power have gotten to the point where it's no longer necessary to do the things we've always traditionally done… [and] as the tools get better, faster and less expensive… [what] it allows for is the image intent of the director and director of photography to be preserved in a way that we've never been able to control before" (Kaufman, 2014b). When discussing *Avatar*,

James Cameron stated that his goal is to render the technological presence of cinema invisible. "[Ideally], the technology is advanced enough to make itself go away. That's how it should work. All of the technology should wave its own wand and make itself disappear" (cited in Isaacs, 2013, P. 246).

Moviegoers of the future might look back on today's finest films as quaint, just as silent movies produced a century ago seem laughably imperfect today (Hart, 2012). Cinematographer David Stump, noting the positive changes brought about by the transition from analog to digital, states that "[the] really good thing that I didn't expect to see… is that the industry has learned how to learn again…. We had the same workflow… for making images for 100 years. Then we started getting all these digital cameras and workflows and… [now] we have accepted that learning new cameras and new ways of working are going to be a daily occurrence" (Kaufman, 2014a).

Gen Z

For Gen Z, the consumption of entertainment and the use of social media is an intrinsic part of their daily life rather than an occasional activity. Gen Z now consume TV shows and movies on their smartphones, tablets, and laptops for about 3.4 hours every day (Baron, 2019).

Globally, in 2018 the number of subscriptions to online video services such as Netflix and Amazon Prime increased by 27% and, for the first time, surpassed cable subscriptions (MPAA, 2019). Streaming services—and there are more than 140 online services providing movies and TV shows to U.S. consumers (MPAA, 2019)—provide Gen Z the opportunity to actively engage in the movie experience on their terms and in ways that are not conducive to sitting in a dark movie theater facing forward and watching a screen. But, for Gen Z, streaming and theatrical viewing can co-exist, it's not an either/or choice. "…movie-polling service PostTrak… found people ages 18 to 24 make up the largest segment of moviegoers" in the USA (Lindahl, 2019).

Most Gen Z moviegoers come to the cinema with friends to socialize before, during and after the film. They prefer to stay in groups and stay constantly connected to their larger web of friends through smartphones and social media apps. Typical movie theaters risk alienating Digital Natives just by asking them to turn off their electronic devices and sit quietly without interaction during the film's screening (Kobzeva-Pavlovic, 2015). Gen Z, like the movie going generations before them, "like to be told stories, stories that make a real impact and stories that they can pass on and talk about [with] their peers. This applies to all their media consumption" (Jones, 2014).

Sustainability

Film productions can also have a hefty carbon footprint. "The average film is estimated to produce 500 [tons] of CO_2 emissions (equivalent to running 108 cars for a year), but this scales up with its budget; a $50m film can produce 4,000 [tons of CO_2]" (Hoad, 2020). With a budget of $356 million, one can only guess what the CO_2 emissions were for *Avengers: End Game* (2019). But carbon dioxide emissions are only part of the global climate change equation, a benchmark two-year study by UCLA, released in 2006, also found that the film industry in the five-county Los Angeles, CA area produced 140,000 tons of ozone and diesel particulate emissions per year (Corbet & Turco, 2006). "But it's not just Hollywood. According to the British Academy of Film and Television Arts (BAFTA), one hour of UK television… produces 13 metric tons of carbon dioxide. This is about what the average American [generates] every year" (Geiger, 2019). When one takes into consideration that most of the big studio's "tent-pole" projects now have crews in the thousands, "from set-builders to [electricians], masseurs and makeup artists, to high-end caterers and server-hungry special effects farms—there is no reason to think Hollywood is any less resource-hungry these days" (Hoad, 2020) than they have been in the past.

Still, the film industry's constituency is mostly eco-friendly by nature and they want to do their part. In fact, the film production industry has made great strides since the 2006 UCLA report and have endeavored to incorporate sustainable initiatives and practices into the physical production process with the aim of reducing its environmental footprint. Early on, most measures to address environmental concerns were not a priority until a significant component of the film—such as a director, an A-list actor, or the producer(s)—demanded action. This was the case during the production of the natural-disaster movie, *The Day After Tomorrow* (2004), when the film's director, Roland Emmerich, sought to ensure that the production would not contribute to global warming by paying $200,000 to offset the carbon footprint of the film and by taking other measures to reduce its environmental impact (Aftab, 2007). Today, "[most] major film studios now have sustainability drives to encourage the use of clean energy on set and the recycling of production materials… Carbon-neutral productions, achieved through emissions offsetting, are now relatively common" (Hoad, 2020).

The American University's Center for Media and Social Impact, in 2009, released the first attempt to identify and define a code of environmental best practices for the film industry. The code emphasizes calculating energy use, consuming less where possible, reducing travel, and compensating for consumption by using carbon offsets (Engel & Buchanan, 2009). The report also provides checklists for filmmakers to use in assessing and reducing environmental impact throughout each phase of production. In 2010, with seed funding and support provided by several major Hollywood studios, the Producers Guild of America released the *Green Production Guide* "…to provide a new resource for film, television and commercial professionals looking to reduce the environmental impacts and carbon emissions of film and television production." At the 2020 Academy Awards, three of the ten nominees for Best Picture received EMA "Green Seal" recognition for their sustainable production practices—namely, *Joker* (2019, Warner Brothers, EMA Gold Seal), *Little Women* (2019, Sony, EMA Gold Seal), and *Jojo Rabbit* (2019, Fox, EMA Green Seal) (EMA, 2020).

Bibliography

Adorno, T. (1975). Culture industry reconsidered. *New German Critique*, (6), Fall. http://libcom.org/library/culture-industry-reconsidered-theodor-adorno.

Agnew, P. (2015, May 19). Research: The rapid rise and even more rapid fall of 3D movies. *Brandwatch Blog* https://www.brandwatch.com/2015/05/4-reasons-why-3d-films-have-failed/.

Aftab, K. (2007, November 16). Emission impossible: Why Hollywood is one of the worst polluters. *Independent*. https://www.independent.co.uk/arts-entertainment/films/features/emission-impossible-why-hollywood-is-one-of-the-worst-polluters-400493.html.

Andersen, A. (2017, August 2). Behind the spectacular sound of *Dunkirk* — With Richard King. *A Sound Effect*. https://www.asoundeffect.com/dunkirk-sound/.

Andrae, T. (1979). Adorno on film and mass culture: The culture industry reconsidered. *Jump Cut: A Review of Contemporary Media*, (20), 34-37. http://www.ejumpcut.org/archive/onlinessays/JC20folder/AdornoMassCult.html.

Bankston, D. 2005, April). The color space conundrum part two: Digital workflow. *American Cinematographer*, 86, (4). https://www.theasc.com/magazine/april05/conundrum2/index.html.

Baron, J. (2019, July 3). The key to Gen Z is video content. *Forbes*. https://www.forbes.com/sites/jessicabaron/2019/07/03/the-key-to-gen-z-is-video-content/#1d1ecde83484 .

Belton, J. (2013). *American cinema/American culture* (4th Edition). McGraw Hill.

Blotky, R. (2014 March 24). Special delivery: DCDC network promises to revolutionize cinema content distribution. *Film Journal International*. http://www.filmjournal.com/content/special-delivery-dcdc-network-promises-revolutionize-cinema-content-distribution.

Bludov, S. (2017, August 16). Is artificial intelligence poised to revolutionize Hollywood? *Medium.* , https://medium.com/dataart-media/is-artificial-intelligence-poised-to-revolutionize-hollywood-e088257705a3.

Bomboy, S. (2015, May 4). The day the Supreme Court killed Hollywood's studio system. *Constitution Daily* [Blog]. http://blog.constitutioncenter.org/2015/05/the-day-the-supreme-court-killed-hollywoods-studio-system/.

Bordwell, D., Staiger, J. & Thompson, K. (1985). *The classical Hollywood cinema: Film style & mode of production to 1960.* Columbia University Press.

Borruto, A. (2017, February 15). IMAX'S first virtual reality theater is transforming the film industry. *Resource.* http://resourcemagonline.com/2017/02/imaxs-first-virtual-reality-theater-is-transforming-the-film-industry/76083/.

Boxoffice Pro (2020). *Plus Report.* https://pulse.boxofficepro.com/year/2025.

Bradley, E.M. (2005). *The first Hollywood sound shorts, 1926-1931.* McFarland & Company.

Bredow, R. (2014, March). Refraction: Sleight of hand. *International Cinematographers Guild Magazine*, 85(03), p. 24 & 26.

Busch, A. (2016, August 25). Wave of the future? How a smart computer is helping to craft a horror film. *Deadline Hollywood*. http://deadline.com/2016/08/concourse-productivity-media-horror-film-impossible-things-smart-computer-1201808227/.

BW Online Bureau (2015, November 17). Battle of the box office. *BW Businessworld*. http://businessworld.in/article/Battle-Of-The-Box-Office/17-11-2015-82087/.

Cieply, M. (2010, January 13). For all its success, will "Avatar" change the industry? *The New York Times*, C1.

Classification and Rating Administration (CARA). (2013). *How: Tips to be "Screenwise."* http://filmratings.com/how.html.

Clark, T. (2020, January 14). A chart of Netflix's Oscar nominations each year since 2014. *Business Insider*. https://www.businessinsider.com/number-of-netflix-oscar-nominations-per-year-chart-2020-1.

Cook, D.A. (2004). *A history of narrative film* (4th ed.). W.W. Norton & Company.

Corbet, C.J. & Turco, R.P. (2006, November). Sustainability in the Motion Picture Industry. *University of California Los Angles (UCLA) Institute of the Environment.* https://www.ioes.ucla.edu/wp-content/uploads/mpisreport.pdf.

Dcinema Today (2017, December 7). THX and China film investment company announce all THX certified cinema. https://www.dcinematoday.com/dc/pr?newsID=5027.

De La Merced, M. (2012, January 19). Eastman Kodak files for bankruptcy. *The New York Times*. http://dealbook.nytimes.com/2012/01/19/eastman-kodak-files-for-bankruptcy/.

Denison, C. (2013, March 1). THX wants to help tune your home theater, not just slap stickers on it. *Digital Trends*. http://www.digitaltrends.com/home-theater/thx-wants-to-help-tune-your-home-theater-not-just-slap-stickers-on-it/.

Dirks, T. (2009, May 29). Movie history—CGI's evolution From *Westworld* to *The Matrix* to *Sky Captain and the World of Tomorrow. AMC Film Critic*. http://www.filmcritic.com/features/2009/05/cgi-movie-milestones/.

Digital Cinema Distribution Coalition (DCDC). (2016). *About us.* http://www.dcdcdistribution.com/about-us/.

Dolby. (2010a). *Dolby Digital Details.* http://www.dolby.com/consumer/understand/playback/dolby-digital-details.html.

Dolby. (2010b). 5.1 Surround sound for home theaters, TV broadcasts, and cinemas. http://www.dolby.com/consumer/understand/playback/dolby-digital-details.html.

Edwards, G. (2014, January 4). C is for composite. *Cinefex Blog*. https://cinefex.com/blog/tag/williams-process/.

Engber, D. (2016, October 20). It looked great. It was unwatchable. *Slate*. http://www.slate.com/articles/arts/movies/2016/10/billy_lynn_s_long_halftime_walk_looks_fantastic_it_s_also_unwatchable.html.

Engel, L., Buchanan, A. (2009, February). Code of best practices in sustainable filmmaking. *American University, Center for Media & Social Impact*. https://cmsimpact.org/resource/code-best-practices-sustainable-filmmaking/.

Environmental Media Association (EMA). (2020). EMA Green Seal for production. https://www.green4ema.org/ema-green-seal-production.

EPS Geofoam raises Stockton theater experience to new heights. (n.d.). http://www.falcongeofoam.com/Documents/Case_Study_Nontransportation.pdf.

Epstein, A. (2020, January 13). Netflix has taken over the Oscars. *Quartz*. https://qz.com/1784161/netflixs-leads-all-2020-oscar-nominations-thanks-to-the-irishman-and-marriage-story/.

Epstein, A. (2017, March 24). Americans are over 3D movies, but Hollywood hasn't got the memo. *Quartz Media*. https://qz.com/940399/americans-are-over-3d-movies-but-hollywood-hasnt-got-the-memo/.

Fleming Jr., M. (2016, January 7). Picture this: Hateful 8 caps strong year for movies for Kodak. *Deadline Hollywood*. http://deadline.com/2016/01/kodak-the-hateful-eight-star-wars-the-force-awakens-bridge-of-spies-1201678064/.

Freeman, J.P. (1998). Motion picture technology. In M.A. Blanchard (Ed.), *History of the mass media in the United States: An encyclopedia,* (pp. 405-410). Routledge.

Fuchs, A. (2016, January 21). Delivering on the promise: DCDC connects content and cinemas large and small. *Film Journal International.* http://www.filmjournal.com/features/delivering-promise-dcdc-connects-content-and-cinemas-large-and-small.

Fulford, R. (1999). *The triumph of narrative: Storytelling in the age of mass culture.* House of Anansi Press.

Geiger, J. (2019, October 17). An inconvenient truth: Hollywood's huge carbon footprint. *Oilprice.com.* https://oilprice.com/Energy/Energy-General/An-Inconvenient-Truth-Hollywoods-Huge-Carbon-Footprint.html.

Gelt, J. and Verrier, R. (2009, December 28) "Luxurious views: Theater chain provides upscale movie-going experience." *The Missoulian.* http://www.missoulian.com/business/article_934c08a8-f3c3-11de-9629-001cc4c03286.html.

Giardina, C. (2020, January 30). 1917: Inside the war drama's seamless VFX stitchery. *The Hollywood Reporter.* https://www.hollywoodreporter.com/behind-screen/1917-inside-war-dramas-seamless-vfx-stitchery-1273626.

Giardina, C. (2016, October 29). James Cameron promises innovation in *Avatar* pequels as he's feted by engineers. *The Hollywood Reporter.* https://www.hollywoodreporter.com/behind-screen/james-cameron-promises-innovation-avatar-sequels-as-hes-feted-by-engineers-942305.

Haines, R. W. (2003). *The Moviegoing Experience, 1968-2001.* McFarland & Company, Inc.

Hancock, D. (2014, March 13). Digital cinema approaches end game 15 years after launch. *I Technology.* https://technology.ihs.com/494707/digital-cinema-approaches-end-game-15-years-after-launch.

Hancock, D. (2015, April 17). Advancing the cinema: Theatres reinforce premium status to ensure their future. *Film Journal International.* http://www.filmjournal.com/features/advancing-cinema-theatres-reinforce-premium-status-ensure-their-future.

Hart, H. (2012, April 25). Fast-frame Hobbit dangles prospect of superior cinema, but sill theaters bite? *Wired.* http://www.wired.com/underwire/2012/04/fast-frame-rate-movies/all/1.

Heuring, D. (2016, June 15). Experimental light-field camera could drastically change the way cinematographers work. *Variety.* , http://variety.com/2016/artisans/production/experimental-light-field-camera-1201795465/.

Hess, J. P. (2017, January 6). Early traveling mattes – The black and blue screens. *Filmmaker IQ.* https://filmmakeriq.com/lessons/black-blue-screens/.

Higgins, S. (2000). Demonstrating three-colour Technicolor: Early three colour aesthetics and design. *Film History,* 12, (3), Pp. 358-383. https://www.jstor.org/stable/3815345.

Hoad, P. (2020, January 9). Vegan food, recycled tuxedos—and billions of tonnes of CO_2: Can Hollywood ever go green? *The Gardian.* https://www.theguardian.com/film/2020/jan/09/vegan-food-recycled-tuxedos-and-billions-of-tonnes-of-co2-can-hollywood-ever-go-green.

Hogg, T. (2020, January 17). VFX Jedi tricks still impress in "Rise of Skywalker." *Animation Magazine,* (298). https://www.animationmagazine.net/vfx/vfx-jedi-tricks-still-impress-in-rise-of-skywalker/.

Hollywood Reporter. (2005, September 15). *Future of Entertainment.* http://www.hollywoodreporter. com/hr/search/article_display.jsp?vnu_content_id=1001096307.

Horkheimer, M. & Adorno, T. (1969). *Dialectic of enlightenment.* Herder & Herder.

Hull, J. (1999). *Surround sound: Past, present, and future.* Dolby Laboratories Inc. http://www.dolby.com/uploadedFiles/zz-_Shared_Assets/English_PDFs/Professional/2_Surround_Past.Present.pdf.

International Data Corporation (2017, August 7). *Worldwide Spending on Augmented and Virtual Reality Expected to Double or More Every Year Through 2021, According to IDC.* , https://www.idc.com/getdoc.jsp?containerId=prUS42959717.

Isaacs, B. (2013). *The Orientation of future cinema: Technology, aesthetics, spectacle.* Bloomsbury.

Jackson, P. (2014). *See It in HFR 3D: Peter Jackson HFR Q&A.* http://www.thehobbit.com/hfr3d/qa.html

Jacobson, B. (2011, June). The Black Maria: Film studio, film technology (cinema and the history of technology). *History and Technology,* 27, (2), 233-241. https://doi.org/10.1080/07341512.2011.573277.

Jones, B. (2012, May 30). New technology in Avatar—Performance capture, fusion camera system, and simul-cam. AVATAR. http://avatarblog.typepad.com/avatar-blog/2010/05/new-technology-in-avatar-performance-capture-fusion-camera-system-and-simulcam.html.

Jones, Z. (2014, December 2). Want to reach Millennials? Don't forget the draw of the silver screen. *MediaPost.* https://www.mediapost.com/publications/article/239198/want-to-reach-millennials-dont-forget-the-draw-o.html#.

Jowett, G. (1976). *Film: The democratic art.* Little, Brown & Company.

Karagosian, M. & Lochen, E. (2003). *Multichannel film sound.* MKPE Consulting, LLC. http://mkpe.com/publications/d-cinema/misc/multichannel.php.

Kaufman, D. (2011). Film fading to black. *Creative Cow Magazine*. http://magazine.creativecow.net/article/film-fading-to-black.

Kaufman, D. (2014a). Technology 2014 | Production, Post & Beyond: Part ONE. *Creative Cow Magazine*. http://library.creativecow.net/kaufman_debra/4K_future-of-cinematography/1.

Kaufman, D. (2014b). Technology 2014 | Production, Post & Beyond: Part TWO. *Creative Cow Magazine*. http://library.creativecow.net/kaufman_debra/4K_future-of-cinematography-2/1.

Kaufman, D. (2017, April 18). Artificial intelligence comes to Hollywood. *Studio Daily*. , http://www.studiodaily.com/2017/04/artificial-intelligence-comes-hollywood/.

Kindersley, D. (2006). *Cinema Year by Year 1894-2006*. DK Publishing.

Kobzeva-Pavlovic, I. (2015, June). the young generation and future of cinema. *TK Architects International, Inc.* https://tkarch.com/theyounggenerationandfutureofcinema/.

Lackey, R. (2015a, November 24). Full frame and beyond—Large sensor digital cinema. *Cinema5D*. https://www.cinema5d.com/full-frame-and-beyond-large-sensor-digital-cinema/.

Lackey, R. (2015b, September 9). More resolution, higher dynamic range, larger sensors. *Digital Cinema Demystified*. http://www.dcinema.me/2015/09/more-resolution-higher-dynamic-range-larger-sensors/.

Lang, B. (2017, April 11). Lytro's Latest VR light-field camera is huge, and hugely improved. *Road To VR*. https://www.roadtovr.com/lytro-immerge-latest-light-field-camera-shows-major-gains-in-capture-quality/.

Light-Field Forum (2015, November 7). *Lytro Immerge: Company focuses on cinematic virtual reality creation*. http://lightfield-forum.com/2015/11/lytro-immerge-company-focuses-on-cinematic-virtual-reality-creation/.

Lindahl, C. (2019, August 14). Yes, young people still go to the movies: Report finds 18-24 is biggest moviegoing segment. IndieWire. https://www.indiewire.com/2019/08/young-people-millenials-movie-ticket-sales-1202166084/.

Lindbergh, B. (2019, February 7). The rise of the iPhone auteur. *The Ringer*. https://www.theringer.com/movies/2019/2/7/18214924/steven-soderbergh-high-flying-bird-iphone-tangerine-unsane-netflix.

Lubin, G. (2014, July 3)' Here's when 3D movies work, when they don't, and what the future holds. *Business Insider*. http://www.businessinsider.com/are-3d-movies-worth-it-2014-7.

Maltz, A. (2014, February 21). Will today's digital movies exist in 100 years? *IEEE Spectrum*. http://spectrum.ieee.org/consumer-electronics/standards/will-todays-digital-movies-exist-in-100-years.

Martin, K. (2014, March). Kong to lift-off: A history of VFX cinematography before the digital era. *International Cinematographers Guild Magazine*, 85(03), p. 68-73.

McMahan, A. (2003). *Alice Guy Blaché: Lost visionary of the cinema*. Continuum.

Media Insider (2018). The future trends in VFX industry. *Insights Success*. http://www.insightssuccess.com/the-future-trends-in-vfx-industry/.

Melville, A. & Simmon, S. (1993). *Film preservation 1993: A study of the current state of American film preservation*. Report of the Librarian of Congress. https://www.loc.gov/programs/national-film-preservation-board/preservation-research/film-preservation-study/current-state-of-american-film-preservation-study/.

Mendelovich, Y. (2020, January 16). The cameras behind Oscar 2020: Film made a comeback and RED enters the list. *Y.M. Cinema Magazine*. https://ymcinema.com/2020/01/16/the-cameras-behind-oscar-2020-film-made-a-comeback-and-red-enters-the-list/.

Mondello, B. (2008, August 12). Remembering Hollywood's Hays Code, 40 years on. *NPR*. http://www.npr.org/templates/story/story.php?storyId=93301189.

Motion Picture Association of America (MPAA). (2020). *Film ratings: Informing parents since 1968*. http://www.mpaa.org/ratings/what-each-rating-means.

Motion Picture Association of America (MPAA) (2019, March). *2018 THEME Report*. https://www.motionpictures.org/wp-content/uploads/2019/03/MPAA-THEME-Report-2018.pdf.

Mondello, B. (2020, February 7). In an age of streaming, "Oscar Bounce" at the box office is... less bouncy. *NPR*. https://www.npr.org/2020/02/07/803641508/in-an-age-of-streaming-oscar-bounce-at-the-box-office-is-less-bouncy

MPAA (2019). 2019 Theme Report. https://www.motionpictures.org/research-docs/2019-theme-report/.

Muñoz, L. (2010, August). James Cameron on the future of cinema. *Smithsonian Magazine*. http://www.smithsonianmag.com/specialsections/40th-anniversary/James-Cameron-on-the-Future-of-Cinema.html.

Neale, S. (1985). *Cinema and technology: Image, sound, colour*. Indiana University Press.

New Jersey Hall of Fame (2018). *2013 Inductees*. https://njhalloffame.org/hall-of-famers/2013-inductees/.

Newitz, A. (2016, June 9). Movie written by algorithm turns out to be hilarious and intense. *Ars Technica*, https://arstechnica.com/gaming/2016/06/an-ai-wrote-this-movie-and-its-strangely-moving/.

Ng, R., Levoy, M., Brüdif, M., Duval, G., Horowitz, M. & Hanrahan, P. (2005, April). *Light field photography with a hand-held plenoptic camera*. http://graphics.stanford.edu/papers/lfcamera/.

Nolfi, J. (2016, July 26). Artificial intelligence writes 'perfect' horror script, seeks crowdfunding. *Entertainment Weekly*. , http://www.ew.com/article/2016/07/26/artificial-intelligence-writes-perfect-movie-scri't/.

Ogden, M. (2019). The Next Innovation in Immersive [Actuality] Media Isn't Technology—It's Storytelling. *The Asian Conference on Media, Communication & Film 2019: Official Conference Proceedings*. https://papers.iafor.org/submission52910/.

Pahle, R. (2019, March 1). Digital delivery dynasty: DCDC celebrates five years. *Boxoffice Pro*. https://www.boxofficepro.com/digital-delivery-dynasty-dcdc-celebrates-five-years/.

Parce qu'on est des geeks! [Because we are geeks!]. (July 2, 2013). *Pleins feux sur—Georges Méliès, le cinémagicien visionnaire* [Spotlight—Georges Méliès, the visionary cinema magician]. http://parce-qu-on-est-des-geeks.com/pleins-feux-sur-georges-melies-le-cinemagicien-visionnaire/.

pattap-21567 (2018). Highest grossing blockbusters of all time adjusted for inflation. *IMDb*. https://www.imdb.com/list/ls026442468/.

Poster, S. (2012, March). President's letter. *ICG: International Cinematographers Guild Magazine*, 83(03), p. 6.

PR Newswire. (2005, June 27). The Walt Disney Studios and Dolby bring Disney Digital 3-D™ to selected theaters Nationwide With CHICKEN LITTLE. http://www.prnewswire.co.uk/cgi/news/release?id=149089.

Producers Guild of America (PGA) (2020). *Green Production Guide*. https://www.greenproductionguide.com .

Rizov, V. (2020, January 21). The 27 movies (more or less) shot on 35mm in 2019. *Filmmaker Magazine*. https://filmmakermagazine.com/108805-26-movies-shot-35mm-2019/#.Xj6ohC2B3UI.

Renner, B. (2020, January 30). 2020 movies: Complete list of all new 2020 movies released in theaters. *Movie Insider*. https://www.movieinsider.com/movies/2020?view=list&scope=&genre=150&plot=.

Renner, B. (2019, December 28). 2019 movies: Complete list of all new 2019 movies released in theaters. *Movie Insider*. https://www.movieinsider.com/movies/2019?view=list&scope=&genre=150&plot=.

Renner, B. (2016, December 28). 2016 movies: Complete list of all new 2016 movies released in theaters. *Movie Insider*. https://www.movieinsider.com/movies/2016?view=list&scope=&genre=150&plot=.

Renner, B. (2015, December 28). 2015 movies: Complete list of all new 2015 movies released in theaters. *Movie Insider*. https://www.movieinsider.com/movies/2015?view=list&scope=&genre=150&plot=.

Renner, B. (2013, December 19). 2013 movies: Complete list of all new 2013 movies released in theaters. *Movie Insider*. https://www.movieinsider.com/movies/2013?view=list&scope=&genre=150&plot=.

Sanyal, R. (2016, April 20). Lytro poised to forever change filmmaking: debuts cinema prototype and short film at NAB. *Digital Photography Review*. , https://www.dpreview.com/news/6720444400/lytro-cinema-impresses-large-crowd-with-prototype-755mp-light-field-video-camera-at-nab.

Savov, V. (2016, October 17). Razer acquires THX, the audio company George Lucas founded in 1983. *The Verge*. https://www.theverge.com/2016/10/17/13309346/razer-buys-thx-lucasfilm.

Science & Technology Council. (2007). *The digital dilemma*. Hollywood, CA: Academy of Motion Picture Arts & Sciences. http://www.oscars.org/science-technology/council/projects/digitaldilemma/index.html.

Science & Technology Council. (2012). *The digital dDilemma 2*. Hollywood, CA: Academy of Motion Picture Arts & Sciences. http://www.oscars.org/science-technology/council/projects/digitaldilemma2/.

Schank, R. (1995). *Tell me a story: Narrative and intelligence*. Northwest University Press.

Sohail, H. (2015, March 27). The state of cinema. *Qube: Events & news*. http://www.qubecinema.com/events/news/2015/state-cinema.

Stray Angel Films (2014, January 21). What cameras were the 2013 best picture Oscar nominees shot on? *Cinematography*. http://www.strayangel.com/blog/2014/01/21/what-cameras-were-the-2013-best-picture-oscar-nominees/.

The Numbers (2020a). *All time worldwide box office*. https://www.the-numbers.com/box-office-records/worldwide/all-movies/cumulative/all-time.

The Numbers. (2020b). *3D–shot in 3D movies*. https://www.the-numbers.com/movies/keywords/3-D-Shot-in-3-D.

The Numbers. (2020c). *Market share for each MPAA rating in 2019*. https://www.the-numbers.com/market/2019/mpaa-ratings.

The Numbers (2020d). *Domestic movie theatrical market summary 1995 to 2020*. https://www.the-numbers.com/market/.

The Numbers (2020e). *Market share for each genre for 2019*. https://www.the-numbers.com/market/2019/genres.

Thompson, A. (2010, January). How James Cameron's innovative new 3-D tech created Avatar. *Popular Mechanics* http://www.popularmechanics.com/technology/digital/visual-effects/4339455.

THX. (2020). *THX certified cinemas*. http://www.thx.com/professional/cinema-certification/thx-certified-cinemas/.

Verhoeven, B. (2017, July 17) *War for the Planet of the Apes*: How the VFX team created the most realistic apes yet. *The Wrap*. https://www.thewrap.com/war-for-the-planet-of-the-apes-vfx-team-more-real/.

VFXMovies (n.d.). *S/VFX Shots Race*. http://www.upcomingvfxmovies.com/svfx-shots-race/.

Vreeswijk, S. (2012). A history of CGI in movies. *Stikkymedia.com*. http://www.stikkymedia.com/articles/a-history-of-cgi-in-movies.

Volpe, A. (2017, September 17). How VR is changing the film industry. *Backstage*. https://www.backstage.com/advice-for-actors/inside-job/how-vr-changing-film-industry/.

WETA Digital (2019, May 20). *WETA and Thanos come full circle in Avengers: Endgame*. https://www.wetafx.co.nz/articles/weta-and-thanos-come-full-circle-in-avengers-endgame/.

Section III

Computers & Consumer Electronics

11

Computers

Glenda Alvarado, Ph.D.[*]

Abstract

Modern societies can barely function without some type of computer. Bulky, awkward, slow, and ugly boxes have been replaced by sleek designs that fit in your pocket. For many consumers, there would be no record of their life if not for the zeros and ones of their digital footprint. The discontinuation of support for some operating systems caused a spike in unit sales in 2019, but overall sales trends continue to show a decline. Desktop and laptop computers are still commonplace in business settings, but smartphones are taking the place of personal systems at home.

Introduction

The pace of technological advancement has been rapid, but the robot revolution expected to permeate our daily life by this point has been somewhat delayed (Picheta, 2020). Although artificial intelligence (AI) is a hot trend at the annual consumer electronics show, most of us do not have robotic servants, although "computerized colleagues have infiltrated some workplaces" according to Picheta (2020). Ackerman (2018) reported from CES 2018 that the "long-predicted PC-phone convergence is happening, but rather than phones becoming more like computers, computers are becoming more like phones." Just two years later, phones and computers are featuring foldable screens (Ackerman, 2020; Goldman, 2020) allowing both to function as multi-purpose devices.

Background

Blissmer (1991) traces the roots of computers back to the 1600s with the invention of logarithms that led to the development of slide rules and early calculators. Blaise Pascal and Gottfried von Leibniz led the computing pack with "mechanical calculators" but the industrial revolution was the driving force toward development of a system that could deal with large amounts of information (Campbell-Kelly, et al., 1996).

---

[*] Marketing Instructor, School of Business, Midlands Technical College (Columbia, South Carolina).

Increasingly elaborate desk calculators, and then punch-card tabulating machines, enabled the U.S. government to process the census data of 1890 in months instead of years. The punch-card system gained wide commercial application during the first few decades of the 20th century and was the foundation of International Business Machines (IBM) (Campbell-Kelly et al., 1996).

Herman Hollerith won the contract for analyzing the 1890 census data and created the Tabulating Machine Company (TMC) in 1896 (Swedin & Ferro, 2005). TMC leased or sold its tabulating machines to other countries for census taking, as well as making inroads to the private sector in the U.S. These computational machines were then put to use to compile statistics for railroad freight and agriculture (Swedin & Ferro, 2005). Through a series of mergers and a shift to focus on big leasing contracts over small office equipment, IBM was established by Thomas Watson in 1924 (Campbell-Kelly et al., 1996). Early computing machines were built with specific purposes in mind—counting people or products, calculating coordinates, breaking codes, or predicting weather.

Toward the end of World War II, general purpose computers began to emerge (Swedin & Ferro, 2005). The Electronic Numerical Integrator and Calculator (ENIAC) became operational in 1946 and the Electronic Discrete Variable Automatic Computer (EDVAC) concept was published the same year (Blissmer, 1991). The first commercially available computer was the UNIVAC (UNIVersal Automatic Calculator). UNIVAC gained fame for accurately predicting the outcome of the 1952 presidential election (Swedin & Ferro, 2005). IBM and UNIVAC battled for commercial dominance throughout the 1950s. Smaller companies were able to participate in the computer business in specialized areas of science and engineering, but few were able to profitably compete with IBM in the general business market.

Until the 1970s, the computer market was dominated by large mainframe computer systems. The Altair 8800 emerged in 1974. It graced the cover of *Popular Electronics* in 1975, touted as the "world's first minicomputer kit to rival commercial models" (Roberts & Yates, 1975). The Altair had to be assembled by the user and was programmed by entering binary code using hand switches (Campbell-Kelly et al.,

1996). This do-it-yourself model was the forerunner to the machines most people associate with computers today. Paul Allen and Bill Gates owe the success of Microsoft to the origins of the Altair 8800 (Swedin & Ferro, 2005). Steve Wozniak and Steve Jobs founded Apple on April Fool's Day in 1976, and their Apple II microcomputer was introduced in 1977. IBM did not join the microcomputer market until 1981 with the IBM PC (Swedin & Ferro, 2005).

Fast-forward to the twenty-first century and personal computing devices are commonplace in business and the private sector. In 1996, a home computer was considered by most a luxury item. By 2006, 51% of the adult public considered a home computer a necessity (Taylor, et al., 2006). In 2019, Holst (2019b) reported that 74% of Americans own a laptop or desktop computer, a drop from the 80% Blumberg and Luke reported in 2017. Anderson (2019) explained this change as part of a broader shift to mobile technology. People do everything from getting their news to applying for a job on a smartphone. Double-digit increases in "smartphone only" connectivity occurred between 2015 and 2019 (Anderson, 2019). Primary usage of desktops, laptops, or tablets for going online has fallen from 53% in 2013 to 30% in 2019. Functionality of smartphones and financial factors connected to broadband service are the most frequently cited reasons for the switch.

Figure 11.1

Desktop/Laptop Ownership Among U.S. Adults 2008–2019

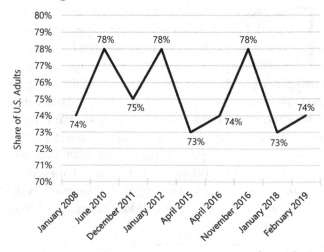

Source: Statista

Figure 11.2

Technology Adoption by Generation

Millennials lead on some technology adoption measures, but Boomers & Gen Xers are also heavy adopters

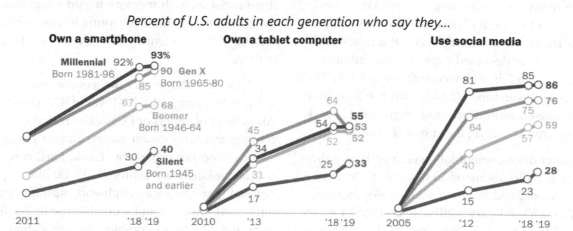

Percent of U.S. adults in each generation who say they...

Own a smartphone

Millennial 92% — 93%
Born 1981-96 — 90 Gen X
85 — Born 1965-80
67 — 68
Boomer
Born 1946-64
40
30 — Silent
Born 1945
and earlier

2011   '18 '19

Own a tablet computer

64
54 — 55
52 — 53
45 — 52
34
25 — 33
31
17

2010   '13   '18 '19

Use social media

81 — 85 — 86
— 76
75
64 — 59
57
40
28
23
15

2005   '12   '18 '19

Note: Those who did not give an answer are not shown.
Source: Survey conducted Jan. 8 - Feb. 7, 2019.

Source: Pew Research Center

Mobile computing gained ground in the early 1990s with a hand-held device introduced by Hewlett-Packard (MacKenzie & Soukoreff, 2002). The first-generation models were primary operated with a stylus on a touchscreen or slate and had limited applications. Apple's Newton (aka Apple MessagePad) emerged in 1993, but one of the most significant entrants to the mobile computing market was the release of the Palm™Pilot in 1996 (Swedin & Ferro, 2005). Microsoft launched the Pocket PC in 2000, with Bill Gates using the term "Tablet PC" during its trade show introduction (Atkinson, 2008). Apple's iPad came to the consumer market in 2010 and changed the landscape of personal computing devices (Gilbert, 2011). For many school children in the United States, iPads are issued instead of textbooks, and homework is done online rather than with pen and paper. Both online and traditional colleges are providing free tablets or laptops with tech support and repair services for the entirety of a student's college career (Collegeconsensus.com, n.d.).

Computer ownership levels have remained roughly the same for more than a decade; however all-purpose devices are taking the place of specialized components (music-players, game systems, electronic readers) (Anderson, 2015). Although tech usage has become the norm, adoption rates have continued to slow because, for some generations, there are few remaining non-users (Schaeffer, 2019). Moving into a new decade, the trend continues to be combining features of devices that were previously sold as separate items. These hybrids, convertibles, and 2-in-1s are emerging as crossover devices, which share traits of both tablets and laptops. Combined features include a touchscreen display designed to allow users to work in a tablet mode as well as a keyboard that can be removed or folded out of the way.

How a Computer Works

The technology ecosystem model presented in Chapter 1 starts with breaking down the hardware of a computer to its basic components. In simplest terms, a computer has four fundamental components: input, output, processing, and storage. Hardware elements or the physical components of which a computer is comprised have become smaller and more compact, but are essentially the same across platforms. These main pieces are the central processing unit (CPU), power circuits, memory, storage, input, and output. Software or application components determine how information or content is used and manipulated.

The CPU is the brain of the computer—a series of switches that are essentially either on or off—a binary system or string of ones and zeroes. Early processors were constructed using vacuum tubes that were hot, fragile, bulky, and expensive. This meant that it took and entire room (or rooms) to accommodate a single computer. The 1940s-era ENIAC weighed 30 tons and contained 18,000 vacuum tubes with the tubes being run at a lower than designed capacity in an attempt to lessen their rate of failure (Swedin & Ferro, 2005). The EDVAC used fewer tubes (5,000), but the heat they generated caused technicians and engineers to work in their underwear during the summer months.

Computer development followed a path very similar to that of radio discussed in Chapter 8. As can be seen there, science and technological developments are often tied to war—military battling to best their foes. In 1947, Bell Telephone scientists invented "point-contact" transistors made out of a semiconductor material—much smaller, lighter and more durable than tubes (Swedin & Ferro, 2005). Military funding sped up this development of transistors that were key components in computers built throughout the 1950s. By the 1960s "integrated circuits" contained multiple tiny transistors on a single silicon "chip." Continued miniaturization has enabled complex processing systems to be reduced from the size of a room to a single chip, often smaller than a pinkie fingernail.

Hardware

Modern computers are built with a motherboard that contains all the basic electrical functions. A motherboard is the physical foundation for the computer. Additional "boards" can be tied into the mother to provide supplementary components—extra memory, high-resolution graphics, improved video, and audio systems. Memory of a computer determines how much information it can store, how fast it can process that information, and how complex the manipulations can be.

For a computer to operate, the CPU needs two types of memory, random access memory (RAM) and storage memory. Most forms of memory are intended for temporary storage and easy access. RAM space is always at a premium—all data and programs must be transferred to random access memory before any processing can occur. Frequently run program data and information are stored in temporary storage known as cache memory; users typically access the same information or instructions over and over during a session. Running multiple programs at the same time creates a "RAM cram" (Long & Long, 1996) situation where there is not enough memory to run programs. That's the reason many troubleshooting instructions include clearing "cache" memory before re-attempting an operation.

New PCs and similar devices are being manufactured with a range of 4GB–16GB of RAM (Martindale, 2019). Budget tablets or PCs have around 2GB. 8GB is recommended for casual gamers and general-purpose users. Computer professionals and hard-core gamers will be looking for machines with 16GB that provide optimal performance. High-end, specific purpose-built workstations can be obtained with 32GB RAM. Typically, the more complex an application or program, the more RAM that is used. However, more RAM does not necessarily mean better performance. Martindale (2019) wrote that "wasted money" on RAM would be better spent on CPUs or graphics cards. RAM should be considered for immediate access, but not long-term storage. For large amounts of data, a hard drive, solid state memory, or external storage device is more appropriate.

Most computers have built-in internal hard drives that contain its operating system and key software programs. External storage systems typically have high capacity and are used to back up information or serve as an archive. These systems connect to a computer using USB, thunderbolt, or firewire connection. Increasingly, additional memory is part of a "cloud," or virtual network system that is digitally stored in a remote location.

Small-scale storage devices called flash drives or thumb drives allow users to share or transport information between computers or systems, especially when network connections are unreliable or not available. Other memory or storage devices include SD cards, CDs, DVDs, and Blu-ray discs. SD cards are most often associated with digital cameras or other small recording devices. Information is easily deleted or can be transferred to a permanent location before the contents are erased and written over. CDs and DVDs are considered "optical discs" and are commonly associated with movies and music files—although some versions have read/write capacity, most are used for a single purpose or one time only. Information in this format is accessed using an optical disc drive that uses laser light to read and write to or from the discs. For many years, software programs were produced as a CD-ROM (Read Only Memory) version.

Manufacturers are building machines without optic drives to save space and lessen battery drain. Headphone jacks, USB sockets, and other input/output ports are also being eliminated as laptops are being constructed less than 15mm in thickness. USB-C ports are emerging as the industry standard connector for power and data. These ports are 2.6mm tall versus the 7.5mm of a standard USB. They have no up/down or in/out orientation offering greater bandwidth for data transfer and reduced power consumption. USB-C ports are now being found on a variety of devices from external hard drives to smartphones, but the slot doesn't serve the same function on every device (Brant, 2019). Be sure to check the details on each device as manufacturers can vary on what hardware is connected to the ports.

Input and output devices are the last pieces of the computer anatomy. Input devices include keyboards, mice, microphones, scanners, and touchpads. Sophisticated computer operators may also have joysticks, hand-held controllers, and sensory input devices. Output options include printers, speakers, monitor, and increasingly, virtual reality headsets.

Most current computers or personal computing devices are also equipped with Bluetooth, a form of technology that uses short-range wireless frequencies to connect and sync systems. Users looking for a wireless experience enjoy the "uncabled" functionality of such systems. Wireless options are also available for input devices (mice and keyboards), often using radio frequency (RF) technology whereby the input item contains a transmitter that sends messages to a receiver that plugs into a USB port.

Software

For all the sophistication and highly technical components of a computer, it is essentially useless without software. These are the programs that enable the computer to perform its operations. At the core of each computer is the operating system—the Windows versus Mac debate continues to rage, but user preference remains a personal choice. These systems have more similarities than differences at this point in time, however Windows is the most used operating system worldwide. According to Liu (2019), Windows holds a 72% share of the desktop, tablet, and console (gaming) operating systems. As of mid-2020, Windows 10 and macOS Catalina are the latest versions. If mobile devices are included in the count, Android is the market-share leader at 40.5% over Windows at 34.2% (Operating Systems Worldwide, 2019).

Operating systems control the look and feel of your devices—and, to a certain extent, how you connect your multiple-device households. One reason for the dominance of Windows is the ability to connect devices from a variety of manufacturers. Mac/Apple products tend to only connect with each other—although there are "compatible" technologies that adapt across platforms. Unfortunately for users, frequent updates or system changes can render software obsolete. MacOS allows users free and continual updates to its operating systems—Catalina, Mojave, High Sierra, Sierra and El Capitan, from newest to oldest, respectively. Windows users lose the ability to get systems support or updates after operating systems are phased out. Microsoft ended mainstream support for Windows 7 in 2015 and extended support was discontinued in January 2020. An unofficial loophole allowed users to upgrade to Windows 10 if they had a license key (Neild, 2020). Windows 8 will cease offering support in 2023 and users of that operating system are encouraged to upgrade to Windows 10 for $139 before they lose the ability to get updates and patches.

The delivery of on-demand computing resources via connection to the internet, known as cloud computing, has eliminated the need for many people to purchase hardware and individual software programs. Storage and access are available for free or on a pay-as-you-go basis. Common cloud computing examples include Microsoft's Office 365 and OneDrive that marry local and online services; Google Drive, with Docs, Sheets and Slides that are purely online; and Apple's iCloud, primarily offering storage, but also offering cloud-based versions of popular programs (Griffith, 2016). Hybrid services such as Box, DropBox, and SugarSync are considered cloud-based because they store synchronized information. Synchronization is the key to cloud computing, regardless of where or how information is accessed.

Consumers are continuously tempted with shareware and trial software that offers a chance to test a program before purchasing a code for continued use. Open source software is also available for users to modify programs for specific usage. There are hundreds of thousands of individual programs and applications that can run on each operating system. Users can write, calculate, hear, watch, learn, play, read, edit, control, and produce…virtually anything.

Recent Developments

Dual-screen folding devices debuted at the 2019 *Consumer Electronics Show* and continued to be popular exhibits at the 2020 show (Goldman, 2020). Lenovo's ThinkPad X1 Fold was deemed one of the best products showcased in 2020, but many people remain skeptical about the durability of a foldable screen (Verge Awards, 2020). In 2020, Intel will retail a NUC (Next Unit of Computing) 9 Extreme Platform that contains swappable components in a convenient cartridge, allowing computer enthusiasts to easily build a PC at home (Portnoy, 2020). Buzzi and Burek (2020) tout the space saving features of all-in-one desktop options for students and professionals with a need for functionality, but limited work area. Large screens offer high resolution and allow multiple windows to be displayed side by side. Curved screens are an option from some manufacturers, as are touch screen systems that will recline to a horizontal position for ease of use (Buzzi and Burek, 2020).

Hardware

Computer prices have remained relatively stable over the past few years. Users can purchase a basic tablet computer for less than $100—most tablets more closely resemble a cell phone than a computer and are a popular choice for those with toddlers and preschoolers. High-end tablets with more versatile and sophisticated operations are available for $900–1200, but the most popular models cost $250–400. Entry-level laptops—perfect for younger families to use for homework, web surfing, and movie watching—can often be found in the same price range. Hybrid

systems offering laptop and tablet functionality in one device can be found for as little as $300 (Brant, 2020).

Acer brand's Swift and Chromebook laptops are top rated by *PC Magazine* in 2020. These budget-friendly options fall in a $350-$400 range (Brant, 2020). Brant (2020) claims that fewer compromises are necessary than in previous years, and most options in this price range include at least 4GB of memory and a minimum of five hours of battery life. Multiple USB ports, HDMI outputs, and SD card slots that previously only appeared on higher-priced computers are now available on basic versions. Dell tops the list of the ultra-portables that are thin, light. and last all day on a single charge. These units are great for general purpose usage and with an average price of $1,000, are a little cheaper than their HP and Lenovo counterparts that received similar ratings, but with a price tag of up to $1,500 (Buzzi, 2019).

Gamers and high-end users will be pleased with the sleek design and long battery life of an Asus ROG Zephus for $2,000 or high frame rates and slick graphics that are touted on a Razer Blade 15 topping out at $3,000 (Buzzi 2020). For Apple enthusiasts, the latest model Mac Pro can reach the $50,000 mark for the highest-end model with all the upgrades (Mihalcik, 2019). The pricey system is designed for a professional market rather than general consumers and features Intel's Xeon processers which aren't found on any standard computers. The base model for the 16″ MacBook Pro actually starts around $2,800. The pre-Christmas 2019 release was the first major update from Apple since 2013. Entry level versions of the iMac desktop start at $1,499 for a 27″ model while the powerful Mac Pro desktop computer starts around $6,000.

Software

Cloud computing has seen enormous growth over the past several years on both a professional and casual scale. While adoption of new systems has slowed, migration to a remote infrastructure has taken off. Cloud services have allowed businesses to develop virtual IT structures and deliver software independent of a user's operating system (Drake & Turner, 2019). Cloud brokers enable companies to mix and match services from different providers, allowing maximum efficiencies and improved redundancies.

Easy access and scalability mean that businesses of any size can take advantage of the services. Amazon (AWS), Google, and Microsoft provide cloud services for free or nominal fees. Large corporations are better served with Oracle or IBM that offer highly customizable options to support a wide range of workloads (Drake & Turner, 2019).

According to Swant (2020), innovations coming out of *CES 2020* focused on making the world better over wowing the crowd. Partnerships with brands and tech companies showcased ways to improve consumer experiences and customer service. Devices for people to monitor their health, along with using data management systems to customize AI to a specific user, are on the rise (Swant, 2020). Hiner (2020) echoed the focus on improved health monitoring with wearable devices, but highlighted the introduction of technology becoming less-obtrusive and blending into the environment. Companies are using sound waves to use a plain surface as a touch interface. Sound wave buttons don't wear out as quickly and haptic feedback can be used to help users locate and interact with these surfaces (Hiner, 2020).

Microsoft relaunched its newest browser, Edge, in early 2020. The updated Edge is compatible with Mac and Windows operating systems and will replace Internet Explorer for many users (Lardinois, 2020). Information security and protection from online trackers are key options. For users who want improved visual appearance, AMD's new Adrenaline 2020 provides some of the best graphics options for gamers and video streamers (Stobing, 2019). The updates provide a refined resolution to improve the visual quality of older games that use flat, pixelated software.

Current Status

Worldwide shipments of traditional PCs (desktops, notebooks, and workstations) grew 2.7% in 2019, the first full year over year growth since 2011 (Chou, et al., 2020). A report from International Data Corporation (IDC) attributes the growth to business sector purchases as commercial users upgraded to Windows 10 when support for older operating systems ended. HP, Dell, and Lenovo branded devices enjoyed market share increases from the upgrade-driven push

(Chou et al. 2020). Overall, Euromonitor International (2019) forecasts the continued decline in sales volume for computers through 2025.

Consumer desire for performance and mobility has fueled innovation in laptop computers. Retail volume sales of laptops rose 4% in the U.S. during 2019 (Country Report, 2019). Innovation in the area of multiple displays and touch screens add versatility to traditional laptops. Declining unit prices make them more attractive for a business' computing needs, and detachable keyboards allow easy conversion for retail and sales situations.

Figure 11.3

Sales of Computers & Peripherals— 2005-2024

Retail Volume – '000 units
361,120

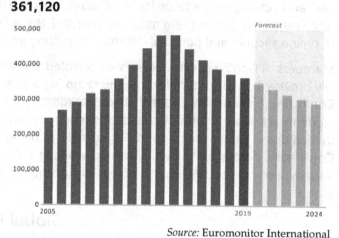

Source: Euromonitor International

Factors to Watch

Technological advances continue to target seniors. Baby Boomers (55 to 73) and the Silent Generation (74 to 91) face unique challenges when it comes to technology (Vogels, 2019). Products aimed at these age groups who often face physical challenges have increased font sizes, voice activation, speech recognition, and hearing aid compatibility. Health tech targeting seniors is a $30 billion industry (Karl, 2019). Devices remind seniors about doctors' appointments and when to take medication. Wearable devices can monitor for falls and medical conditions, alerting emergency services when the need arises. Senior care residences are boosting network infrastructures to accommodate the growing reliance on technological devices (Vogels, 2019). While technology aids greatly inpatient diagnoses and treatment, there are great risks when it comes to information security and HIPAA concerns. Network and server geo-fencing is expected to be further developed as one way to protect patients' sensitive information (Vogels, 2019). (For more on digital health, see Chapter 17.)

The idea of quantum computing has been around for decades, but workable technology is starting to catch up to the theory (Hunter, 2020). According to Greene (2020), both Google and IBM were able to demonstrate application of quantum technology in 2019. If existing computers cannot get faster, the ever-increasing amount of data in the world will not be able to be processed. A quantum bit, or qubit, can store an exponential number of bits, making a quantum computer faster and more powerful than current digital versions (Coughlin, 2020).

Artificial intelligence (AI), augmented reality (AR) and virtual reality (VR) continue to be categories that are expanding. Autonomous vehicles are also gaining traction with automakers and tech firms actively pursuing the technology.

Gen Z

Generation Z is poised to become the largest U.S. consumer population. They are "digital natives" and spend ten or more hours a day with technology, mostly mobile (Molla, 2019). Gen Z is not as optimistic about technology as might be imagined—more than any other generation, they are concerned about their own inability to disconnect and actually desire to be less reachable. Molla (2019) surmised that their, albeit short, lifetime of experience with computers makes them more experienced, less enamored with technology and warier of the consequences of online security and personal information getting into the wrong hands.

Members of Gen Z are extremely visual-oriented and strongly influenced by peers. For marketers, this generation will browse a physical store for trial and sampling and then search online for the best deal (Herosmyth.com, 2018). Conflicting information about their preparedness for a career abounds. *Inside Higher Ed* reports many Gen Z students are weak in soft skills needed for engaging in face to face communication (Bauer-Wolf, 2018). Proficiency in tech doesn't translate to good writing skills or being able to spend time in a traditional working environment. Conversely, *Forbes* touts Gen Z as hard-working entrepreneurs who are eager to learn (Wingard, 2018). They eagerly embrace technology to access information, work collaboratively, and connect with others around the globe.

Global Perspective

Growing competition from mobile phones has eroded the computer market. IBIS World (Gonzales, 2019) forecasts that the global computer industry will decline at a rate of 1.1% over the next five years. Additionally, on a global level, the price for computers and peripherals is predicted to fall at a rate of 2.8%. There has been a shift away from countries with high manufacturing costs, and Taiwan is currently home to the two largest players in computer production—Quanta and Compal (Gonzales, 2019). On average, almost half of the private households around the world had a computer in 2018 (Holst, 2019a, August 19). In developing countries, that number is barely 30%, for developed countries it's more than 80%. As telecommunication technology increases mobility, and wireless network expansion continues, demand for computers and peripherals is lessened. IBIS projects that global internet usage will increase at a rate nearing 4% over the next five years, but the majority of that access will be via smartphones (Gonzales, 2019).

Bibliography

Ackerman, D. (2018, January 12). PC-phone convergence is happening, but not how you think. *CNET*. https://www.cnet.com/news/pc-phone-convergence-is-happening-not-how-you-think-ces-2018/.

Ackerman, D. (2020, January 17). Don't get too hyped about folding or dual-screen laptops just yet. *CNET*. https://www.cnet.com/news/dont-get-too-hyped-about-folding-or-dual-screen-laptops-just-yet/.

Anderson, M. (2019, June 13). Mobile Technology and Home Broadband. Pew Research Center. https://www.pewresearch.org/internet/2019/06/13/mobile-technology-and-home-broadband-2019/.

Anderson, M. (2015). *Technology Device Ownership: 2015*. Pew Research Center. http://assets.pewresearch.org/wp-content/uploads/sites/14/2015/10/PI_2015-10-29_device-ownership_FINAL.pdf.

Atkinson, P. (2008). A bitter pill to swallow: the rise and fall of the tablet computer. *Design issues*, 24(4), 3-25.

Bauyer-Wolf, J. (2018, February 23). *Overconfident Students, Dubious Employers. Inside Higher Ed*. https://www.insidehighered.com/news/2018/02/23/study-students-believe-they-are-prepared-workplace-employers-disagree.

Blissmer, R. H. (1991). *Introducing computers: concepts, systems, and applications* (1991–92 ed.). Austin, TX: John Wiley & Sons, Inc.

Blumberg, S., & Luke, J. (2017). Wireless substitution: Early release of estimates from the National Health Interview Survey, July–December 2016. http://www.cdc.gov/nchs/nhis.htm.

Brant, T. (2019, June 14). What is a USB-C? An explainer. *PC Magazine*. https://www.pcmag.com/how-to/what-is-usb-c-an-explainer.

Brant, T. (2020, January 14). The best budget laptops for 2020. *PC Magazine*. https://www.pcmag.com/picks/the-best-budget-laptops.

Buzzi, M. (2019, December 9). The best gaming laptops for 2020. *PC Magazine*. https://www.pcmag.com/picks/the-best-gaming-laptops.

Buzzi, M. (2020, January 14). The best ultraportable laptops for 2020. *PC Magazine*. https://www.pcmag.com/picks/the-best-ultraportable-laptops.

Buzzi, M. & Burek, J. (2020, January 3). The best all-in-one computers for 2020. *PC Magazine*. https://www.pcmag.com/picks/the-best-all-in-one-computers.

Campbell-Kelly, M., Aspray, W., & Wilkes, M. V. (1996). Computer A History of the information Machine. *Nature*, 383(6599), 405-405.

Chou, J., Ubrani, J., Reith, R., & Shirer, M. (2020). Traditional PC Volumes Close Out an Impressive 2019 with Fourth Quarter Growth of 4.8%, According to IDC. [Press release]. https://www.idc.com/getdoc.jsp?containerId=prUS45865620.

Collgegeconsensus.com, n.d. (No Date). 25 Colleges that offer free laptops or iPads. College Consensus. https://www.collegeconsensus.com/rankings/free-laptops-for-college-students/.

Coughlin, T. (2020, January 14). Memory for quantum computing. *Forbes*. https://www.forbes.com/sites/tomcoughlin/2020/01/14/memory-for-quantum-computing/#2598dd393d77.

Country Report (2019, August). Computers and peripherals in the US. Euromonitor International. https://www.portal.euromonitor.com/portal/analysis/tab.

Drake, N., & Turner, B. (2019, December 20). Best cloud computing services of 2020: for Digital Transformation. *Tech Radar*. https://www.techradar.com/best/best-cloud-computing-services.

Euromonitor International (2019, November). Computers and peripherals in World. *Euromonitor International*. https://www.portal.euromonitor.com/portal/analysis/tab.

Gilbert, J. (2011, December 6, 2017). HP TouchPad bites the dust: Can any tablet dethrone the IPad? *Huffington Post*. https://www.huffingtonpost.com/2011/08/19/hp-touchpad-ipad-tablet_n_931593.html.

Goldman, J. (2020, January 6). With Concept Duet and Ori, Dell brings dual-screen, foldable experience to laptops. *CNET*. https://www.cnet.com/news/concept-duet-concept-ori-dell-brings-dual-screen-foldable-experience-laptops/.

Gonzales, E. (2019, May). Global Computer Hardware Manufacturing Industry Report. *IBIS World*. https://www.ibisworld.com/global/market-research-reports/global-computer-hardware-manufacturing-industry/.

Greene, T. (2019). IBM thinks outside of the lab, puts quantum computer in a box. *The Next Web*. https://thenextweb.com/artificial-intelligence/2019/01/10/ibm-thinks-outside-of-the-lab-puts-quantum-computer-in-a-box/.

Griffith, E. (2016). What is cloud computing? *PC Magazine*. https://www.pcmag.com/article2/0,2817,2372163,00.asp.

Herosmyth.com (2018). 75 Eye-Opening Statistics on How Each Generation Uses Technology. *Herosmyth*. https://www.herosmyth.com/article/75-eye-opening-statistics-how-each-generation-uses-technology.

Hiner, J. (2020, January 10). The 5 biggest tech trends from CES 2020. *CNET*. https://www.cnet.com/news/the-5-biggest-tech-trends-ces-2020/.

Holst, A. (2019a, August 19). Computer penetration rate among households worldwide 2005-2018. Statista. https://www.statista.com/statistics/748551/worldwide-households-with-computer/.

Holst, A. (2019b, December 3). Desktop/laptop ownership among adults in the United States from 2008 to 2019. Statista. https://www.statista.com/statistics/756054/united-states-adults-desktop-laptop-ownership/.

Hunter, M. (2020, January 16. The quantum computing era is here. Why it matters—and how it may change our world. *Forbes*. https://www.forbes.com/sites/ibm/2020/01/16/the-quantum-computing-era-is-here-why-it-mattersand-how-it-may-change-our-world/#23fca25a5c2b.

Karl, J. (2019, December 10). Why tech designers must keep seniors' needs in mind. *Health Tech*. https://healthtechmagazine.net/article/2019/12/why-tech-designers-must-keep-seniors-needs-mind.

Lardinois, F. (2020, January 15). Here is the first stable release of Microsoft's new Edge browser. *Tech Crunch*. https://techcrunch.com/2020/01/15/here-is-the-first-stable-release-of-microsofts-new-edge-browser/.

Liu, S. (2019, September 3). Market share held by the leading computer (desktop/tablet/console) operating systems worldwide from January 2012 to August 2019. Statista. https://www.statista.com/statistics/268237/global-market-share-held-by-operating-systems-since-2009/.

Long, L. E., & Long, N. (1996). *Introduction to Computers and Information Systems*. Upper Saddle River, NY: Prentice Hall PTR.

MacKenzie, I. S., & Soukoreff, R. W. (2002). Text entry for mobile computing: Models and methods, theory and practice. *Human–Computer Interaction, 17*(2-3), 147-198.

Martindale, J. (2019, October). How much RAM do you need? *Digital Trends*. https://www.digitaltrends.com/computing/how-much-ram-do-you-need/.

Mihalcik, C. (2019, December 10). Apple's Mac Pro costs more than $50,000 if you get all the upgrades. *CNET*. https://www.cnet.com/news/apples-mac-pro-costs-more-than-50000-if-you-get-all-the-upgrades/.

Molla, R. (2019, October 15). Generation Z doesn't always want to hear from you. Vox. https://www.vox.com/recode/2019/10/15/20915352/generation-z-technology-attitudes-optimism-always-reachable-survey-gfk.

Neild, D. (2020, January 14). What to do when your OS becomes obsolete—and how to save money in the process. *Popular Science*. https://www.popsci.com/story/diy/obsolete-os-guide/.

Operating System Market Share Worldwide. (2019, December). Statcounter Global Stats. https://gs.statcounter.com/os-market-share#monthly-201809-201908-bar.

Picheta, R. (2020). Nanobots, ape chauffeurs and flights to Pluto. The predictions for 2020 we got horribly wrong. CNN Business. https://www.cnn.com/2020/01/01/tech/2020-predictions-we-got-wrong-scli-intl/index.html.

Portnoy, S. (2020, January 8). CES 2020: Intel launches NUC 9 Extreme, NUC 9 Pro Workstation mini-PC kits. *ZDNet*. https://www.zdnet.com/article/ces-2020-intel-launches-nuc-9-extreme-kit-nuc-9-pro-workstation/.

Roberts, H. E., & Yates, W. (1975). Altair 8800 minicomputer. *Popular Electronics, 7*(1), 33-38.

Schaffer, K. (2019, December 20). U.S. has changed in key ways in the past decade, from tech use to demographics. Pew Research Center. https://www.pewresearch.org/fact-tank/2019/12/20/key-ways-us-changed-in-past-decade/.

Stobing, C. (2019, December 10). Preview: AMD pumps up Radeon software to 'Adrenalin 2020' with brand new tech. *PC Magazine*. https://www.pcmag.com/news/preview-amd-pumps-up-radeon-software-to-adrenalin-2020-with-brand-new-tech.

Swant, M. (2020, January 14). Shift to more minimal—and more useful—tech catches CMOs' attention at CES 2020. *Forbes*. https://www.forbes.com/sites/martyswant/2020/01/14/shift-to-more-minimaland-more-usefultech-catches-marketers-attention-at-ces-2020/#73d9f74e7c32.

Swedin, E. G., & Ferro, D. L. (2005). *Computers: the life story of a technology*, Westport, CT: Greenwood Publishing Group.

Taylor, P., Funk, C., & Clark, A. (2006). Luxury or necessity? Things we can't live without: The list has grown in the past decade [Press release]. http://assets.pewresearch.org/wp-content/uploads/sites/3/2010/10/Luxury.pdf.

Verge Awards (2020, January 10). The Verge Awards at CES 2020. *The Verge*. https://www.theverge.com/2020/1/10/21058850/ces-2020-verge-awards-best-laptop-tv-tech-pc-concept-gadget.

Vogels, E. A. (2019, September 9). Millennials stand out for their technology use, but older generations also embrace digital life. Pew Research Center. https://www.pewresearch.org/fact-tank/2019/09/09/us-generations-technology-use/.

Wingard, J. (2018, November 21). Training Generation Z. *Forbes*. https://www.forbes.com/sites/jasonwingard/2018/11/21/training-generation-z/#454ce5d7bde0.

12

Automotive Telematics

Jeffrey S. Wilkinson, Ph.D. & Angeline J. Taylor, M.S.[*]

Abstract

Advances in Automotive Telematics are transforming road-based travel throughout the world. Automotive Telematics underlies why and how our cars, trucks, and other vehicles are becoming smarter, safer, and heading toward independence. Technology innovations inside and outside the vehicles are rapidly transforming transportation industries as well as commuter driving habits. Inside each vehicle, the emphasis on elaborate infotainment systems are giving way to surveillance and safety, sparking significant changes in the insurance industry. Experiments continue with autonomous driving, hydrogen fuel-cell systems, and electric hybrid models with implications for sustainability and reducing the environmental impact of automobiles worldwide.

Introduction

Our love affair with the automobile is quickly becoming co-dependent, and we like it. Amazing innovations in our four-wheeled friends are making it increasingly difficult to live without them. Many of these improvements come under the heading Automotive Telematics, which is the umbrella term referring to equipping vehicles with networking capabilities combined with advanced computer information technology systems. These innovations are making significant changes to the vehicles we use to move from one location to another.

The automobile, or motor vehicle commonly referred to as "the car" (from old Anglo-French "carre" or "wheeled vehicle") continues to evolve in form as well as function. In 2018, 86 million cars were sold in

[*] Wilkinson is Associate Professor and Taylor is Visiting Instructor, School of Journalism and Graphic Communication, Florida A&M University (Tallahassee, Florida).

the 54 largest markets worldwide (Bekker, 2019). Although still mostly gasoline-dependent, the trend toward electric continues, with a 75% increase reported in the sale of electric cars or hybrids (Bekker, 2019). While all nations rely on motor vehicles, perhaps the most car-dependent culture continues to be the United States (Buehler, 2014). Because of the legacy of fossil-fuels, cars and trucks continue to have a dramatic impact on the global environment (International Organization of Motor Vehicle Manufacturers, n.d.).

While the most popular and widely sold vehicle is still the car, this chapter highlights innovations that are being applied across the spectrum of transportation vehicles, including trucks, semi-tractor trailers, RVs (recreational vehicles), ATVs (all-terrain vehicles) motorcycles, and golf carts. Personal transportation devices such as scooters, bikes, skateboards, and hoverboards have also been marketed with advanced communication technology (although the hydrogen-powered skateboard does not seem to be a thing yet).

Background

More than 50 years ago, noted science fiction author Isaac Asimov attended the 1964 New York World's Fair and submitted his prediction for future cars in a New York Times essay. Asimov predicted then that in 2014 we would see cars with "robot-brains:"

> Much effort will be put into the designing of vehicles with "robot-brains"—vehicles that can be set for particular destinations and that will then proceed there without interference by the slow reflexes of a human driver. (Bogost, 2014).

Asimov and others thought that cars would travel suspended on compressed air. A widely popular children's animated sitcom at the time, *The Jetsons*, featured family patriarch George Jetson flying his car to his office. With a flick of a button, the automobile folded up into a briefcase he carried inside (Duan, 2018). Although wearable technology is common today, fold-up cars flying on compressed air remain a cartoonist's dream.

Today's vehicles are a far cry from the first cars invented in 19th century Europe. In America, the

Model Ts that rolled off Henry Ford's assembly lines in the early 1900s helped cement the symbiotic relationship between human and machine. And, from the beginning, electronics were integral to automobiles. Historically, the first battery-based electronic automotive components provided combustion and lighting, followed by radio. The first successful AM car radio was invented by William Lear (of Learjet fame) in 1930 (Berkowitz, 2010). Lear sold the device to the Galvin brothers who then sold the basic unit under the name Motorola for the then-unheard-of-price of $130 (a Model A Deluxe coupe cost a staggering $540). The first FM car radio appeared in 1952 from German maker Blaupunkt, and the next year automakers could install a dual-band AM/FM radio. Other communication innovations include car-telephones (1940s but primitive and experimental), cruise control (late 1950s) on-board entertainment systems (1960s, eight-track tapes; 1970s, cassettes; 1980s, compact discs; and 2000s, DVDs) (Gray, n.d.). As analog turned to digital, tape was replaced by MP3, USB ports, satellite radio, Bluetooth, HDMI input, Wi-Fi, and voice activation technology which are now common features.

Table 12.1

Key Developments in the Evolution of Automotive Telematics & GPS

Year	Development
1978	Navstar 1, first Global Positioning System (GPS) satellite launched
1983	GPS made available for public use
1985	First car navigation system for consumers launched
1996	GPS is made an international utility
2010	First smartphone application (app) that combine telematics launched
2014	Era of full connectivity begins
2016	Europe's new navigation satellite program Galileo activated
2018	GPS is now a constellation of 31 satellites
2025	Worldwide projection that 88% of new cars will feature integrated telematics

Source: Wise (2018) telematics.tomtom.com

Today, automotive telematics use Global Positioning System (GPS) technology to ascertain vehicle location and digital cellular networks to transmit data from the vehicle to other software and hardware systems (Wise, 2018). Data sent includes location, vehicle diagnostics, and driving style.

For perspective, back in 2001, there were no electric or even semi-electric cars in production. The most advanced vehicle powertrain then was in the first-generation Toyota Prius hybrid (Richardson, 2018). Cars did not have sensors to detect other vehicles. Cameras were carried on a strap and used film. Car radios came with tape decks or CD players. Texting was not a problem in 2001, and social media platforms like Twitter, Instagram, and Snapchat did not exist.

Recent Developments

The auto industry is changing in several significant ways but there are two particularly clear overarching trends. First is a concerted industry-wide move toward autonomous vehicles, and second, a move away from fossil fuels. Although hard numbers are difficult to find, many predict that by 2040 electric cars will outnumber internal combustion cars (Valdes-Dapena, 2019), and by 2025 a significant number of those cars will be driving themselves (Richardson, 2018).

Autonomous Vehicles

The recent move to self-driving cars has been staggeringly aggressive. But the developments are uneven and the range of capabilities must balance innovation with what can be marketed and sold right now. Every major car and truck maker has a model utilizing high levels of autonomy as well as others operating with limited capabilities. To help guide these efforts, the National Highway Traffic Safety Administration (NHTSA) has defined different levels of autonomous driving, from zero to five (Figure 12.2).

Smart car capabilities include knowing how to avoid traffic jams, knowing where to park, opening the door for the passengers, playing favorite music, turning on the lights when needed, choosing optimal speed, driving safely, and being efficient when dangerous moments happen (Enero, 2018).

As of mid-2020, only one company, Waymo, has developed a Level 5, fully autonomous, self-driving automobile. This vehicle division of Alphabet, parent company to Google, has been testing cars without a driver in the driver's seat since 2018 on the roads of Chandler, Arizona. Each car is equipped with sensors offering 360-degree views. The vehicles drive within a 100-square mile area, collecting data, acquiring more experience, and becoming "smarter" (Marr, 2018). The team started in 2009 in secret and used Toyota Prius cars. The goal for fully autonomous vehicles is to improve human mobility, safety, and free up time for people to do something besides attending to driving behind the wheel (Marr, 2018).

Figure 12.2

Levels of Autonomous Driving

Level 0 – No Automation	The driver controls all steering, brakes and power.
Level 1 – Driver Assistance	Most functions are still controlled by the driver but specific functions (steering or accelerating) can be done automatically by the car.
Level 2 – Partial Automation	At least one driver assistant system is automated, such as cruise control or lane-centering. The driver has hands off of the steering wheel and foot off pedal at some time.
Level 3 – Conditional Automation	Drivers are still in control in level three cars; however, a transfer from human to car and vice-versa for "safety critical functions" can take place.
Level 4 – High Automation	Vehicles at level 4 are designed to perform safety-critical driving functions and monitor road conditions.
Level 5 – Full Automation	Vehicles at level 5 are expected to perform at the same level of a human driver in all scenarios.

Source: Car and Driver, 2017; Reese, 2016).

As of mid-2020, Waymo has reported that collectively millions of miles had been driven by the fleet's vehicles. In Fall 2019, the company began tests taking actual passengers around Phoenix using their driverless robotaxis. The driverless rides in Arizona are only part of Waymo's overall transportation strategy. The company also signed a multiyear contract with Transdev North America to provide bus drivers and other transportation workers to airports and cities (Boudway & Brustein, 2020).

There is plenty of competition for these markets. A subsidiary of Intel, Mobileye, is also in the robotaxi business (Korosec, 2020). Mobileye supplies automakers with computer vision technology that powers advanced driver assistance systems and reported it had been installed in up to 54 million vehicles in 2019 (Korosec, 2020).

In the United States, Tesla has become a brand known for innovation and change. Tesla's line of smartly-designed cars and trucks regularly make news for its self-driving cars and experiments with cars that can be summoned. U.S. automakers are also pushing into electric and hydrogen-powered cars. For example, General Motors announced plans to have 20 new all-electric vehicle models by 2023 and hoped to have one million electric GM vehicles on the road by 2025 (Richardson, 2018). Overseas, all major car makers like Honda, Toyota, Mercedes Benz, and Renault are running tests and experimenting with ways that their models can stand out. For example, Renault's Symbioz features a "lounge mode" for passengers. When the self-driving feature is activated, the front seats swivel 180 degrees to face the back seats, turning the car interior into a cozy interactive space (Renault, n.d.).

There are at least 20 or more models of cars known for their self-parking or parking assistance abilities. Manufacturers making self-parking cars include BMW, Mercedes Benz, Ford, Volkswagen, Tesla, Volvo, Ram, Jaguar, and Lincoln (Loveday, 2018). Some of these cars are equipped with features like automatic emergency braking, lane departure warnings, and self-parallel parking (Driverless cars, 2019).

Figure 12.1

Tesla using Networked Charging Station

Source J. Wilkinson

Developing a true self-driving car has become more tricky, however. Bad weather affects road conditions, streets and highways are not always correctly mapped, and there are limitless variables and unknowns at any given time. The smart system must continually learn to anticipate and adjust at a moment's notice. Waymo software engineer Andrew Chatham said a critical part of their testing is selecting where to test-drive. "We are not driving in the dense traffic of Mumbai, where you have really hairy traffic conditions. We are not driving in the worst blizzard you can imagine" (Driverless cars, 2019).

Alternative Fuel Systems

The second major trend is a move away from fossil fuels. Several automakers are testing alternatives such as electric hybrids and alternative fuels such as hydrogen. Significant investments are being made to develop hydrogen fuel cell vehicles (HFCVs). HFCVs will significantly reduce emissions because hydrogen burns cleaner than gasoline and is far more efficient. Unlike gasoline- and diesel-powered vehicles, they do not create any greenhouse gas emissions during vehicle operation.

There are two big barriers to adoption. One barrier is how to set up the infrastructure for refueling, and the other is getting consumers to switch over to the new pricier models of HFCVs. It would take a massive re-structuring for gas stations to begin offering hydrogen, and the cheapest pricing models still make hydrogen more expensive than gasoline (Kaslikowski, 2019. Enabling factors are high-quality products, changing lifestyle of consumers, and rising spending power (Kenneth Research, 2020).

As of late 2019, the three most popular hydrogen fuel cell cars available for lease or purchase were the Toyota Mirai, the Honda Clarity Fuel Cell, and the Hyundai Nexo (Kaslikowski, 2019). All three cars are listed at almost identical prices, between $58,300 and $58,500.

The 2020 Honda Clarity Fuel Cell was tested in California in 2019 and travels about 360 miles on a tank of hydrogen. (Szymkowski, 2019). A massive outage of hydrogen fuel availability in California during 2019 made it difficult for owners to drive the fuel cell powered car (Szymkowski, 2019). Toyota reiterated their commitment to the Mirai in Fall 2019, announcing they would launch the second-generation hydrogen fuel-cell in 2020 (Buckland, 2019).

The outlook for hydrogen fuel cell cars is positive despite the infrastructure problems, and it is projected to have annual growth of more than eight percent (Kenneth Research, 2020). In the U.S., California is the acknowledged leader for testing and experimenting with HFCVs (Kenneth Research, 2020).

The President of Mercedes-Benz Canada, Brian Fulton, said the company's CASE initiative is Connected, Autonomous, Shared, and Electric. Fulton said they were working to end key fobs and rely solely on telematics to analyze driver behavior and adapt to the customer's mood, for example, setting the appropriate temperature and programming appropriate music at a given time. Fulton said the top cars will be electric and "mostly" autonomous, and could be powered by hydrogen fuel-cell. Mercedes has already announced that at least 25% of all the cars it sells in 2025 will be either pure electric or plug-in hybrids (Richardson, 2018).

Other Developments

Fleet Management Systems.

While smart cars are regularly featured in news reports, a seismic shift in fleet management has also been occurring. Large trucks and vans are also moving toward autonomous operation and using electrification and hydrogen as fuel. Samsara, Fleetio, and other companies provide software to help trucking firms track and coordinate vast truck fleets simultaneously across the country. Car companies like Uber and Lyft do the same. Fleet management software enables companies to adapt and adjust truck routes and deliveries in real-time. Automated fleets are fast becoming a standard and integral part of business. Robert Brown, head of government relations at TuSimple, which is developing self-driving truck technology, said startups and truck makers are focusing more on this technology, and today the number of companies doing truck automation is in the "high teens" (Gilroy, 2020).

There are several next-generation trucks being built. For example, Tesla's all-electric Semi truck, unveiled in 2017, is expected to go into limited production at the end of 2020. The Semi is expected to be priced starting at $150,000 for the 300-mile model and $180,000 for the 500-mile model (Gilroy, 2020). Other companies are developing innovations to improve truck and fleet performance. Nikola Corporation announced it will unveil battery technology at the end of 2020 that doubles the range of heavy-duty trucks that are either fully electric or use a hydrogen fuel cell (Gilroy, 2020).

Smart Roads

Smart car telematics are forcing governments to upgrade roads to work with autonomous and semi-autonomous vehicles. For example, the Colorado Department of Transportation has been working extensively to develop hundreds of miles of smart roads to accommodate self-driving autonomous vehicles (Roberts, M., 2019). Research indicates around 90% of all traffic accidents involve reaction time and concentration. The roads are equipped with sensors to provide real-time information to assist the long-haul trucks going through the state's most challenging mountain roads. Experts predict that smart cars will make roads much safer, and accident numbers could decline by as much as 90% by 2050 (Hulsey, 2017). Problems, however, come to play when considering road hazards, weather and security. Smart technology relies on sensors on the car and in roadside equipment.

Health, Safety, and Insurance

Automotive telematics have been designed to protect us in very sophisticated ways, sometimes in ways we don't realize. Reliance on GPS, which is standard in most new cars, must identify the vehicle's exact location at all times. Many car buyers may not be aware this feature is built-in. The details of tracking are often buried in the fine print of the dealer agreement and warranty documentation which are routinely signed when a consumer buys a new car. Carmakers give themselves the right to collect personal data that transforms "the automobile from a machine that helps us travel to a sophisticated computer on wheels that offers more access to our personal habits and behaviors than smartphones do" (Holley, 2018).

"The thing that car manufacturers realize now is that they're not only hardware companies anymore—they're software companies," said Lisa Joy Rosner, Chief Marketing officer of Otonomo—a company selling car-connected data.

> The first space shuttle contained 500,000 lines of software code, but compare that to Ford's projection that by 2020 their vehicles will contain 100 million lines of code. These vehicles are becoming turbocharged spaceships if you think of them from a purely horsepower perspective (Holley, 2018).

Automakers defend themselves noting that—like digital home assistants—we agree to the practice when we purchase the device. The car makers state they only collect customer data for benign research purposes (Holley, 2018), and it's only used to improve vehicle performance and enhance customer safety.

Telematics and in-car electronic systems have evolved and are now analogous to airplane event data recorders, so-called "black boxes" for automobiles (Holley, 2018). The information collected today is far more detailed and can be extracted and connected to the internet (Holley, 2018). For example, users of the OnStar system can have a dealer remotely diagnose problems with the car and then suggest the closest repair shop.

In 2018, there were at least 78 million cars on the road equipped with some means to monitor the performance of the driver (Holley, 2018). By 2021, automotive experts expect 98% of all new cars sold in the United States and Europe will automatically include this function (Hicks, 2018). Smart systems monitor your driving behavior, and if you weave or show signs of "instability" the car will recommend you pull over and rest (Lewis, 2019).

According to the National Association of Insurance Commissioners (NAIC), telematics devices measure several elements that interest underwriters including miles driven, time of day, where the vehicle is driven, rapid acceleration, hard braking, hard cornering, and air-bag deployment (NAIC, 2019).

This interest is reflected in the trend toward Usage-Based Insurance, pioneered by Ohio-based Root Insurance, but quickly picked up by others (Juang, 2018). Insurance companies are using your driving data to assess your rates. The goal is simply to provide safe drivers with lower rates and charge unsafe drivers higher rates. Many if not all insurers now offer performance-based programs, and legally, the insurers are required to disclose what data is used in adjusting those rates (Hicks, 2018).

The sophistication of telematics even enables monitoring inside the car, if the driver allows it. Currently, companies are trying to get policyholders to opt into initiatives like Progressive Insurance's "Snapshot" program. Customers can earn discounts based on safe driving performance, confirmed by an app that tracks your driving behavior (Juang, 2018). For example, a driver who drives long distance at a high speed will be charged a higher rate than a driver who drives short distances at a slower speed (NAIC, 2019). Insurance companies are also experimenting with how to integrate inside-the-car dashboard cameras. Some legal issues are involved, and consumers often feel this is a bit too intrusive.

Current Status

It is estimated that globally we surpassed one billion cars worldwide in 2009, then surpassed 1.2 billion in 2015 (OICA, n.d.). Today there are probably more than 1.6 billion cars and trucks travelling somewhere on earth. By comparison, the movement toward electric cars seems tiny. The International Energy Agency (IEA) reported that the global electric car fleet grew substantially from 2017 to 2018, from around 3 million to just above 5.1 million (IEA, 2019). The world's largest electric car market is the People's Republic of China, followed by Europe, and the United States. By the year 2040, projections are for sales of new electric vehicles to outnumber internal combustion engine cars (Valdes-Dapena, 2019).

Smart car technologies are estimated to save drivers between $60 and $266 each year, totaling up to $6.2 billion in annual fuel cost savings (Stevens Institute, 2018). These efficiencies and economies of scale are expected to increase with improved autonomous systems.

Governments are also encouraging the move toward sustainability, and trying to anticipate where the flood of innovation is taking us. In early 2020, the U.S. federal government established regulations for automated vehicle technologies. The United States Department of Transportation's (USDOT) "AV 4.0" regulations would unify standards for automated vehicles across 38 federal departments, independent agencies, commissions and the office of the U.S. President (USDOT, 2020). USDOT Secretary Elaine Chao said the regulations were needed to help America stay competitive in AV technology development and integration (USDOT, 2020).

Factors to Watch

It is difficult for reality to match the optimism pushing for change. Five years ago, you could find articles predicting there would be millions of self-driving cars in operation today. As mentioned earlier, we remain in the earliest stages of testing and development. Still, updated forecasts predict that once autonomous cars are made available, consumers will rush to buy. In the midst of all this change, the main areas to watch will continue to be telematics and smart systems.

Safety

The issue of safety remains of paramount concern. The safety of the driver, the passengers, and others undergirds much of the developments in adding sensors and extensive software monitoring features to the vehicle telematics. The powerful assumption is that computers and robots do not sleep, nor do they drink or take medications. Advocates of driverless cars say the country could be saved from the 37,000 traffic deaths a year by using autonomous vehicles. Experts, however, say, no one can be certain about the safety potential of self-driving cars until they are a part of daily life (Driverless cars, 2019).

There were a small but significant number of deaths recorded in 2019 involving autonomous cars. The incidents ranged from Tesla's AutoPilot, which is nominally a Level 3 autonomous vehicle to instances with Uber's self-driving car which is somewhere in the area of Level 4 or Level 5 autonomy (Schmelzer, 2019).

The single U.S. case where a pedestrian was killed by a level 3 autonomous vehicle occurred March 18, 2018. That first recorded pedestrian death occurred in Tempe, Arizona when a self-driving Uber car accidentally struck and killed a pedestrian who strayed into the road (Gonzales, 2019). The National Transportation Safety Board (NTSB) concluded that multiple sensors missed or misread the situation and could not avoid hitting and killing pedestrian Elaine Herzberg who was walking her bike outside of a crosswalk (Gonzales, 2019). The NTSB report identified three places where Herzberg was not detected. The SUV had "a fusion" of three sensor systems—radar, lidar (light detection/ranging system), and a camera—designed to detect an object and determine its path. However, the system could not determine whether Herzberg was a pedestrian, vehicle, or a bicycle, and it failed to correctly predict her path (Gonzales, 2019).

There are often risks associated with innovation and it is logically impossible to prepare for all possible tragedies. The nature of machines underscores the seriousness of research and development. Undoubtedly accidents will occur, but overall, telematics reduces the overall accident and death rates involving motor vehicles.

Hacking

Another key concern plaguing automobiles with advanced technology, is hacking. In 2015, Charlie Miller and Chris Valasek were hired to research hacking into a Jeep Cherokee through the internet. The researchers successfully hacked the system and from a remote location, were able to turn the steering wheel, disable the brakes and shut down the engine (Ilunin, 2017). The researchers were also able to access thousands of other vehicles that used the wireless entertainment and navigation system called Uconnect (Ilunin, 2017). Dodge, Jeep, and Chrysler issued a major recall of 1.4 million vehicles after the hack (Ilunin, 2017). When they announced their findings, the risks of smart car hacking became a reality for many. Afterward, German researchers released a report noting where they were also able to unlock and start cars with wireless key fobs. They reported that hackers could even drive off in a smart car when the key fob remained at the owner's house (Ilunin, 2017). "If it's a computer and it connects to the outside world, then it is hackable," said Yoni Heilbronn, vice president of marketing at automotive security firm Argus Cyber Security (Ilunin, 2017).

According to several leading car-makers, between now and 2025 we will begin to see the first Level 4 autonomous vehicles on the road (GreyB Research, 2019), and sometime after 2030 we may begin seeing Level 5 autonomous cars. The delay in Level 5 will come from the obvious risk factors in turning over complete control to autonomous systems. Several layers of regulation and infrastructure will be needed before these can begin to happen. For all the tests being conducted, each time there is an accident or death, things slow down exponentially.

Regulation

In light of this, the regulatory structure is trying to keep up with the proposed changes. Governments around the world wrestle with how quickly to unleash self-driving cars on an unwitting populace? Few officials fully understand how these changes will ultimately look in the future. Unveiling a new system too soon could wreak havoc and result in tragedy and disaster. Governments will continue to monitor tests as well as facilitate an orderly transition in the marketplace. City roads and infrastructure must be improved and adapted to these anticipated changes. Ultimately it will be the regulatory issues at the federal, state and local levels that will hinder the launch of autonomous vehicles (GreyB Research, 2019). Autonomous vehicles are the hottest topic gripping the auto industry. How well the machines are designed and will respond to us will determine our future.

Gen Z

For individuals born after 1997, driving a car is not the rite of passage it used to be. In a trend first noticed among older groups, the "always-pugged-in" generation is increasingly comfortable ordering Uber or Lyft to take them somewhere without the hassle of insurance or finding a parking space. According to research from the University of Michigan, the overall percentage of young people with driver's licenses has steadily declined for over a generation (Bomey, 2016). This trend holds for first-time-drivers aged 16 as well as older cohorts aged 20-24.

For example, among 20–24 years olds, the percentage holding a driver's license slid from 92% in 1983 to 77% in 2014. The numbers are similar for 16-year-olds, typically the first time a young person can apply for the license. According to the same UM study, the percentage of 16-year-olds applying for a license has continually dropped over the past few decades, from 46% in 1983 to 24.5% in 2014.

Since World War II, cars have symbolized adulthood, freedom, and independence. Cars enabled teenagers to forge their identities through cultural activities like dating and cruising with friends. Clearly the meaning and function of cars has changed for Gen Zers from those older generations.

One factor is the cost of a new car. The average cost of a new car in 2019 was $37,000 (Frio, 2019). Opting to exclusively use ride services like Uber and Lyft can cost under $200 a month, far below a typical new car payment (Passy, 2019). A second factor is the time and money involved with insurance, maintenance and repair, and fuel. A third factor is sustainability—the environment and "being green." Cars pollute and young people recognize peers who use smart alternative forms of transportation like buses, trains, or getting around locally using a bicycle, scooter, or skateboard.

Still, the move toward smart cars is popular with Gen Zers, and half—50%—of car-owning millennials prefer to buy a "green" car over a traditional car (Brown, 2017). Generation Zers find self-driving vehicles appealing due to the safety factor that could be offered (Manufacturing Group, 2016). 61% of the Gen Zers believe the roads will be much safer with autonomous vehicles (Manufacturing Group, 2016).

Sustainability

Cars and trucks have been a major cause of pollution worldwide, and manufacturers are trying to reduce the adverse effects on our environment. As we approach 1.7 billion or more cars and trucks worldwide, we must acknowledge how the production, use, and disposal of vehicles worldwide leave a massively bad impact on the environment.

Throughout the lifespan of every vehicle, there are adverse effects from beginning to end. For example, the initial production process leaves a damaging footprint because we build our cars using materials like steel, rubber, glass, plastics, and paints—all of which produce emissions in the production process.

Once completed, our vehicles are placed on the open road where they may last years and travel countless thousands of miles. During this time the vehicles do the greatest amount of harm to the air. Most of the environmental damage—80–90%—is attributed to fuel consumption and emissions of air pollution and greenhouse gases (National Geographic, 2019). But besides polluting the air, the infrastructure needed for our vehicles to operate does damage to our fields and forests. Roads, parking lots, rest areas and gas stations all scar the natural beauty of the planet. In other words, to "get your motor runnin", [to] head out on the highway, we had to pave paradise and put up a parking lot" (credits to Mars Bonfire and Joni Mitchell).

Most cities and towns have authorized locations to dispose of cars and trucks. Abandoning cars illegally does serious damage to the earth because plastics, battery acids, and oil can leach into the soil, placing deadly toxins into the ground. Since it is estimated that 85% of the material in junked cars and trucks can be recycled, it is best for them to be disposed of properly at a licensed auto salvage yard (New Hampshire Department of Environmental Services, n.d.).

Bibliography

Bekker, H. (2019, February 23). 2018 (Full Year) International: Worldwide Car Sales and Global Market Analysis. Best-Selling-Cars.com. https://www.best-selling-cars.com/global/2018-full-year-international-worldwide-car-sales-and-global-market-analysis/.

Berkowitz, J. (2010, October 25). The History of Car Radios. Caranddriver.com. https://www.caranddriver.com/features/a15128476/the-history-of-car-radios/.

Bogost, I. (2014, November). The Secret History of the Robot Car. *The Atlantic*. https://www.theatlantic.com/magazine/archive/2014/11/the-secret-history-of-the-robot-car/380791/.

Boudway, I., and Brustein, J. (2020, January 18). Waymo's long-term commitment to safety drivers in autonomous cars. SFgate.com. https://www.bloomberg.com/news/articles/2020-01-13/waymo-s-long-term-commitment-to-safety-drivers-in-autonomous-cars.

Bomey, N. (2016, January 19). Millennials spurn driver's licenses, study finds. *USA Today*. https://www.usatoday.com/story/money/cars/2016/01/19/drivers-licenses-uber-lyft/78994526/.

Brown, M.. (2017, August 22). Where will millennials take the car industry? Lendedu. https://lendedu.com/blog/millennials-cars/.

Buckland, K. (2019, September 25). Toyota preparing next-gen Mirai fuel-cell car for 2020 launch: chairman. Reuters.com. https://www.reuters.com/article/us-toyota-hydrogenidUSKBN1WA0DM.

Buehler, R. (2014, February 4). 9 Reasons the U.S. Ended Up So Much More Car-Dependent Than Europe. Citylab.co. https://www.citylab.com/transportation/2014/02/9-reasons-us-ended-so-much-more-car-dependent-europe/8226/.

Car and Driver. (2017, October 3). Autonomous self driving car levels. Caranddriver.com. https://www.caranddriver.com/features/a15079828/autonomous-self-driving-car-levels-car-levels/.

Duan, N. (2018, January 4). Everything about our post-apocalyptic future was predicted on the Jetsons. *Quartz*. https://qz.com/quartzy/1170999/the-iconic-jetsons-television-show-predicted-the-future/.

Driverless Cars are taking longer than we expected. Here's Why (2019, July 14). *New York Times*. https://www.nytimes.com/2019/07/14/us/driverless-cars.html.

Enero, D.A. (2018, March 20). Smart cars - driving the innovation. Apiumhub.com. https://apiumhub.com/tech-blog-barcelona/smart-cars-driving-innovation/.

Frio, D. (2019, December 10). Cheapest new cars: most affordable cars for 2020. Edmunds.com. https://www.edmunds.com/vehicles/cheapest-new-cars/.

Gilroy, R. (2020, January 2). Truck Orders to Plunge, GDP to Fall Lower, Diesel Prices to Rise in 2020. *Transport Topics*. https://www.ttnews.com/articles/truck-orders-plunge-gdp-fall-lower-diesel-prices-rise-2020.

Gonzales, R. (2019, November 7). Feds say self-driving Uber SUV did not recognize jaywalking pedestrian in fatal crash. NPR. https://www.npr.org/2019/11/07/777438412/feds-say-self-driving-uber-suv-did-not-recognize-jaywalking-pedestrian-in-fatal-crash.

Gray, C. (n.d.). Technology in automobiles. R6 Mobile. http://support.r6mobile.com/index.php/knowledgebase/152-technology -in-automobiles.

GreyB Research (2019, September 5). Autonomous Vehicle Market Report: Size, Forecast, Companies, and Research Trends of the Self-Driving Industry. GreyB.com. https://www.greyb.com/autonomous-vehicle-market-report/.

Hicks, K. (2018, February 20). Tracking Technology: Your car is definitely watching you – but that might not be a bad thing. *Zebra*. https://www.thezebra.com/insurance-news/5576/tracking-tech-good-and-bad/.

Holley, P. (2018, January 15). Big Brother on wheels: Why your car company may know more about you than your spouse. *Washington Post*. https://www.washingtonpost.com/news/innovations/wp/2018/01/15/big-brother-on-wheels-why-your-car-company-may-know-more-about-you-than-your-spouse/.

Hulsey, L. (2017, September 21). 'Smart Car' Technology May Make Roads Safer, but some fear data hacks. *DaytonDaily News.com*. https://www.daytondailynews.com/news/transportation/smart-car-technology-may-make-roads-safer-but-some-fear-data-hacks/ccfhLefEmuL0wF6n4MmonK./

Ilunin, I. (2017, October 31). Smart Car Hacking: A Major Problem for IoT. *Hackernoon*. https://hackernoon.com/smart-car-hacking-a-major-problem-for-iot-a66c14562419.

International Energy Agency (2019, May). Global EV Outlook 2019: Scaling up the transition to electric mobility. IEA.org. https://www.iea.org/reports/global-ev-outlook-2019.

International Organization of Motor Vehicle Manufacturers, (n.d). 2017 Production Statistics. http://www.oica.net/category/production-statistics/.

Juang, M. (2018, October 6). A new kind of auto insurance technology can lead to lower premiums, but it tracks your every move. CNBC.com. https://www.cnbc.com/2018/10/05/new-kind-of-auto-insurance-can-be-cheaper-but-tracks-your-every-move.html.

Kaslikowski, A. (2019, October 23). Hydrogen was the fuel of tomorrow, so what happened? Digitaltrends.com. https://www.digitaltrends.com/cars/hydrogen-cars/.

Kenneth Research (2020, January 10). Hydrogen Fuel Cell Vehicle Market Forecast To 2023. Marketwatch.com. https://www.marketwatch.com/press- release/hydrogen-fuel-cell-vehicle-market-forecast-to-2023-trends-industry-size-sale-2020-01-10.

Korosec, K. (2020, January 15). Mobileye takes aim at Waymo. Techcrunch.com. https://techcrunch.com/2020/01/15/mobileye-takes-aim-at-waymo/.

Lewis, N. (2019, December 27). Car camera system could help keep drivers awake at the wheel. CNN Business. CNN.com. https://www.cnn.com/2019/12/27/business/technology-detects-drowsy-drivers/index.html

Loveday, S. (2018, December 18). 20 Best Self-Parking Cars in 2018. *U.S. News & World Report Best Cars*. https://cars.usnews.com/cars-trucks/best-self-parking-cars.

Marr, B. (2018, September 21). Key Milestones of Waymo – Google's Self-Driving Cars. *Forbes*. https://www.forbes.com/sites/bernardmarr/2018/09/21/key-milestones-of-waymo-googles-self-driving-cars/#7a3840625369.

Manufacturing Group (2016, March 16). Gen-Z more excited about driving than Millennials: Study shows excitement, greater willingness to get licenses, cars. TodaysMotorVehicles.com https://www.todaysmotorvehicles.com/article/gen-z-millennials-driving-autotrader-kelley-blue-book-031616/.

Mitchell, J. (1970). Big yellow taxi. On *Ladies of the Canyon* [LP]. New York: Reprise.

Mars Bonfire (Edmonton, Dennis) (1968). Born to be wild. On *Steppenwolf* [LP]. ABC Dunhill RCA.

National Association of Insurance Commissioners. (2019, May 19). The Center for Insurance Policy and Research. https://content.naic.org/cipr_topics/topic_usage_based_insurance.htm.

National Geographic Staff (2019, September 4). The environmental impacts of cars, explained: automobiles have a big footprint, from tailpipe emissions to road infrastructure. Nationalgeographic.com. https://www.nationalgeographic.com/environment/green-guide/buying-guides/car/environmental-impact/.

New Hampshire Department of Environmental Services (n.d.). Where junk cars go. https://www.des.nh.gov/organization/commissioner/pip/publications/wmd/documents/where_junk_cars_go.pdf.

Organization of Motor Vehicle Manufacturers (OICA) (n.d.). World vehicles in use. OICA.net. http://www.oica.net/wp-content/uploads//Total_in-use-All-Vehicles.pdf.

Passy, J. (2019, April 12). Will Uber and Lyft make 'car cutting' the new cord cutting? Marketwatch.com. https://www.marketwatch.com/story/will-uber-and-lyft-make-car-cutting-the-new-cord-cutting-2018-04-20.

Reese, H. (2016, January 20). Autonomous driving levels 0 to 5: Understanding the differences. *Tech Republic*. https://www.techrepublic.com/article/autonomous-driving-levels-0-to-5-understanding-the-differences/.

Renault (n.d.). Renault SYMBIOZ concept. Renault.co.uk. https://www.renault.co.uk/concept-cars/symbioz.html.

Richardson, (2018, January 12). The ride of the future: Auto makers look into their crystal balls – and their shops – to get an idea as to what's fueling the car of 2025. TheGlobeandMail.com. https://www.theglobeandmail.com/globe-drive/.culture/commentary/auto-makers-predict-what-the-car-of-2025-will-looklike/article37537602/.

Roberts, A. (2019, April 20). Driving? The kids are so over it. *Wall Street Journal*. https://www.wsj.com/articles/driving-the-kids-are-so-over-it-11555732810.

Roberts, M. (2019, May 17). Inside the Smart Highway System That Could Revolutionize Travel in Colorado. Westword.com. https://www.westword.com/news/inside-the-smart-highway-system-v2x-that-could-revolutionize-travel-in-colorado-10692021.

Schmelzer, R. (2019, September 26). What Happens When Self-Driving Cars Kill People? Forbes.com. https://www.forbes.com/sites/cognitiveworld/2019/09/26/what-happens-with-self-driving-cars-kill-people/#4d7ea6c9405c.

Stevens Institute of Technology. (2018, November 20). Smart Car technologies save drivers $6.2 billion on fuel costs each year. https://www.sciencedaily.com/releases/2019/07/190729111337.htm.

Szymkowski, S. (2019, December 18). 2020 Honda Clarity Fuel Cell will work better in the cold now: Not that Southern California — the biggest market for the fuel cell model — gets that cold. Cnet.com. https://www.cnet.com/roadshow/news/2020-honda-clarity-fuel-cell-price-features/.

U.S. Department of Energy. (2014, September 14). The History of the Electric Car. https://www.energy.gov/articles/history-electric-car.

U.S. Department of Transportation. (2020, January 8). Trump administration releases "Ensuring American leadership in automated vehicle technologies: Automated vehicles 4.0." https://www.transportation.gov/briefing-room/trump-administration-releases-%E2%80%9Censuring-american-leadership-automated vehicle?utm_source=morning_brew.

Valdes-Dapena, P. (2019, September 6). By 2040, more than half of new cars will be electric. CNN.com. https://www.cnn.com/2019/05/15/business/electric-car-outlook-bloomberg/index.html.

Wise, B. (2018, March 29). A journey through the history of telematics. Tomtom.com. https://www.webfleet.com/en_gb/webfleet/blog/history-of-telematics/.

13

Virtual & Augmented Reality

Rebecca Ormond, M.F.A.[*]

Abstract

Virtual and augmented reality (VR/AR) systems have been used in specialized professional applications for decades. However, when Oculus released the Oculus Rift SDK 1 system in 2012 and Facebook purchased Oculus for $2 billion in 2014, a consumer VR entertainment industry was born. Many early inhibitors to consumer adoption, such as latency issues that caused VR sickness, are now being resolved, while new markets, such as Enterprise AR and RPG (role playing games) VR gaming are emerging through new head mounted displays (HMD), most notable the all-in-one HMD. As of 2020, VR and AR use has expanded across numerous markets, and the combined AR and VR markets are predicted to continue to grow.

Introduction

Virtual Reality (VR), Augmented Reality (AR), Mixed Reality (MR), and Extended Reality (XR), are differentiated by software, hardware, and content. VR artist and author Celine Tricart (2018) describes VR content as "an ensemble of visuals, sounds and other sensations that replicate a real environment or create an imaginary one," while AR content "overlays digital imagery onto the real world." When VR and AR are used together, we call it "mixed reality." VR, AR, and MR are all forms of XR. Another notable distinction is when the XR user interacts with physical objects or has a sense of "touch" usually via a controller which is referred to as haptics, in other words "haptic systems serve as input devices" (Jerald, 2016).

[*] Assistant Professor, Department of Media Arts, Design, and Technology, California State University, Chico (Chico, California)

Typically, VR is watched on a stereoscopic (a distinct image fed to each eye) head mounted display (HMD) that employs inertial measurement unit tracking (IMU) that "combine gyroscope, accelerometer and/or magnetometer hardware, similar to that found in smartphones today, to precisely measure changes in rotation" (Parisi, 2016). HMDs fall into three broad categories: smartphone-based, tethered (requiring a computer and software development kit or SDK), and all-in-one (also referred to as stand-alone, similar in function to a tethered HMD, but not tethered to a computer). Tethered HMDs allow for active haptics, 6 degrees of freedom (DoF), which means you can move around, and agency, which means you can guide the experience. Playstation VR and Oculus Rift are examples of tethered HMDs.

The all-in-one HMD is relatively new, with many HMDs, including Oculus Go, VIVE Focus, and others (Conditt, 2017), first announced for release in 2018. First generation all-in-one HMDs were 3DoF without haptics, but later releases such as the Oculus Quest include 6DoF and active haptics (Dujmovic, 2019). Smartphone-based HMDs use a smartphone compatible with the HMD, but have only 3DoF and no agency, which translates into limited movement and user control of the experience. Google Cardboard and Samsung Gear VR are examples of smartphone HMDs. Alternatively, VR can be projected with a Cave Automatic Virtual Environment (CAVE) system, "a visualization tool that combines high-resolution, stereoscopic projection and 3D computer graphics that create the illusion of a complete sense of presence in a virtual environment" (Beusterien, 2006). What all VR systems have in common is they block out the real world and are therefore considered immersive.

AR is viewed through glasses or smart devices such as a smartphone (not housed in an HMD), which do not occlude the real world as seen by the wearer and are therefore not immersive. The 2015 release of the Microsoft HoloLens, which uses "a high-definition stereoscopic 3D optical see-through head-mounted display (and) uses holographic waveguide elements in front of each eye to present the user with high-definition holograms spatially correlated and stabilized within your physical surroundings" (Aukstakalnis, 2017) is an AR HMD. AR Apps can be viewed in an HMD or a smartphone; for example, the popular *Pokémon Go* app is an example of a smartphone AR app.

Apps are created using software development kits (SDKs). "A software development kit (or SDK) is a set of software development tools that allows the creation of applications for a certain software package, software framework, hardware platform, computer system, video game console, operating system, or similar development platform" (Tricart, 2018).

VR content can be "live" action, computer generated (CG), or both. Live action VR content is captured via synchronized cameras or a camera system using multiple lenses and sensors, with individual "images" being "stitched" together to form an equi-rectangular or "latlong" image (similar to how a spherical globe is represented on a flat map). VR experiences are usually posted in a non-linear edit system (NLE) or Game Engine and Digital Audio Workstation (DAW). Initially, there were only professional 3D (stereoscopic) 360° VR cameras used by big studios, such as the Ozo 360° stereoscopic camera (Kharpal, 2015), but beginning in 2016, prosumer and consumer VR cameras, available in 3D 180°, 2D 360° and 3D 360°, became available, and popular streaming sites such as Facebook, YouTube, Vimeo and others began expanding their services to include "magic window" and HMD support for VR both business and consumer streaming, including some live streaming.

Figure 13.1

Insta360 One X

Source: J Meadows.

Table 13.1

Non-Entertainment VR/AR Companies & Products

Industry	Company/Organization	VR/AR Application
Real Estate	Matthew Hood Real Estate Group & Matterport, Inc.	Real estate viewable as MR on HMD
Automotive Engineering	Ford Motor Co. & NVIS	Design Analysis & Production Simulation
Aerospace Engineering	NASA & Collaborative Human Immersive Laboratories (CHIL)	CHIL's CAVE to assist in engineering
Nuclear Engineering and Manufacturing	Nuclear Advanced Manufacturing Research Center (NAMRC) & Virtalis	Active Cube /Active Wall 3D visualization
Medicine (non-profit)	HelpMeSee (non-profit organization and global campaign) uses a number of organizations including Moog, Inc., InSimo, and Sense Graphics	The Manual Small Incision Cataract Surgery (MSICS) simulator
Medicine (education)	Academic Center for Dentistry & Moog Industrial Group	The Simodont Dental Trainer
Heath (PTSD treatment)	The Geneva Foundation & Virtually Better, Inc., Navel Center	Bravemind: MR using stereoscopic HMD (sensor& hand-held controller)
Health (phobias)	Virtually Better & VRET software suites	MR using stereoscopic HMD (sensor & hand-held haptics)
Medicine (operation room)	Vital Enterprises	Vital Stream (medical sensor & imaging)
Defense (flight)	NASA's Armstrong Flight Research Center & National Test Pilot School	Fused Reality (flight simulator)
Defense (planning and rehearsal)	U.S. Military developed The Dismounted Soldier Training System (DSTS)	Fully portable immersive environment MR training system
Defense (rescue)	U.S. Forest Service with Systems Technology, Inc.	VR Parachute Simulator (PARASIM)
Advanced Cockpit Avionics	US Military & Lockheed Martin	f-35 Joint Strike Fighter Helmet Mounted Display system
Education (vocational training)	Lincoln Electric with VRSIM, Inc.	VTEX (a virtual welding system)
Education (theory, knowledge, acquisition & concept formation)	Multiple educational institutions using multiple Augmented Displays	Multiple Building and Information Modeling (BIM) applications, AR display & interactive lessons
Education (primary)	Multiple Schools with the Google Expeditions Pioneer Program	Google's Expeditions Kit
Big Data	Big Data VR challenge launched by Epic Games & Welcome Trust Biomedical Research Charity	LumaPie (competitor) built a virtual environment where users interact with data
Telerobotics/ telepresence	Developed through a space act agreement between NASA &General Motors	Robonaut 2 (remote robot controlled for use in space via stereoscopic HMD, vest and specialized gloves)
Architecture	VR incorporated into computer-assisted design programs	Allow visualization of final structure before or during construction.
Art	Most VR tools can be repurposed by artists to create works of art	Virtual art extends our understanding of reality

Source: R. Ormond

VR games are built-in game engines, for example, Unity 3D or Unreal. Since game engines include "video" importing and exporting capabilities, non-gaming VR content can also be created in game engines. These game engines are free and have been very popular with both professional and amateur content creators. Today, most of the major Non Linear Editors (NLE) will also work alone or via plugin with VR's latlong aspect ratio and common over-under or side-by-side stereoscopic video formats.

Increasingly, VR, AR, MR and all forms of XR are being used beyond entertainment. Table 12.1 lists many examples of VR/AR companies and products for non-entertainment industries. These examples were sourced from *Practical Augmented reality: A guide to the technologies, applications, and human factors for AR and VR* (Aukstakalnis, 2017), where full descriptions can be found. In particular, note that real estate, architecture, and flight simulators are robust markets for VR.

Background

"Virtual reality has a long and complex historical trajectory, from the use of hallucinogenic plants to visual styles such as Trompe L'Oeil" (Chan, 2014). Even Leonardo da Vinci (1452-1519) "in his Trattato della Pittuara (Art of Painting) remarked that a point on a painting plane could never show relief in the same way as a solid object" (Zone, 2007), expressing his frustration with trying to immerse a viewer with only a flat frame.

Then in 1833 Sir Charles Wheaton created the stereoscope, a device that used mirrors to reflect two drawings to the respective left and right eye of a viewer, thus creating a dual-image drawing that appeared to have solidity and dimension (Figure 13.2) Wheaton's device was then adapted into a handheld stereoscopic photo viewer by Oliver Wendall Holmes, and "it was this classic stereoscope that millions used during the golden age of stereography from 1870 to 1920" (Zone, 2007). "Google's Cardboard as an inexpensive mass-produced VR viewer is almost identical in purpose and design to its analog ancestor, Holmes' patent-free stereoscope, from over 150 years ago" (Tricart, 2018).

Figure 13.2

Charles Wheaton's Stereoscope (1833)

Source: Wikimedia Commons-public domain

Between the mid-1900s and 1960, there were a few "mechanical" VR devices, most notably Edwin Links' flight simulator created in 1928, with 10,000 units being sold principally to the Army Air Corps by 1935 (Jerald, 2016), but the next real big innovation involved the computer.

In 1968, Ivan Sutherland and Bob Sproull constructed what is considered the first true VR head-mounted display, nicknamed the Sword of Damocles, which was "somewhat see through, so the user could see some of the room, making this a very early augmented reality as well" (Tricart, 2018). By the 1980s, NASA developed a stereoscopic HMD called Virtual Visual Environment Display (VIVED), and Visual Programming Language (VPL) laboratories developed more HMDs and even an early rudimentary VR glove (Virtual Reality Society, 2017). During this same time, VR creators, most notably VPL laboratories, turned their attention to the potential VR entertainment market. By 1995, the New York Times reported that Virtuality managing director, Jonathan Waldern, predicted that the VR market would reach $4 billion by 1998. However, this prediction ended with "most VR companies, including Virtuality, going out of business by 1998" (Jerald, 2016). At this time limited graphic ability and latency issues plagued VR, almost dooming the technology from the start. Still, "artists, who have always had to think about the interplay between intellectual and physical responses to their work, may play a more pivotal role in the development of VR than technologists" (Laurel, 1995). During this same time, artist

Char Davies and her group at SoftImage (CG creator for Hollywood films) created Osmose, "an installation artwork that consists of a series of computer-generated spaces that are accessed by an individual user wearing a specialized head mounted display and cyber-vest" (Chan, 2014). Osmose is an early example of the "location based" VR experiences and haptic suits that are re-emerging today.

In 2012, Oculus released the Oculus Rift SDK I, a breakthrough tethered stereoscopic HMD (software and hardware) with a head-tracking sensor built into the lightweight headset (Parisi, 2016). Oculus Rift SDK1 and SDK 2 (both tethered HMDs) were released to the public, but between the cost of the hardware and the required computer, the final cost remained well over $1,000 and required that the user have some computer knowledge. AR was also introduced via Google Glass in 2012, but many issues from limited battery life to being "plagued by bugs"—and even open hostility by users (because of privacy concerns)—led to them being quickly pulled from the market (Bilton, 2015). Then in 2014, Facebook bought Oculus for $2 billion (Digi-Capital, 2017a) while Google cardboard was introduced at the Google I/O 2014 developers' conference, with SDK's both Android and iOS operating systems with simple cardboard instructions available on the web (Newman, 2014). In 2015, Microsoft announced HoloLens, "a high-definition stereoscopic 3D optical see-through head-mounted display which uses holographic waveguide elements in front of each eye to present the user with high-definition holograms spatially correlated and stabilized within your physical surroundings" (Aukstakalnis, 2017). Meanwhile, Valve, which makes Steam VR, a VR system, and HTC collaborated to create HTC VIVE, a tethered HMD (PCGamer, 2015) and Oculus partnered with Samsung to create Samsung Gear VR, a smartphone HMD in November 2015 (Ralph & Hollister, 2015.)

By 2016, Sony released the Playstation VR, which, while tethered, could be used with the existing Playstation 4, selling the most tethered HMD units in 2016 (Liptak, 2017) while Google Cardboard was biggest seller in the VR and AR market, outselling all other HMDs (tethered and smartphone) as well as AR glasses combined, with an estimated 10 million sold in 2017 (Vanian, 2017).

Figure 13.3

VR Demonstration using the HTC VIVE

Source: H.M.V. Marek, TFI

Soon after Google cardboard, the market was flooded with many HMDs for phones, including Google's own Daydream, a higher end smartphone HMD announced at the Google I/0 16 developers' conference (Hollister, 2016). Simultaneously, AR apps such as *Pokémon Go*, intended for smart phones (not housed in an HMD) outperformed expectations, and early pioneers such as Magic Leap announced the first generation of all-in-one consumer AR glasses (Conditt, 2017). However, there were clear, long-term inhibitors to early AR and VR HMDs. According to Digi-Capital (2017b): "5 major challenges must be conquered for them [AR glasses] to work in consumer markets: (1) a device (i.e., an Apple quality device, whether made by Apple or someone else), (2) all-day battery life, (3) mobile connectivity, (4) app ecosystem, and (5) telco cross-subsidization." While the major inhibitors to VR (viewing in HMD) were resolution, field of view (FOV), and latency, as well as the need for a specific app, headset, and browser. In a 2017 interview, Antti Jaderholm, co-founder and Chief Product Officer at Vizor, said, "...85% consume

WebVR experiences from our site without any head-set" (Bozorgzadeh, 2017).

In 2017, Google announced advances to WebVR and the release of ARcore, while Apple announced ARKit (i.e., SDK's), and we saw increased investment in app creation. Meanwhile, as a solution to the cost and complexity of the tethered VR HMD and resolution and latency issues of the smartphone HMD, the all-in-one HMD arrived. Examples include the Oculus Go priced under $200 and HTC's VIVE Stand Alone (Greenwald, 2017).

Entertainment VR evolved from early suggestions that it might be the new "film" to its own category, Cinematic VR, with most major film festivals (Sundance, Tribecca, Sundance, etc.) having a VR category. The Academy of Motion Picture and Sciences awarded its first VR award "in recognition of a visionary and powerful experience in storytelling," and Obama's White House tour, *The People's House* by Felix & Paul's VR studio, won an Emmy (Hayden, 2017).

Distribution also diversified through transmedia, i.e., reselling the same story franchise across multiple experiences in venues like theme parks or VR arcades. For example, Star Wars was also available as Star Wars VR experience, Star Wars: Secrets of the Empire, at Void locations (The Void, 2017).

Recent Developments

VR and AR were active industries in 2018 and 2019 with five major shifts inhibiting or enabling VR, AR, MR, and XR (all of them under the broader concept of Extended Reality). The first shift was the decline of the smartphone VR HMD, inhibiting sales in late 2018. According to Strategy Analytics, "Smartphone based VR account[ed] for the majority of shipments" in 2018 (Macqueen, 2019), but early in 2018 the all-in-one HMD, most notably the Oculus Go at a $199 price point, was released.

Very soon after, Samsung stopped bundling VR headsets, with their smartphones, and Google stopped manufacturing its own smartphone HMD, Daydream View, explaining that it had "clear limitations" (Hruska, 2018).

Figure 13.4

Watching Cinematic VR on the Oculus Go Headset

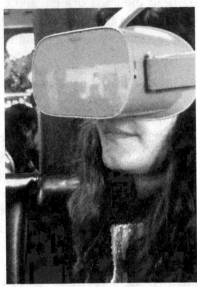

Source: R. Ormond

While Oculus Go sales were strong, they could not offset the decline in smartphone sales. However, in 2019 the all-in-one HMD got some major upgrades (SuperData, 2019) most notably with Oculus Quest, which featured 6DoF and active haptics, both of which are important for role playing games (RPG), for only $399. For the first time, all-in-one HMDs started to close the sales gap with the PlayStation VR.

Figure 13.5

Virtual Reality Headset Shipments

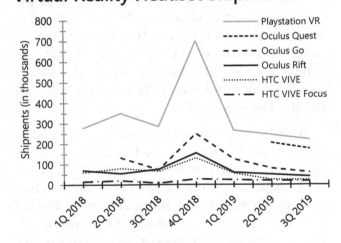

Source: SuperDATA, A Nielsen company

While gaming is more than 50% of the VR market, VR represents only a small portion of the gaming market. For more details on gaming, see Chapter 14.

Since VR gaming represents over half of VR sales (Milijic, 2019), the arrival of a game-ready all-in-one HMD drove sales. In September 2019, at Oculus Connect 6, "Mark Zuckerberg announced that the company had surpassed $100 million in revenue in the Oculus store...20% of that revenue was for Quest titles" (Matney, 2019). In October, *AsGard's Wrath*, a more than 40-hour epic VR RPG adventure game (most games are around 2 hours), was released through the Oculus store for the Oculus Rift (high-end tethered HMD) for $39.99. While this is just one game, reviews heralded it as the next generation of RPG VR, a "staggering masterpiece with some of the most intricately detailed environments I've seen in any VR game" (Jagneaux, 2019) with "more than 1,000 ratings in just a few days" (Reddit, 2019). Finally, in November, Oculus released Oculus Link, which essentially allowed the Quest to play any Rift game, including *Asgard's Wrath*. By December, "Retailers like Amazon and Best Buy sold out" (Fink, 2019a). According to a SuperData (2019) XR Dimensions report, there was "widespread consumer adoption of the Oculus Quest standalone VR device."

In the third shift, AR glass manufacturing and sales found a new market with the move from consumer to Enterprise AR, which also enabled sales in 2019. According to Tom Mainelli, group vice president of Consumer and Devices Research at IDC, "Enterprise AR is evolving quickly and many companies are actively looking for hardware solutions they can use to improve existing business processes and drive new ones" (Mainelli and Ubrani, 2019). Google, for example, may have canceled the Daydream smartphone VR HMD, but it has relaunched and upgraded an expensive version of its AR glasses, "Enterprise Edition 2," exclusively for business (Diamandis, 2019). HTC "announced the formation of a new business unit, VIVE Enterprise Solutions" (Fink, 2019b), and Magic Leap launched Enterprise Suite that includes a Magic Leap 1 headset (an update to the One Creator edition) (Engadget, 2019b).

In the fourth and fifth shifts, increased data rates via 5G and the return of AR glasses re-enabled investment in AR apps, most notably those would also work in 3D space, including new startups, so while there was major slump in investing in 2018 (Venture Scanner, 2019), by 2019 funding was returning (Palumbo, 2019). For example, Google also reinvested in "AR experiences like Google Lens, AR walking navigation in Maps, and AR search" (Engadget, 2019a). "Scape is focusing on enterprise-class, city scale mapping, as is Facebook, which is working to map both the world and the interiors of buildings within it. By Contrast, Snapchat is mapping individual landmarks for consumer applications" (Horowitz, 2019).

Three-dimensional depth mapping (stereoscopic) is data intensive. However, 5G connectivity (discussed in Chapter 20) increases speeds up to 20 times that of 4G, and "as of Q3 2019, there are now 4 million 5G connections globally; an increase of 166% over the second quarter," (Shein, 2019). We also see smaller companies such as HumanEyes announcing at the Consumer Technology Association's CES 2020: "In anticipation of the new 5G era, HumanEye's Cloud-based Suite utilizes high-performance Edge computing resources to simplify and streamline the Capture-Create-Share VR workflow" (GlobalNewswire, 2020), and Tilt Five raising $1.7 million through a kickstarted campaign to build an AR board game that allows you to see stereoscopic AR overlays, via its AR HMDs, of animated characters and objects over a board with other players in a room or players anywhere (Boland, 2019).

In the 2019 *Forbes* article "The Important Risks and Dangers of Virtual and Augmented Reality," author Bernard Marr (2019) wrote about the risks in both the business and consumer market where by "using AR applications we can expose vast amounts of data about ourselves... filters bused by Snapchat and Facebook can gather biometric data when we show them our face. This includes facial expressions, speech data, and even retina patterns that can be used to uniquely identify us." Lt. Gen. Susan Lawrence (2019) addressed military and government safety issues, noting "... data underlying XR is vulnerable to bias, misuse and cyber attacks. Both virtual and augmented reality can be manipulated for adverse purposes that could threaten our individual, mental and societal well-being."

Current Status

In looking at total VR headsets sales in 2019, Futuressource Consulting reports that the "VR headset Market increased by five percent in 2019," which was slightly lower than projections because of smartphones where there was "an 11 percent drop in market volume in 2019" (Meier, 2020). All-in-one HMDs and gaming are driving market increases, according to market researcher SuperData's 2019 year-end report entitled "Stand alone headsets accounted for 49% of VR Shipments," where we see 2019 end revenue at "$6.3 billion for XR, or extended reality games based on technologies such as virtual reality or augmented reality" (Takahashi, 2020). Despite growth, especially all-in-one headsets with active haptics and 6DoF, "it remains extremely difficult to get a handle on the actual size of market opportunity for a VR project… (as) some percentage of the most popular VR headsets remain unused most of the time" (Hamilton, 2020).

Factors to Watch

Major enablers currently being developed include eye tracking, hand tracking, and advances to passthrough as well as hints of new consumer AR glasses.

VR eye tracking, already available on HTC's VIVE Pro, "allows you to control things in VR just by looking at them, but can also produce higher fidelity VR images with less processing power—by only rendering the parts of a scene that you're actually looking at in high resolution, instead of wasting those resources on your peripheral vision too. It's a technique called foveated rendering" (Hollister, 2019). Hand tracking is coming to Oculus Quest, which "unlocks mechanics for

VR developers" and "reduces barriers of entry to VR for people who are not familiar or comfortable with gaming controllers" (Oculus, 2019), while Oculus Passthrough+ (already on Rift but coming to Quest) lets you toggle on and off a full spatial view through the glasses, reducing safety risks and disorientation (Oculus, 2019.)

Meanwhile the 2019 announcement of NReal consumer AR glasses at $499 had speculators noting: "Once Nreal Light is widely available for consumers and enterprises to purchase, the pressure will be on makers of enterprise only AR hardware to justify their existence (Horowitz, 2019). Other players such as Apple have announced that they have both VR and AR HMDs with major advances planned for upcoming years.

Gen Z

How Gen Z interacts with the world around them is likely to be very different from previous generations, both as workers and consumers of goods. "Virtual and augmented reality are set to disrupt the digital workplace" (Roe, 2018), while businesses are increasingly replacing "the traditional shop with virtual showrooms" (Rotter, 2019). As a result, Gen Z will likely interact less with the "live world." Yet, ironically, this same technology will likely increase the global interactions among Gen Z since they are increasingly likely to perform international tasks remotely using VR and AR.

Globalization

VR and AR are increasingly being used to perform international and multinational tasks remotely, thanks to ultra-fast wireless networks such as 5G. Susan Lund and Jacques Bughin (2019) noted that even global trade flows could be affected: "Already today, services trade is growing 60% faster than goods trade overall," and "5G wireless networks, virtual reality and augmented reality…opens new possibilities for delivering services globally." As we move forward, we will likely see ongoing increases in globalized services in connection with VR and AR.

Bibliography

Aukstakalnis, S. (2017). Practical augmented reality: A guide to the technologies, applications, and human factors for AR and VR. Pearson Education, Inc.

Beusterien, J. (2006). Reading Cervantes: A new virtual reality. Comparative Literature Studies, 43(4), 428-440. http://www.jstor.org/stable/25659544.

Bilton, N. (2015). Why Google Glass broke. *New York Times*. https://www.nytimes.com/2015/02/05/style/why-google-glass-broke.html.

Boland, M. (2019). The AR show: Race cars, rocket ships and reinventing AR. *AR Insider*. https://arinsider.co/ 2019/12/27/the-ar-show-race-cars-rocket-ships-and-reinventing-ar/.

Bozorgzadeh, A. (2017). WebVR's Magic Window is the gateway for pushing VR to billions of people. *Upload VR*. https://uploadvr.com/webvrs-magic-window-gateway-pushing-vr-billions-people/.

Chan, M. (2014). *Virtual Reality: Representations In Contemporary Media*. Bloomsbury Publishing, Inc.

Conditt, J. (2017). Worlds collide: VR and AR in 2018. *Engadget*. https://www.engadget.com/2017/12/20/vr-and-ar-in-2018/.

Diamandis, P. (2019). Augmented 2030: the apps, headsets, and lenses getting us there. *SingularityHub*. https://singularityhub.com/2019/09/13/augmented-2030-the-apps-headsets-and-lenses-getting-us-there/.

Digi-Capital. (2017a). After a mixed year, mobile AR to drive $108 billion VR/AR market by 2021. https://www.digi-capital.com/news/2017/01/after-mixed-year-mobile-ar-to-drive-108-billion-vrar-market-by-2021/#.Wlj7O0tG1mo.

Digi-Capital. (2017b). The four waves of augmented reality (that Apple owns). https://www.digi-capital.com/news/2017/07/the-four-waves-of-augmented-reality-that-apple-owns/#.Wi63-7Q-fOQ).

Dujmovic, J. (2019). Opinion: Here's why you will be hearing more about virtual reality. *Market Watch*. https://www.marketwatch.com/story/heres-why-you-will-be-hearing-more-about-virtual-reality-2019-07-15.

Engadget. (2019a). Google's Daydream VR experiment is over. https://www.engadget.com/2019/10/15/google-daydream-view-smartphone-vr-pixel-4/.

Engadget. (2019b). Magic Leap shifts focus to business with an updated AR headset. https://www.engadget.com/2019/12/10/magic-leap-enterprise-suite/.

Fink, C. (2019a). The biggest XR stories of the year. *Forbes*. https://www.forbes.com/sites/charliefink/2020/12/30/the-biggest-xr-stories-of-the-year/#7f5f0a7b70ab.

Fink, C. (2019b). HTC doubles down with VIVE VR enterprise solutions. *Forbes*. https://www.forbes.com/sites/charliefink/2019/07/01/htc-doubles-down-with-vive-vr-enterprise-solutions/#228b0dd13fbd.

GlobalNewswire. (2020). HumanEyes Technologies launches new cloud-based suite and presents immersive virtual reality experience at CES 2020. https://www.globenewswire.com/news-release/2020/01/05/1966202/0/ en/HumanEyes-Technologies-Launches-New-Cloud-Based-Suite-and-Presents-Immersive-Virtual-Reality-Experience-at-CES-2020.html.

Greenwald, W. (2017). The best VR (virtual reality) headsets of 2018. *PCMag*. https://www.pcmag.com/.article/342537/the-best-virtual-reality-vr-headsets.

Hamilton, I. (2020). VR in 2020: Small development teams are building runways to the future. *Upload*. https://uploadvr.com/editorial-vr-devs-market-size/.

Hayden, S. (2017). Felix & Paul's The People's House 360 video wins an Emmy. *Road to VR*. https://www.roadtovr.com/felix-pauls-360-video-peoples-house-inside-white-house-wins-emmy/.

Hollister, S. (2016). Daydream VR is Google's new headset – and it's not what you expected. *CNET*. https://www.cnet.com/news/daydream-vr-headset-partners-google-cardboard-successor/.

Hollister, S. (2019). THC's VIVE Pro Eye VR headset with eye-tracking arrives for $1,599. *The Verge*. https://www.theverge.com/2019/6/6/18655871/htc-vive-pro-eye-vr-headset-eye-tracking-price-release-us-canada.

Horowitz, J. (2019). 4 enterprise AR trends to watch in 2020. *VentureBeat*. https://venturebeat.com/2019/12/27/4-enterprise-ar-trends-to-watch-in-2020/.

Hruska, J. (2018). VR market expected to improve despite sharp declines in sales. *Extreme Tech*. https://www.extremetech.com/gaming/276470-vr-market-expected-to-improve-despite-sharp-decline-in-sales.

Jagneaux, D. (2019). Best VR of 2019 nominee: Asgard's Wrath delivered a true VR epic. *Upload*. https://uploadvr.com/best-vr-of-2019-nominee-asgards-wrath/.

Jerald, J. (2016). *The VR book: Human centered design for virtual reality*. Association for Computing Machinery.

Kharpal, A. (2015). This is why Nokia is betting big on virtual reality. *CNBC*. https://www.cnbc.com/2015/11/12/nokia-ozo-virtual-reality-big-bet.html.

Laurel, B. (1995). Virtual reality. *Scientific American*, 273(3), 90-90. http://www.jstor.org/stable/24981732.

Lawrence, S. (2019). Incoming: Extended reality offers the military boundless opportunity, as well as risk. https://www.afcea.org/content/incoming-extended-reality-offers-military-boundless-opportunity-well-risk.

Liptak, A. (2017) Sony's Playstation VR sales have been stronger than expected. *The Verge*. https://www.theverge.com /2017/2/26/14745602/sony-playstation-vr-sales-better-than-expected.

Lund, S., and Bushin, J. (2019). Next-generation technologies and the future of trade. *Vox*. https://voxeu.org/article/next-gen-eration-technologies-and-future-trade.

Macqueen, D. (2019). VR headset forecast by device type 2014-2024 Nov 2019 publication. *Strategy Analytics*. https://www. strategyanalytics.com/access-services/media-and-services/virtual-and-augmented-reality/market-data/report-detail/ vr-headset-forecast-by-device-type-2014-2024-nov-2019-publication?langredirect=true.

Mainelli, T., and Ubrani, J. (2019). AR/VR headsets return to growth in Q1. *Digitaltvnews*. http://www.digitaltvnews.net/ ?p=33123.

Marr, B. (2019). The important risks and dangers of virtual and augmented reality. *Forbes*. https://www.forbes.com/ sites/bernardmarr/2019/07/17/the-important-risks-and-dangers-of-virtual-and-augmented-reality/ #7435333e3d50.

Matney, L. (2019). Oculus eclipses $100 million in VR content. *TechCrunch*. https://techcrunch.com/2019/09/25/oculus-eclipses-100-million-in-vr-content-sales/.

Meier, D. (2020). VR headset technology "will be commercially viable within the next five years. *TVB Europe*. https://www.tvbeurope.com/data-centre/vr-headset-technology-will-be-commercially-viable-within-the-next-five-years.

Milijic, M. (2019). 29 Virtual reality statistics to know in 2020. *Leftronic*. https://leftronic.com/virtual-reality-statistics/.

Newman, J. (2014). The weirdest thing at Google I/O was this cardboard virtual reality box. http://time.com/2925768/ the-weirdest-thing-at-google-io-was-this-cardboard-virtual-reality-box/.

Oculus. (2019). Introducing hand tracking on Oculus Quest – Bringing your real hands into VR. [Oculus Blog]. https://www.oculus.com/blog/introducing-hand-tracking-on-oculus-quest-bringing-your-real-hands-into-vr/ ?locale=en_US.

Palumbo, A. (2019). VR hardware revenue soars in 2019as Oculus Quest and Valve Index remain in high demand. https://wccftech.com/vr-hardware-revenue-soars-in-2019-as-oculus-quest-and-valve-index-remain-remain-in-high-demand/.

Parisi, T. (2016). *Learning virtual reality: Developing immersive experiences and applications for desktop, web and mobile*. O'Reilly Media, Inc.

PCGamer. (2015). SteamVR, everything you need to know. http://www.pcgamer.com/steamvr-everything-you-need-to-know.

Ralph, N., & Hollister, S. (2015). Samsung's $99 Gear VR launches in November, adds virtual reality to all new Samsung phones (hands-on). *CNET*. https://www.cnet.com/products/samsung-gear-vr-2015/preview/.

Reddit. (2019). Asgard's Wrath. https://www.reddit.com/r/oculus/comments/dkhxyi/asgards_wrath_has_ more_than_ 1000_ratings_in_just/.

Roe, D. (2018). 6 ways businesses are using virtual and augmented reality today. *CMSWire*. https://www.cmswire.com/ digital-workplace/6-ways-businesses-are-using-augmented-and-virtual-reality-today.

Rotter, M. (2019). How will AR and VR change industries in 2020. *ARPost*. https://arpost.co/2019/12/11/how-will-ar-and-vr-change-industries-in-2020/.

Shein, E. (2019). 5G connections increase 166% to 4 million globally in Q3. *TechRepublic*. https://www.techrepublic.com/ article/5g-connections-increase-166-to-4-million-globally-in-q3/.

SuperData. (2019). SuperData XR Q3 2019 update. https://www.superdataresearch.com/superdata-xr-update/.

Takahashi, D. (2020). SuperData: Games hit $120.1 billion in 2019, with Fortnite topping $1.8 billion. *Venture Beat*. https://venturebeat.com/2020/01/02/superdata-games-hit-120-1-billion-in-2019-with-fortnite-topping-1-8-billion/.

The Void. (2017). Star Wars: Secrets of the Empire. https://www.thevoid.com/dimensions/starwars/secretsoftheempire/.

Tricart, C. (2018). *Virtual reality filmmaking: Techniques & best practices for VR filmmakers*. Routledge.

Vanian, J. (2017). Google has shipped millions of cardboard reality devices. *Fortune*. http://fortune.com/2017/03/01/google-cardboard-virtual-reality-shipments/.

Venture Scanner. (2019). Virtual reality in 2019 projected to significantly exceed 2018. https://www.venturescanner.com/ 2019/09/05/virtual-reality-funding-in-2019-projected-to-significantly-exceed-2018/.

Virtual Reality Society. (2017). Profiles. https://www.vrs.org.uk/virtual-reality-profiles/vpl-research.html.

Zone, R. (2007). *Stereoscopic cinema and the origins of the 3-D film, 1838-1952*. University Press of Kentucky.

14

Video Games

Isaac Pletcher, M.F.A.*

Abstract

A video game is an interactive digital entertainment that you "play" on some type of electronic device. The medium has existed in some form since the 1950s, and since that time, has become a major worldwide industry, spawning technology battles and being birthed in new iterations that become ubiquitous parts of modern life. This chapter will look at the development of video games from the beginning, through the golden age, and into the era of the console wars. Further, it will point out the enormous revenues video games make, and show how the #metoo movement and unions are working to change the face of the industry.

Introduction

Video games have grown up! Luminaries such as Ray Bradbury once thought of the entertainment medium as "a waste of time for men with nothing else to do" (Ray Bradbury, n.d.). Yet games have proved to be robust motivators for growth and development, as instruments of social change, powerhouse enterprise, and even military application. It is time for denizens of the Age of Interconnectivity to realize that video games and gamers are no longer on the margins, living on Mountain Dew and Cheetos in their parents' basement. Video games are an indelible part of the future of the entertainment industry. Actor Andy Serkis has perhaps correctly stated that, "Every age has its storytelling forum, and video gaming is a huge part of our culture" (Andy Serkis, n.d.). Shigeru Miyamoto, creator of gaming icons like Mario and Donkey Kong, has stated that, "the obvious objective of video games is to entertain people by surprising them with new experiences" (Sayre, 2007). This chapter seeks to explore the intersection of known and new experiences by presenting a brief history of video games, moving into an analysis of the current industry, especially addressing social/technological evolution and cultural shift. It then continues with a look at the industry's massive economic footprint, and concludes with a vision for

---

* Assistant Professor of Film/TV at Louisiana State University (Baton Rouge, Louisiana)

the next few years with a brief section on breaking into the industry.

Background

In 1958 Willy Higinbotham of Brookhaven National Laboratory developed the first truly interactive electronic game, *Tennis For Two*. It was a tennis match between two players controlled with a single knob, programmed on an oscilloscope and presented at an open tour day of the lab. Despite the massive popularity of the game Higinbotham would later claim, "It never occurred to me that I was doing anything very exciting. The long line of people I thought was not because this was so great but because all the rest of the things were so dull" (Tretkoff & Ramlagan, 2008). Over the next two years, his game would get a bigger screen and upgrades that allowed players to simulate the gravities of the moon and Jupiter. While Higinbotham was simply trying to make something amusing for lab visitors, historian Steven Kent (2002) has stated that early iterations of games leading to the marriage of video display and interactive controls were obvious evolutions from the "well-established amusement industry." He even goes so far as to trace these amusements back to the 17th century game of Bagatelle.

While Higinbotham's contribution was significant, it is difficult to say exactly when the video game industry began. Egenfeldt-Nielsen, Smith, & Tosca (2016) have pragmatically pointed out that, "No trumpets sounded at the birth of video games, so we must choose what constitutes the beginning." Kent (2002) is careful to point out the evolutionary forebears of modern game consoles as novelty paddle and mechanical games from the 1940s and 50s that "simulated horse racing, hunting, and Western gunfights." But once the television became popular, the race to put these mechanical games into an electronic version began.

In 1952 Sandy Douglas, a Cambridge PhD candidate developed an electronic version of *Noughts and Crosses* (Tic-Tac-Toe) played against Cambridge's first programmable computer, the Electronic Delay Storage Automatic Calculator (EDSAC). A telephone dial was used to input moves, and the game results were displayed on a cathode-ray-tube. While it was the first graphics-based computer game, the dependency on programmed punch cards gives ammunition to the argument against it being the first true video game in the sense of *Tennis for Two*.

Once Higinbotham's game showed the possibility of controllable video entertainment, a number of early gaming figures starting making their marks. In 1961, MIT student Steve Russell developed *Spacewar*, which started life as a hack for the PDP-1 computer. Two other students, Alan Kotok and Bob Sanders, developed remote controls that could be wired into the computer to control the game functions, and thus the controller was born. By the early 1960s then, the display, ability to play a game, and ability to control the game functions had been achieved. However, all these developments had happened in academic or government environments, and video game monetization was still in the future.

By 1966 Ralph Baer, an employee at defense contractors Sanders Associates Inc., began work on game systems with colleagues Bill Harrison and Bill Rusch. Their breakthrough came in the form of the "'Brown Box" a prototype for the first multiplayer, multi game system" (The Father of the Video Game, n.d.). Essentially, the trio had managed to create a video game system whereby the user could change between games, and the player had a way to connect it to their home television. American company Magnavox saw the value of the system and licensed it, marketing it in May 1972 as the *Magnavox Odyssey*. Though the console wars of the 1990s and 2000s were still a generation off, once Magnavox put their system on the market Atari, Coleco, URL, GHP, and others followed suit within four years. Once the video game console existed, cabinet games quickly emerged, and developers began to look for a game that would carry their product to the emerging home and arcade markets.

Through the late 1970s and into the 1980s, a series of games and systems came to prominence with blazing speed in what a number of video game historians have dubbed the "second generation" of video games. Kent (2002) has called it simply, "The Golden Age," illustrated in Figure 14.1.

Table 14.1

Golden Age

1978
- Space Invaers (Taito Corporation)
- Computer Othello (Nintendo Co. Ltd.)

1979
- Asteroids (Atari, Inc.)
- Galaxian (Namco, Ltd.)
- Cosmos Handheld System (Atari, Inc.)
- Microvision Handheld System (Milton Bradley)

1980
- Pac-Man (Namco, Ltd.)
- Game & Watch Handheld System (Nintendo Co. Ltd.)

1981
- Frogger (Konami Holdings Corp.)
- Ms. Pac-Man (Midway Games, Inc.)
- Donkey Kong (Nintendo Co. Ltd.)

1982
- Burgertime (Data East Corp.)
- Q*bert (D. Gottlieb & Co.)
- Commodore 64 Home Computer (Commodore Intl.)

1983
- Super Mario Brothers (Nintendo Co. Ltd.)

Source: S. Kent

This "Golden Age" came to an end in the United States in 1983 as the game/console market crashed due to a combination of "oversaturated market, a decline in quality, loss of publishing control, and the competition of Commodore, PC, and Apple computers" (Cohen, 2016). In Japan, however, a different landscape meant the home console market had not felt the same effects, allowing Nintendo to release the Famicom (short for Family Computer) Home Entertainment System. Atari initially had the chance to license Famicom for American release, but when the deal fell through, Nintendo decided to set up shop in the U.S. In 1985, Nintendo of America—which had been established in 1980—began selling the system in North America as the Nintendo Entertainment System (NES). The wild success of this system touched off the third generation of video games and cemented Nintendo as industry titan, spelling doom for Atari and forcing other companies into breakneck development just to keep pace.

This mad dash to maintain market superiority led to the Console Wars—in essence, a video game arms race. System technologies were being developed, not just for companies to match their competitors, but also to provide gamers with more sophisticated experiences. Technologists Mike Minotti (2014) and Jordan Minor have traced the first five iterations of console wars, which started in 1989 and have lasted to the present time, pointing out the winners and losers in Figure 14.2.

Table 14.2

Console Wars

1989–1994 (The 16-bit graphic revolution)

Super Nintendo Entertainment System	49.10M units sold
vs	
Sega Genesis	30.75M units sold

1995–1999 (Video games now in 3-D)

Sony PlayStation	102.49M units sold
vs	
Nintendo 64	32.93M units sold
vs	
Sega Saturn	9.26M units sold

2000–2006 (Sega bows out, Microsoft enters the ring)

Sony PlayStation 2	155M units sold
vs	
Microsoft Xbox	24M units sold
vs	
Nintendo GameCube	21.74M units sold
vs	
Sega Dreamcast	9.13M units sold

2007–2013 (Nintendo reclaims the crown)

Nintendo Wii	101.63M units sold
vs	
Microsoft Xbox 360	84M units sold
vs	
Sony PlayStation 3	80M units sold

2013–Current for Jan. 2020 (4K Gameplay)

Sony PlayStation 4	102.8M units sold
vs	
Microsoft Xbox One	41M units sold
vs	
Nintendo Wii U	102.53M units sold
vs	
Nintendo Switch	15M units sold

Source: S. Kent

Over the next three decades, *Nintendo, Microsoft,* and *Sony* have become the three key industry players. Whether this market balance can be maintained is yet to be seen, especially as all three of these companies have consoles that compete with PCs as a platform.

However, with improvements in television display quality, the consoles are clearly in the forefront, as the main advantage of PC gaming was the potential for improved graphics. In fact, the number of games shared between PC and console players is vast, with only a handful of titles like *World of Warcraft* and *League of Legends* being exclusive to the PC platform.

The development of increasingly realistic games led to some to call for regulation of the industry. By 1987 the Software Publishers Association of America, a body of congressional lobbyists for the video game industry, had decided to not institute an MPAA style ratings system and instead let developers police themselves and put warnings on their games when they deemed it necessary. Executive Director, Ken Wasch said, "Adult computer software is nothing to worry about. It's not an issue that the government wants to spend any time with…They just got done with a big witchhunt in the music recording industry, and they got absolutely nowhere… (Williams, 1988)." This laissez-faire attitude, however, was not enough for certain members of the U.S. Congress, and by 1993, senators Joe Lieberman and Herb Kohl told the association that if they did not set up a rigorous self-regulatory body, within one year the industry as a whole would be subject to governmental regulation. This prompted the formation of the Entertainment Software Ratings Board (ESRB).

While every game released on main consumer consoles since the 1990s has received an ESRB rating, these ratings are relatively new (since 2015) on the Google Play Store, and still wholly absent on the Apple App Store, which still prefers to use their own internal review process. Unfortunately, the ESRB has been recently accused to misrating games as it has come to light that raters, "don't actually play video games for more than an hour rate them. Instead they're given a premade video package of the game to review and then give a rating based on that" (Zamora, 2019). This has led to an increase in games with microtransactions and loot boxes (explained later in the chapter), as the developers become more aware that, at least in the United States, there will be no ratings repercussions for having them in the game, especially when labeled as "surprise mechanics."

Recent Developments

Since 2015, three of the most important developments in the gaming industry were the, "distribution revolution…the rise of social games…and the body as interface (Egenfeldt-Nielsen, et al., 2016)." Since that writing, the body as interface has progressed and is discussed briefly later in this chapter and at length in Chapter 13. Social games like *Fortnite* and *League of Legends* have replaced the early 2010s *Farmville* and *Words with Friends,* which in their own ways, seemed to be the distant descendants of the Massive Multiplayer Online (MMO) games still played with the wide interconnectivity of online objects. It seems that almost any game won't survive or make money unless it has the capacity for wide social structures. However, the distribution models and methods for games have seen an acute change since 2017.

In 2017, the packaging of a video game on a solid-state object like a disc and selling as a physical product in a store was already passé. There are now app stores that sell games for both Apple and Android devices as well as specific app stores for Sony PlayStation and Microsoft Xbox users. In addition, if a gamer is using a Mac or PC, "platforms such as Steam offer direct purchase, data storage, downloadable content (DLC), and other community features" (Egenfeldt-Nielsen et al., 2016). In fact, by the end of April 2019, Steam had reached "one billion accounts and 90 million active users per month" (Arif, 2019). In addition, NewZoo's estimate of 45% market share for mobile gaming in 2019 far outstrips the 25% and 21% percent of the gaming market that goes to console games and downloaded computer games, respectively (Warman, 2019). Clearly, the video game world is seeing a significant market shift to mobile gaming, developers and game conglomerates can no longer make the same revenues with boxed games. Therefore, since 2015, these developers have not produced nearly as many boxed units of game titles and instead are opting for downloads and microtransactions. Monetization service company Digital River released a report in early 2017 which states that because of the digitalization of video games, the average gamer is expecting more from the developer for less money. Consumers are less willing to pay $60 for a boxed game and instead choose titles with a steady stream of new content. Publishers seek to meet these expectations and have adopted a "games as service" model, releasing fewer titles over time while keeping players engaged longer with regular updates and add-ons (SuperData, 2017).

Of course, if the game is cheap or "free-to-play" (costing nothing for the download and to start gameplay), these updates and add-ons are what cost money, and players must pay for them or risk never being able to finish the game or see the premium content. These are charged as small amounts throughout the gameplay and are the micro-transactions that add up and, according to some estimates, have almost tripled the game industry's value. After all, why charge $60 for a game once when over the development life of a game, a player may pay upwards of $100-$500? This model has gained traction and, as journalist Dustin Bailey (2017) mentions, while these "modern business practices are failing in the court of public opinion, they are profitable. Extremely, *obscenely* profitable." It is this very profitability that ultimately drives the developments in the industry. While many gamers and critics bemoan this practice, it seems to show no signs of slowing down.

Another source of profitability is the digital distribution mechanism that games use to get to their end consumers. As of early 2020, Steam is seeing competition in the form of Epic's PC game store and Google's Stadia Service. Unfortunately for Google, Bradley Gastwirth (2019) of investment firm Wedbush Securities stated in November 2019 that Google Stadia was rushed out "ahead of the new console launches next year from Microsoft and Sony. So, it came a little bit early, it doesn't offer all the features and functions that Google has promised." It seems that more companies, while aware of the necessity of digital platforms for distribution and marketing, might still have a way to go to catch up to the industry leader. And it is precisely this race that has led to the rise of one of the most controversial aspects of an industry practically built on controversy, the loot box.

In non-gamer terms, the loot box is synonymous with roulette or a slot machine in the real world. They are a "virtual bundle of random items exchanged for real-world money"—paid for through microtransactions (Busby, 2019). Wright (2017), points out that, "Videogame loot boxes can trace their origins back to decorative cards packed in with cigarettes more than a hundred years ago," and are the direct descendants of rare cards from packs of sports cars and card game decks. However, the "random" nature of the reward packets used in games such as *Overwatch, Fifa,* and *Shadow of War* has been considered by many world governments to amount to illicit lotteries and contribute to problem gambling among other addictive behaviors. In fact, the practice of selling loot boxes ingame has been decided to be problematic enough that it has been banned in Belgium, The Netherlands, China, and Sweden. As we look forward to the next generation of gaming, it will be interesting to see if loot boxes will be determined to be problem enough to be universally banned. And if they are, common sense dictates that developers will find another way to stretch the profitability of games for as long a period as possible.

One of the most interesting developments in video games since 2018 has been the emergence of esports as a billion-dollar industry with players who make millions of dollars per year. For the uninitiated, esports is competitive video gaming and it has millions of fans all around the world and as CNN's AJ Willingham (2018) points out, it can turn "casual gamers into serious stars who can sometimes rake in seven-figure earnings and massive brand endorsements." Wijman (2019) points out that in 2018, revenue growth of esports grew by 32% and in 2019, it jumped again by 26.7% to $1.1 billion. For more on esports and the global trends that affect sponsorship, media rights, and advertising, see Chapter 18.

In addition to esports, the last few years has seen some video game developers attempting to turn games into a way to bridge social gaps. Hideo Kojima, a former developer at Konami who created the *Metal Gear* series, is on the forefront of this movement with his independently produced game *Death Stranding,* which premiered in November 2019. The BBC's Steffan Powell (2019) points out that, "Gaming communities have been dogged by claims of toxic behavior by some players for a number of years" (i.e., Gamergate, described later in this chapter). The objective of Kojima's game is to deliver messages to others in a post-apocalyptic wasteland and is programmed in such a way that it is only possible to trade positive messages with others in networked gameplay. Kojima describes his game as, "using bridges to represent connection—there are options to use them or break them. It's about making people think about the meaning of connection (Powell, 2019)" It is, perhaps, very appropriate to think that may be the first sign of gaming coming full circle and using the Age of Interconnectivity to actually build connections instead of competition between humans.

Some of the most noteworthy recent developments in the gaming industry are about the industry itself. Since 2018, the industry has seen fascinating advances in its own #metoo movement, unionization, and financial demographics. In 2014, independent game developer Zoe Quinn created a game called *Depression Quest* that was distributed for free on Steam and is a first-person text-based game where the player is a young adult suffering from depression. Quinn developed the game not to be an amusement but to, "broaden the medium's subject matter with depictions of life's darker aspects" (Parkin, 2014). The controversy came when a former boyfriend of Quinn's wrote a rambling blog post that alleged a relationship between Quinn and a journalist who had merely mentioned—not even reviewed—the game in an article. Opponents of Quinn decried the perceived lack of journalistic integrity and used it as an excuse to launch a cavalcade of malevolent personal attacks and subject her to doxing (when an individual's personal details are made known online so anyone is able to harass them) as well as launch arbitrary attacks on other female gaming luminaries like critic Anita Sarkeesian. This online mobbing led the industry to have to address two specific subjects; namely, how women are treated in gaming circles and the ethics of game journalism. Unfortunately, many gaming critics feel like the situation was addressed on internet message boards and even through some legal channels but never adequately resolved. The scandal was dubbed "Gamergate" and as late as August 2019, Marie-Claire Isaaman, (in Orr, 2019) CEO of Women in Games stated that, "After GamerGate in 2014 and a lack of resolution from that intensive moment, I sensed another wave would come." That wave seems to be moving slowly, yet steadily, forward in gaming

culture. In September 2019, at the Women in Games European Conference, "Senior women in the games industry took the opportunity afforded by the two-day conference to discuss workplace abuse and the wider issue of toxicity in gaming culture (Orr, 2019)." While this is an ongoing problem in the gaming community, the conference delegates decided that the best way forward is through education, "not just in the workplace, but much earlier, in a school setting" (Orr, 2019). Only time will tell if this necessary movement in the gaming industry and culture will have the desired effect.

Not limited to game journalists and developers, women and under-represented minorities are often the subject of harassment in online game play. The Anti-Defamation League conducted a study which found that about 65% of online game players experienced severe harassment (Smith, 2019) 53% of those who experienced harassment said they were targeted for their gender, gender identity, race, ethnicity, religion, or sexual orientation. Game and console makers are tackling the problem with revised community standards, easier reporting structures, and easy ways to block players. Nonetheless, many female gamers say they mute their microphone when playing to avoid harassment (Smith, 2020).

Another current movement in the gaming industry is unionization. Since 2018, Game Workers Unite (GWU) has worked to create a labor movement in the game industry for the purpose of ending "the institutionalized practice of unpaid overtime, improve diversity and inclusion at all levels, support abused or harassed workers, and secure a steady and living wage for all" (Kerr, 2018). While many may see the gaming industry as a good way to marry their ambitions towards art and technology, the reality, according to Declan Peach, a founding member of the Independent Workers Union of Great Britain's (IWGB) GWU branch, is that it is normal "for games workers to endure zero-hours contracts, excessive unpaid overtime and even sexism and homophobia as the necessary price to pay for the privilege of working in the industry (Lomas, 2018)." In other words, GWU is working to bring attention and solutions to the unregulated nature of the gaming industry that allows developers to pay minimal wages to employees working uncompensated overtime hours to meet development and delivery deadlines and then reducing overhead with mass layoffs. This cyclical hiring and firing has, according to GWU, been virtually holding game workers hostage in situations of overwork and abuse such as was reported at Rockstar Games in the lead up to the release of *Red Dead Redemption 2* (Hall, 2018). Their goal is legal representation for game workers and the establishment of a certification for games that would signify "games developed by studios with healthy labor practices (Johnson, 2018)." This is a movement that seems maybe inevitable and perhaps analogous to the labor movements in Hollywood during the 1930s and 40s. In fact, the similarity seems especially apropos given the financial value of the gaming industry and its emergence as one of the pre-eminent worldwide entertainment media.

Current Status

Between April 2017 and June 2019, analytics firm NewZoo tracked the segmented and overall growths of the gaming market and showed that worldwide, gamers spent $108.9 billion on games in 2017, and that number in 2019 skyrocketed to $152.1 billion, coming from 2.5 billion gamers (Warman, 2019). (For perspective, remember that the U.N. currently estimates the world population at 7.8 billion). What remains interesting for the gaming ecosystem is that the most lucrative segment of gaming is still the mobile market. Mobile gaming's market share jumped from 42% to 45% of the global market between 2018 and 2019, to an estimated $68.5 billion. In addition, the American market for the first time outpaced the Chinese market as the world's largest revenue-producer, taking advantage of China's nine-month licensing freeze on new games between 2018 and 2019, part of an effort to reduce the amount of screen time spent by children in that country. NewZoo now predicts a global gaming market value of $196 billion by 2022 (Wijman, 2019).

In September 2017, Epic Games released *Fortnite*, a battle-royale shooter built on the company's signature Unreal Engine that has since earned Epic revenues of over $3 billion, and increased the company's value by more than $7 billion between 2017 and 2019. By March 2019, the game had over 250 million players on PlayStation, Xbox, Microsoft, and macOS. While these numbers are some of the largest for a single game in history, it is perhaps the cultural capitol that *Fortnite* has built that has most surprised industry analysts. This capitol comes in both relatively unintentional ways and through purpose, market-driven means. 63% of the game's players are aged between 18-24, a demographic that also includes many renowned world sports figures. In fact, the signature dances of the *Fortnite* characters have been seen as celebrations by players in the NFL, MLB, NBA, NHL, Premier League, Serie A, La Liga and many others, creating a cultural connection that cannot be denied—after all, if a sports hero plays a game, then how "nerdy" can it really be?

In addition, Epic has capitalized on its cultural prominence by partnering with other brands for creative, targeted marketing. In February 2019, EDM music producer Marshmello was contracted to perform a short, exclusive, in-game dance music set. Ten million players of the game went to the specified map area to watch the 10-minute set. *Fortnite* also seems to be a trendsetter in the new field of using games as promotional media. With built-in audiences, marketers are beginning to see the benefits of dynamic brand insertion and alternate revenue streams. In December 2019, director JJ Abrams and Disney worked out a deal with Epic that saw the release of exclusive clips and a preview for *Star Wars: The Rise of Skywalker* available only to the 250 million people who have a game account. While the concept of the "influencer" is a topic for Chapter 23 of this book, Epic and *Fortnite* seem close to defining what that term means in the gaming world. And it is far beyond the stereotype that dogged gamers for decades. It seems to be, in fact, the sports star and the celebrity, the musician and the movie maker—a symbiosis that seeks to combine the call of culture with the rhythms of youthful rebellion.

Consider that for 2019, NewZoo numbers gamers at 2.5 billion worldwide, a 300 million person increase from the 2017 number of 2.2 billion. And the 2017 projection for 2020 overall revenue was surpassed in 2019 when revenues hit $152.1 billion, far exceeding the $128 billion estimate made just two years earlier (Warman, 2019). However, the gaming industry has also experienced scandal. Analytics firm Digital River, in 2017, stated that because the digital marketplace is only going to become more robust, publishers must be more concerned than ever about fraud. Currently, when a game is purchased online, the buyer is given a digital key that unlocks gameplay. Because it is easy enough to fraudulently buy the key, it is also easy to sell the ill-gotten key on a gray market. And the company G2A is exactly that—a company that, among other activities, resells keys for games on pretty much every single major platform. Since 2018, G2A has come under fire from major developers for "fraudulently facilitating the sale of illegally obtained game keys" (Brown, 2019). The feud heated to the point where in summer 2019, developers were taking to social media to encourage players to actually pirate games instead of using the keys bought from G2A. The problem inherent is that only G2A has the list of keys bought on its site and only the developers of each separate game have the list of fraudulently obtained game keys. While G2A has offered to have an outside party do audits to reconcile these lists for three smaller companies, they have stated that any audits done after that must have the cost split between their coffers and the developers. Unsurprisingly, this has proved to be the bridge too far for large developers and so it seems as though these gray market transactions will continue apace.

Factors to Watch

The factors to watch for video games are the continued growth of market share, the integration of games with ever more advanced technology, and perhaps ironically, the video game as envied career and social connector. In terms of gaming platforms, we are still in the fifth iteration of the console wars, waiting for new generations of the PlayStation and Xbox which are both slated for release in time for 2020 holiday shopping. The new systems will have improved speed and core processors, with graphics forecasted to support 8K video. Interestingly, neither system seems to be banking on VR for the next console, and there is, as of yet, no indication when Sony or Microsoft will go all in on that technology either (King, 2019). Of the three main gaming platforms—console, mobile, and PC—PC gaming continues to lose ground to mobile and console gaming. Worldwide, PC gaming accounted for 23% of gaming revenue, down 4% from its 27% stake in 2017. And while the overall revenues are up—PC gaming improved from $29.4 billion in

2017 to $35.7 billion in 2019—the overall growth of the industry still reads that growth as the lowest share of the market.

In 2018, for the first time, video game sales eclipsed movie global box office revenue by a margin of $43.8 billion to $41.7 billion (Shieber, 2019). The video game market has been so robust that Reed Hastings, CEO of Netflix, stated in his 2019 first quarter letter to shareholders that "We compete with [and lose to] Fortnite more than HBO" (Perez, 2019). Realistically, video games shouldn't be the majority of Hastings' concerns as global streaming revenues trail both video games and box office, making only $28.8 billion in 2018. However, the message seems clear: video games seem to be moving ahead of traditional entertainment media as the choice of many people. In fact, analytics firm Statista (2018) has pointed out that the three biggest video game openings of all time—opening being three days from launch on all platforms—have dramatically outperformed the three biggest film openings of all time, all of which opened to worldwide audiences, as illustrated in Figure 14.1.

FIGURE 14.1

Video Games vs. Blockbuster

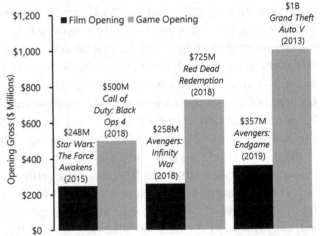

Source: Box Office Mojo

If there is a potential pitfall to the current growth of the video game industry it may be in the pricing structure itself. NBC's Tom Loftus (2003) explained that when the next generation of game consoles are in development, the current generation is in the downward part of its lifecycle. Historically, it made sense that developers let the current generation of games fade away without updates when the new consoles were on the horizon. Yet because of the current "games as service" model, this is not happening. The games keep getting updated and fans keep making their own updates when developers stop. A major title game that retailed between $40–65 in 2015 still retails for between $40–50. A standard PlayStation game cost less than $1 million to develop in 1998, but now can cost upward of $200 million, while the life cycle of the game needs to be far longer. However, the game costs to the consumers aren't changing, and so as mentioned, the trend of microtransactions continues unabated.

In terms of technology, virtual reality (VR) and augmented reality (AR) are beginning to impact the industry. One simply needs to attend any of the major video game or technology gatherings to confirm this. The coming release of the next generation of consoles is poised to take another step in that direction. However, as Statista (2018) points out, while the VR gaming market is projected to hit $23 billion by the end of 2020, "60% of gamers worldwide said they were not interested in purchasing a virtual reality headset." Quite simply, it seems that the technology has simply not yet caught up to the storytelling possibilities of VR. This makes for a market situation where gamers are hesitant to spend their money on expensive VR systems (even for existing consoles) that they might hate. Conversely, the marketing for the VR systems is, for some companies, cautious (to use a kind word) and outright non-existent for others. Morris (2019) analysts phrase it simply: "Perhaps more damning is the fact that there hasn't been a definitive software product that captures people's imagination on a widespread basis." Chapter 13 of this book offers a much more in-depth analysis of this phenomenon and possible market solutions that may be on the horizon.

Whether *Death Stranding* is successful enough to inspire other developers to follow in Kojima's altruistic footsteps, and the possibility of gaming remaining as the top money earning relaxation media, remains to be seen. But it is clear that video games are no longer the purview of the lonely and disaffected. Those who dismiss them are summarily dismissing an entire entertainment industry with billions of participants and revenues that rival, and even surpass, other traditional media. And they may even be used as a force for positivity in this, our Age of Interconnectivity. As Khan (2020) in the game *Metro 2033* states, "Even in dark times, we cannot relinquish the things that make us human." Video games have grown up and are finally one of those things.

Gen Z

As end-to-end (birth to death) technology "haves," Gen Z is easy to distinguish by their broad knowledge base and familiarity with video games. Forbes identifies Gen Z-ers as being born between 1995 and 2010 and, as stated by Deep Patel, have "always lived in a connected world…. Switching between different tasks and paying simultaneous attention to a wide range of stimuli comes naturally to them" (2017). This is the generation that, as stated previously, makes up the majority of accounts on *Fortnite*. It is also the generation that has grown up watching game play-throughs on platforms like Youtube and Twitch. Team Whistle—an online sports and entertainment hub for this generation, has done research that shows that 77% of Gen Z males are video game watchers on the above-mentioned platforms, and 68% would say that online social gaming is an important aspect of their very lives (Research Services, 2019). As we move further into the future, it seems inevitable Gen Z will continue to drive the video game marketplace, and that the children of Gen Z—named by many media outlets as Generation Alpha—will surpass even their parents and grandparents in video game proficiency.

Sustainability

Currently, the sustainability of video games seems to fall into two camps: the environmental efforts of the industry competitors and games that work to further eco-consciousness in the players themselves. The UN Environment Programme has pointed out that efforts have been made by companies like Sony, Microsoft, Google, Ubisoft, and Twitch, among others, to bring their corporate cultures and product development methods more in line with current environmental and climate concerns. Keishamaza Rukikaire has written that the promised efforts of these companies combined will "result in a 30 million tonne reduction of CO_2 emissions by 2030 [and] see millions of trees planted" (2019). Likewise, writers like Lindsey Murri (2018) have pointed out that games like Eco, Aven Colony, and Raft carry heavy themes of sustainability that are intended to change the thought patterns of players. Further, Andrew Urevig (2019) points out that the UN itself has identified Gen Zers as potential climate and environmental shapers through video games via a sevenfold strategy that includes: nudging players towards action, sustainability campaign organization, raising money for global causes, e-waste and efficiency awareness, sustainable game spotlighting, game and personality promotion for environmental concerns, and parental involvement.

Bibliography

Andy Serkis Quotes. (n.d.). https://www.brainyquote.com/authors/andy-serkis-quotes.

Apple changes rules on app 'loot boxes.' (2017, December 22). http://www.bbc.com/news/technology-42441608

Arif, S. (2019, April 30). The number of Steam accounts has just hit one billion. https://www.vg247.com/2019/04/30/steam-one-billion-accounts/.

Bailey, D. (2017, October 11). The game industry's value has tripled because of DLC and microtransactions. https://www.pcgamesn.com/games-as-service-digital-river-report.

Brown, F. (2019, July 12). G2A proposes a key-blocking tool as a solution to fraud. https://www.pcgamer.com/g2a-proposes-a-key-blocking-tool-as-a-solution-to-fraud/.

Busby, M. (2019, November 22). Loot boxes increasingly common in video games despite addiction concerns. https://www.theguardian.com/games/2019/nov/22/loot-boxes-increasingly-common-in-video-games-despite-addiction-concerns.

Cohen, D. (2016, October 19). The History of Classic Video Games - The Second Generation. https://www.lifewire.com/the-second-generation-729748.

Current World Population. (n.d.). https://www.worldometers.info/world-population/.

Deckard Cain - Stay Awhile and Listen. (2011, October 28). https://www.youtube.com/watch?v=tAVVy_x3Erg.

Egenfeldt-Nielsen, S., Smith, J. H., & Tosca, S. P. (2016). *Understanding video games: the essential introduction*. New York, NY: Routledge.

Gastwirth, B. (2019, November 20). Google Stadia is not that exciting, says Wedbush Securities. https://www.cnbc.com/video/2019/11/20/google-stadia-is-not-that-exciting-says-wedbush-securities.html.

Hall, C. (2018, December 27). Why 2019 could be the year video game unions go big. https://www.polygon.com/2018/12/27/18156687/game-workers-unite-game-developer-union-us-uk-france.

Hall, C. (2019, July 5). Repeatedly accused of fraud, G2A offering independent audits of its catalog. https://www.polygon.com/2019/7/5/20683026/g2a-controversy-audit-petition.

Johnson, A. (2018, August 8). The games industry needs unions - and these are the people trying to make it happen in the UK. https://www.eurogamer.net/articles/2018-08-07-the-games-industry-needs-unions-and-these-are-the-people-trying-to-make-it-happen-in-the-uk.

Kent, S. L. (2002). *The ultimate history of video games*. [Kindle Version]. Amazon.com

Kerr, C. (2018, December 14). The first union for game industry workers has launched in the UK. https://www.gamasutra.com/view/news/333009/The_first_union_for_game_industry_workers_has_launched_in_the_UK.php.

Kahn, (n.d.). https://www.quora.com/What-is-a-great-philosophical-quote-from-a-video-game.

King, J. (2019, September 20). PS5 vs Xbox Series X (Xbox 2): Which next-gen console is right for you? https://www.trustedreviews.com/news/ps5-vs-xbox-2-3697353.

Loftus, T. (2003, June 10). Top video games may soon cost more. http://www.nbcnews.com/id/3078404/ns/technology_and_science-games/t/top-video-games-may-soon-cost-more/#.WkTY7raZPOQ.

Lomas, N. (2018, December 14). UK video games workers unionize over 'wide-scale exploitation' and diversity issues. https://techcrunch.com/2018/12/14/its-about-ethics-in-the-games-industry/.

Mansoor, I. (2019, December 13). Fortnite Usage and Revenue Statistics (2019). https://www.businessofapps.com/data/fortnite-statistics/.

McDonald, E. (2017, April 20). The Global Games Market 2017 | Per Region & Segment. https://newzoo.com/insights/articles/the-global-games-market-will-reach-108-9-billion-in-2017-with-mobile-taking-42/

Minor, J. (2013, November 11). Console Wars: A History of Violence. http://uk.pcmag.com/game-systems-reviews/11991/feature/console-wars-a-history-of-violence

Minotti, M. (2014, August 20). Here's who won each console war. https://venturebeat.com/2014/08/20/heres-who-won-each-console-war/

Morris, C. (2017, June 13). Gamers aren't buying the VR hype, and game makers are quietly hedging their bets. https://www.cnbc.com/2017/06/13/e3-virtual-reality-isnt-really-catching-on-with-gamers.html.

Murri, L. (2018, September 3). Top 3 Video Games About Sustainability. https://medium.com/the-wild-thoughts-blog/top-3-video-games-about-sustainability-f0d53a195c0.

Orr, L. (2019, September 17). 'This industry has a problem with abuse': dealing with gaming's #MeToo moment. https://www.theguardian.com/games/2019/sep/17/gaming-metoo-moment-harassment-women-in-games.

Owen, P. (2016, March 09). What Is A Video Game? A Short Explainer. https://www.thewrap.com/what-is-a-video-game-a-short-explainer/

Parkin, S. (2017, June 20). Zoe Quinn's Depression Quest. https://www.newyorker.com/tech/annals-of-technology/zoe-quinns-depression-quest#.

Pasquarelli, A., & Schultz, E. J. (2019, January 22). Move over Gen Z, Generation Alpha is the one to watch. https://adage.com/article/cmo-strategy/move-gen-z-generation-alpha-watch/316314.

Patel, D. (2017, September 21). 8 Ways Generation Z Will Differ From Millennials In The Workplace. https://www.forbes.com/sites/deeppatel/2017/09/21/8-ways-generation-z-will-differ-from-millennials-in-the-workplace/#731df6076e5e.

Perez, S. (2019, January 18). Netflix thinks 'Fortnite' is a bigger threat than HBO. https://techcrunch.com/2019/01/18/netflix-thinks-fortnite-is-a-bigger-threat-than-hbo/.

Powell, S. (2019, November 4). Death Stranding: Hideo Kojima explains his new game. https://www.bbc.co.uk/news/newsbeat-50172917.

Ray Bradbury Quotes. (n.d.). https://www.brainyquote.com/quotes/ray_bradbury_626594.

Richter, F. (2018, November 6). Infographic: Video Games Beat Blockbuster Movies Out of the Gate. https://www.statista.com/chart/16000/video-game-launch-sales-vs-movie-openings/.

Ross, M. (1947). Labor Relations in Hollywood. *The Annals of the American Academy of Political and Social Science, 254*, 58–64.

Rukikaire, K. (2019, September 23). Video games industry levels up in fight against climate change. https://www.unenvironment.org/news-and-stories/press-release/video-games-industry-levels-fight-against-climate-change.

Sayre, C. (2007, July 19). 10 Questions for Shigeru Miyamoto. Time, Inc. http://content.time.com/time/magazine/article/0,9171,1645158-1,00.html

Research Services. (2019, September 9). Two-Thirds of Gen Z Males Say Gaming is a Core Component of Who They Are. https://www.aaaa.org/gen-z-males-say-gaming-core-component-who-they-are/.

Shieber, J. (2019, January 22). Video game revenue tops $43 billion in 2018, an 18% jump from 2017. https://techcrunch.com/2019/01/22/video-game-revenue-tops-43-billion-in-2018-an-18-jump-from-2017/.

Smith, D. (2019). Most people who play video games online experience "severe" harassment, new study finds. Business Insider. https://www.businessinsider.com/online-harassment-in-video-games-statistics-adl-study-2019-7

Smith, N. (2020). Three women dreamed of joining the NBA 2K League. The road to the draft was a gauntlet of abuse and adversity. Washington Post. https://www.washingtonpost.com/video-games/esports/2020/02/24/three-women-dreamed-joining-nba-2k-league-road-draft-was-a-gauntlet-abuse-adversity/

Statista Research Department. (2016, June 14). Global VR gaming market size by segment 2020. https://www.statista.com/statistics/538801/global-virtual-reality-gaming-sales-revenue-segment/.

Statt, N. (2018, December 7). Why Epic's new PC game store is the Steam competitor the industry needed. https://www.theverge.com/2018/12/7/18129563/epic-games-store-fortnite-valve-steam-competition-pc-gaming-distruibition.

Stuart, K. (2019, December 13). Star Wars: The Rise of Skywalker trailer to be revealed in Fortnite. https://www.theguardian.com/games/2019/dec/13/star-wars-the-rise-of-skywalker-footage-fortnite.

SuperData Research Holding, Inc. 2017. *Defend Your Kingdom: What Game Publishers Need to Know About Monetization & Fraud.* New York: Digital River.

Taylor, H. (2017, October 10). Games as a service has "tripled the industry's value." http://www.gamesindustry.biz/articles/2017-10-10-games-as-a-service-has-tripled-the-industrys-value

The Father of the Video Game: The Ralph Baer Prototypes and Electronic Games. (n.d.). http://americanhistory.si.edu/collections/object-groups/the-father-of-the-video-game-the-ralph-baer-prototypes-and-electronic-games

Tretkoff, E., & Ramlagan, N. (2008, October). October 1958: Physicist Invents First Video Game (A. Chodos & J. Ouellette, Eds.). https://www.aps.org/publications/apsnews/200810/physicshistory.cfm

Urevig, A. (2019, May 15). How 2.4 billion gamers might help save the planet. https://ensia. com/notable/video-games-save-environment-improve-sustainability/.

VanDerWerff, E. T. (2014, October 13). Why is everybody in the video game world fighting? #Gamergate. https://www.vox.com/2014/9/6/6111065/gamergate-explained-everybody-fighting.

Warman, P. (Ed.). (2019, August). Newzoo's Global Games Market Report. https://newzoo.com/solutions/standard/market-forecasts/global-games-market-report/.

What is open gaming? (n.d.). https://opensource.com/resources/what-open-gaming

Wijman, Tom. (2019, June 18). The Global Games Market Will Generate $152.1 billion in 2019 as the US Overtakes China as the Biggest Market. https:// newzoo.com/insights/articles/the-global-games-market-will-generate-152-1-billion-in-2019-as-the-u-s-overtakes-china-as-the-biggest-market/.

Williams, J. (1988, January). Goodbye "G" Ratings (The Perils of "Larry"). *Computer Gaming World, 43,* 48-50.

Willingham, A. J. (2018, August 27). What is eSports? A look at an explosive billion-dollar industry. https://edition.cnn.com/2018/08/27/us/esports-what-is-video-game-professional-league-madden-trnd/index.html.

Wright, S. T. (2017, December 8). The evolution of loot boxes. https://www.pcgamer.com/uk/the-evolution-of-loot-boxes/.

Zamora, B. (2019, October 18). ESRB's lazy rating system turns a blind eye to simulated gambling. https://www.arcurrent.com/opinion/2019/10/18/esrbs-lazy-rating-system-turns-a-blind-eye-to-simulated-gambling/.

15

Home Video

Matthew J. Haught, Ph.D.*

Abstract

Home video encompasses a range of formats: broadcast, cable, satellite, internet protocol television (IPTV), over-the-top media services (OTT), and online video distributors (OVD). The three web-based formats have spurred change to television sets, with internet connectivity built in (smart TVs), added via multi-platform streaming connectors (Google Chromecast, Amazon Fire TV Stick), or through an external device (Blu-ray player, gaming system, Apple TV, Roku). The escalation of cord-cutting—dropping traditional cable or satellite service—and the rise of popular niche programming by OTT and OVD providers, has shifted the business model for the entertainment and service provider industries, a trend that continues to develop.

Introduction

For decades, home video technology centered on receiving and displaying broadcast, cable, and satellite signals. Second to that purpose was recording content via videocassette recorder (VCR) and digital video recorder (DVR) and displaying content from personal video recorders on video home system cassette tapes (VHS), laserdiscs, digital video discs (DVDs), and Blu-ray discs. However, the gradual shift toward internet and web-based home video has eroded home video's legacy formats in favor of internet protocol television (IPTV), over-the-top media services (OTT), and online video distributors (OVD).

Today's home video receivers can receive signals over the air (via an external antenna); via cable, satellite, or the internet; from DVDs and Blu-rays; and from streaming service providers. Tube-based televisions have largely disappeared from stores, replaced by plasma, OLED, LCD, and LED flat-panel displays. However, some video gamers have re-embraced tube displays for their better connectivity with classic video game platforms (Neikirk, 2017; Robertson 2018).

Home video technology exists primarily for entertainment; but it also has utility as a channel for information and education. Broadcast television delivers a broad national audience for advertisers, and cable, OTT, and OVD platforms deliver niche demographics.

* Associate Professor, University of Memphis, Department of Journalism and Strategic Media (Memphis, Tennessee)

Background

Today's home video ecosystem sits at the intersection of three technologies: television, the internet, and the film projector. Home video encompasses the many paths video content has into users' homes, as well as the devices used to consume that content.

Broadcast television began in 1941 when the Federal Communications Commission (FCC) established a standard for signal transmission. Up to that point, Philo T. Farnsworth, an independent researcher, as well as engineers and researchers at RCA and Westinghouse, had been working separately to develop the mechanics of the television set. Farnsworth is credited with inventing the first all-electric television, while RCA developed the cathode ray tube, and Westinghouse developed the iconoscope as a component of early video cameras. These inventions coalesced into the broadcast television system, where an audio-video message captured, encoded into an electromagnetic signal, sent through the airwaves, received via antenna, and reproduced on a television screen (Eboch, 2015).

In 1953, issues with limited spectrum created a second tier of broadcast frequencies; the initial space was categorized as very high frequency (VHF), and the additional space as ultra-high frequency (UHF). Consumers were slow to adopt UHF, but eventually, in 1965 the FCC mandated UHF compatibility for television set manufacturers.

As discussed in Chapter 7, cable television originated in 1948 when rural communities in Arkansas, Oregon, and Pennsylvania separately erected community antennae to receive distant broadcast signals. By the late 1950s, these antennae could receive local and distant signals, providing subscribers choices for content. Following a few regulation issues, cable television began to expand rapidly in the 1970s, and some networks began using satellites to send signal to cable systems. In the mid- to late-1990s, cable systems began to transition from antenna and satellite distribution to a fiber optic and coaxial network—capable of supporting television signals, telephone calls, and internet (CalCable, n.d.). Satellite television began as users, primarily in rural areas, installed large dish receivers to receive signals sent via satellite. In 1990, DirecTV was the first direct broadcast satellite service available to users, followed soon by USSB, Dish, and others. Over the next two decades, the industry consolidated to two primary providers: AT&T's DirecTV and Dish.

Consumer internet access began in the 1980s through telephone dial-up service. However, by the mid-1990s, the slow speed of telephone modems was evident, and cable companies began to provide internet access using the same coaxial cables that distributed television signals, supplanting telephone companies and third-party internet Service Providers (ISPs). Eventually, telephone ISPs increased in speed to compete against cable internet. As internet connectivity improved, video began to migrate to the platform.

Finally, the home film projector began as a way to show films and recorded home movies. Its use was expanded with the introduction of the consumer VHS and Betamax VCRs in the 1970s. After VHS won the standard war with Betamax, home video recording became quite popular through the VCR and through portable video camera recorders. The landmark *Sony vs. University City Studios* case in 1984, commonly called the Betamax case, established that home VCR users could record television programming for non-commercial use without infringing on copyright. In the 1990s and 2000s, DVDs replaced VHS as the primary means of storing recorded video, and portable video camera recorders transitioned to flash media for storage. In the late-2000s, Sony's Blu-ray format beat out Toshiba's HD DVD as the format of choice for high definition video storage. Further, the digital video recorder, available through most cable and satellite systems, replaced the VCR as a means of recording and time-shifting/storing television programming.

Recent Developments

The recent developments in home video extend ongoing evolutions in the industry. Four key areas: Smart TVs, cord cutting via the expansion of OVD and OTT services, evolving video platforms, and changes to Net Neutrality laws, shape the industry's present state. And, for the foreseeable future, these issues should remain at the forefront of home video.

Figure 15.1

Video Subscription Services

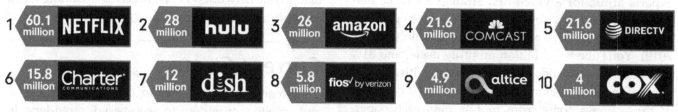

1 60.1 million NETFLIX 2 28 million hulu 3 26 million amazon 4 21.6 million COMCAST 5 21.6 million DIRECTV
6 15.8 million Charter COMMUNICATIONS 7 12 million dish 8 5.8 million fios by verizon 9 4.9 million altice 10 4 million COX.

Netflix leads all video subscription services with Amazon Video and Hulu also drawing considerable audiences. All three offer popular original programming.

Source: NCTA, 2019

Bigger & Brighter TVs

The *2020 Consumer Electronics Show*, as typical, featured a bevy of new television features, increases in size and quality, and incremental improvements to features in standard consumer models. Samsung continued to promote its gigantic 292-inch television set, deemed The Wall; when the Korean electronics giant launched The Wall in 2018, it was a quarter of its current size. LG showed off its roll up and roll down TV models while expanding on its OLED display technologies. With Sony and Vizio in the OLED marketplace now, LG no longer stands alone with the advanced LED display (Katzmaier, 2020).

Consumer television displays have improved from HDTV to 4K (also known as Ultra HD television). The next step in quality is 8K technology offered by Samsung, LG, Sony, TCL, and others. However, little video is being recorded or captured in 8K. Disney's Marvel Studios produced Guardians of the Galaxy 2 in 8K, and Japan hopes to have 8K technology in place for the 2020 Tokyo Olympics, but, as of mid-2020, 4K is the upper limit for most video production. As is typical with televisions, as prices drop and technology becomes more accessible, adoption grows; so, look for the quality evolution in ultra-high definition to continue (Withers, 2018).

Cord cutting

Over-the-top services and online video distributors, such as cable-like providers Verizon FIOS, AT&T U-verse, and Dish's Sling app, plus streaming video services Netflix, Hulu, Amazon Prime, Disney+, YouTube Red, CBS All Access, and Crunchyroll, are disrupting the traditional television ecosystem. Nearly two thirds of U.S. homes have at least one OTT service. For many OTT consumers, streaming services have displaced traditional television viewing; users watch OTT content for 2.5 hours per day, on average (Salkowitz, 2019).

OTT services have become a mainstream media channel. Millennials who subscribe to at least one OTT service typically watch more than twice as much OTT content than live TV, and fewer than 40% of millennials plan to keep traditional cable. And they're not going back: nearly two-thirds of cord cutters say there's nothing that could bring them back to traditional cable (Salkowitz, 2019). Advertising dollars are slow to follow, but most OTT subscribers don't mind ads on their services. EMarketer projects that by 2023, the number of households without pay TV will rise to 56.1 million, while those with cable will drop to 72.7 million (Fitzgerald, 2019).

Platforms

Home video, once defined by the television set, has broadened to include a range of devices and platforms. Thanks to the Internet of Things (IoT) trend, home video has a presence on Facebook, Instagram, and other social media, in picture frames and even on the refrigerator. With this expansion of platforms and consumption has come an increase in video advertising. Facebook, the largest social network, has pivoted hard toward video, with user-generated newsfeed video, newsfeed video advertising, streaming user-generated Facebook Live, and even streaming live sports, including college football (Mediakix, 2018).

Facebook is not alone with its video pivot. Social network video is popular on Twitter, LinkedIn, TikTok, Instagram, and, of course, YouTube. Advertising experts argue for a video-first strategy in social network marketing, as social video is cheap to produce and engages passive audiences to consume (Duarte, 2019). YouTube has 2 billion monthly users, and 81% of U.S. 15–25-year-olds use the platform, with all users averaging about 11.5 minutes of watching a day; 70% of those views come from mobile devices (Cooper, 2020). YouTube also claims its mobile usage numbers show it reaches more 18–34-year-olds than any TV network.

Net Neutrality

The 2017 FCC decision to rollback prohibitions on throttling speeds on certain websites or charging users for premium access to websites has begun to affect home video. Carriers are slowing down streaming video over mobile data (Li, et al, 2019). Now, many leading U.S. mobile phone carriers are imposing speed limits for popular streaming video platforms.

Table 1:

How Slow Can You Go?

Mobile Provider	Streaming Site		
	Netflix	YouTube	Amazon
AT&T	1.4	1.4	NA
Sprint	1.9	2.1	2.1
T-Mobile	1.8	1.5	1.5
Verizon	3.3	3.98	3.95

Bandwidth speed limits in Mbps for selected streaming video sites on popular U.S. carriers.

Source: Li, et al, 2019.

With Net Neutrality struck down, carriers have been able to reduce the speed of home video delivered over mobile internet. This speed throttling, as it is known, leaves users with blurry video, lags in sound, and pauses for buffering—limitations that were on the way out with 4G and 5G mobile networks (see Chapter 20).

What remains to be seen, however, is the effect on terrestrial internet. While some have anecdotally noted throttling (McCue, 2019), no study has found broad limits on speeds on wired internet access. Yet, many users are noting these slowdowns and have turned to virtual private networks (VPNs) to access streaming video content and to bypass any slowdowns.

Current Status

Home video continues to pivot from terrestrial and satellite providers to web-based providers. In 2017, pay TV market penetration was down to 75%, from a high of 88% in 2010 (Watson, 2019). Yet, 96.5% of U.S. homes with a television receive a video signal of some form—paid or otherwise (Nielsen, 2017). Of the 25% of households that do not have pay television, two-thirds had previously been pay TV subscribers; however, many among Generation Z are "cord nevers," having never subscribed to pay television. The oldest members of Generation Z in 2020 are about 23 years old, so as this generation ages and moves out of their parents' homes and become economically independent, the pay television industry could see further shifts (Fitzgerald, 2019).

Home video technology now engages mobile devices alongside television sets. Users, particularly younger users, are increasingly watching home video content on tablets and mobile devices (Eck, 2017). As such, content creators for home video are shifting; Amazon, Netflix, and Hulu are producing original (critically acclaimed and award-winning) content for their platforms, in addition to content from traditional video creators. And specialized platforms are playing a role to provide content previously inaccessible to consumers; for example, Crunchyroll, an online video distributor specializing in anime, reports having 2 million paid subscribers and 50 million total users (Crunchyroll, 2020). Specialized web networks with premium ad-free subscriptions or free-to-watch with advertising options are providing less-visible content to their niche audience.

Moreover, creatives developing new shows might first test them on YouTube or another low-end platform, and larger content companies can pluck new talent from these engaged creatives (Eck, 2017). Or, shows on traditional television can extend their brands with web shows.

Consider first the example of World of Wonder (WOW). The production company's flagship show, RuPaul's *Drag Race*, airs on VH1. However, because fans of the show are deeply engaged with its content, WOW has multiple active free programs on YouTube. Then, in 2017, it launched WOW Presents Plus, a premium YouTube channel to air additional content from former RuPaul's *Drag Race* contestants and other specialized content for the LGBTQ community.

Second, the example of Issa Rae's *The Misadventures of Awkward Black Girl* YouTube series, exemplifies online video as an incubator of new talent. Rae launched the web series in 2011; its first episode, "The Stop Sign," earned 2.4 million views. Rae's success translated to an HBO series, *Insecure*, in 2016, which itself landed her deals for two more series with the premium network (Crucchiola, 2017).

Both these examples show how home video has democratized the television landscape. WOW's programming caters to an LGBTQ audience, and Issa Rae's work is popular among African-American audience. Networks have typically been hesitant to embrace programming for these limited audiences; yet, the alternative networks offered by streaming platforms create space for experimental shows and niche content.

Niche audiences are not the only targets, though. Quibi, a streaming platform launched in April 2020, limits content to 10 minutes (Wood, 2020). Users must subscribe to access content, and can pay more to avoid ads. But the kicker is that the content is mobile only, and is presented in both portrait and landscape video formats. The platform plans to offer 175 shows during its first year; major television and film studios have already bought in, producing content in the short format specifically for Quibi (Wood, 2020).

The power of home video today is the ability reach diverse audiences with diverse content. No longer are audiences limited to the content created by studios and networks or by the channels offered by their local cable companies. Users are able to find content that suits their tastes—and watch it on whatever device they have, whenever they want—and often, paying to avoid advertising, or watching free content with ads specific to their demographics.

Technology Ecosystem

Hardware

The hardware for home video encompasses a range of devices, including television sets, computer monitors, tablets, and mobile phones (Eck, 2017). Similarly, these displays have associated terrestrial, satellite, or web-based connectivity through a receiver, and ultimately, the sender mechanism for encoding the video signal.

Television sets continue to evolve through internet of things (IoT) technology (Spangler, 2018). What used to be a purely home video terminal, now ranks among twenty or so other smart devices in the home capable of displaying video from OVD, as well as monitoring other smart home systems. However, the key feature of a television set is its size. Mobile device screens are rarely more than 12 inches wide, but television screens often measure up to 60 or more inches (Pierce, 2018).

Software

Software needed for home video viewing is limited. Typically, the viewing device itself has built-in viewing capabilities. However, mobile applications for OVD are common. Netflix, Amazon, and Hulu are the most popular apps, but more niche programming apps follow.

Organizational Infrastructure

Home video infrastructure can be broken down into two broad categories: Viewing device manufacturers and service providers. Devices include television sets, computer monitors, mobile devices, Blu-ray and DVD players, and smart television connectors. Providers include cable and satellite services, internet service providers, IPTV services, and OVD.

With more than 20% market share, Samsung leads all television set manufacturers. The second most popular, LG Electronics, has about 12%, with Sony, Hisense, TCL, Panasonic, and Vizio with strong presences (Technavio, 2019). Higher-end Sony and LG organic light emitting diode (OLED) and Samsung quantum dot light emitting diode (QLED) are popular and deliver higher definition, but at higher costs (Katzmaier, 2017). Samsung, Sony, LG, and Panasonic also dominate the list of Blu-ray and DVD player manufacturers.

Top computer manufactures are led in market share by Lenovo (24.2%), HP (22.4%), Dell (15.9%), Apple (7.1%), and ASUS (6.1%) (Gartner, 2019). Mobile device operating systems are dominated by Apple's iOS and Google's Android OS, together accounting for nearly all of the market share (Vincent, 2017). Globally, Android serves 87% of users, while Apple serves 13% (Holst, 2019). In the U.S., however, the market is more even; Apple leads smartphone manufacturers with about 42% of the market with Samsung (28.3%), LG (10%) and Lenovo following (Counterpoint, 2019). Apple and Samsung also dominate the tablet market share.

Comcast and Charter are the largest cable providers in the U.S. AT&T's DirecTV leads satellite with nearly 20 million subscribers, with Dish following; however, with AT&T's announcement of the launch of its last satellite, it seems the tide is turning away from satellite television (Rodriguez, 2018). Comcast, Charter, and AT&T lead internet service providers, with Verizon following (Rutherford, 2019).

Netflix, Amazon, Hulu, and Disney+ boast high subscriptions costs. However, some OVD platforms offer free subscriptions. YouTube, for example, has 1 billion active users per month, while YouTube Red has more than 2 million paying subscribers; those 2 million can watch videos on the network ad-free, while the remainder can use the service, but must watch ads before most videos. This free-with-ads/paid-ad-free model operates for several platforms.

Social Systems

The NCTA—The Internet & Television Association, is the lead trade association for the U.S. broadband and television industries. The NCTA was one of the lead lobbying groups pushing for the FCC's 2017 rollback of Net Neutrality rules. However, the association claims it supports an open internet, including no blocking of legal content, no throttling, no unfair discrimination, and transparency in customer practices (NCTA, n.d.). The 2017 Net Neutrality rollback removes rules blocking extra charges or slowing of content for specific sites, which is just what has happened— particularly on mobile video.

Individual User

Service penetration for television remains high; Nielsen estimated there were 119.6 million TV homes in the U.S. for the 2017-18 season, which is about 80% of households (Nielsen, 2017). Meanwhile, 77% of U.S. adults own a smartphone, 51% own a tablet, and 78% own a desktop or laptop—all capable of home video viewing (Pew, 2017).

As of May 2019, 65% of U.S. households have at least one internet-enabled device capable of streaming home video content (Nielsen, 2019). Users access video content through gaming systems, smart televisions, and multimedia devices with some overlap (Nielsen, 2019). In all, these streamers consume 2 hours and 4 minutes of streaming content per day.

Ultimately, individual user adoption can vary widely, from full cable and satellite with premium channels and multiple OTT, OVD platforms via multiple enabled devices to no television service with only a free OVD subscription on a single web-enabled device.

Sales & Subscriptions

Overall, the home video industry generates about $23 billion annually in revenue. Streaming services account for the lion's share (55.5%, $12.9 billion). Electronic sales and rentals account for $4.6 billion. DVD and Blu-ray disc sales are down to $4 billion; video rentals from kiosks and stores continues to fall, accounting for about $1.4 billion. (Richter, 2019).

The number of home cable and satellite subscribers continues to decline as cord-cutters drop cable services in favor of OTT and OVD services. About 65% of U.S. homes have cable or satellite television, down from a high of 91% in 2011. Conversely, about 69% of U.S. homes have internet streaming video (Guzior, 2019).

In advertising, television is no longer king. Brands spent about $70.3 billion on domestic television in 2019, while digital advertising up to $129.3 billion (eMarketer, 2019a). Companies are concentrating television spending on fewer brands and products, and they are even launching new products without national television campaigns. Experts predict spending will continue to dip through the early 2020s (eMarketer, 2019b).

Factors to Watch

As cord-cutting and online transitions escalate, home video could transition further away from its traditional tether to a television set (Eck, 2017). With new OVD platforms emerging in 2020 including HBO Max and Comcast's Peacock, and OVD platforms Hulu, YouTube Red, Amazon Prime, Netflix, Disney+, and others continue to generate original content, expect more mobile video consumption.

The caveat to this mobile pivot comes through the FCC's 2017 decision to overturn Net Neutrality rules that prohibited ISPs from throttling speeds or charging uses more for streaming services. Considering the bulk of ISPs are cable, satellite, or IPTV companies, users who drop mainline video service in favor of OVD platforms outside the ISP's home brand might see slower speeds or higher costs, particularly as mobile internet users are already having streaming video services slowed.

Regardless, the ability of OVD and legacy video distributors to deliver an audience to advertisers remains strong. Big data tracking of users will be easier as more make the switch to online video. Thus, brands will be able to target specific users and demographics more easily and will be able to more effectively measure the return on investment for ad spending. Further, demographic-specific creative advertising could be channeled to users on the fragmented OVD platform—all driven by big data.

Home video is a commodity of the media convergence economy. Television set manufacturers have switched to making mobile phones and tablets. Computer manufacturers are making mobile phones and tablets and smart television systems. Movie and television studios, independent content creators, and cable companies have their own streaming platforms or are investing in others. Home video is evolving far beyond channels and receivers to a new era of content and glass.

Gen Z

For Generation Z, home video technology is the bread and butter of their media consumption. When watching video, they're much more likely to do it via a streaming service than by a paid cable or satellite connection (Fitzgerald, 2019). Their main source of television-style video content is Netflix, but they spend more time on YouTube. Because many users access the technology plans shared with family and friends, their ultimate adoption of home video routines could change in the next few years.

However, what we do know is that, by consumption patterns, video is a regular part of Gen Z's routines: 65% of them check Instagram, with its mix of photo and video content, and 62% use YouTube every day (Green, 2019). On these ad-driven platforms, Gen Z users aren't just consuming content—they're actively curating a unique mix of channels that suit their individual tastes (Baron, 2019). They're creating content, too, and unlike their Millennial predecessors who valued perfection in their content, Gen Z influencers are striving for authenticity, showing their real, unfiltered lives, and experiences (Lorenz, 2019).

Global Concerns & Sustainability

As home video shifts from tangible means to web-based signals, the ability for users worldwide, particularly those in rural and low-income countries and regions could see their ability to consume media diminished. The United Nations' *State of Broadband* 2019 report found that only 54.3% of homes worldwide have a broadband internet connection (UNESCO, 2019). Further, as some counties choose to limit access to websites and media from some countries, some users could be limited in their media—counterintuitive in an era of expanding access. Privacy also remains a concern, as privacy laws like the *General Data Protection Regulation* in the European Union and the *California Consumer Privacy Act* shape the types of data home video providers can collect and store about their consumers. These changes in privacy are largely seen as good for consumers, but could dramatically change the ways home video providers conduct business.

Home video products' sustainability factors are both positive and negative. With video being delivered via the internet, physical copies no longer have to be made, packaged, and shipped. Watching home video can have a significant carbon footprint, though, because of the energy consumed by the device and the need to power the data center where content is stored (BBC, 2018). However, this footprint can be decreased as data centers transition to greener power methods, The carbon impact of home video all depends on the fuel used to power devices, the fuel used to power data centers, and the type of internet used to consume the media (Chandaria, Hunter, & Williams, 2011). Larger screens require more energy, and 3G and 4G mobile internet take more energy than Wi-Fi connections (BBC, 2018).

Finally, the proliferation of devices and their planned obsolesce creates some sustainability concerns. The precious metals used in device construction are difficult to obtain and to recycle, and electronic waste can be harmful to the environment (Ding, et al, 2019).

Bibliography

Baron, J. (2019, July 3). The Key To Gen Z Is Video Content. https://www.forbes.com/sites/jessicabaron/2019/07/03/the-key-to-gen-z-is-video-content/.

BBC. (2018, October 12). Climate change: Is your Netflix habit bad for the environment? https://www.bbc.com/news/technology-45798523.

CalCable. (n.d.). History of Cable. https://www.calcable.org/learn/history-of-cable/.

Chandaria, J., Hunter, J., & Williams, A. (2011, May). The carbon footprint of watching television, comparing digital terrestrial television with video-on-demand. In *Proceedings of the 2011 IEEE International Symposium on Sustainable Systems and Technology* (pp. 1-6). IEEE.

Cooper, P. (2020, January 13). 23 YouTube Statistics that Matter to Marketers in 2020. https://blog.hootsuite.com/youtube-stats-marketers/.

Counterpoint. (2019, December 4). US Smartphone Market Share: By Quarter. https://www.counterpointresearch.com/us-market-smartphone-share/.

Crucchiola, J. (2017, December 18). Issa Rae Is Working on Two New Shows for HBO. http://www.vulture.com/2017/12/issa-rae-is-working-on-two-new-shows-for-hbo.html.

Crunchyroll. (2020). About Crunchyroll. https://www.crunchyroll.com/about.

Ding, Y., Zhang, S., Liu, B., Zheng, H., Chang, C. C., & Ekberg, C. (2019). Recovery of precious metals from electronic waste and spent catalysts: A review. *Resources, conservation and recycling, 141,* 284-298.

Duarte, J. (2019, April 25). Opinion: The new realities of social media. https://adage.com/article/cmo-strategy/opinion-new-realities-social-media/2166521.

Eboch, M. M. (2015). *A History of Television.* Minneapolis: ABDO Publishing.

Eck, K. (2017, April 20). Is Mobile About to Overtake Traditional TV Viewing? www.adweek.com/tv-video/why-more-people-than-ever-are-watching-tv-on-mobile/.

eMarketer. (2019a, February 19). US Digital Ad Spending Will Surpass Traditional in 2019. https://www.emarketer.com/content/us-digital-ad-spending-will-surpass-traditional-in-2019.

eMarketer. (2019b, November 13). TV Will Drop Below 25% of Total US Ad Spending By 2022. https://www.emarketer.com/content/tv-will-drop-below-25-of-total-us-ad-spending-by-2020.

Fitzgerald, T. (2019, August 6). Cord Cutting Is About To Go Way, Way Up. https://www.forbes.com/sites/tonifitzgerald/2019/08/06/cord-cutting-is-about-to-go-way-way-up/.

Gartner. (2019, January 10). Gartner Says Worldwide PC Shipments Declined 4.3 Percent in 4Q18 and 1.3 Percent for the Year. https://www.gartner.com/en/newsroom/press-releases/2019-01-10-gartner-says-worldwide-pc-shipments-declined-4-3-perc.

Green, D. (2019, July 3). The most popular social media platforms with Gen Z. https://www.businessinsider.my/gen-z-loves-snapchat-instagram-and-youtube-social-media-2019-6/.

Guzior, B. (2019, April 23). Streaming households could overtake traditional TV homes in 2019. https://www.bizjournals.com/bizwomen/news/latest-news/2019/04/streaming-households-could-overtaketraditional-tv.html.

Holst, A. (2019, September 13). Android vs iOS market share 2018. https://www.statista.com/statistics/272307/market-share-forecast-for-smartphone-operating-systems/.

Katzmaier, D. (2017, May 24). QLED vs. OLED: Samsung's TV tech and LG's TV tech are not the same. https://www.cnet.com/news/qled-vs-oled-samsungs-tv-tech-and-lgs-tv-tech-are-not-the-same/.

Katzmaier, D. (2020, January 10). TVs of CES 2020: OLED, 8K, giant screens and other trends of absolute excess. https://www.cnet.com/news/tv-ces-2020-oled-8k-giant-screens-trends-absolute-excess/.

Li, F.; Niaki, A.A.; Choffnes, D.; Gill, P.; and Mislove, A. (2019). A Large-Scale Analysis of Deployed Traffic Differentiation Practices. SIGCOMM '19: 2019 Conference of the ACM Special Interest Group on Data Communication, August 19–23, 2019, Beijing, China.

Lorenz, T. (2019, April 24). The Instagram Aesthetic Is Over. https://www.theatlantic.com/technology/archive/2019/04/influencers-are-abandoning-instagram-look/587803/.

McCue, T. J. (2019, June 29). How Can I Tell If My Internet Provider Is Throttling (Slowing) My Connection Speed? https://www.forbes.com/sites/tjmccue/2019/06/27/how-can-i-tell-if-my-internet-provider-is-throttling-slowing-my-connection-speed/.

Mediakix. (2018, October 31). The Facebook Video Statistics You Need To Know In 2018. https://mediakix.com/blog/facebook-video-statistics-everyone-needs-know/.

NCTA. (2019). Top 10 Video Subscription Services: NCTA - The internet & Television Association. https://www.ncta.com/industry-data/top-10-video-subscription-services.

Neikirk L. (2017, March 09). The tube-based CRT TV has officially run out of time. http://televisions.reviewed.com/features/the-tube-based-crt-tv-has-officially-run-out-of-time.

Nielsen. (2017, August 25). Nielsen Estimates 119.6 Million TV Homes in the U.S. for the 2017-18 TV Season. http://www.nielsen.com/us/en/insights/news/2017/nielsen-estimates-119-6-million-us-tv-homes-2017-2018-tv-season.html.

Nielsen. (2019, August). Nielsen Local Watch Report. https://www.nielsen.com/wp-content/uploads/sites/3/2019/08/local-watch-report-aug-2019.pdf.

Pew. (2017, January 12). Mobile Fact Sheet. http://www.pewinternet.org/fact-sheet/mobile/.

Pierce, D. (2018, January 10). The Future of TV Is Just Screens All The Way Down. https://www.wired.com/story/ces-2018-the-future-of-tvs/.

Richter, F. (2019, January 24). Infographic: Streaming Dominates U.S. Home Entertainment Spending. https://www.statista.com/chart/7654/home-entertainment-spending-in-the-us/.

Robertson, A. (2018, February 6). Inside the desperate fight to keep old TVs alive. https://www.theverge.com/2018/2/6/16973914/tvs-crt-restoration-led-gaming-vintage.

Rodriguez, A. (2018, December 3). It's the beginning of the end of satellite TV in the US. https://qz.com/1480089/att-just-declared-the-end-of-the-satellite-tv-era-in-the-us/.

Rutherford, S. (2019, June 24). The 5 Biggest US ISPs Actually Think They're Ready for Streaming Games. https://gizmodo.com/the-5-biggest-us-isps-actually-think-theyre-ready-for-s-1835737417.

Salkowitz, R. (2019, April 10). How Much Streaming Can We Take? New Data Sheds Light On The OTT Revolution. https://www.forbes.com/sites/robsalkowitz/2019/04/10/how-much-streaming-can-we-take-new-data-sheds-light-on-the-ott-revolution/.

Sony Corp. of America v. Universal City Studios, Inc., 464 U. S. 417 (1984).

Spangler, T. (2018, January 10). Comcast Wants to Turn Xfinity Into an 'internet of Things' Smart-Home Platform. http://variety.com/2018/digital/news/comcast-xfinity-internet-of-things-home-automation-1202657823/.

Technavio. (2019, September 5). Largest TV Manufacturers by Market Share Worldwide 2019: Best TV Brands: Global TV Market. from https://blog.technavio.com/blog/largest-tv-manufacturers-by-market-share.

UNESCO. (2019, September 25). New report on global broadband access underscores urgent need to reach the half of the world still unconnected. https://en.unesco.org/news/new-report-global-broadband-access-underscores-urgent-need-reach-half-world-still-unconnected.

Vincent, J. (2017, February 16). 99.6 percent of new smartphones run Android or iOS. https://www.theverge.com/2017/2/16/14634656/android-ios-market-share-blackberry-2016/

Watson, A. (2019, November 22). Pay TV penetration in the U.S. 2019. https://www.statista.com/statistics/467842/pay-tv-penetration-rate-usa/.

Wood, M. (2020, January 14). Quibi is spending more than a billion to enter streaming wars. https://www.marketplace.org/shows/marketplace-tech/quibi-spending-more-than-a-billion-wading-into-streaming-wars-luring-subscribers-will-be-key/

Withers, S. (2018, August 31). What is 8K TV and do we need it? https://www.wired.co.uk/article/what-is-8k-tv.

16

Digital Imaging & Photography

Michael Scott Sheerin, M.S.*

"no time for cameras ... we'll use our eyes instead"

—*from the song* Cameras *by Matt and Kim*

Abstract

or the past couple of decades, digital photography has been by far the most popular visual medium. But the way we capture the images has changed. This chapter looks back at the history of photography from its camera obscura analog beginnings, through the advent of digital photography, and into the current state of the medium. Light-field photography, lens-less cameras and artificial intelligence (AI) software are some of the newer innovations that are starting to play major roles in the future direction of image capture. These advancements change the way a digital image is captured, processed and shared. So, exactly what is a digital image? It is a numerical representation of a two-dimensional image. The smallest element of the image, the pixel (derived from picture element), holds the values that represent the brightness of a color at any specific point on the image. Thus, the digital image is a file that contains information in the form of numbers. The way we capture this information, the software processes that can be applied to it, the way we share it, and future trends in the industry are explored in this chapter.

---

* Associate Professor, School of Journalism and Mass Communications, Florida International University (Miami, Florida)

Introduction

Advancements in the field point to a time when we'll only need our eyes to take pictures, as image capture technology, to borrow from the Greek, has become "phusis," defined as a natural endowment, condition or instinct (Miller, 2012). Look around you, and I bet there is a device within three feet that you can use to capture an image; perhaps that device is even in your hand or on your wrist as you read this chapter on your mobile phone. For Gen Z and Millennials (and, let's be honest, older generations as well), the mobile phone is a natural appendage. And an appendage that we use a lot, as it has been estimated that we will take 1.4 trillion photos in 2020 (Papacharissi, 2018), with nearly 85% of them captured on a mobile phone (Cakebread, 2017). These digital images are ultimately shared on social networks like Snapchat (20,000 photos shared/second) (Aslam, 2019a), WhatsApp (4.5 billion photos shared/day) (BT Online, 2017), Facebook (350 million photos uploaded/day) (Aslam, 2019b), or Instagram (100+ million photos & videos uploaded/day) (Aslam, 2019c), for anyone to see, making the digital image ubiquitous in our culture.

George Orwell's novel, *Nineteen Eighty-Four*, published in 1949, fictitiously predicts a dystopian society that has had its civil liberties stripped due to government surveillance, i.e. Big Brother. But as we see today, it's not Big Brother alone that is watching, as the plethora of images shared on social media platforms confirms that everyone is participating in the surveillance, including self-participation. Over 1 million selfies are taken globally every day, with over 54 million images that are tagged #selfie residing on Instagram (Muller, 2019). And, according to one poll, every third image taken by a millennial was a selfie (Zetlin, 2019). And that figure doesn't take into account the newest trend of taking slow-motion selfies, or "slofies" (Kastrenakes, 2019). Thus, Orwell's imagined surveillance takes place with complete compliance by its subjects. As Fred Ritchin, professor of Photography and Imaging at New York University's Tisch School of the Arts, points out, "we are obsessed with ourselves" (Brook, 2011). Now couple this "social surveillance" trend with facial-recognition technology, and it's easy to see how someone could be "tagged" and tracked very easily. Taking it a step further, and not satisfied with our own "self-surveillance," China has implemented mandatory facial recognition scans for anyone registering a new sim card. The government claims that the deployment of this AI technology will "protect the legitimate rights and interest of citizens in cyberspace" (Kuo, 2019). Many have raised concerns about the dystopian use of this technology. In what is said to be the first lawsuit brought against the Chinese government regarding the use of facial recognition, Guo Bing, a professor at Zhejiang Sci-Tech University, claims that the consumer rights protection law was violated when his face was scanned at a safari park without his consent. In fact, Facebook, which uses a facial recognition algorithm called Photo Review, changed its policy due to user complaints. The default setting now opts new users out of the use of all facial recognition, including tagging. One can still opt in if they want the algorithm to detect and tag. However, users that were automatically opted in when Facebook first implemented the technology still have to change their settings and opt out if they don't want to be facially recognized and tagged (Glaser, 2019).

These findings suggest that the photograph is not just a standalone part of the digital image industry anymore. It's fully converged with the computer and cell phone industries, among others, and has changed the way we utilize our images "post shutter-release." These digital images are not the same as the photographs of yesteryear. Those analog photos were continuous tone images, while the digital image is made up of discreet pixels, ultimately malleable and traceable, to a degree that becomes easier with each new version of photo-manipulating software, GPS, AI, and detection algorithms. And unlike the discovery of photography, which happened when no one alive today was around, this sea change in the industry has happened right in front of us; in fact, we are all participating in it—we are all "Big Brother." This chapter will look at some of the hardware and software inventions that continue to make capturing and sharing quality images easier, as well as the implications on society, as trillions of digital images enter all our media.

Background

Digital images of any sort, from family photographs to medical X-rays to geo-satellite images, can be treated as data. This ability to take, scan, manipulate, disseminate, track, or store images in a digital format has spawned major changes in the communication technology industry. From the photojournalist in the newsroom to the magazine layout artist, and from the social scientist to the vacationing tourist posting to Facebook via Instagram, digital imaging has changed media and how we view images.

The ability to manipulate digital images has grown exponentially, and has become easier to do, with the addition of imaging software. And with the advancements of AI in imaging software apps, the manipulation is starting to be done *for you*. Though apps from the likes of Photoshop and Lightroom already exist, there was a somewhat steep learning curve involved with using them. With the 2020 release of Adobe's Photoshop Camera app for smart phones, high-end AI can do things such as adjust the dynamic range, tonality, and scene type for you automatically (Lee, 2019). Photoshop Camera uses the AI platform Sensei to recognize the subject of the image and suggests in real time which image filters to use. Other AI advancements include the ability to transform a 2D photograph into a 3D image. A beta version of the LucidPix app was released during the 2020 CES convention, with plans for full release in the second quarter of 2020. Unlike previous apps that require dual lens smart phones, or for the user to move the camera around the subject when in capture mode, LucidPix generates a depth map of the captured image as it converts the 2D image into a 3D one. "... the visual medium has become more multidimensional, leading to more portrait photos, 3D content, and AR and VR being created," states Han Jin, Founder and CEO at Lucid (Grigonis, 2020).

Looking back, history tells us photo-manipulation dates to the film period that apps like Faded, with its vintage film and sepia filters, attempt to capture. Images have been manipulated as far back as 1906, when a photograph taken of the San Francisco Earthquake was said to be altered as much as 30% according to forensic image analyst George Reid. A 1984 National Geographic cover photo of the Great Pyramids of Giza shows the two pyramids closer together than they are, as they were "moved" to fit the vertical layout of the magazine (Pictures that lie, 2011). In fact, repercussions stemming from the ease with which digital photographs can be manipulated caused the National Press Photographers Association (NPPA), in 1991, to update their code of ethics to encompass digital imaging factors (NPPA, 2019). Here is a brief look at how the captured, and now malleable, digital image got to this point.

The first photograph ever taken is credited to Joseph Niepce, and it turned out to be quite pedestrian in scope. Using a technique he derived from experimenting with the newly invented lithograph process, Niepce was able to capture the view from outside his Saint-Loup-de-Varennes country house in 1826 in a camera obscura (Harry Ransom Center, 2020). The capture of this image involved an eight-hour exposure of sunlight onto bitumen of Judea, a type of asphalt. Niepce named this process heliography, which is Greek for sun writing (Lester, 2006). It wasn't long after that the first photographic self-portrait, now known as a selfie, was recorded in 1839. (See Figure 16.1) Using the daguerreotype process (sometimes considered the first photographic process developed by Niepce's business associate Louis Daguerre), Robert Cornelius "removed the lens cap, ran into the frame and sat stock still for five minutes before running back and replacing the lens cap" (Wild, 2020).

Figure 16.1

First "Selfie"

Source: Library of Congress

The next 150 years included significant innovation in photography. Outdated image capture processes kept giving way to better ones, from the daguerreotype to the calotype (William Talbot), wet-collodion (Frederick Archer), gelatin-bromide dry plate (Dr. Richard Maddox), and the now slowly disappearing continuous-tone panchromatic black-and-white and autochromatic color negative films. Additionally, exposure time has gone from Niepce's eight-hour exposure to 1/500th of a second or less.

Cameras themselves did not change much after the early 1900s until digital photography came along. Kodak was the first to produce a prototype digital camera in 1975. The camera, invented by Steve Sasson, had a resolution of .01 megapixels and was the size of a toaster (Zhang, 2010). In 1981, Sony announced a still video camera called the MAVICA, which stands for magnetic video camera (Carter, 2015a). It was not until nine years later, in 1990, that the first digital still camera (DSC) was introduced. Called the Dycam (manufactured by a company called Dycam), it captured images in monochromatic grayscale and had a resolution that was lower than most video cameras of the time. It sold for a little less than $1,000 and had the ability to hold 32 images in its internal memory chip (Aaland, 1992).

In 1994, Apple released the Quick Take 100, the first mass-market color DSC. The Quick Take had a resolution of 640 × 480—equivalent to a NTSC TV image—and sold for $749 (Kaplan, 2008). Complete with an internal flash and a fixed focus 50mm lens, the camera could store eight 640 × 480 color images on an internal memory chip and could transfer images to a computer via a serial cable. Other mass-market DSCs released around this time were the Kodak DC-40 in 1995 for $995 (Carter, 2015b) and the Sony Cyber-Shot DSC-F1 in 1996 for $500 (Carter, 2015c).

The DSCs, digital single lens reflex (DSLR) and mirrorless (MILC) cameras work in much the same way as a traditional still camera. The lens and the shutter allow light into the camera based on the aperture and exposure time, respectively. The difference is that the light reacts with an image sensor, usually a charge-coupled device (CCD) sensor, a complementary metal oxide semiconductor (CMOS) sensor, or the newer, live MOS sensor. When light hits the sensor, it causes an electrical charge.

The size of this sensor and the number of picture elements (pixels) found on it determine the resolution, or quality, of the captured image. The number of thousands of pixels on any given sensor is referred to as the megapixels (MP). The sensors themselves can be varied sizes. A common size for a sensor is 18 × 13.5mm (a 4:3 ratio), now referred to as the Four Thirds System (Four Thirds, 2017). In this system, the sensor area is approximately 25% of the area of exposure found in a traditional 35mm camera. Many of the sensors found in DSLRs are full frame CMOS sensors that are 35mm in size (Canon, 2018), whereas are most of the MILC sensors are either Four Thirds or Advanced Photo System Type-C (APS-C), which are slightly larger than Four Thirds sensors (Digital Camera, 2020).

The pixel, also known in digital photography as a photosite, can only record light in shades of gray, not color. To produce color images, each photosite is covered with a series of red, green, and blue filters, a technology derived from the broadcast industry. Each filter lets specific wavelengths of light pass through, according to the color of the filter, blocking the rest. Based on a process of mathematical interpolations, each pixel is then assigned a color. Because this is done for millions of pixels at one time, it requires a great deal of computer processing. The image processor in a DSC must "interpolate, preview, capture, compress, filter, store, transfer, and display the image" in a very short period (Curtin, 2011).

This image processing hardware and software is not exclusive to DSCs, MILCs, or DSLRs, as this technology has continued to improve the image capture capacities of the camera phone. Starting with the first mobile phone that could take digital images—the Sharp J-phone released in Japan in 2000—the lens, sensor quality, processing power, ease, and convenience have made the mobile phone the go-to-camera for most of the images captured today. An example of improved image capture technology in mobile phones can be found on the Apple iPhone 11. With a triple lens array and an image capture sensor of 48 MP, the iPhone 11 also offers time-of-flight (ToF) technology, which can measure the distance of several objects from the camera's sensor, aiding in depth-of-field AI calculations (Kronfli, 2019).

Recent Developments

According to Saffo's 30-year rule, the digital imaging industry has reached full maturity (1990–2020). Thus, it can be expected that growth in some areas has slowed. One example of this is seen in the lack of new camera models (DSC, DSLR, and MILCs) coming to market. There has been a steady decline since 2014, with less than 40 new models introduced in the last couple of years (there were 98 new models in 2014) (Sylvester, 2019). But that doesn't imply that it is not still a dynamic industry. New areas in light-field photography, mobile phone apps, and AI continue to push the technological boundaries.

Change has always been a part of the photographic landscape. Analog improvements, such as film rolls (upgrades from expensive and cumbersome plate negatives) and Kodak's Brownie camera (sold for $1.00) brought photography to the masses at the start of the 1900s. The advent of the digital camera and the digital images it produced in the late 20th century was another major change, as the "jump to screen phenomenon" gradually pushed the film development and printing industry off to the sidelines (Reis, 2016). And when this new digital imaging phase converged with the telephone industry at the start of the 21st century, the industry really underwent a holistic change. The rapid rise in the use of the mobile phone as camera is the real technological advancement that made the largest impact on the digital imaging and photography industry up to now.

But we don't stop there. Technological advancements in the field continue, and they usually happen in one of two ways. The first is by incremental improvements, and this can be illustrated with a few examples. One is digital sensor improvement, while another is lens speed, as determined by its aperture. Currently, the Experimental Optics lens, coming in with an aperture at f0.75 is the fastest lens on the market (though it lacks built-in auto-focusing capabilities) (Artaius, 2020) Perhaps stretching the use of the term incremental is the advent of the gigapixel camera, such as the one being developed in Chile at the Large Synoptic Survey Telescope (LSST) Project. This 3.2 billion-pixel camera will be the world's largest camera and is expected to be completed in 2022.

Each panoramic snapshot that the LSST captures will be of an area 40 times the size of a full moon (LSST, 2020). A triple array lens camera phone is another example of incremental improvement, as is the implementation of a 10X lossless hybrid zoom on the Reno camera phone by Oppo. It has a 48MP main camera lens, an 8MP wide-angle camera lens and a 13MP telephoto lens (Brown, 2019). Two other innovations that deal with camera lenses begin to pull the technological advancements from the incremental innovation column to the disruptive innovation column, the second of the two types of innovation.

The disruptive innovation that digital imaging is currently undergoing is going to have the biggest impact on the future of the industry. From advances in light-field and computational photography, to lens-less camera innovation, this is the area to watch. It's not just the ubiquitous use of today's mobile phones for capturing images, coupled with the sharing of these images due to the rise of social media and the internet of things (IoT), but it's also the new ways that these images are captured that contribute to this disruptive innovation. A leader in this field is the company Light, who released the L16 camera in 2017. It is a 52 MP, 16-lens camera prototype the size of a smartphone. Reviews have been less than impressive, but Wired writer David Pierce's prediction that the algorithms used in the L16 "will power your next smartphone, or maybe the one after that" (Pierce, 2017) has come to fruition in 2020. Nokia partnered with Light, to create the PureView9 smartphone, which has Zeiss optics and a five-camera array that uses the same light-field technology found on the L16.

Figure 16.2

Nokia 9 PureView

Source: GizChina

Light has also partnered with Sony and Qualcomm among others, so expect a slew of future smartphones deploying this technology (Monckton, 2020). The way light-field cameras work is they capture data rather than images and process the data post-shutter release to produce the desired photo, including the ability to reinterpret it by focusing on any part of the image field afterwards. Light field, defined by Arun Gershun in 1936 and first used in photography by a team of Stanford researchers, led by Lytro inventor Ren Ng, in 1996, is part of a broader field of photography known as "computational photography," which also includes high-dynamic range (HDR) images, digital panoramic images, as well as the aforementioned time-of-flight and light-field camera technology. The images produced in computational photography are "not a 1:1 record of light intensities captured on a photosensitive surface, but rather a reconstruction, based on multiple imaging sources" (Maschwitz, 2015). To clarify, optical digital image capture in general is not a 1:1 record of the light entering thru the camera lens. Only about one third of the sensor's photosites are used, as "two thirds of the digital image is interpolated by the processor in the conversion from RAW to JPG or TIF" (Mayes, 2015). The way we see with our own eyes, in fact, can be considered more like a lossy JPG than a true 1:1 recording, as "less than half of what we think of as 'seeing' is from light hitting our retinas and the balance is constructed by our brains applying knowledge models to the visual information" (Rubin, 2015).

To take it a step further, computational photography can be done with light captured by a lens-less camera. The optical phased array (OPA) camera that Ali Hajimiri of CalTech has developed is an example of this type of innovation. He states, "It can mimic a regular lens but can switch from a fish-eye to a telephoto lens instantaneously—with just a simple adjustment in the way the array receives light" (Perkins, 2017). A paper put out by researchers at The National Center for Biotechnology Information proposes a thin lens-less camera that realizes a "super-field-of-view imaging without lenses or coding masks and therefore can be used for rich information sensing in confined spaces. This work also suggests a new direction in the design of CMOS image sensors in the era of computational imaging" (Nakamura, 2019).

What computational photography really does is allow the final image to be reinterpreted in ways never seen before in the industry. We are starting to see this under the X Reality (XR) umbrella, where the X stands for any number of variables such as Augmented Reality (AR) or Virtual Reality (VR). Earlier advances such as Apple's ARKit, as well as ARCore for Android phones, toolkits that allow third-party developers to work on AR apps, have primed our smart phones for that killer AR app that breaks the dam. According to Digi-Capital, a consultancy specializing in AR and VR, XR is expected to be a $108 billion business by 2021 (Kahney, 2018), while Tim Cook, CEO of Apple, says, "AR is going to take a while, because there are some really hard technology challenges there. But it will happen, it will happen in a big way, and we will wonder when it does, how we ever lived without it. Like we wonder now how we lived without our phone today" (Marr, 2018). For more on AR, VR, and XR, see Chapter 13.

The disconnect between optical capture, which itself was an interpretation of the actual object being photographed—a "willful distortion of fact," as eloquently stated in 1932 by Edward Weston—and the new data capture methods has opened a realm of possibilities (Weston, 1986). The old camera obscura method that has been in place for upwards of 160 years and used in both analog and digital image capture methods, with light passing thru the lens aperture onto a light recording mechanism, has changed. And digital photography has changed with it, as it enters "a world where the digital image is almost infinitely flexible, a vessel for immeasurable volumes of information, operating in multiple dimensions and integrated into apps and technologies with purposes yet to be imagined" (Mayes, 2015).

In the past, photography represented a window to the world, and photographers were seen as voyeurs—outsiders documenting the world for all of us to see. Today, with the image capture ability that a mobile phone puts in the hands of billions, photographers are no longer *them*—outsiders and talented specialists. Rather, they are *us*—insiders that actively participate in world events and everyday living, and, thanks to social media sites such as Snapchat, WhatsApp, Facebook, and Instagram, we share this human condition in a way that was never seen before.

This participation also falls under the disruptive innovation column. In *Bending the Frame*, Fred Ritchin (2013) writes, "The photograph, no longer automatically thought of as a trace of a visible reality, increasingly manifested individuals' desires for certain types of reality." This argument about what is or isn't a photograph, in part spurred on by technologic advances in the medium, has played out before, starting with George Eastman's mass production of the dry plate film in 1878 (prior to this innovation, photographers made their own wet-collodion plates). Lewis Carroll, noted early photographer and author, upon examination of a dry plate film, stated, "Here comes the rabble (Cicala, 2014)." "No one likes change," freelance photographer Andrew Lamberson says. "People who shot large format hated on the people who shot medium format, who in turn hated on the people who shot 35mm, who in turn hated on people who shot digital" (McHugh, 2013).

There is already a rumble stating that digital photography is dead. But the reality is, the rules have just changed. As we've seen in the past, "waves of innovation always disrupt the status quo generation after generation" (Hall, 2019). Based on the exciting innovations in light-field and computational photography, the use of AR, MR, and AI, and the possibilities that lens-less cameras bring, it seems that the industry is undergoing another mediamorphosis and is not actually dying. To drive that point home, an estimated 147,000 digital images where just uploaded to Facebook while you read the last two sentences (Aslam, 2019b)!

Most post-shutter release developments that are not a function of the camera itself, but have still contributed to this innovative disruption, have occurred mainly in the mobile phone arena. Apps for image manipulation continue to evolve, and though there are many great advances (Photoshop Camera is one), the dark side of technology also comes into play. Deepfake, a technology that allows for seamless replacement of one person's face with another on either still or video imagery, has been used for harmless entertainment purposes (Snapchat's Cameo), but also by bad actors attempting to sow disinformation campaigns (Shwayder, 2019). The use of new technology for nefarious purposes is not new and technology to combat deepfakes is already in the works. RefaceAI, a

company that released the photo-deepfake app Reflect, is working on a web service that will detect deepfakes, and Amber Video has already released Amber Authenticate, which runs on the blockchain platform Ethereum. Amber Video works by adding "hashes" to video being recorded. These hashes are "cryptographically scrambled representations of the data—that then get indelibly recorded on a public blockchain. If you run that same snippet of video footage through the algorithm again, the hashes will be different if anything has changed in the file's audio or video data (Newman, 2019). Roman Mogylnyi, CEO of RefaceAI, said, "We saw from the very beginning that this technology was being misused. We thought we could do it better" (Shwayder, 2019).

Current Status

The number of digital images taken each year continues to grow at an exponential rate, increasing from 350 billion in 2010 (Heyman, 2015) to an estimated 1.4 trillion in 2020 (Papacharissi, 2018). This means that every two minutes, humans take more photos than ever existed in total during the first 150 years of photography (Eveleth, 2015). It is estimated that 85% of the 1.4 trillion images taken in 2020 will be taken on mobile phones (Cakebread, 2017). In fact, it's estimated that "more than 90% of all humans who have ever taken a picture, have only done so on a camera phone, not a stand-alone digital or film-based 'traditional' camera" (Ahonen, 2013). So, what happens to all these captured digital images? With close to 190 billion images in total uploaded to Facebook and Instagram combined, we see that most of these images will not be printed. Instead they will "jump to screen," as we take, view, manipulate, transfer, and post these images via mobile phones onto social media sites. Some of these images will only "jump to screen" for a limited time, due to the increased use of photo apps such as Snapchat, making them somewhat "ephemeral" in nature.

The increased use of the mobile phone as camera has caused the number of digital camera purchases to drop. But sometimes what is old is new again. According to Don Franz, Publisher of Photo Imaging News, "The worldwide unit shipment of instant printers and camera/printer combos is estimated to rise from 8 million in 2016 to more than 12 million in 2021.

And Fujifilm sold more than 10 million Instax cameras last year" (Digital Imaging Reporter, 2019).

As we can see, the field of digital imaging and photography has matured since that first DyCam recorded a black and white low-resolution image back in 1992. With the new innovations discussed, it seems that it has now entered a new transformative period. The infinite flexibility of the digital image, the information sources that it can carry, and the possibilities for future usage in AR apps and yet-to-be developed technologies all are indicative of how robust and dynamic the field is. And the exciting thing is that we are all able to take part in this growth, as the only tool we need to participate is in our hands.

Factors to Watch

Image Manipulation Detection

With the increased release of fake digital images, Adobe, a name synonymous with image manipulation, has teamed with scientists from UC Berkeley to create an AI tool that detects manipulated images. By feeding thousands of both unedited and manipulated images into a Convolutional Neural Network (CNN), they were able to use machine learning to detect even the most subtle changes. When testing the validity of the algorithm, human volunteers were able to spot only 53% of the manipulations, while the algorithm

spotted 99% of them (Vincent, 2019). Adobe researcher Richard Zhang adds, "The idea of a magic universal 'undo' button to revert image edits is still far from reality. But we live in a world where it's becoming harder to trust the digital information we consume, and I look forward to further exploring this area of research" (Adobe, 2019).

Advances in mobile phone battery technology.

One element of the industry that has struggled to keep pace with all the advances in the hardware and software technology of our mobile phones is battery technology, but this is about to change. Developments in this area will come fast and furious, as we will see things like air charging, lightning fast re-charging, and longer-life batteries. In the works is a molybdenum disulfide-based rectenna (radio wave harvesting antenna) that harvests AC power from Wi-Fi in the air, converting it to DC power, while the StoreDot charger uses biological semiconductors made from peptides that can completely recharge a smartphone in a minute (Langridge, 2020). A plethora of battery substrates, aimed at increasing battery life, can be found being tested in labs around the world. From a new IBM battery, made from materials extracted from seawater, to a solid-state, lithium-ion battery with sulfide-superionic conductors, to both lithium-sulfur and graphene batteries, scientists are working with new ways to extend battery life.

Gen Z

Gen Zers are the first generation to be completely raised in the digital age, so it makes sense that Gen Zers feel at home taking and sharing digital photos. When editing these photos, a survey conducted by JWT Intelligence and SnapChat shows that 55% of them find that social media apps and the internet offer more creative space than found offline (Marketing Charts, 2019). And they are acutely aware of how this online connectivity affects their lives, as a study by the research firm, The Center for Generational Kinetics, states that 42% of them—a number higher than reported by any other generation—said "social media affects how other people see you" and that same percentage also said "social media has a direct impact on how they feel about themselves" (Seymour, 2019). Interestingly, along the lines of what was old is new again, one upward trend in the digital image industry predicts an expected 150% increase in sales of instant printers and camera/printer combos from 2016 to 2021. This is due to the popularity of these devices with both Gen Zers and Millennials. It shows that they still relate to "the instant gratification of a print and the retro experience" that prior generations enjoyed (Digital Imaging Reporter, 2019).

Sustainability

Environmental sustainability can be looked at in two ways. One is how digital imaging can help in our ability to monitor the environment. The other is how the industry works to make sure their policies and business practices are "green." A good example of the first factor comes from Rochester Institute of Technology (RIT). Led by Anthony Vodacek, part of RIT's Landstat project uses spectral imaging to determine concentrations of waterbody components, such as those related to harmful cyanobacteria blooms (Vodacek, 2019). The second way that digital imaging is involved in sustainability is illustrated by a report from Stockholm's Water Authority, which reported that "silver pollution had halved in the city's sewers over a five-year period, citing the increase in digital photography as a major factor" (Breakwell, 2019). (Silver is an important ingredient in analog film.) Also helpful is the fact that companies like Nikon, Samsung, and Sony are removing toxic chemicals from their camera production lines, while Canon, which eliminated the use of lead components in all new phones, is using components from older camera phones to manufacture new ones.

Bibliography

Aaland, M. (1992). *Digital photography*. Avalon Books, CA: Random House.

Adobe. (2019). Adobe Research and UC Berkeley: Detecting Facial Manipulations in Adobe Photoshop. *Adobe Blog*. https://theblog.adobe.com/adobe-research-and-uc-berkeley-detecting-facial-manipulations-in-adobe-photoshop/.

Ahonen, T. (2013). The Annual Mobile Industry Numbers and Stats Blog. *Communities Dominate Brands*. http://communities-dominate.blogs.com/brands/2013/03/the-annual-mobile-industry-numbers-and-stats-blog-yep-this-year-we-will-hit-the-mobile-moment.html.

Artaius, J. (2020). The 7 fastest lenses in 2020 – break the speed limit with glass as fast as f/0.75. *Digital Camera World*. https://www.digitalcameraworld.com/features/the-7-fastest-lenses-in-2019-break-the-speed-limit-with-glass-as-fast-as-f075.

Aslam, S. (2019a). Snapchat by the Numbers: Stats, Demographics & Fun Facts. *Omnicore*. https://www.omnicoreagency.com/snapchat-statistics/.

Aslam, S. (2019b). Facebook by the Numbers: Stats, Demographics & Fun Facts. *Omnicore*. https://www.omnicoreagency.com/facebook-statistics/.

Aslam, S. (2019c). Instagram by the Numbers: Stats, Demographics & Fun Facts. *Omnicore*. https://www.omnicoreagency.com/instagram-statistics/.

Breakwell, A. (2019). Is Analogue or Digital Photography More Environmentally Friendly? *Gobe Magazine*. https://mygobe.com/explore/analogue-versus-digital-photography-eco-friendly/.

Brook, P. (2011). Raw Meet: Fred Ritchin Redefines Digital Photography. *Wired*. http://www.wired.com/rawfile/2011/09/fred-ritchin/all/1.

Brown, A. (2019). Hands on: Oppo Reno 10x Zoom camera review. *Digital Camera World*. https://www.digitalcameraworld.com/reviews/hands-on-oppo-reno-10x-zoom-camera-review.

Business Today Online. (2017). WhatsApp users share 55 billion texts, 4.5 billion photos, 1 billion videos daily. *Business Insider*. http://www.businesstoday.in/technology/news/whatsapp-users-share-texts--photos-videos-daily/story/257230.html.

Cakebread, C. (2017). People will take 1.2 trillion digital photos this year — thanks to smartphones. *Business Insider*. http://www.businessinsider.com/12-trillion-photos-to-be-taken-in-2017-thanks-to-smartphones-chart-2017-8.

Canon. (2018). Technology Used in Digital SLR Cameras. *Canon*. https://global.canon/en/technology/support14.html.

Carter, R. L. (2015a). DigiCam History Dot Com. http://www.digicamhistory.com/1980_1983.html.

Carter, R. L. (2015b). DigiCam History Dot Com. http://www.digicamhistory.com/1995%20D-Z.html.

Carter, R. L. (2015c). DigiCam History Dot Com. http://www.digicamhistory.com/1996%20S-Z.html.

Chojecki, P. (2019). Moore's law is dead. *towards data science*. https://towardsdatascience.com/moores-law-is-dead-678119754571.

Cicala, R. (2014) Disruption and Innovation. *PetaPixel*. http://petapixel.com/2014/02/11/disruption-innovation/.

Curtin, D. (2011). How a digital camera works. http://www.shortcourses.com/guide/guide1-3.html.

Digital Camera (2020). Mirrorless Camera Sensor Sizes. Database. https://www.digicamdb.com/mirrorless-camera-sensor-sizes/.

Digital Imaging Reporter (2019). Digital Imaging Reporter's 2019 State of the Industry. *Digital Imaging Reporter*. https://www.direporter.com/state-of-the-industry/digital-imaging-reporters-2019-state-of-the-industry.

Eveleth, R. (2015). How Many Photographs of You Are Out There In the World? *The Atlantic*. http://www.theatlantic.com/technology/archive/2015/11/how-many-photographs-of-you-are-out-there-in-the-world/413389/.

Four Thirds. (2017). Overview. Four Thirds: Standard. http://www.four-thirds.org/en/fourthirds/whitepaper.html.

Glaser, A. (2019). Facebook Will Tell You How to Turn Off Facial Recognition. Why Wait? *Slate*. https://slate.com/technology/2019/09/turn-off-facial-recognition-facebook-default-how-to.html.

Grigonis, H. (2020). This A.I. app converts any photo into 3D, no dual lenses required. *Digital Trends*. https://www.digital-trends.com/photography/this-a-i-app-converts-any-photo-into-3d-no-dual-lenses-required/.

Hall, P. (2019). Is Photography as We Know It Dying? *Fstoppers*. https://fstoppers.com/business/photography-we-know-it-dying-425728.

Harry Ransom Center-The University of Texas at Austin. (2020). The First Photograph. *Exhibitions*. https://www.hrc.utexas.edu/niepce-heliograph//.

Heyman, S. (2015). Photos, Photos Everywhere. *N.Y. Times*. http://www.nytimes.com/2015/07/23/arts/international/photos-photos-everywhere.html?_r=1.

Kahney, L. (2018). Your smartphone is ready to take augmented reality mainstream. *Wired*. https://www.wired.co.uk/article/augmented-reality-breakthrough-2018.

Kaplan, J. (2008). 21 Great Technologies That Failed. *Features*. http://www.pcmag.com/article2/0,2817,2325943,00.asp.

Kastrenakes, J. (2019). Apple is trying to trademark 'Slofie'. *The Verge*.

Kronfli, B. (2019). The best camera phone in 2020: which is the best smartphone for photography? *Digital Camera World*. https://www.digitalcameraworld.com/buying-guides/best-camera-phone.

Kuo, L. (2019). China Brings in mandatoru facial recognition for mobile phone users. *The Guardian*. https://www.theguardian.com/world/2019/dec/02/china-brings-in-mandatory-facial-recognition-for-mobile-phone-users.

Langridge, M. (2020). Future batteries, coming soon: Charge in seconds, last months and power over the air. *Gadget news*. https://www.pocket-lint.com/gadgets/news/130380-future-batteries-coming-soon-charge-in-seconds-last-months-and-power-over-the-air.

Lee, D. (2019). Adobe is launching a free AI-powered Photoshop Camera app. *The Verge*. https://www.theverge.com/2019/11/4/20938121/adobe-ai-photoshop-camera-app-photos-ios-android-sensei.

Lester, P. (2006). Visual communication: Images with messages. Belmont, CA: Wadsworth.

LSST. (2020). The Large Synoptic Survey Telescope. *Public and Scientists Home*. http://www.lsst.org/lsst.

Marketing Charts. (2019). Why Gen Z Is A Generation of Creativity. *marketing charts*. https://www.marketingcharts.com/demographics-and-audiences/teens-and-younger-109216.

Marr, B. (2018). 16 Fascinating Augmented Reality Quotes Everyone Should Read. *Forbes*. https://www.forbes.com/sites/bernardmarr/2018/09/05/16-fascinating-augmented-reality-quotes-everyone-should-read/#219a38df5107.

Maschwitz, S. (2015). The Light L16 Camera and Computational Photography. *Prolost*. http://prolost.com/blog/lightl16.

Mayes, S. (2015). The Next Revolution in Photography Is Coming. *Time*. http://time.com/4003527/future-of-photography/.

McHugh, M. (2013). Photographers tussle over whether 'pro Instagrammers' are visionaries or hacks. *Digital Trends*. http://www.digitaltrends.com/social-media/are-professional-instagrammers-photographic-visionaries-or-just-hacks/.

Miller, A.D. (2012). A Theology Study of Romans. USA: Showers of Blessing Ministries International Publishing.

Monckton, P. (2020). Radical Nokia Handset Reveals Future For Smartphone Cameras. *Forbes*. https://www.forbes.com/sites/paulmonckton/2019/02/28/radical-nokia-handset-reveals-future-for-smartphone-cameras/#c18a23843bc1.

Muller. M. (2019). Selfie 'Statistics'. *Infogram*. https://infogram.com/selfie-statistics-1g8djp917wqo2yw.

Nakamura, T. (2019). Super Field-of-View Lensless Camera by Coded Image Sensors. *Sensors*. https://www.ncbi.nlm.nih.gov/pmc/articles/PMC6471751/.

National Press Photographers Association. (2019). NPPA Code of Ethics. https://nppa.org/node/5145.

Newman, L. (2019). A New Tool Protects Videos From Deepfakes and Tampering. *Wired*. https://www.wired.com/story/amber-authenticate-video-validation-blockchain-tampering-deepfakes/.

Papacharissi, Z. (2018). A Networked Self and Platforms, Stories, Connections. Routledge.

Perkins, R. (2017). Ultra-Thin Camera Creates Images Without Lenses. *News*. https://www.caltech.edu/about/news/ultra-thin-camera-creates-images-without-lenses-78731.

Pictures that lie. (2011). *C/NET News*. https://www.cnet.com/pictures/pictures-that-lie-photos/24/.

Pierce, D. (2017). REVIEW: LIGHT L16. *Gear*. https://www.wired.com/review/light-l16-review/.

Reis, R. (2016). Making the Best of Multimedia: Digital Photography. *Writing and Reporting for Digital Media*. Dubuque, IA: Kendall Hunt.

Ritchin, F. (2013). Bending the Frame: Photojournalism, Documentary, and the Citizen. Aperture Foundation.

Rubin, M. (2015). The Future of Photography. *Photoshop Blog*. https://theblog.adobe.com/the-future-of-photography/.

Schwayder, M. (2019). Why deepfakes will soon be as commonplace as Photoshop. *Digital Trends*. https://www.digital-trends.com/news/why-deepfakes-will-soon-be-as-commonplace-as-photoshop/.

Seymour, E. (2019). Gen Z: Born to Be Digital. *VOA*. https://www.voanews.com/student-union/gen-z-born-be-digital.

Sylvester, C. (2019). 2018 Camera Wrap-Up. http://blog.infotrends.com/2018-camera-wrap-up/.

Vincent, J. (2019). Adobe's prototype AI tool automatically spots Photoshopped faces. *The Verge*. https://www.theverge.com/2019/6/14/18678782/adobe-machine-learning-ai-tool-spot-fake-facial-edits-liquify-manipulations.

Vodacek, A. (2019). Modeling the impact of future Landsat spectral capabilities for monitoring water quality and harmful algal blooms. *Research*. https://www.rit.edu/dirs/research/modeling-impact-future-landsat-spectral-capabilities-monitoring-water-quality-and-harmful.

Weston, E. et al. (1986). Edward Weston: Color Photography. Tucson, AZ: Center for Creative Photography.

Wild, C. (2020). 1839. The First Selfie. *Mashable*. http://mashable.com/2014/11/07/first-selfie/#NMKmtmAcnsqw.

Zhang, M. (2010). The World's First Digital Camera by Kodak and Steve Sasson. *PetaPixel*. https://petapixel.com/2010/08/05/the-worlds-first-digital-camera-by-kodak-and-steve-sasson/.

Zetlin, M. (2019). Taking Selfies Destroys Your Confidence and Raises Anxiety, a Study Shows. Why Are You Still Doing It? INC. https://www.inc.com/minda-zetlin/taking-selfies-anxiety-confidence-loss-feeling-unattractive.html.

17

eHealth

Sharon Baldinelli, M.P.H.*

Abstract

This chapter focuses on eHealth-related technology, usages, and data gathering by professionals and individuals. eHealth technology has enhanced the health record accessibility by professionals and patients. Evaluation of big data uses, along with how the U.S. compares globally with this technology, is included. The impact of health-related apps on wellness and livelihood rounds out the chapter.

Introduction

James (Jim) Caridi, M.D., is a self-described interventional radiologist "blue collar doc," chair of the department of radiology at Tulane University, and a professor of pediatrics and radiology. He thinks in terms of patients, their care and their outcomes; and he works to balance patient needs with the options that are the best practices for their diagnosis. Dr. Caridi is also husband to a supportive wife, Rhonda, and a father to five children. He is a man that when speaking of his family, he is as proud of his grown children as he is of his minor child. He loves his family and works to balance all these roles.

Yet, sometimes, life takes a person into unexpected spaces. For Dr. Caridi, this space came when he became the patient himself. He became Jim. Diagnosed on August 10, 2011, this event affected him as a professional and within his roles of husband and father. Feeling chronic neck pain while treating a patient, he soon realized that this pain could not be simply attributed to "old age." After having his neck scanned and learning of a neck tumor nearly replacing his C4 vertebrae, undergoing additional testing then finally, he received the diagnosis: multiple myeloma, a blood cancer originating from his bone marrow. In plain terms Jim underwent numerous treatments including drug regiments and a bone marrow transplant; however, he has never been in remission. His prognosis was poor, and he was

* Doctoral candidate in the College of Communication & Information Sciences, University of Alabama (Tuscaloosa, Alabama)

told to expect three years of life. Jim chose to channel his inner "Rocky," and no matter how many times he was knocked down, he believed that he must get up and keep working to move forward.

The work ethic and beliefs that Jim values is credited with helping to keep him motivated within his professional and family roles. He did not quit working, he did not quit exercising, he did not quit being a husband; nor did he quit being a father to his grown children or to his young son. Instead, he opted for working hard and playing when he could, all while living within the parameters of a disease.

Now that you have met Dr. Caridi, or Jim, whichever you prefer, you may be asking, "What does this introduction have to do with eHealth?" The answer is a simple: eHealth ties together all of his illness characteristics in his full and active life. eHealth is the very glue, through information and communication technology (ICT) gathering, that binds healthcare innovations into usable and accessible forms for both healthy individuals as well as those fighting with illness (*eHealth*, 2020).

eHealth has provided many individuals more power and control over their health status than at any point in history. This power has allowed health decisions to be made with greater confidence by patients and family members. People are experiencing this empowerment by connecting with others who may be experiencing similar situations, their own digital record keeping for weight loss, and even access to their own health facility's electronic medical records. The speed of technology has created greater options for how the internet can be used in health choices, as well as prevention, and how individuals will choose to utilize available options. For example, individuals can listen to podcasts to gain health information, use smartphone apps to record meal plans and provide reminders on the importance of breathing to relax, and even track their exercise through shoes to map and record their distance running mileage. Innovations in eHealth have created an arena where everyday people can make health decisions while feeling, investing, and being informed in their own health outcomes. As you read this chapter, consider how each section applies to professionals (Dr. Caridi) and to the personal lives of people just like Jim.

More about Dr. Caridi
One in a Million

When Dr. Caridi was diagnosed with cancer, his son gave him a piece of advice: To approach his disease like he approached becoming a doctor. The words resonated with Dr. Caridi, in the same tone as a movie poster of Rocky worked as his mantra to make it to and through medical school to become a physician. Dr. Caridi's background and upbringing did not provide the same advantages that other students in his cohort carried with them. So, as a 1 in a million chance to become a physician and a 1 in a million chance to live with the cancer diagnosis, Dr. Caridi decided make better information available. He started a Facebook page called 'oneIAM club.' This page is reviewed by Dr. Caridi regularly and offers information on topics such as updates on cancer research, diet, exercise, supplements and most importantly a way for anyone diagnosed with any cancer to feel connected.

You may check out his Facebook page at: https://www.facebook.com/OneIAMClub/

Background

For twenty years, what started as a seed soon grew into a tree with roots and branches, from the term *eHealth* that was coined in 1999 (Schreiweis et al., 2019). A term was needed for researchers to describe the phenomenon of health technology and related interests. These innovations were soaring in medical and private sectors. Research was paramount to the success of many of the technologies being invented and utilized (Schreiweis et al., 2019). How wide of a market was the technology making? Was the data from the instrumentation reliable? How long would the item work before the user(s) became bored or believed the piece to be outdated? The answers to these questions assisted in creating the foundation for continuing to use eHealth to allow better health choices and potential health outcomes by medical professionals and private individuals.

When eHealth technology was introduced, many individuals were intrigued by the idea of just how far innovations could expand. Ideas about electronic medical records (EMR), health data availability, and concerns for privacy all circled in conversations. Medical providers were concerned about their observations, findings, and medical notes being made available to the wrong audience(s). Additional concerns for all involved included confusion over differing terminology. For instance, electronic health record (EHR) verses electronic medical record (EMR) seem like interchangeable names for the same item, but the following definitions are offered within the healthcare field:

Electronic Medical Record: Used mostly intra-practice for diagnoses and treatments of patients

Electronic Health Record: Moves with patients between different organizations for extended views of health

(Differences between EHR and EMR, 2020)

Overall, both resources provide valuable protection for the medical provider and patients: fewer medical errors, timely information, better medication prescribing history, and ultimately improved outcomes in patient participation for healthier lifestyles and preventative care habits (Differences between EHR and EMR, 2020). Would electronic medical records be enough for the general public? Was the idea of an electronic medical record too much for medical professionals and/or the general public to accept as trustworthy and wanted information? Was there enough interest by the medical industry to invest in advanced technology?

As some medical providers were articulating concerns about the privacy of their own records and observations for their patient loads, some patients were also voicing anxieties about their private health information potentially making it into the wrong hands; general public members should not have access to anyone's health records aside from those who needed access or were directly invested in a patient's care. For Americans, the desire to have medical information

available (Atherton, 2011), while keeping concerns of confidentiality and integrity in handling of the sensitive information in mind, was addressed through the *Health Insurance Portability and Accountability Act* (HIPAA)(Al-Issa et al., 2019) . This act was signed into law on August 21, 1996 (Committee on Health Research and the Privacy of Health Information: The HIPAA Privacy Rule et al., 2009). Since being passed into law, there have been several updates to HIPAA as the needs for American society and concerns of privacy with technology have continued to evolve, the latest as the *Omnibus Rule* of 2013 (Goldstein & Pewen, 2013).

We now know that implementation of EMRs and EHRs have become a common practice for many medical providers. When patients arrive for an initial visit with a practitioner, they are often handed a handheld device to answer questions about their own medical history. With a few clicks to confirm or deny medical history from drop-down menus, lists of prior surgeries, labs or tests, and geographical locations of one's medical information, the newly arrived patient can provide an electronic signature and hand their information to the receptionist. In real time, the software "talks" to the centralized computer system of the medical practice, and the medical team can access patients' information within moments without having the congestion of loose papers, misinformation from illegible handwriting, and sources who may not be available if additional information is needed from prior care.

With a click, medical offices can often request electronic health records from other locations since the patient's signature often provides the key to access. The patient no longer must wait for his/her medical history to be relayed verbally or scanned from loose papers and then sent via fax, mail or overnight delivery service (depending on the urgency). Patients like Jim can travel to give lectures with peace of mind knowing that his EHR is available should he experience a life-threatening issue, no matter how far he is away from home.

Table 17.1

Timeline of EHR's

Year	Event
1960's–1970's	• Academic medical centers develop their own EHR's • First EHR systems were known as Clinical Information Systems (CIS) • University of Utah collaborates with 3M to create medical record system • Lockheed creates medical record keeping system: Example of processing speed, supports multiple users, flexibility • 1968: Massachusetts General Hospital in collaboration with Harvard begins COSTAR, example for needing usage of wider vocabulary across institutions • Department of Veteran Affairs (VA) implements VISTA
1980's	• Leaders see benefits for industry-wide standards • Institute of Medicine (IOM) initiates analysis of paper health records, improvement sought • HL7: International nonprofit standards-developing organization (SDO) of EMR's founded
1990's	• 1991: Results of initial analyses published • 1997: Second analysis conducted; IOM argues case for EMR to decrease medical errors
2000's	• 2000: Study of medical errors released by IOM, *To Err is Human* • 2004: President Bush speaks on topic in State of the Union Address • 2009: President Obama incorporates EHR as part of *American Recovery and Reinvestment Act* of 2009 as part of the *Health Information Technology for Economic and Clinical Health Act* (HITECH)

Source: S. Baldinelli

Recent Developments

For many individuals, eHealth is not considered as a major factor in their lives. Instead it may be ignored, taken for granted, or simply not understood that it is intertwined into our day-to-day living. One such occurrence illustrating how much eHealth has permeated into our society is through Apple. When we think of Apple, we think of MacBooks, iPads, AirPods, iPhones, and even the Apple Watch. What do all these pieces have to do with eHealth?

According to Apple CEO, Tim Cook, "Our active install base of devices has now surpassed 1.5 billion, up over 100 million in the last 12 months" (Neely, 2020). These devices have the capability to be used in conjunction with eHealth and personal health practices. One can use a MacBook to look up a website for access to an exercise community, iPads may be loaded with apps to peruse your running times based on your preference of brand, AirPods might help pass the time away as you listen to your favorite podcast for health advice, while iPhones and the Apple Watch can be used for tracking individual fitness habits, maps, or reminders of scheduled health appointments or fitness classes.

Apple is one of many companies that has taken the quest for availability of health data to a new level by offering information through their products and platforms. For example, when visiting the Apple webpage, at https://www.apple.com/healthcare/, individuals can learn different ways in which to use Apple products to create eHealth access for themselves and for career medical research (Healthcare, 2020). The Apple Watch, for example, carries access to apps which provide feedback on heartrate notifications, including irregular heart rhythms (an ECG app), wearer fall detection, medical ID information access (Apple Watch, 2020), fitness trends and even noise pollution notifications (Apple Research, 2020).

Another recent development focuses on three-dimensional (3D) printing, also known as additive manufacturing, combined with nanotechnology, which can create porous conductors that are highly flexible (Davoodi et al., 2020). The conductors can be manufactured through 3D printing and are easily sized and contoured to fit into the best of spaces. The rubber-graphene material is hardy and robust enough to be used in areas with high humidity and extreme temperatures (Tough, flexible sensor invented for wearable tech, 2020). Examples of usages in these environments would be inside of athletic shoes, part of a running singlet or even sewn in as a part of a rower's unisuit. These sensors will enhance the usage of wearable devices by "monitoring everything from vital signs to athletic performance" (Tough, flexible sensor invented for wearable tech, 2020).

Tools, Gadgets & eHealth

Using eHealth technology, it is easy to imagine the possibilities of better health outcomes for individuals, insurance companies, and government revenue economics. One such program, the Diabetes Prevention Recognition Program (DPRP), has added an approved digital (eHealth) curriculum as a choice for participants in their diabetes prevention program, based on findings from a digital diabetes prevention program (DPP), and a national DPP was created (Almeida et al., 2020). The need for this style of curriculum is apparent when considering that approximately 84 million Americans have been diagnosed as prediabetic and 30.3 million people have been diagnosed as diabetic, per the Centers for Disease Control and Prevention (CDC). This program's goal is to reduce body weight and HbA1c lab values while working towards sustainable behaviors in their participating populations.

In a clinical trial, the primary aim was on intervention effectiveness while assessing the implementation of the digital curriculum. During the study, additional measures (blood pressure, additional blood labs, body measurements, quality of life, psychological state, health behavior, and other potential cardiovascular risks) of effectiveness were taken and compared with an in-person small group diabetes prevention class (Almeida et al., 2020). The digital curriculum included weekly online lessons, health coaching, online social forums, and internet-enabled tools for support of socialization, health education, weight tracking, and activity monitoring. Overall, the findings for the program suggest that "digitally-based DPP and weight loss programs using evidence-based behavioral strategies combined with automated self-monitoring tools and remote health coaching can be successfully used to help individuals reduce their risk for type 2 diabetes" (Almeida et al., 2020).

The findings support individuals working with eHealth technologies to keep from being handed a type 2 diabetes diagnosis, T2D. For those with limited mobility access or time restraints, eHealth tools seem to be an acceptable course to reach wider populations to participate in such behavior modification programs. From a financial view, studies using these types of eHealth components are providing valuable data for such instruments and programs to be supported through insurance benefits and healthcare payment models. eHealth tools used through this study included: weight scales, pedometers, and Fitbit products (Almeida et al., 2020). With a small investment in eHealth tools, individuals can lower their risk of being diagnosed as a T2D.

Examples of health tools for monitoring cardiac chronic health conditions are the KardiaMobile and the KardiaMobile 6L. The FDA has cleared this personal electrocardiogram (EKG or ECG) apparatus to tell if a someone is experiencing irregular heartbeat patterns. Individuals place their fingers on the small device to detect their heartbeat. The monitor can detect an abnormal beat, also known as arterial fibrillation, or AFib. The device can also detect whether an individual is experiencing a normal cardiac rhythm. The results are then displayed on a configured smartphone which has been updated with the appropriate eHealth app. These products offer peace of mind for many individuals by storing the individual's history of EKG recordings that are accessible to share as needed, providing a personalized heart report with 30 days' worth of data and even medication reminders while tracking medication history all through a portable device that is slightly larger than a USB drive (AliveCor: KardiaMobile, 2020).

Gamification

Because of the current climate of preventive healthcare, individuals have become more motivated to focus on their health. One advancement in this arena is the rise of gamification products. What motivates one individual is different than what may motivate another individual. Since 2010, the concept of gamification has created heightened awareness of how to motivate individuals for better health outcomes (Sardi et al., 2017). Sebastian Deterding offers the most basic and applicable definition for the purposes of this chapter, "Gamification is the use of game design elements in non-game contexts" (Sardi et al., 2017, p. 32). Gamificaiton may provide an incentive for consumers to continue to use eHealth apps and products, and to seek achievement through "badges." If you have a smartphone, you know there are many apps you can choose to purchase or freely download allowing you to earn badges as different activities are performed. One example is the Apple Activity app. You can challenge a friend to a fitness contest, you can compete with yourself to achieve personal goals, or even set a new exercise record with your family. All these activities strive for the common goal of increasing activity. As humans, many of us are motivated to continue to perform simply by receiving feedback that our efforts have been acknowledged. Providers of these technologies receive data notifying them which apps are being utilized and how frequently.

Not all products and platforms are true eHealth enhancers, some are simply available because enough people enjoy technology. One such product that could arguably fall into this category is the Ember mug or tumbler, for beverages. Through a Bluetooth connection with a smartphone, you can control the temperature of your favorite beverage (*Ember: Turn up the heat*, 2020). For this type of product, there may be some uncertainty regarding the level of contribution to health. Rest assured, this product has been used in scientific research when a controlled temperature is needed for a liquid (Jiang et al., 2018). This product also pairs with Apple Health for collecting personal caffeine intake data (*Ember: Turn up the heat*, 2020). If scientific research does not constitute reason enough to invest in a Bluetooth, app-controlled travel mug, perhaps the benefits of quality of life measure may persuade you.

Figure 17.2

Ember Tumbler

Source: S. Baldinelli

Figure 17.1

Badges

Source: S. Baldinelli

Current Status

Healthcare Statistics

The apps and tools discussed earlier become important when thinking about the top chronic conditions/diseases in the United States of America, which include heart disease, stroke, cancer, diabetes, obesity, arthritis, Alzheimer's disease, epilepsy, and dental cavities (*Health and Economic Costs of Chronic Diseases*, n.d.). To understand the financial burden of chronic disease on American society, consider that 75% of the total costs of healthcare is paid to cover these chronic conditions (Raghupathi & Raghupathi, 2018) which is broken down to approximately $5,300 being paid out per individual annually (Raghupathi & Raghupathi, 2018). At the public insurance level, approximately 96 cents out of every Medicare dollar and 83 cents of every Medicaid dollar is spent on chronic medical conditions (Raghupathi & Raghupathi, 2018). In 2016, $1 trillion was paid to cover medical costs for chronic diseases (Levine et al., 2019). And when taking into account lost work productivity, the amount of money lost increases to a staggering $3.7 trillion (Levine et al., 2019) tagged to chronic diseases.

Globalization

When thinking about healthcare, individual perceptions and experiences are the guide. Often, changes to our viewpoints come from a specific experience or a friend who shares a new idea with us for consideration. Sometimes a broader view is needed to help guide future innovations and research endeavors. This happens to be the case when the realization is made that eHealth resources and prospective research are benefitted by considering a worldwide view, also referred to as the globalization of healthcare. Fahmida Hossain explains the phenomenon best, "Globalization and multiculturalism have forced fundamental and, truly, paradigmatic changes to the entire breadth of healthcare practices" (Hossain, 2020, p.159). eHealth is included in the scope of healthcare practices.

Table 17.2, obtained from 2015 World Health Organization (WHO) data, outlines how the United States compares with the world reporting of utilizing eHealth for medical care (Organización Panamericana de la Salud, 2017).

Table 17.2

Comparison of U.S.A. to Global Participation with eHealth Resources

eHealth Policy Data Item	U.S. Participation	Global Participation %
National eHealth policy or strategy	Yes	58%
Funding through public	Yes	40%
Private or commercial funding	Yes	40%
Government-supported internet sites in multiple languages	Yes	48%
Addresses patient safety and quality of care data	Yes	46%
Protects privacy of personally identifiable data of individuals	Yes	78%
Governs sharing of personal and health data between research entities	Yes	39%
Allows individuals electronic access to their own health-related data housed in EHR	Yes	29%
Allows individual(s) to demand their own health data be corrected when known to be inaccurate in EHR	Yes	32%
Governs civil registration and vital statistics	Yes	76%
Usage of Social Media for Health by Organizations and/or Individuals	**U.S. Participation**	**Global Participation %**
Promote health messages as part of health promotion campaigns	Yes	78%
Run community-based health campaigns	Yes	62%
Help decide what health services to use	Yes	56%
Participate in community-based health forums	Yes	59%

Source: S. Baldinelli

Preventative health care is defined as health maintenance that keeps emergency health care services from being required. Examples of preventative health care services include annual check-ups, lab screenings, mammograms, monitoring for healthy weight parameters for all ages, and even mental health screenings for conditions such as depression (*Prevention Services*, n.d.). The above listed services serve as a pathway to help prevent chronic medical conditions, or a condition that "is a physical or mental health conditions that lasts more than one year and causes functional restrictions or requires ongoing monitoring or treatment" (Raghupathi & Raghupathi, 2018). Sadly, preventative services are underutilized. In 2015, of American adults age 35 years or older, only 8% received all appropriate and recommended clinical preventative services (Levine et al., 2019).

Big Data

This interest in records, record keeping, privacy and information has put a focus on all kinds of data. Personal data, medical entity data, health provider data, and even government health data, are now available instantaneously for consumers and practitioners. For the purposes of this chapter, we will take a moment to consider big data and what this means in the eHealth industry.

eHealth is bolstered by big data. One example of how big data is improving eHealth is through the synthesis of information around the specialties of psychiatry and mental health. While big data is providing the quantitative markers, eHealth is improving the outcomes in mental health medicine. By using the resources of the collective data, focus groups have been formed by professionals to aid them in providing better services. Recent studies suggest that the mental healthcare system can benefit from eHealth device opinions; however, these are far from being agreed upon, and the management of these tools in care environments should be further developed and regulated (Morgiève et al., 2019). These enhancements could potentially yield improvements in both therapeutic and preventative practices once agreements can be made (Morgiève et al., 2019).

But, what are *big data* in healthcare? To some professionals, the preferred term is *health analytics,* which is defined as "the transformation of data for the purpose of providing insight and evidence for decision and policy-making" (Peterson et al., 2016). To provide consistency, the term we will use for our discussion is big data. Big data is a broadening of the above concept and refers to "data sets that are much larger and/or complex than what traditional data processing can accommodate" (Peterson et al., 2016). Through innovations in eHealth, the data sets have become increasingly larger as the generated information is compared on individual, societal, country and world-wide population levels. Reports garnered from eHealth sources are accessed by medical providers and provide a view of health outcomes, as retrospective analyses, for populations of interest (Peterson et al., 2016). (For a broader discussion of big data, see Chapter 24.)

A prime innovative force in creating, accumulating, and sorting all of this created data is artificial intelligence (AI). According to *Encyclopedia Britannica,* AI is "…the ability of a digital computer or computer-controlled robot to perform tasks commonly associated with intelligent beings… .such as the ability to reason, discover meaning, generalize or learn from past experience." One well-known AI product is IBM's supercomputer, Watson. This incredible piece of equipment is performing groundbreaking work in the fight against chronic diseases by creating and following algorithms for patients and their health care options (Zhou & Myrzashova, 2020). AI is used to monitor the condition of patients who are suffering with chronic disease, and then data is structured to provide input from patients, refer them to specialists, or provide other timely health related recommendations such as taking medicine or following up with a medical provider. The American Heart Association is already working with Watson on behalf of chronic heart disease patients (Zhou & Myrzashova, 2020). Through eHealth, better health outcomes are ahead.

Factors to Watch

Some may wonder why technological gadgets and products are tied to eHealth? The answer is so basic; yet, quite complex. The answer is HOPE. Individuals generally wish to live healthy, happy and long lives. They hope that with better habits, they can enjoy these benefits while enticing friends and family members to improve their health outcomes, as well. For many people, having the instant feedback from eHealth products and services provides a tangible record of instant gratification and accomplishment.

While "hope" sounds altruistic in nature, an alternative viewpoint to caring about eHealth and eHealth products is through the understanding of preventative health care versus treatment for individuals with health conditions and ailments. One of the most important potential outcomes from all of this data that is being generated through eHealth is the basic question:

How will this information be used? Medical professionals like Dr. Caridi will have greater information and data available for making and offering health choices for decisions regarding his patients and the health entities his work supports. To others, people like Jim will have greater, potentially more reliable, information on which to base health choices. Either viewpoint creates conversations in the healthcare community for better health outcomes that will be available to wider populations. Advances in eHealth will only continue to strengthen medical, social, and personal communities. Future eHealth advances will include wider availability of health outcome data, needs for professional technological training and support, offerings of more informative health conversations due to increased health literacy by the general public, and most importantly, security in your own decision making which will ultimately improve our population's quality of life while potentially decreasing healthcare costs.

Gen Z

For Gen Z, eHealth offers a vast opportunity for self-efficacy, or belief in one's own capabilities (Buckworth, 2017), in many areas: health outcomes for self and family members, professional development, options of increased access to data about and how one compares to others participating in similar hobbies, activities or health choices, and even ready-made communities to support health goals. While this may appear to be an overwhelming responsibility, it is important to consider that GenZ has been living with eHealth their entire lives!

Bibliography

Al-Issa, Y., Ottom, M. A., & Tamrawi, A. (2019). eHealth cloud security challenges: A survey. *Journal of Healthcare Engineering, 2019*, 1–15. https://doi.org/10.1155/2019/7516035.

AliveCor: KardiaMobile. (2020). https://www.alivecor.com/kardiamobile.

Almeida, F. A., Michaud, T. L., Wilson, K. E., Schwab, R. J., Goessl, C., Porter, G. C., Brito, F. A., Evans, G., Dressler, E. V., Boggs, A. E., Katula, J. A., Sweet, C. C., & Estabrooks, P. A. (2020). Preventing diabetes with digital health and coaching for translation and scalability (PREDICTS): A type 1 hybrid effectiveness-implementation trial protocol. *Contemporary Clinical Trials, 88*, 105877. https://doi.org/10.1016/j.cct.2019.105877.

Apple Research. (2020). Apple Research App. https://www.apple.com/ios/research-app/.

Apple Watch. (2020). Watch-Apple. https://www.apple.com/watch/.

Atherton, MD, J. (2011). History of medicine: Development of the electronic medical record. *Virtual Mentor: American Medical Association Journal of Ethics, 13*(3), 186–189. https://journalofethics.ama-assn.org/article/development-electronic-health-record/2011-03.

Buckworth, J. (2017). Promoting self-efficacy for healthy behaviors: *ACSM's Health & Fitness Journal, 21*(5), 40–42. https://doi.org/10.1249/FIT.0000000000000318.

Committee on health research and the privacy of health information: The HIPAA privacy rule, Board on health sciences policy, Board on health care services, & Institute of Medicine. (2009). *Beyond the HIPAA Privacy Rule: Enhancing Privacy, Improving Health Through Research* (S. J. Nass, L. A. Levit, & L. O. Gostin, Eds.). National Academies Press. https://doi.org/10.17226/12458.

Davoodi, E., Montazerian, H., Haghniaz, R., Rashidi, A., Ahadian, S., Sheikhi, A., Chen, J., Khademhosseini, A., Milani, A. S., Hoorfar, M., & Toyserkani, E. (2020). 3D-Printed ultra-robust surface-doped porous silicone sensors for wearable biomonitoring. *ACS Nano, 14*(2), 1520–1532. https://doi.org/10.1021/acsnano.9b06283.

Differences between EHR and EMR. (2020). USF Health. https://www.usfhealthonline.com/resources/key-concepts/ehr-vs-emr/.

EHealth. (2020). World Health Organization. https://www.who.int/ehealth/en/.

Ember: Turn up the heat. (2020). Ember Technologies. https://ember.com/?gclid=CjwKCAiAg9rxBRADEiwAxKDTugs7LoP2KLWKW16z4HkB6JdiaZC6fWc-btAMO28-t7ygcgp4BSDPpRoCpjQQAvD_BwE.

Goldstein, M. M., & Pewen, W. F. (2013). The HIPAA Omnibus Rule: Implications for public health policy and practice. *Public Health Reports, 128*(6), 554–558. https://doi.org/10.1177/003335491312800615.

Health and Economic Costs of Chronic Diseases. (n.d.). Centers for Disease Control and Prevention: National Center for Chronic Disease Prevention and Health Promotion (NCCDPHP). https://www.cdc.gov/chronicdisease/about/costs/index.htm.

Healthcare. (2020). Apple Healthcare. https://www.apple.com/healthcare/.

Hossain, F. (2020). Narrative authority: A key to culturally competent healthcare. In J. Gielen (Ed.), *Dealing with Bioethical Issues in a Globalized World* (Vol. 14, pp. 157–194). Springer International Publishing. https://doi.org/10.1007/978-3-030-30432-4_12.

Jiang, X., Loeb, J. C., Manzanas, C., Lednicky, J. A., & Fan, Z. H. (2018). Valve-enabled sample preparation and RNA amplification in a coffee mug for Zika virus detection. *Angewandte Chemie International Edition, 57*(52), 17211–17214. https://doi.org/10.1002/anie.201809993.

Levine, S., Malone, E., Lekiachvili, A., & Briss, P. (2019). Health care industry insights: Why the use of preventive services is still low. *Preventing Chronic Disease, 16*, 180625. https://doi.org/10.5888/pcd16.180625.

Morgiève, M., Sebbane, D., De Rosario, B., Demassiet, V., Kabbaj, S., Briffault, X., & Roelandt, J.-L. (2019). Analysis of the recomposition of norms and representations in the field of psychiatry and mental health in the age of electronic mental health: Qualitative study. *JMIR Mental Health, 6*(10), e11665. https://doi.org/10.2196/11665.

Neely, A. (2020, January 28). *Apple has more than 1.5B active devices in the wild, up 100M in last year.* AppleInsider. https://appleinsider.com/articles/20/01/28/apple-has-more-than-15b-active-devices-in-the-wild-up-100m-in-last-year.

Organización Panamericana de la Salud. (2017, May). PAHO EHealth: United States of American eHealth country profile. https://www.who.int/goe/publications/atlas/2015/usa.pdf?ua=1&ua=1.

Peterson, C. B., Hamilton, C., & Hasvold, P. (2016). *From innovation to implementation: EHealth in the WHO European region.* WHO Regional Office for Europe..

Prevention Services. (n.d.). [Glossary index]. https://www.healthcare.gov/glossary/preventive-services/.

Raghupathi, W., & Raghupathi, V. (2018). An Empirical study of chronic diseases in the United States: A visual analytics approach to public health. *International Journal of Environmental Research and Public Health, 15*(3), 431. https://doi.org/10.3390/ijerph15030431.

Sardi, L., Idri, A., & Fernández-Alemán, J. L. (2017). A systematic review of gamification in e-Health. *Journal of Biomedical Informatics, 71*, 31–48. https://doi.org/10.1016/j.jbi.2017.05.011.

Schreiweis, B., Pobiruchin, M., Strotbaum, V., Suleder, J., Wiesner, M., & Bergh, B. (2019). Barriers and facilitators to the implementation of eHealth services: Systematic literature analysis. *Journal of Medical Internet Research, 21*(11), e14197. https://doi.org/10.2196/14197.

Tough, flexible sensor invented for wearable tech. (2020, March 6). *Waterloo News.* https://uwaterloo.ca/news/news/tough-flexible-sensor-invented-wearable-tech?fbclid=IwAR0G92gb1ZeLfA6M2sThDIEf7Wy5Wg8niur1KOO_uhBbKAI-qgBKMBmUCFng.

Zhou, H., & Myrzashova, R. (2020). AI based systems for diabetes treatment: A brief overview of the past and plans for the future. *Journal of Physics: Conference Series, 1453*, 012063. https://doi.org/10.1088/1742-6596/1453/1/012063.

18

Esports

Max Grubb, Ph.D.*

Abstract

From its beginnings as simulations for human computer interaction research, video games have evolved in the last 60 years from simple entertainment for individuals and competition among friends to esport leagues and tournaments. Now the realm of professionals and collegiate players and teams, esports is quickly evolving into a billion-dollar industry with no end in sight to its growth. This chapter examines the development of esports, its technology, and its future.

Introduction

In less than ten years, esports has grown into one of the most exciting forms of entertainment for both participants and spectators. In 2019, over 450 million viewers globally tuned in to watch esports events (Root, 2019; Newzoo, 2019). As part of the sports industry esports earned more than $1 billion (Fischer, 2019). Mat Bettinson first coined the term "eSport" during the 1999 launch of the Online Gamers Association (The OGA, 1999). Esports has been defined several ways but at its core is professional competitive gaming with the specific goal of a championship title or prize money (Newzoo, 2019). Popular esports are often team-based multiplayer games played in tournaments for large audiences both in person and online (Dwan, 2017).

Historically, esports can trace their heritage to the 1950s when computer scientists created simple games and simulations for their research. Esports can be classified into two periods: the arcade and internet eras (Lee and Schoensted, 2011). As computer technologies evolved, so did video games, leading first to competitive gaming in arcades and player experiences at home to esports today; all in conjunction with the growth of home computer technologies and the internet.

What technology is needed for esports? Basically, players need a high-powered gaming computer—usually a desktop computer with a powerful graphics card, CPU, and lots of memory. Peripherals needed include controllers, displays, microphones, headphones, and speakers. Players also need a broadband internet

* Grubb is Senior Lecturer in the College of Creative Arts and Communication, Youngstown State University; Youngstown, Ohio

connection and of course, the game! Esports includes a multitude of video game genres from traditional sports video games such as *Madden 20,* but mostly focuses on first person shooters (FPS), real-time strategy (RTS), and multiplayer battle arena games such as *Dota 2, League of Legends, Fortnite, Counter-Strike: Global Offensive,* and *Overwatch.*

Background

Competitive gaming has been around as long as video games. A 1962 MIT newspaper, *Decuscope,* reported that students competed at *Spacewar,* a two-player game that was programmed on a data processor. This inspired students to tinker with the code which involved variations of the *Spacewar* game. Eventually this competition led to a small 24 student tournament, the Intergalactic Spacewar Olympics, held at the Stanford University Artificial Intelligence Laboratory on October 19, 1972 (Baker, 2016).

But it was the economic incentive that motivated entrepreneur Nolan Bushnell, founder of Atari, that led to the growth of video games in arcades. Atari's *Pong* is considered the first video arcade game which quickly grew in popularity. At its peak, there were over 35,000 *Pong* machines in the United States, earning a staggering $200 a week per machine (Goldberg, 2011).

Arcade gaming continued to grow in popularity and with that popularity came competitive arcade gaming. For example, a nationwide competition was launched to see who could get the highest score on the arcade game *Donkey Kong.* Scores were conveyed through word of mouth and published in trade magazines. Atari sponsored the first arcade competition, for *Space Invaders* in 1980 with more than 10,000 players (Edwards, 2013).

Arcade gaming competitions, though, pale against today's esports which really began to take off with the development of internet enabled PC and console gaming in the 1990s. Improvements in internet connectivity speeds contributed to the rise in multiplayer online role-playing games—the precursor to today's prolific online gaming environment (Overmars, 2012).

Where entrepreneurs saw the opportunities in video games, both in arcades and at home, gamers moved from simple competition in the arcade and tournaments for the highest scores to a more robust gaming experience online. Thus, the social aspects of competition were evolving, stirring the beginnings of electronic sports (esports).

While there had been tournaments since the 1970s, the first high-stakes tournament took place in 1997 when *Quake* hosted "Red Annihilation." It attracted over 2,000 participants, and the winner was awarded the Ferrari then owned by the developer of the game, John Carmack (Edwards, 2013). That year, the Cyberathlete Professional League (CPL) was created leading to the emergence of the first esport professionals. Later that year CPL held its first tournament, and the next year it offered $15,000 in prize money (Edwards, 2013). One professional to appear during this time was Johnathan "Fatal1ty" Wendel who reportedly won about a half-million dollars throughout his esports career (Wendel, 2016). It was also during this time that esports was judged to be an emerging spectator sport

During this time most games were FPS, arcade style, and sports games. Edwards (2013) observed Real-Time Strategy (RTS) games were then developed in the late 1990s. Similar to chess, RTS games require thought and long-term planning. As a RTS game, *Starcraft* was a significant factor in the growth of esports with its unlimited strategic possibilities reaching its pinnacle of popularity after 2000.

As the world entered the new millennium, the world of esports experienced seismic growth in popularity. In 2000, two major international tournaments were introduced, the World Cyber Games and the Electronic Sports World Cup. Both became annual events, serving as international platforms for esports competitors and providing the template for major esports tournaments (Edwards, 2013). These, and the growing number of competitions, gave gaming companies that hosted them the opportunity to feature the two principal game types: FPS and RTS. Games such as *Quake, Halo,* and the *Call of Duty* series are FPS. These and other FPS games require quick reaction and reflexes with the need for fast implementation and use of buttons.

Gamers are attracted to RTS games because they can choose from a variety of characters with different skills and abilities. In addition, RTS competitions normally involve 5-member teams competing against each other. Collaboration among team members is a must for success. RTS games attract competitors from around the world with thousands attending championship events. Games such as *Starcraft*, *World of Warcraft* and *League of Legends (LoL)* are RTS games that contributed to the rapidly expanding esport industry.

Esports tournaments and leagues began to take off in South Korea in the early 2000s with the game *Starcraft*. When Riot Games' *League of Legends* came out in 2009, the sport exploded (Wingfield, 2014). Leagues for esports have since been established around the world. The largest is the ESL, formerly known as the Electronic Sports League, formed in 2000. Major League Gaming was formed in New York City in 2002 for professional gamers. Leagues are also formed around specific games such as *Overwatch*. There is a great deal of money in professional esports. For example, 10 million people watched the Overwatch League in January 2018. Blizzard charged teams a total of $20 million to participate in the league, with the fee expected to grow to $60 million next year (Conditt, 2018)! Traditional sports has taken notice of esports with Patriots owner Robert Kraft and Mets COO Jeff Wilpon being among the first owners of franchises ("Robert Kraft Calls Upstart" 2017). Figure 18.1 shows the Florida Mayhem team entering Blizzard Arena for an Overwatch League Event.

Esports isn't just for the players, as the number of spectators for esports has grown exponentially. Millions of people watch videos of game play on YouTube and the live streaming video game service Twitch. Tournaments can fill stadiums and arenas with devoted fans. As esports has grown since 2015 years, so has the construction of arenas for spectators. The largest esports arena in North America opened in November 2019. At a cost of $10 million the Esports Stadium in Arlington, Texas, has an 85 foot-long LED wall with flexible seating and a high tech broadcast studio. The Fusion Arena is being constructed in Philadelphia in 2020 with 3500 seats and a training facility. Other arenas are being developed for esports by renovating existing facilities (Hughes, 2019).

Figure 18.1

Florida Mayhem Entering Blizzard Arena for an Overwatch League Event

Source: Blizzard Entertainment

Recent Developments

Esports continue to grow in popularity, and new developments in esports include continued growth and investment in professional esports, televised competitions on cable channels, arenas built solely for esports, collegiate esports teams, and even talk of including esports in the Olympics.

The popularity of esports hasn't escaped the owners and managers of professional sports including FIFA, NHL, MLB, NBA, Formula 1, and NFL. The NHL entered the arena with the 2018 NHL Gaming World Championship. Using EA's *NHL* on the PS4 and Xbox One, players competed in 18 rounds leading to the championship in Las Vegas during the NHL Awards in June 2018 (Associated Press, 2018). The NBA is going so far as to form their own esports league, NBA 2K. Formed in 2017, the league had 17 participating NBA teams in its inaugural season in 2018 using *NBA 2K*, a game from Take Two Interactive Software (NBA, 2017). In 2020 the league has grown to 21 teams. The NBA drafts over 100 professional players who receive 6-month contracts with benefits and compensation around $35K (Harrison, 2018).

In its effort to embrace digital media, Formula 1 launched an esports series. Based on the F1 video game, Formula 1 uses the series as a marketing tool for both the traditional sport and video game. Formula 1 pursues it for fan engagement, but also to explore possible financial opportunities. Commercial managing director Sean Bratches observed that esports is a way for Formula 1 to reach a new and younger audience (Richards, 2017). Team Vitality CEO, Nicolas Maurer, noted that the skill-set used for the real Formula 1 sport is the same for the F1 esports game. Unlike League of Legends and Dota 2, F1 reaches a small online niche audience. The game is designed to resemble as much as possible the actual sport (MacInnes, 2019).

Want to watch all these gamers? Fans can watch online, on television, or go to live events. Twitch was introduced in 2011, purchased by Amazon in 2014, and has become the most popular online service used to watch live gaming. Calling itself "the world's leading social video service and community for gamers," Twitch allows users to watch live gaming with audio (About Twitch, 2018). Twitch is also available as an app for a wide range of devices including smartphones, game consoles, Google's Chromecast, and Amazon FireTV. Users can just watch other gamers but the draw of Twitch is the opportunity to interact with chat and cheers. Popular gamers have huge followings on Twitch. For example, Ninja had an average of almost 80,000 viewers and DrDisRespectLIVE had about 27,000 viewers in March, 2018. Professional leagues also have large followings on Twitch. For example, the Overwatch League has an average of 60,000 viewers (The Most, 2018; About Twitch, 2018). To watch competitive esports on Twitch, users use official league channels. Other popular ways to watch online include Mixer, MLG.tv, and YouTube Gaming.

The concept of watching esports on television was scoffed at just a few years ago. Now ESPN2 carries esports and has a deal with MLG. The network has broadcast the finals of a collegiate tournament of Blizzard's *Heros of the Storm* and major events such as the International *Dota 2* Championships and BlizzCon. TBS has ELeague—a partnership between IMG and Turner Sports. ELeague has esports events for games including *Streetfighter V*, *Counter Strike: Global Offensive*, and *Overwatch* (ELeague, n.d.).

Technologically savvy fans have higher expectations as esports spectators than the traditional broadcast audience. Broadcasts feeds consist of a picture-in-picture inset with the player's interfaces. There are dedicated player cams on each player's stations that provide an immersive experience that young fans desire. Viewers are able to see the intense esports players' expressions and movements during competitions. Thus, with player cams and more than 50 video sources in an arena broadcasts provide viewers a different experience compared to traditional sports (Root, 2019).

Esports fan argue that nothing is better than seeing an event live. 12,000 fans watched Team Newbee win $1 million at the International *Dota 2* Championship in Seattle in 2014 (Wingfield, 2014). Since then, the numbers have grown. Over 174,000 spectators filled a stadium in Poland for the IEM (Intel Extreme Masters) World Championships in 2017 (Elder, 2017). In the U.S., regular events are held in arenas in major cities. For example, ESL One, a *Counter Strike: Global Offensive* competition, was held at Brooklyn's Barclay Center in 2017 with full capacity and over five million streams (Kline, 2017).

There are now dedicated esports arenas. The first Esports Arena was built in Orange County, CA in 2015. The Luxor Hotel and Casino in Las Vegas has the flagship Esports Arena Las Vegas. The arena opened on March 22, 2018 in a 30,000 square-foot former nightclub space and is part of Allied Esports International's network of esports arenas. The Las Vegas arena features a competition stage and state of the art streaming and television production studios. (See Figure 18.2) The company has plans to build 10-15 more Esports Arenas across the United States in the next few years (Esports Arena, 2018).

Are esports just another sport like basketball and track? Across the world, high schools, colleges, and universities are featuring esports teams. Often beginning as student clubs and then club sports, esports teams are now fully-funded athletics teams at many universities. Robert Morris University Illinois was the first higher education institution to offer scholarships for esports athletes in 2015. The initial team focused on *League of Legends* and recruited from the League of Legends High School Starleague (HSEL) (RMU Becomes, 2014). As of March 2020, there were at least 175 varsity esports programs at universities in the U.S. up from just 50 in 2018.

Figure 18.2

Esports Arena Las Vegas

Esports Arena, the first of its kind to open in North America, specializes in large Esports events including the very first tournament for Blizzard Entertainment's newly released game *Overwatch*.

Source: Esports Arena

The governing body for collegiate esports is the National Association of Collegiate Esports (NACE) (Morrison, 2018). Founded in 2016, NACE members institutions are "developing the structure and tools needed to advance collegiate esports in the varsity space" (About NACE, n.d.). NACE has partnered with the High School Esports League (HSEL), the National Junior College Athletic Association (NJCAA), and the National Association of Intercollegiate Athletics (NAIA). What's missing? The NCAA of course. The NCAA announced in December 2017 that it was investigating collegiate esports. In 2019 the NCAA Board of Governors voted to table the idea of governing esports indefinitely, basically killing the idea (Hayward, 2019). However, student esport athletes

are getting NCAA-like support at some universities. For example, NACE member the University of California at Irvine fully supports their esports team with dedicated practice facilities, an esports arena, and scholarships. UC, Irvine supports teams for *Overwatch* and *League of Legends*. Collegiate teams generally play for scholarship money (Conditt, 2018).

With amateur collegiate and professional esports exploding, inclusion in the Olympics could be in the future. The International Olympic Committee began to explore the possibility in 2017 for inclusion in the Paris summer games of 2024 and approved demonstrations of esports for the games (Schaffhauser, 2019). The IOC is interested in esports for its appeal to youth but state that the sport must conform to Olympic values

(Esports:International, 2017). This standard could be an issue for some games considering their violent content. Indeed, in March 2017, the IOC clearly stated that it will not consider any games that contain violence. Games will probably be related to traditional sports rather than popular esport games like *Call of Duty* and *Overwatch* (Orland, 2018).

Nielsen (2019) provides insight giving a holistic overview of the stakeholders into the esports ecosystem including game publishers, leagues, teams, event operators, streaming/platforms/broadcasters/ and gaming personalities. As seen in Figure 18.3 Esports has evolved into a major industry with each stakeholder contributing to its growth.

Current Status

Newzoo released its 2019 Global Esports Market Report in February 2019. The report listed seven key takeaways:

- Global esports revenues will reach $1.1 billion in 2019, a year-on-year growth of +26.7%. North America will account for $409.1 million of the total, and China for $210.3 million.

- Brands will invest $897.2 million in the esports industry, 82% of the total market (media rights, advertising and sponsorship). This spending will grow to $1.5 billion by 2022, representing 87% of total esports revenues.

- The number of esports enthusiasts worldwide will reach 201.2 million in 2019, a year-on-year growth of +15.%. The total esports audience will reach 453.8 million this year.

- The global average revenue per esports enthusiasts will be $5.45 this year, up 8.9 % from $5.00 in 2018.

- In 2018, there were 737 major esports events that generated an estimated $54.7 million in ticket revenues, down from $58.9 million in 2017.

- The total prize money of all esports events held in 2018 reached $150.8 million, an increase from $112.1 million in 2017. The *League of Legends* World Championship was the most watched event on Twitch in 2018 with 53.8 million viewing hours. It also generated $1.9 million in ticket revenues. Overwatch League was the most-watch league by live viewership hours on Twitch, generating 79.5 million hours (NewZoo, 2018).

Figure 18.3

Esports Ecosystem

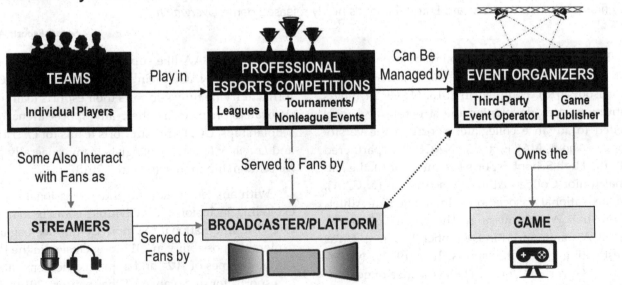

Source: Nielsen (2019)

Nielsen (2017) reports that the average age of the esports audience member is 26, with male fans more likely to stream esports. However, Nielsen (2019) found that globally 22% of esports fans are female, with Korea (32%) and China (30%) leading in female fans. However, only 17% of esports fans in the U.S. are female. Activate (2019) found that in the U.S. 68% of esports viewers were age 18-34. Esports fans watch comparatively little linear television but are avid streamers. For example, U.S. esports fans watch an average of just 4.4 hours of television per week (Nielsen, 2017).

As evidence concerning a dedicated and growing audience for esports, Nielsen (2019) found that globally 60% of Twitch viewers had been esports fans for at least four years. This was not the case with the U.S. where only 20% had been following for this long, with 23% new viewers this year. This report also found that 70% of Twitch viewers gave more time to esports that traditional sports. Further, in 2019, one in three Twitch viewers watched five hours of *Fortnite* live in the last year.

The Nielsen (2017) study found that, excluding Twitch, most fans followed both players and games via YouTube. The most popular team in the U.S. and U.K. was Cloud9 while Fnatic was the most popular in France and Germany. Top broadcasters in the U.S. for esports for the first six months of 2017 were Disney XD with 13 shows, TBS with seven, ESPN2 with five, NFL Network with three, and ESPNU, the CW and ESPN with one each. Interestingly this study also asked whether esports should be considered an actual sport (53%), a university sport (41%), and an Olympic sport (28%) (Nielsen, 2017). However, in 2020, YouTube negotiated exclusive rights to stream Activision Blizzard's Esports leagues, *Call of Duty* League, the *Overwatch* League, and *Hearthstone*. This programming is YouTube's effort to become the new home of esports, a title held by Twitch (Alexander, 2020).

According to esportsearnings.com (2020), total prize money for esports was $219 million with 4,685 tournaments and 24,352 players. The top earning games for 2019 were *Fortnite* with $64 million in prize money, over 350 tournaments, and 2,273 players.

In second place, *Dota 2* awarded $47 million in prize money, 205 tournaments, and 1,286 players. *Counter Strike: Global Offensive* came in at third place with $22 million in prize money, over 782 tournaments, and 3,828 players. The classic *League of Legends* fell from third to sixth place with $9 million, 162 tournaments, and 1,747 players. Nielsen (2019) reported that *Fortnite* had in 2017 become a major game disrupter, skyrocketing to the most watched game on Twitch in 2019.

Factors to Watch

Expect continued growth in the esport audience, prize money, leagues, and revenues. Television broadcasts should continue to proliferate as networks see the possibilities of esports programming and its tough-to-reach audience of young men.

Expect collegiate esports to be embraced by the governing organizations which will help further standardize the sport for both institutions and players. As of this writing, this is difficult because esports teams range from NACE supported varsity teams to small student clubs. There is even a wide disparity in which university division these teams are located on a campus. Some are in athletics, while others are in student affairs or other divisions.

The big players in the game industry such as Blizzard and Valve will continue to dominate but look for smaller games to make inroads. The social aspect of how audiences consume games through sites such as Twitch and YouTube means that smaller games can easily go viral and stand out.

One hot aspect of gaming may soon become part of the esports ecosystem, virtual reality (VR) and augmented reality (AR). Matthew McCauley, Misfits social media and marketing director, observed that VR and AR could revolutionize esports. He believes that the key to esports growth is in new technologies. Both VR and AR are still developing and expensive, yet, they have the potential to enhance user experience. Just think about how VR could provide a participant the ability to roam in-the-game. Audiences could have a virtual experience interacting with players (Ayles, 2019).

Gen Z

Born between 1995 and 2014, Generation Z is perfect for esports. Nielsen (2019) research notes that this generation has high and different expectations for entertainment. They consume sports differently than their parents and grandparents. Gen Z values quick paced sporting events. In addition, Gen Z are more loyal to individual athletes than teams or leagues. Gen Z rarely watch traditional broadcast platforms, preferring instead streaming in viewing sports. Traditional sports has taken notice and are intergrating esports competitions into the entertainment experience. The NBA, NHL, and MLB have entered the esports arena. An example is Formula 1 which introduced its esports league in 2017 in an effort to engage Gen Z audiences. The growth in esports and its majority Gen Z audience had not gone unnoticed by advertisers. Thus, esports has seen tremendous growth in sponsorships by various brands. Esports is in a wonderful position as it becomes mainstream to cater to the Gen Z audience.

Bibliography

About NACE (n.d.). National Association of Collegiate Esports. https://nacesports.org/about/.

About Twitch (2018). Twitch. https://www.twitch.tv/p/about/.

Activate Technology and Media Outlook 2020. (2019, October 23). https://activate.com/outlook/2019/.

Alexander, J. (2020, January 27). Call of Duty League, the Overwatch League, and Hearthstone. https://www.theverge.com/2020/1/27/21082612/youtube-blizzard-activision-esports-leagues-twitch-live-streaming.

Associated Press (2018). NHL takes esports on ice with gaming tournament. *The New York Times*. https://www.nytimes.com/aponline/2018/03/09/business/ap-hkn-starting-esports.html.

Ayles, J. (2019, October 17). How Technology Can Help Esports Continue Its Astonishing Growth. https://www.forbes.com/sites/jamesayles/2019/10/17/how-technology-can-help-esports-continue-its-astonishing-growth/#66f658be4219.

Baker, C. (2016) Game Tournament. *Rolling Stone*. https://www.rollingstone.com/culture/news/stewart-brand-recalls-first-spacewar-video-game-tournament-20160525.

Conditt, J. (2018). College esports is set to explode, starting with the Fiesta Bowl. *Engadget*. https://www.engadget.com/2018/02/22/college-esports-is-set-to-explode-starting-with-the-fiesta-bowl/.

Dwan, H. (2017). What are esports? A beginner's guide. *The Telegraph*. https://www.telegraph.co.uk/gaming/guides/esports-beginners-guide/.

Edwards, T. (2013). Esports: A Brief History. *Adani*. http://adanai.com/esports/.

Elder, R. (2017). The esports audience is escalating quickly. *Business Insider*. http://www.businessinsider.com/the-esports-audience-is-escalating-quickly-2017-3.

ELeague (n.d.) TBS. http://www.tbs.com/sports/eleague.

Esports Arena (2018). Esports Arena Las Vegas. https://esportsarena.com/lasvegas/.

Esports: International Olympic Committee considering esports for future games. BBC. http://www.bbc.com/sport/olympics/41790148.

Esportsearnings.com. (2020, January). Overall Esports Stats for 2019. https://www.esportsearnings.com/history/2019/games

Fischer, J. (2019). White Paper: The Esports Phenomenon Brings New Challenges and Opportunities in Media Tech. https://www.sportsvideo.org/2019/11/15/white-paper-the-esports-phenomenon-brings-new-challenges-and-opportunities-in-media-tech/.

Goldberg, F. (2011). The origins of the first arcade video game: Atari's Pong. *Vaniety Fair*. https://www.vanityfair.com/culture/2011/03/pong-excerpt-201103.

Harrison, S. (2018). The Big Breakaway. *The Outline*. https://theoutline.com/post/3616/the-big-breakaway-nba-esports-league-nba-2k.

Hayward, A. (2019). NCAA votes to not govern collegiate esports. *The Esports Observer*. https://esportsobserver.com/ncaa-nogo-collegiate-esports/.

Hughes, C. J. (2019, May 28). As E-sports Grow, So Do Their Homes. *The New York Times*. https://www.nytimes.com/2019/05/28/business/esports-arenas-developers.html.

Kline, M. (2017). Gaming fans are crowding into stadiums to watch esports events. Here's why. *Mashable.* https://mashable.com/2017/09/22/esports-events-are-filling-stadiums/#iDJLIPMP_Oq7.

Lee, D. and Schoensted, L. (2014). Comparison of eSports and traditional sports consumption motives. *Journal of Research,* v6, 2. pp 39-44.

MacInnes, P. (2019, December 3). F1 Esports season crosses line before top gamers steer out of virtual world. https://www.theguardian.com/sport/2019/dec/03/f1-esports-season-jarno-opmeer-renault.

Morrison, S. (2018). List of varsity esports programs spans America. *ESPN.* http://www.espn.com/esports/story/_/id/21152905/college-esports-list-varsity-esports-programs-north-america.

NBA (2017). 17 NBA Teams to Take Part in Inaugural NBA 2K esports league in 2018. http://www.nba.com/article/2017/05/04/nba-2k-esports-league-17-nba-teams-participate-inaugural-season.

NewZoo (2018). *Newzoo 2018 Global Esports Market Report.* https://resources.newzoo.com/hubfs/Reports/Newzoo_2018_Global_Esports_Market_Report_Excerpt.pdfc.

NewZoo (2019). *Newzoo 2019 Global Esports Market Report* https://newzoo.com/solutions/standard/market-forecasts/global-esports-market-report/.

Nielsen. (2019, October 23). Activate Technology and Media Outlook 2020. https://activate.com/outlook/2019/

Nielsen (2017). *The Esports Playbook.* http://www.nielsen.com/content/dam/corporate/us/en/reports-downloads/2017-reports/nielsen-esports-playbook.

Orland, K. (2018). Violent video games not welcome for Olympic esports consideration. *Arstechnica.* https://arstechnica.com/gaming/2018/03/olympic-committee-open-to-esports-but-only-without-violence/.

Overmars, M. (2012). A brief history of computer games. *Stichingspel.* https://www.stichtingspel.org/sites/default/files/history_of_games.pdfbritis.

Richards, G. (2017). F1 enters eSport arena with official championships to start in September. https://www.theguardian.com/sport/2017/aug/21/f1-esports-championships-abu-dhabi-grand-prix.

RMU Becomes First University to Offer Gaming Scholarships With the Addition of eSports to Varsity Lineup (2014). Robert Morris University Illinois. www.rmueagles.com/article/907.

Robert Kraft Calls Upstart Overwatch League The "future Of Sport In Many Ways" https://www.sportsbusinessdaily.com/Daily/Issues/2017/07/13/Leagues-and-Governing-Bodies/Overwatch.aspx.

Root, J. (2019, September). Esports Productions Up Their Game With Dedicated 'Player Cams.' *TVB EUROPE,* 58-59.

Schaffhauser, D. (2019). Esports joining Olympics in 2024. The Steam Universe. https://steamuniverse.com/articles/2019/07/30/esports-joining-olympics-in-2024.aspx.

The Most Popular Twitch Streamers, March 2018 (2018). TwitchMetrics. https://www.twitchmetrics.net/channels/popularity.

The OGA (1999). Eurogamer.net. http://www.eurogamer.net/articles/oga.

Wendel, J. (2016). The original. *The Players' Tribune* https://www.theplayerstribune.com/fatal1ty-esports-the-original/.

Wingfield, N. (2014). In E-Sports, video games draw real crowds and big money. *The New York Times.* https://www.nytimes.com/2014/08/31/technology/esports-explosion-brings-opportunity-riches-for-video-gamers.html?ref=bits.

19

Ebooks

Steven J. Dick, Ph.D. [*]

Abstract

Ebooks are digital publications applying technology to transmit fiction and nonfiction works to computer screens and handheld devices. Ebooks at first disrupted traditional formats by affording the means of electronic access to authors' works. Assisted by groups like Project Gutenberg, ebooks also preserved the library of publications in electronic format for popular distribution. Independent authors now have the capacity to circumvent traditional publishing outlets and reach new audiences through ebook technology. It allows access to audio volumes, text-to-speech works, interactive narratives, and an enhanced experience with multimedia content.

This format may have disrupted the publishing industry, but was itself disrupted by new marketing strategies and access options. Amazon, for example, the market leader for electronic reading, revamped the way it distributes ebooks moving to capture the market with its powerful Prime services while competitors strive to follow its lead. The ebook future promises innovations for both authors and a wider readership.

Introduction

The idea of distributing books in digital form is almost as old as computer programming itself. Even with slower transmission speeds, book-length data files have been distributed without delay or unnecessary expense since 1971. The goal at first was to free the publishing world from paper, and expand the archival storage for all books. Ebooks soon became an important part of the publishing industry, and even though popular adoption has fallen short of expectations amid competing formats and devices, new models have revolutionized the business and raised questions.

This book is available to you in both electronic and paper formats. Why did you choose the format you are reading now? Some like the portability of carrying

[*] Analyst, Modern Metrics Barn, Youngsville, LA

several digital books without additional weight, or appreciate the advantage of acquiring a book quickly, downloading it without worrying about physically returning it. Traditional readers enjoy the feel of book pages in their hands, the "under the tree" experience of reading anywhere they want without having to buy new software or navigate unfamiliar hardware. Some readers like to disconnect from technology while preferring an easy-to-share book.

What factors influence your book choice? Do price comparisons enter in your decision of selection? How about the type of content—would a Bible or another religious book feel the same to you if viewed in digital text? The relationship readers have with their books affects the market for ebooks. That relationship has, in some ways over time, changed demonstrably. This chapter will explore the factors leading to the evolution of ebooks beginning with the component parts of this important innovation.

Background

Like most media, ebooks are not a single technology. There has been a combination of content, software, and hardware elements designed and reformulated to create the industry that exists today.

Content

The *content* is the first necessary component traced to its digital evolution on a college campus around 1971. University of Illinois prodigy, Michael S. Hart, enjoyed access to ARPANET computers (the predecessor to the internet). He used his access to key in the words and distribute great documents like the *Declaration of Independence*, followed by the *Bill of Rights*, the *American Constitution*, and the Christian *Bible* (The Guardian, 2002). That effort continues in 2020 as Project Gutenberg, which offers more than 60,000 free ebooks (www.Gutenberg.org). Project Gutenberg and like-minded groups moved swiftly to transfer analog text to digits through labored keyboard strokes or by just scanning works into computer format. Independent authors joined in the movement releasing their books through creative commons minus copyright restrictions, escaping the travail of traditional publication channels. These publicly accessible volumes formed the backbone of ebooks and soon were produced by commercial publishers entering the field. Three main companies, Ereader, Bibliobytes, and Fictionwise, initiated the sale of ebooks (The Telegraph, 2018). Their content was text, but audio recordings soon appeared on the horizon.

In 1932, the tests of audio recorded books featured a chapter from Hellen Keller's book *Mainstream* and Edgar Allan Poe's *The Raven* (Audio Publishers Association, n.d.). "Talking Books" from the Library of Congress originally were intended for the blind. Their start was slow but got a strong push forward from the audio cassette's introduction in 1963. From there books on tape rose in popularity from the 1970s through 1980s, while the digital evolution in hardware and software forged new links in this chain.

One company, Audible, created a digital audio player in 1997 and joined Apple's iTunes library in 2003. Not long thereafter digital downloads began to surpass CD audiobooks around 2008. Audiobook software is naturally reliant on computer technology and yet carefully crafted to afford the same reading experience across media platforms. Thus, the iOS version and the Android versions share a similar look and feel. Perhaps most significant in economic terms was the corporate giant Amazon's entry in the field, buying Audible in 2008 while merging sales across book formats.

Google created its book search engine in 1994 (later Google Books), and by that time, market forces had moved to convince authors to distribute their content in electronic format. Public domain books were enhanced through the presentation of clearer formatting, annotations, and illustrations producing works sold at a low, yet profitable rate. Independent authors, aspiring to become successful professionals, became the first ones to offer their compositions through this digital form of content.

Amazon joined the ebook market in earnest when it introduced its Kindle E-reader (2007) and began selling content from major authors. Barnes & Noble, Apple, and Google (among others) followed Amazon, and the book market grew more attractive. Major publishers joined in the throng ramping up to meet the demand of new sales. In 2011, Amazon announced ebooks had outsold its traditional works (Miller & Bossman, 2011). The evolution in technology produced yet another popular feature.

Software

The ebook technology capitalized on two innovations in software—file formats and display. Over the years, multiple versions of each were marketed. Popular file formats contained markers leading to key features of the ebook experience, while display software afforded simple data like chapter markers, page turn locations, display of images, and interaction with dictionaries. File formats evolved from text storage (ASCII or American Standard Code for Information Interchange) and word processing (Adobe PDF). Early abandoned formats like LIT and MOBI were still found in online ebook resources, and could still be read on certain devices.

Display software applications have moved to afford readers a satisfying channel to communicate across devices. Interactivity of this sort invites a reader to move from device to device and return to it at the same point with a seamless flow between text and audiobooks. Interactivity can extend across channels as well, including shared personal notes, highlights, and social media posts. Religious books such as Bibles supplied by Olive Tree enable users to compare passages across translations, immediately find commentary, maps, dictionaries, and other study aids. Dedicated apps like WebMD and cookbooks like Tasty are also classified as interactive ebooks.

The final stage of software development protects the author's publication from being shared without sale or rent, or otherwise used inappropriately. Digital Rights Management (DRM) software—once common to music property—remains a legal safeguard for ebooks. Such DRM applications can link an ebook to a physical device or a loan time, but they also provide the security necessary for authors or owners needing to enforce their license.

Hardware

The first handheld devices for ebooks arrived in an array of personal digital assistants (PDA) debuting in the 1990s. As a precursor to the modern smartphone, these devices were especially useful for appointments, contacts, and similar data. Complimenting their utility were new apps for reading. Apple entered first in 1993 with its short-lived Newton, followed by the Palm Pilot in 1996, and the Microsoft Reader in 2000. Such handheld devices invited viewers to read text and write on an interactive screen about the size of the modern smartphone. That visual experience left something to be desired, however, until a digital screen innovation was implemented.

The most visible ebook hardware was the ereader with the innovation of E-Ink. E-Ink is a form of electronic paper (Primozic, 2015) appearing with the Kindle in 2007 (Wagner, 2011). This special screen appearing on early Kindles was invented to resemble paper's natural reflective quality. Since pages reflect light rather than producing it, ereaders illuminated the text with less intensity, a longer battery life, and were more relaxing to the eyes.

Table 19.1

Ebook File Formats

Format	Creator	Introduced	Notes
ASCII	American National Standards Institute	1963	Many have built on this standard to add additional capability such as "rich text."
ePub	International Digital Publishing Forum	2007	Widely used by Google, Android, and Apple. An open format but modified by some uses.
AZW	Amazon	2007	DRM added to MOBI for Amazon Kindle. Replaced by KF8.
LIT	Microsoft	2000	Microsoft discontinued support in 2012.
ODF	OASIS and OpenOffice	2005	Used in many alternatives to Microsoft Office. XML
MOBI	Mobipocket	2000	Altered original Palm format then purchased by Amazon.
PDF	Adobe	1993	Designed for page layout.

Source: S. Dick

Sales of E-Ink devices, however, were impacted by conflicting market forces. First, the advantage of extended battery life seemed to be unimportant as cell phone and tablet batteries were vastly improved. Second, the economic advantage of this dedicated device became diminished when ebook sellers allowed the volume sales price to approach the higher cost of paper books. Finally, the penetration of tablets and mobile phones with ebook apps undercut the market because ereaders meant carrying a second handheld device. And despite their promising start, E-Ink devices soon lost consumer sales to alternative technologies (See Figure 19.1). E-Ink reached its zenith in 2012 about the time ebook prices began to rise substantially, and smartphone or computer tablets competed rather successfully (Haines, 2018).

The rise in popularity of smartphone and tablet sales (Milliot, 2016) can be attributed as one factor in the decline of sales. Market observers contend E-Readers were not as attractive to prospective buyers—especially young consumers. Handheld electronics were expensive and, in the case of E-Readers, only useful for one purpose. Even the simplest E-Reader costs between $50 and $200. It would soon appear to be smarter to pay for paper-bound books instead of the E-Reader plus the cost of the book. Similar arguments criticized E-Readers for the problem of screen fatigue as well as the reduced capacity to retain the content of digital text (Allcott, 2019).

Figure 19.1

Trend of Mobile Device Ownership

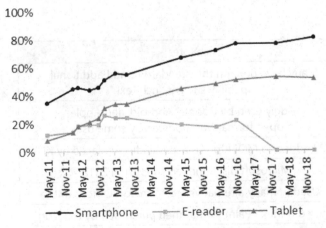

Source: S Dick based on data from Pew Research Center (2019)

Recent Developments

Ebooks have produced mixed sales over the past few years with flat market growth noted for both kinds of books. Pew Research (n.d.) surveys over the past few years have indicated an unchanging percentage of people read physical books (around 90%) and read ebooks (around 36%). It could be argued that between 2016 to 2019 ebook readership has dropped but it may be too early to know definitively. Audiobooks, while still the least popular format, have grown dramatically since 2015 (17-27%).

Certain categories of books remain popular for paper publishing. The latest fad of adult and children's coloring books, and hardback books lead the categories that sell better in paper (Sweney, 2016). People seem to have a natural affinity for relaxing with paper rather than with electronic ink.

Figure 19.2

Book Format for People Reading (One Book Past Year)

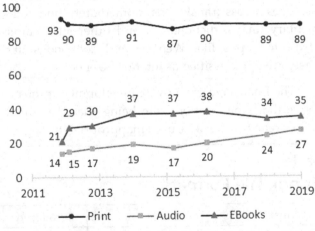

Source: S. Dick based on Pew Research (n.d.)

It is also possible that sales numbers do not accurately reflect ebook sales. Interactive books and games for children, medical texts, and religious publications may not be seen properly in publishing charts, and can quickly cross a line making their data invisible to book market measures. For example, religious publisher Olive Tree produces Bibles and Bible study products outside normal publishing channels. Somewhat similarly, Chu Chu TV (www.youtube.com/user/The ChuChutv) is a YouTube Channel and application. It rewrites traditional children's stories and nursery rhymes. Finally, the original Grey's Anatomy has been repackaged in the app Visual Anatomy. Absent interactive technology, these alternative applications might be produced by the publishing industry, but unlikely figure in ebook sales data.

New pricing models also impact ebook market statistics. First, as major publishers enter the market, calculations are not as accurate for independent publishers and writers concentrating instead on their best-selling books. Second, free or promotional books are easy to distribute (Trachtenberg, 2010), and rather than selling titles outright, a book is given away in exchange for indirect publicity or corporate promotion (Anderson, 2009).

Finally, in July 2014, Amazon introduced Kindle Unlimited—a substantial collection of ebooks available at no additional cost to subscribers of their Prime service. Instead of buying or renting a single book, the pricing was based on a subscription model for a library of titles, and authors were paid a share of subscriptions instead of a per-book royalty. Both Barnes & Noble and Kobo adopted subscription services as well. Because of these systemic changes, an accurate view of ebook sales in such a diverse marketplace is difficult to apprehend (Pierce, 2017).

Figure 19.3

Average Price of Top 10 Books on Amazon

Note: Two books in the new book category were not available in audio. One book, a Kobe Bryant memorial was only available in magazine format and was skipped.

Source: S Dick based on Amazon Prices February 15. 2020

Current Status

Publishers, authors, customers, and distributors continue to search for acceptable pricing models. Amazon has tried to enforce sales contracts with authors to give them the best possible pricing and remain the most important distributor of books in all formats. Looking at the current price of the most popular books (see Figure 19.3), we can see how among all-time top selling books (e.g., *Hunger Games, 50 Shades of Gray,* etc.), ebooks cost slightly more than paperback—probably due to the price pressure felt by used books. Exclusively hardbound new releases, on average, cost about $3.50 more than ebook versions.

There are other choices for authors today. Major models include self publishing, direct to distributor, subscription services, publisher wholesale, or publisher agency models—each holding advantages and disadvantages. Such deals can be shaped by negotiations with the distributors or by the type of book.

Self-publishing gives the author greatest control over content but also the most work in distribution. Promoting a book is one challenge, while DRM is another one if the author chooses a free or creative commons model. Most authors will do some self-promotion regardless of the distribution channel.

Direct to distributor: Several distributors will negotiate directly with authors to sell books for a flat fee or royalty. Such sales are available on a particular website where books can be shipped to readers at a reduced charge offering a discount as a result of eliminating the so-called middle man in the equation.

Subscription-based distribution (discussed below) is promoted by Amazon's Unlimited service placing a particular volume in a selection of books. Subscribers can access any in the set for a flat monthly fee. The author is not paid by the book but by the number of "pages" read. The author must agree to a renewable 90-day window exclusive to Amazon. Major publishers also select between two distribution models—wholesale or agency model.

In the wholesale model, the publisher or author suggests a price but sells the book to the distributor at a much lower wholesale price. The distributor is then free to set whatever price it wishes to put on the book.

So, the price to the distributor may be $6.99 with a suggested price of $12.99, but the distributor may choose to sell the book for $9.99. This makes it more difficult for the publisher to sell the book at comparable prices across platforms.

The agency price model gives more control to the publisher—so it is favored by the larger publishing houses. With agency pricing, the publisher sets the amount for the sale of the book, and the distributor receives a commission on each purchase. For example, the publisher sets the price for $12.99 and gives the distributor 40% on each book sold. This model poses a problem for Amazon based on its principal aim to standardize prices and control sales. Overall, this approach also has the effect of pushing prices higher—more equivalent to printed books. Agency pricing is consequently blamed for raising ebook prices and contributing to a slump in their sales (Miller & Bosman, 2011).

The Amazon Model

As should be clear by now, it is impossible to evaluate the ebook industry without carefully considering the dramatic impact of Amazon. This company has the capacity to set prices and terms, and dominate the market by garnering an estimated 75% of the revenue. Because of its pinnacle position, Amazon enjoys what has been described as "monopsony power," where one buyer/seller dictates the terms over a category of goods (London, 2014). Commercial authors cannot ignore Amazon if their book is to be successful.

For the purposes of this chapter, consider how Amazon promotes its Kindle Direct Publishing (KDP)—a contractual relationship designed for the independent author (Amazon, n.d.). KDP makes certain rigorous demands of the author in exchange for services and preset royalty amounts. There are optional promotional programs such as the Kindle Countdown Deal (limited time promotional discount). The Kindle publishing page helps authors to choose between one royalty arrangement (allowing distribution on other stores) or an exclusive deal with Amazon.

One of the problematic qualities of Amazon's vast influence is its "most favored nation" (MFN) clause. The MFN requires publishers doing business with the company to reveal the terms of book contracts made with competitors. This information puts Amazon in the position of demanding comparable or superior terms. By marshaling MFN power over contracting writers, the corporation can require significant concessions to reduce their costs and approach monopoly status (one effective seller) in terms of cost sharing, promotion, and release date. As a result, the European Community agreed to end MFN for Amazon in exchange for the end of a three-year anti-trust investigation (Vincent, 2017).

Producing an Ebook

The challenge for struggling independent authors is how to produce a marketable book in the first place. Modern ebooks have more in common with a web page than a word processing document, which means the content must be actively resized to meet each reader's preferences so publishers incorporate a display size or build one on cascading style sheets using HTML technologies. This technical concern naturally leads into design elements and display files.

There are two main categories of display files: reflowable and fixed layout. Reflowable files (ePub, AZW, and MOBI) allow the dynamic assignment of page lengths based on user preferences and screen size. Such files allow smooth lines and page breaks when the display changes from a small mobile phone screen to a large tablet. Reflowable content works best with traditional books that have minimal illustrations and are proven best sellers.

Fixed layout files are best for books needing illustrations to be associated with text that cannot be moved. Fixed layout files cover items like instruction manuals, cookbooks, and sales literature. The best known in this category would be Adobe's Portable Document Format (PDF). A PDF is like a computer file in that pages expand or shrink but maintain the same look.

Table 19.2

Comparison of Reflowable and Fixed Layout

	Reflowable	Fixed Layout
Distribution formats	ePub, AZK, and MOBI	PDF, KF8, ePUB3, Pageperfect.
Production Software	Public domain and distributer based software.	Quark and Apple iTunes Producer.
Advantages	Most popular, more selling platforms, easier to update, user control over fonts and screen size.	Maintains look, media capable, interactivity, Java-capable.
Disadvantages	Undependable relationship between text and illustrations. Branding may be a problem.	PDFs cannot be sold on many platforms, larger file size, problems with small screens, more difficult to produce.
Best for	Text only books with few illustrations.	Children's Books, graphic novels, cookbooks, instruction manuals, textbooks, corporate publications.

Source: S. Dick

Fixed layout files can take the next step to true interactivity. The files may include links to online media, quizzes, and interactive content. For example, children's ebooks may allow elements to move around the page, react to clicks (taps) or make sounds. Programs like Quark and Apple iTunes Producer are most popular to produce these books. Fixed layout ebooks are sold at most bookstores but are incompatible with certain platforms (Wilson, 2015).

Once the basic book is created, the author must validate its content as an original work for multiple platforms to incorporate DRM elements (if used). At that point, the major concern becomes one of marketing. A few companies help with all elements of ebook creation and marketing but prices are not yet stable due to multiple options and reliability of their products.

Factors to Watch

Despite the unimpressive growth of ebook sales, publishers prepared to do battle with one of their traditional allies. In 2019, certain publishers began resticting the distribution of ebooks by libraries. They based their decision on four factors. First, customers were going to public libraries for free copies of books they might otherwise purchase. Second, large urban libraries were exasperating the situation by "selling" library cards outside service areas. Third, libraries were creating public expectations for free book ownership, damaging long-term marketing strategies. Commercial publishers did not want to deal with the crisis newspapers faced in desperation for financial stability due to free and unlimited digital access. Finally, unlike a physical book, an ebook never gets damaged, worn out, and it is returned on time. This digital capacity to bypass replacement editions, or need to purchase extra shelf copies, might offer huge savings for libraries but create an enormous revenue challenge for book publishers.

For their part, librarians argue that their distribution is a traditional motivator of sales in promoting books to people who go on to buy copies. However, libraries tend to serve people who cannot afford to buy books. In truth, most library patrons actually would not be traditional book buyers.

The library issue is an example of the evolving market models for ebooks. Wholesale distribution, new independent publishers, and subscription book services all combine to create an ebook marketplace hoping to find its way to consumers and profitability. In the coming years even more business models are likely.

Gen Z

There is some reason to believe the Gen Z reader is, in fact, the major driving force in ebook distribution. Gen Z buys more ebooks than other age groups—partially due to the class needs of high school and college students.

The student market is a major book market but the same young people continue to read ebooks for enjoyment past graduation. Readers between 18-29 are most likely to consume books (Statista, 2019) and visit the library 2.5 times for every trip they make to the movie theater (Mccarthy, 2020).

The changes in publishing are also starting to affect student's course material costs (National Association of College Stores, n.d.). Between 2007-2008 and the 2017-18 school year, the average college student spending decreased from $701 to $484 for required course materials. The reduction in cost was mostly due to increased use of free (downloaded or borrowed) material and ebooks, increasing from 10%-25% of all students.

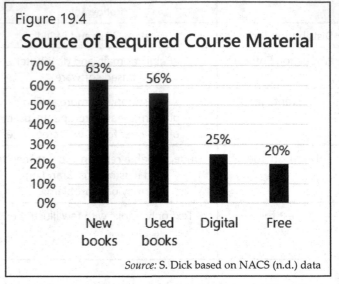

Figure 19.4

Source of Required Course Material

Source: S. Dick based on NACS (n.d.) data

Global Concerns & Sustainability

The market structure for ebooks have a fairly low impact compared to hardbound copies. It was estimated 689 million print books were sold in 2019 (Statista , 2020), but that figure does not cover global distribution or unsold inventory. These physical books require paper, glue, and other materials. Regardless of the amount of paper used, the book may not be recyclable. Books must to be shipped to customers, and unsold inventory shipped back.

Ebooks may be one of the cheapest media products to produce. Other than a computer, ebooks require almost no specialized equipment. The average 300-page Kindle ebook requires only 2.6 mb (Elite Authors, n.d.). Distribution can be accomplished in even low speed networks. To the extent that a country is already using handheld devices, ebook distribution requires no new infrastructure. As such, it is one medium that can be produced almost anywhere in the world.

Bibliography

Allcott, L. (2019, October 19). Reading On-screen vs Reading in Print: What's is the Difference for Learing? National Libary of New Zealand. https://natlib.govt.nz/blog/posts/reading-on-screen-vs-reading-in-print-whats-the-difference-for-learning.

Amazon. (n.d.). Kindle Direct Publishing. Amazon. https://kdp.amazon.com/en_US/help/topic/G201541130

Anderson, C. (2009). *FREE: The Future of a Radical Price*. Hyperion.

Audio Publishers Association. (ND). A History of AudioBooks. Audio Pubishers Association. https://www.audiopub.org/uploads/images/backgrounds/A-HISTORY-OF-AUDIOBOOKS.pdf.

Elite Authors. (n.d.). *The average size of a Kindle E-book*. Elite Authors. https://eliteauthors.com/blog/the-average-size-of-a-kindle-e-book/.

Haines, D. (2018, Feb 25). The Ereader Device is Dying a Rapid Death. Just Publishing Advice. https://justpublishingadvice.com/the-e-reader-device-is-dying-a-rapid-death/.

London, R. (2014, October 20). Big, bad Amazon. *The Economist*. https://www.economist.com/blogs/freeexchange/2014/10/market-power.

Mccarthy, J. (2020, January 24). In U.S., Library Visits Outpaced Trips to the Movies in 2019. *Gallup*. https://news.gallup.com/poll/284009/library-visits-outpaced-trips-movies-2019.aspx.

Miller, C. C., & Bosman, J. (2011, May 19). Ebooks Outsell Print Books at Amazon. *New York Times*. http://www.nytimes.com/2011/05/20/technology/20amazon.html.

Miller, C. C., & Bossman, J. (2011, May 19). E-Books Outsell Print Books at Amazon. *New York Times*. http://www.nytimes.com/2011/05/20/technology/20amazon.html.

Milliot, J. (2016, June 17). As E-book Sales Decline, Digital Fatigue Grows. *Publishers Weekly*. https://www.publishersweekly.com/pw/by-topic/digital/retailing/article/70696-as-e-book-sales-decline-digital-fatigue-grows.html.

National Association of College Stores. (n.d.). Student Watcch Key Findings. *NACS*. https://www.nacs.org/research/studentwatchfindings.aspx.

Pew Research Center. (2019, June 12). Deographics of Mobile Devices. *Internet and Technology*. https://www.pewresearch.org/internet/fact-sheet/mobile/.

Pew Research Center. (n.d.). Methodology. Pew Research. https://www.pewresearch.org/wp-content/uploads/2019/09/FT_19.09.25_BookReadingFormats_Methodology_Topline.pdf.

Pierce, D. (2017, December 20). The Kindle changed the Book Business. Can it change books? *Wired*. https://www.wired.com/story/can-amazon-change-books/.

Primozic, U. (2015, March 5). Electronic Paper Explained. *Visionect*. https://www.visionect.com/blog/electronic-paper-explained-what-is-it-and-how-does-it-work/.

Statista . (2020, January 16). Unit sales of printed books in the United States from 2004 to 2019. Statica. https://www.statista.com/statistics/422595/print-book-sales-usa/.

Statista. (2019). E-book readers in the United States in 2018, by age. Statista. https://www.statista.com/statistics/249767/e-book-readers-in-the-us-by-age/.

Sweney, M. (2016, May 13). Printed Book Sales Rise for the First Time in Four Years as Ebook Sales Decline. *The Guardian*. https://www.theguardian.com/media/2016/may/13/printed-book-sales-ebooks-decline.

The Guardian. (2002, Jan 3). Ebooks Timeline. *The Guardian*. https://www.theguardian.com/books/2002/jan/03/ebooks.technology.

The Telegraph. (2018, February 17). Google Editions. a history of ebooks. *The Telegraph*.http://www.telegraph.co.uk/technology/google/8176510/Google-Editions-a-history-of-ebooks.html.

Trachtenberg, J. A. (2010, May 21). E-Books Rewrite Bookselling. *Wall Street Journal*. https://www.wsj.com/articles/SB10001424052748704448304575196172206855634.

Vincent, J. (2017, May 4). Amazon Will Change Its Ebook Contracts with Publishers as EU ends Antitrust Probe. *The Verge*. https://www.theverge.com/2017/5/4/15541810/eu-amazon-ebooks-antitrust-investigation-ended.

Wagner, K. (2011, Sept 28). The History of the Amazon Kindle So Far. *Gizmodo*. https://gizmodo.com/5844662/the-history-of-amazons-kindle-so-far/.

Wilson, J. L. (2015, March 25). Everything You Need to Know About Digital Comics. *PC Magazine*. https://www.pcmag.com/article2/0,2817,2425402,00.asp.

Networking Technologies

20

Telephony

William R. Davie, Ph.D. *

Abstract

Telephony is the general term describing electronic transmissions of voice and data by means of a handheld device. In contemporary terms, the primary instrument of telephony is the mobile phone, globally linking billions of people through distant exchanges of conversation and transmissions of text, video, audio, and graphics. Driving the current development of telephony is fifth generation (5G) wireless to be used for the Internet of Things (IoT), artificial intelligence (AI), augmented reality (AR), live streaming, and other forms of communication and commerce.

Introduction

Telephony transmits streams of voice and data between distant points by wired and wireless means. In 2018, mobile telephony passed a milestone when the majority of American homes (57%) moved to wireless-only phones (Blumberg & Luke, 2019). Today the idea of living without a mobile phone seems unthinkable for most of the world's population. Telephony is at the center of interpersonal communications, and knowing how this technology evolved over 150 years is helpful to understanding its future.

Background

The principal features of the telephone are the microphone/transmitter, the earphone/receiver, and the ringer/signal—all projects of various telephonic inventors including Johann Philipp Reis, Antonio Meucci, and Charles Bourseul, among others. Most prominent is Alexander Graham Bell, the Scottish visionary who migrated to Boston and worked on his "talking telegraph" to help the hearing impaired, which included his wife Mabel, who was rendered deaf by scarlet fever as a child. Bell's work with Thomas Watson, and his famous race to the U.S. patent office in 1876 carrying specifications for the

* BORSF Regents Chair of Communication, University of Louisiana, Lafayette (Lafayette, Lousiana).

electric telephone beating out his rival Elisha Gray is part of telephony's legend (AT&T, 2010a).

The company Bell and his associates envisioned, called American Telephone and Telegraph, was founded in 1885 (AT&T, 2010b). While inventing the telephone was Bell's vision, the business of telephony relied on Theodore Vail's determination. Vail led an effort to crisscross the country with poles and wires and a consuming passion to grow AT&T into an indomitable firm. Vail's company overtook smaller telephonic enterprises and expanded his transcontinental cables for coast-to-coast service (John, 1999). By the time he retired in 1919, Vail boasted his company controlled every telephone instrument, every switch, and every pole in the nation. AT&T's preeminence even had the blessing of the federal government when the Federal Communications Commission classified it in 1934 as a *protected monopoly* (Thierer, 1994).

Figure 20.1

Traditional Telephone Local Loop Network Star Architecture (POTS Network)

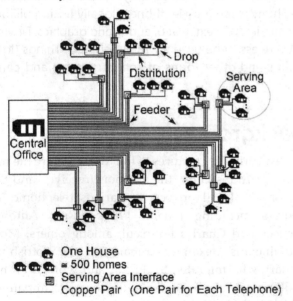

One House
≈ 500 homes
Serving Area Interface
Copper Pair (One Pair for Each Telephone)

Source: Technology Futures, Inc

Plain Old Telephone Service (POTS) transmitted voice calls via copper wires linked to an analog exchange center where multiple calls rapidly switched back and forth between trunk and branch lines to reach their destination. The telephonic system for

transmitting calls via cables was called the Public Switched Telephone Network (PSTN) (Livengood, Lin, & Vaishnav, 2006). (See Figure 22.1) This mechanical PSTN would be replaced one day by a computer-driven system expanding both phone capacity and the speed of transmission.

Over the next half century, telephony evolved by means of Storage Program Control (SPC), where home phone numbers were computed and data were stored. The SPC was a faster way of digitally switching phone calls through a centralized control system. Microprocessors routed vast numbers over "blocks" or series of exchanges (Viswanathan, 1992). Fiber optic glass strands eventually replaced copper wires, and modems began translating calls into binary signals for conversion into sound waves.

Birth of Mobile Phones

The move from wired to wireless telephony was fueled by the automobile boom in the 20th century. Telephone-equipped vehicles were needed to summon aid for fire engines, ambulances, and police cruisers in addition to non-emergency uses such as consumer ride services. At first radio transmissions were used between base stations and vehicles, but linking to the PSTN via a traveling handset was impractical until a project was built in St. Louis to facilitate a Mobile Telephone Service (MTS) and solve the vehicle-in-motion challenge. MTS was modeled after conventional dispatch systems of wireless radio (Farley, 2005). Motorola built the radios while the Bell System installed them showing the potential for consumer uses of wireless telephony.

Texas ranchers and oil company executives also needed a portable means to communicate remotely with workers in the field. Thomas Carter of Dallas in 1955 introduced his two-way radiophone attachment, and the homemade Carterfone acoustically linked mobile calls through AT&T wires and poles. AT&T attempted to put a stop to Carter's usurping transmissions and filed a legal complaint that was referred to the FCC. The Commission's decision on the Carterfone in 1968 affirmed its earlier ruling that held "any lawful device" could connect with phone company facilities (*Carter v. AT&T*, 1968). This FCC ruling set the stage for the cellular boom ensuing years later.

First Handset

Today's mobile invention owes something to the genius of an electrical engineer who worked on car phones in Motorola's research and development lab. Martin Cooper was fascinated by the "communicator" from the TV show Star Trek and hoped to make it a reality (Time, 2007). In 1973, he took to the streets of New York City where he triumphantly placed a call to Joel S. Engel, another New York engineer at competing Bell Labs. Cooper's phone was bulkier than Captain Kirk's model, weighing almost two pounds. It was dubbed "the brick," but it worked (Economist, 2009). Inventing a mobile phone was one feat, but linking antennas to follow mobile phones traveling across distances proved to be another hurdle.

Figure 20.2

Cellular Telephone Network Architecture

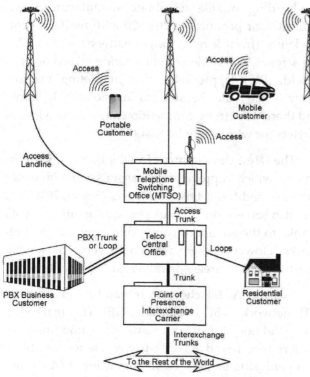

Source: Technology Futures, Inc.

Early land-mobile telephony needed a band of frequencies for networking over a wide area, a prohibitively expensive challenge. For conventional mobile phones, AT&T designed a network of carefully spaced

hubs. Bell engineers Engel and Richard Frenkiel drew up a cellular map with a network of antenna links (Lemels, 2000). Their map caught the attention of FCC engineers in Washington, D.C., who wondered how their design would overcome certain technical obstacles. Instead of using broad coverage antennas, they turned to another option (Oehmk, 2000). They linked cellular signals to small reception spaces so that the same frequencies could be used via a network of non-adjacent "cells." Figure 22.2 shows cellular network architecture.

Smartphone Evolution

The transition to smartphone technology began with the vision of integrating personal computers with a mobile operating system to link voice, text, and Internet data with a cellular network. Personal Digital Assistants or PDAs were the first step along this route. Second came IBM's Simon Personal Communicator introduced in 1994. It used a touch screen to open an address book, a calculator, and a sketchpad. Simon was not a market success but a fleeting harbinger signaling the future of mobile devices (Sager, 2012).

By converging computer and cellular technology, mobile phones evolved. Nokia moved forward in 1996 with its 9000 model, integrating portable phone features with Hewlett-Packard's Personal Digital Assistant (PDA). Motorola's engineers in 2003 next pushed ahead with the launch of the MPx200. This venture with Microsoft used a Windows-based operating system for AT&T wireless services including email and instant messaging, eventually enabling other mobile phone applications or "apps" (Baguley, 2013; CHM, 2020).

Tiers of Smartphone Technology

At the basic level of its architecture, the smartphone is actually a stack of four tiers of technology (Grimmelman, 2011). The first tier is the wireless networks identified by abbreviations—CDMA, GSM, EDGE, EVDO, LTE—and numbered by generation (1G, 2G, 3G, 4G, and 5G). They link the mobile phone to the outside network through internal antennas. The second tier is mobile phone itself such as Apple's iPhone, Samsung's Galaxy, or Google's Pixel. Each comes with an operating system, which is the third tier of the architecture. Apple's iOS, Android, Windows, or another operating system acts as the intermediary between handset and applications. Smartphone apps are the fourth tier pressed into service for all sorts of activities: posting on social media, playing a game, calculating numbers, monitoring fitness, or just brewing coffee in the kitchen. A brief look at its history explains how this system evolved.

From Blackberry to Android

A Canadian firm, Research in Motion (RIM), held the cellular world in the palm of its hands in 2002 when it launched the Blackberry cellular phone. Blackberry's compact features for wireless email, mobile faxing, and its tiny keyboard with raised letters, numbers, and symbols attracted a huge following (Connors, 2012). It achieved its peak in 2008, when urban professionals transfixed by their mobile devices created "Blackberry jams" on city sidewalks and used the wry moniker "Crackberry" to warn of its addictive features. Apple and Android models entered the fray in 2007 and 2008 rendering their users even more mobile-phone dependent. Competition grew fast in the cellular market, and Blackberry soon fell behind its mobile rivals, dropping to quarterly lows that approached a billion dollars in losses (Austen, 2013). RIM and Blackberry fought back with innovations like the model Z10, but Apple's iPhone and the up-and-coming Androids eventually overtook Blackberry in the global market.

Apple's iPhone was an exemplar of inventive engineering at its 2007 debut, and despite its price tag, spectacular for that day ($499), customers lined up to pay the iPhone's pied piper, Steve Jobs. The chief executive already had confirmed brand loyalty through Apple laptops and iPods (Vogelstein, 2008), but it was iPhone's special features that captured the market. Its slide and press icons rendered tiny Blackberry keys or poking at one with a stylus as outmoded. Its operating system (iOS) used a web browser inside Apple's silver case of dark glass and shining icons. Millions were convinced they needed an iPhone and it rose to preeminence, but not for long. Google's open-source Linux-based Android operating system debuted in 2008 when two manufacturers introduced the same basic model with two different names: HTC Dream and TMobile G-1 (Priya, 2008; Shankland, 2008). The Android phones had a slide-out keyboard and posed little

threat to the iPhone's dominance. But soon these competitors offered capacitive touch screens, and Androids surged past the iPhone in sales.

Samsung's smartphone gained an edge through lower pricing and comparable features, but Apple felt certain it had stolen iPhone's patented innovations and took the evidence to court (Lee, 2011). Late in 2017, the U.S. Supreme Court let stand Apple's victory over Samsung for using the slide-to-unlock feature while Apple's other claims stood pending until 2018 when a California jury declared Samsung owed Apple $539 million in damages. Samsung paid $399 million but additional damages were required for infringing on Apple's design and utility patents. Both parties finally reached a settlement in 2018, sparing the two corporate rivals further litigation and court appeals (Kastrenakes, 2018).

Recent Developments

Leading mobile telephone manufacturers enhanced their products in 2019-20 with multiple camera lenses (front & rear), longer battery life, and folding screens, in addition to eliminating screen notches (Nield, 2019). In promotion and marketing, the currency of the realm became 5G for mobile telephony, and there was strong competition among the wireless carriers for who would be first.

The latest development in telephony is 5G technology, which improves data transmission with wider channels and faster links to servers (Segan, 2019). 5G can also serve more devices per square mile than 4G thanks to the advanced computing power of 5G networks. However, there was controversy over what constituted 5G wireless in advertising promotion.

In 2019, AT&T chose to rebrand its existing 4G LTE network as 5G Evolution (5GE). This marketing move did not sit well with some tech community observers who found AT&T's use of the term to be at best confusing because customers using AT&T's network might think they were actually getting actual 5G wireless (Statt, 2019). For mobile phones, terms like 5G evolutionary or just simply 5GE became part of the vocabulary in 2019, but area coverage for this wireless network was limited.

In April 2019, Verizon became the first wireless provider to offer 5G networking by using ultra-high-speed—yet low coverage area—mmWave towers (McCann, Moore, & Lumb, 2020). Other networks soon followed in Verizon's footsteps. While negotiating its merger with rival Sprint, T-Mobile launched its own 5G network, including not only 28 Ghz mmWave frequencies but also covering longer distances to reach sub-600 Mhz frequencies for suburban and rural communities.

5G & Smartphone Features

Smartphone makers showcase technical improvements each year to convince customers to upgrade to a larger screen, a water-resistant model, or new photographic views. Critics in 2019 noted how recent innovations seemed more like iterations on familiar themes (Eadiccio, 2019). The photographic capabilities of smartphones improved through widespread adoption of a multiple camera system combining three or four lenses to record wide-angle views up to 360-degrees, one with telephoto zoom capacity, another with a 3D image, or one to enhance selfies. Lenses would pop out from the smartphone when taking a picture and then withdraw back into hiding. Picture quality improvements came without adding pixels, but with brighter screens.

In design evolution, some smartphone manufacturers eliminated bezels and notch cutouts progressively moving their models to what one writer described as a "seamless slab of glass" (Eadiccio, 2019). New foldable phones were marketed when the Samsung Galaxy Fold, Huawei's Mate X, and Motorola's Razr all added this bendable feature. Of the foldable models, the Mate X from Huawei sold the largest screen: 8 inches at $2,400. Second came Samsung's Galaxy fold with a 7.3-inch screen retailing at around $2,000. The 6.2-inch screen on Motorola's foldable Razr, was the most affordable at $1,500. This revival of the flip phone harkened to an earlier era, but this one could bend in half and snapped shut like the original model.

Speed is a priority for marketing new mobile phones including faster refresh rates such as those advertised by Google's Pixel 4, which at 90Hz offered a higher refresh rate than average but not as fast as the Razr Phone 2 boasting 120Hz. Samsung's Galaxy S10 phone had a 5G option yet only about 30 cities in the United States were fully wired for it when the phone was introduced (Katrenakes, J., 2019). The Samsung Galaxy Note 10 was introduced with a 5G model, while TMobile launched its own OnePlus 7T Pro 5G McClaren capable of connecting to a 600 MHz low band network. Apple, for its part, refrained from rolling out its 5G model in 2019. Instead its iPhone 11 Pro showcased a triple-camera design and enhanced picture though XDR (External Data Representation) along with a four-hour longer battery life. Apple's iPhone 11 only would connect to 4G. There was speculation Apple was holding back on 5G until the iPhone 12 series was ready for market in 2020.

FCC & 5G

After the Federal Communications Commission adopted new licensing, service, and technical rules for deploying spectrum bands above 24 GHz in 2018 it sought comments for mobile services transmitting in the high frequency mmW bands enabling 5G wireless networking. While different standards for 5G were evolving, American consumers were sold on the future of "ultra-fast" connections for web browsing, video streams, AI sensors, monitors, and machines linking to the Internet of Things (IoT).

The FCC used the acronym FAST—Facilitating America's Superiority in Technology—to promote this next generation network. Its FAST campaign emphasized the promise of faster speeds, larger capacity, and lower latency all possible thanks to 5G. Still larger data streams required more electromagnetic spectrum, and so the government had to begin freeing up space in low, medium, and high bandwidths.

The FCC held auctions for wireless bandwidth in 24 and 28 GHz frequencies and then moved to even higher bands of 37, 39, and 47 GHz. A portion of the 2.5 GHz spectrum was designated for auction in 2019 and then a smaller portion of 3.5 GHz was put up for auction in 2020. C-band frequencies falling between 3.7-4.2 GHz frequencies got a nod of approval for 5G. The lower range at 600 MHz was already repurposed for mobile broadband carriers thanks to broadcast television stations giving up a portion of their spectrum in two FCC auctions held in 2018-19 (Pai, 2019).

The ongoing transition in cable from copper to fiber serves high-speed broadband networks. The number of 5G cell sites proliferated in 2019-20 as the U.S. wireless industry moved to make telephony networks denser. Wireless carriers began installing cells to carry millimeter waves causing an explosion in the number of antenne locations. The number of wireless small cells quadrupled from about 13,000 to almost 60,000. The signals reaching the home through internet transmission of text, graphics, audio, and video content required investments in broadband networks that grew by $3 billion in 2018 for the second year in a row, but the need for official approval meant some public entities controlling poles and sites wanted a share of the profits demanding anywhere from $50 to $5000 in fees per site (Pai, 2019).

The problem of robocalls caught the attention of both the FCC and Congress and its connection to the wireless industry was made explicit by CTIA's Nick Ludlum (2019), who noted how wireless providers were given the green light to automatically block illegal robocalls by default. President Trump signed into law the *Telephone Robocall Abuse Criminal Enforcement and Deterrence Act* (TRACED Act) on the last day of 2019 giving the FCC and federal and state-level law enforcement have criminal penalties to use against unwanted and illegal calls.

Current Status

The accelerating growth of mobile commerce (m commerce) has overcome the challenge of small smartphone screens growing to $230 billion revenues in the United States. What was commonplace in U.S. airports, sporting events, and concerts now appears to be more acceptable in Asia than in the U.S. for all sorts of daily transactions. The transfer of money by mobile phone known as the digital wallet began to catch on in 2014 when Apple introduced its Apple Pay service and activated one million credit cards within the first week of its launch (Rubin, 2015). Samsung Pay joined the competition soon thereafter. Android Pay used NFC (near field communication) technology, and the joint venture of AT&T, Verizon, and T-Mobile moved m-commerce forward in 2018 when Android Pay and Google Wallet combined forces to create the single system of Google Pay. The digital transformation of m-commerce through contactless payments using mobile handsets continued to grow, but not quite as fast as expected (Juniper Research, 2016).

Forbes.com (Kolchin, 2019) reported the reasons why the U.S. lagged behind Asia in general and China in particular when it came to mobile wallets. The platforms used either radio-frequency identification (RFID) or near-field communication (NFC) technology and encountered technical issues depending on if they were Apple or Android units. Also, younger demographics found the familiar use of credit cards with payback incentives to be more attractive. One success story was the Starbucks app, which scanned more than 23.4 million users based on its system of pre-paid accounts using a QR code.

M-commerce comprised nearly a quarter of all e-commerce sales in 2019. On "Cyber-Monday" online shoppers purchased $3 billion worth of merchandise using mobile phones (Perez & Lunden, 2019). Mobile phone sales drove 2019 online holiday spending, and technological advances made shopping by mobile phone easier making m-commerce more mainstream. Facebook, Twitter, and Instagram all include "buy now" buttons on their advertisements sending users directly to the product shown, ready for purchasing. Shoppers enter their payment method once, and then are ready to buy.

Wearable Technology

At first the smartphone wristwatch was a tough sell, but Apple and Samsung opened up a retail battlefront to grow their market shares. Their smartwatches caught on with versatile functions for time, temperature, apps, and phone calls through touch screens responding or voice commands (Shanklin, 2014). Industry observers liked the smartwatch for fitness monitoring, day planning, and online access but the challenges of excessive drain on the battery plus added billing costs stifled growth (Hartmans, 2017b).

Health concerns were at the forefront of Apple's latest smartwatch in 2020 offered in various metals or a ceramic case. It had a standard watch face on the normal screen, along with applications for an echocardiogram or heart rhythms. The Apple Watch Series 5 also featured the standard fitness tracking apps and a log for daily activity. Among the myriad of features it offered was Apple Pay, which could be scanned at a coffee kiosk or used to send money to friends. (For more on digital health, see Chapter 17.)

Competing against Apple with Android smartwatches was something of a challenge (Bohn, 2019). One LTE enabled smartwatch was promoted in 2019 to stream music, make calls, and see app notifications. Because of the exorbitant fees incurred by such a device, the less expensive watch, a non-LTE model was recommended. And some Android phone companies, like Samsung, developed watches capable of being paired with other phone brands.

Popular Apps

When it comes to the most popular smartphone applications, Google rose to dominance in 2017 with Google Play, Google Search, GMail, Google Maps, and YouTube (Hartmans, 2017a), but social networking apps like Facebook, Snapchat, and Instagram remain popular. Pandora's music app also attracts smartphone users. It is noteworthy that both voice minutes and traditional text messaging are declining uses for mobile phones. What is taking their place are communication apps such as Skype, FaceTime, Facebook messenger, and WhatsApp (FCC, 2017).

The number of mobile app downloads was rising steadily worldwide having surpassed 178 billion downloads in 2017. By 2022 there is projected to be 258 billion downloads of apps reflecting a 45% increase over five years' time. In dollar terms these downloads generated $935 billion in revenue from both app sales and advertising. Leading the gaming app publishers were King, Supercell, and Bandai Namco entertainment, while Google, Tinder, and Pandora dominated other apps based on revenue (Clement, 2019).

Global Frontiers

Worldwide, more people are fascinated by their mobile phones than ever before and that has led nations around the world to review the evolutionary development of cellular networks. At the end of 2019, 67% of the world's population subscribed to mobile telephony (GSMA, 2019). While nearly two-thirds of the world's population had entered the mobile-global age, Group Speciale Mobile Association (GSMA) predicted by 2020, there would be 5.7 billion mobile

subscribers on the planet ranging from 87% in Europe to 50% of Sub-Saharan Africa. Smartphone marketing shifted geographically to the developing countries of India and Pakistan, where analysts predict 310 million unique subscribers by 2020. Huawei, the mobile giant of China continued its surge in sales thanks to the booming Asian Pacific market. More than half of the mobile users in the world live in Asia, where two of the most populous countries, India and China, add millions of mobile subscribers each year.

The worldwide sale of smartphones had been growing until the fourth quarter of 2018, when the market stalled at just 0.1 percent growth over the fourth quarter of 2017. There was some sour news for Apple, which recorded its worst quarterly decline in iPhone sales (11.8%) since 2016, according to a global research firm (Gartner, 2019). The flagship smartphones from Samsung, Huawei, and Apple models continued to enjoy strong sales of their middle and lower priced models, but "slowing incremental innovation at the high end coupled with price increases, deterred replacement decisions for high-end smartphones," according to Gartner research (2019).

Factors to Watch

Both the number of wireless subscribers and data downloads by mobile carriers are increasing in the United States, according to the CTIA figures. There were 396 million mobile subscribers by the end of 2016, reflecting an uptick of about 5% from the previous year. And in terms of data, Americans used 82% more mobile data in 2018 than in 2017, according to the Cellular Telecommunications and Internet Association (Ludlum, 2019). The vast majority of Americans subscribe to one of four national service providers—Verizon, AT&T, TMobile, or Sprint, with 92% of the American population served by 3G networks. The problem is in rural areas, where only 55% of the population can access all four of the top competitors in broadband delivery. These four service providers account for 411 million connections, but the resellers and Mobile Virtual Network Operators (MVNOs) like TracFone Wireless and Google Fi reach certain segments of the market, where low-income consumers or those with lower data needs are able to buy a cellular plan.

Two-year contracts were once routinely accepted for U.S. wireless customers to buy smartphones. Those customer agreements meant they would pick a phone and pay a subsidized price. In return, a two-year agreement with the carrier came with a monthly service charge. Then the two-year contract began to disappear, and unlimited minutes took its place. TMobile disrupted the two-year format with its "Un-Carrier" campaign where customers bought smartphones in installments or at a single purchase price.

The choices for cellular agreements are defined as either pre-paid or post-paid. The prepaid option offers customers a fixed number of minutes, data, or texts for a particular period, while postpaid billing charges customers at the end of each month and tacks on extra charge for usage above a set figure. The move to unlimited data plans was viewed by some as misleading since for mobile carriers it meant consumers would have to pay more or deal with the "throttling" of data downloads that makes it a slower process. The price ranges for mobile plan were $30.00-100.00 varying considerably by company and plan.

American Frontiers

The Federal Communications Commission evaluates commercial mobile wireless services in the U.S. with an emphasis on competitive market conditions. None of the four largest providers, AT&T, Verizon, TMobile and Sprint, are viewed as dominant in their share of the mobile market, but all hope to meet the demand for wireless access and in the case of TMobile and Sprint intend to merge as one to make it happen. The Cellular Telecommunications and Internet Association (CTIA) collect data for use by public and private firms, and the CTIA report shows steady growth in mobile subscribers along with the spectacular surge in data usage noted above.

The wireless service companies are now focused on fifth generation (5G) cellular networking, In anticipation of 5G networks, the FCC made available 3250 MHz of Extremely High Frequency (EHF) bandwidth that transmits millimeter waves (mmW), which are subject to atmospheric attenuation and limited by distance. What the Next Generation Mobile Networks Alliance predicts is faster data speeds with digital access to the Internet of Things (IoT) through the use of wireless technology. The challenge in technology comes in assigning higher frequencies and wider bandwidth to meet the needs of 5G networks.

Gen Z

Wireless companies building new 5G networks promise super speed downloads for mobile phones. Innovations come at a price for the fastest-growing generation—Gen Z—who find faster downloads a necessity to make streaming video content easily accessible for such past times as binge-watching TV series. One study estimated this generation averaged more time on mobile devices than any other demographic and would continue to do so at an accelerating rate (Rasool, 2019).

Gen Z spends an estimated 4 hours, 15 minutes per day using mobile devices to access social media, text one another, view videos, or search for items online. Close to two-thirds (64%) of this demographic polled in a Snapchat survey (The Youth, 2019) reported they're constantly online, and 78% said mobile devices are their go-to technology. Gen Z members are said to be "always-on," although that's not just because 97% own a smartphone, but because over half (57%) feel insecure not having one ready in hand.

Compared to older demographics, Gen Z appears more inclined to view streaming video on mobile devices. They prefer uninterrupted entertainment, including music (69%), or movies (61%), along with retail shopping, checking news, and weather (Freier, 2019). A majority of Gen Z use ad blockers to stop the intrusion of commercial messages interfering with viewing their programs.

Gen Z also understands their online activity is monitored and archived by commercial and industry observers. They share concerns about personal privacy on their mobile devices as well as other invasions by security cameras and surveillance tech, including facial recognition and traffic sensors. Such worries cause some loss of enthusiasm for new technology—not just for privacy reasons—but also from fatigue for having too many choices for subscriptions, streaming, and the like.

Sustainability

Planned obsolescence is the term used to describe when a company has purposely designed a product to speed up its expiration date prompting users to purchase a new model. This profitable strategy was made infamous by the automobile industry at a time when Detroit was eager for customers to trade in their late model cars and buy newer models once parts began to falter. In computer technology, planned obsolescence has made way for new laptops, workstations, and mobile devices inspiring the government of at least one country, France, to adopt a law, making planned obsolescence a commercial crime. One of this law's targets—older smartphones began operating at a slower speed after a new line of handheld devices hit the market. In France, California, and Illinois courtrooms (Apple faces, 2017), lawyers sought to prove their case for "throttling" the performance of certain mobile phones.

More than 40 refined metals and elements are needed to make a phone, and according to Apple, manufacturing makes up 77% of the firm's carbon footprint. Lithium-ion batteries are only good for a specified number of cycles of charging and recharging, which means either battery replacement or buying a new model phone. Most smartphones are designed in a way that prevents users from replacing parts. Many phones do not offer replacement components, keeping users from maintaining a working older model.

By extending a smartphone's lifespan, owners could substantially decrease the environmental impact of its disposal, and avoid replacing existing handsets with new models without accounting for older technology. One solution is iFixit, a California firm that offers a reputable source of information for free smartphone repairs, while others advocate right-to-repair legislation to force mobile phone makers to provide owners independent access to replacement parts and technical repair information. Short of that solution smartphone users must do whatever is necessary to maintain battery life and protect their handset screen from harmful damage to keep from adding to the landfill sites where outmoded technology like mobile phones are buried.

This sustainability problem has been noted by the CTIA, which has initiated programs to support to extend the lifespan of smartphones and substantially decrease the environmental impact of its disposal. Under the rubric of the Wireless Industry Service Excellence (WISE) program, the industry is moving to put in place new standards for assessing devices and training technicians to repair them (CTIA, 2019).

Bibliography

Apple faces lawsuits over slowed iPhones. (2017, Dec. 22). *BBC.com*. https://www.bbc.com/news/technology-42455285.

AT&T. (2010a). Inventing the telephone. AT&T Corporate History. http://www.corp.att.com/history/inventing.html.

AT&T. (2010b). Milestones in AT&T history. AT&T Corporate History. http://www.corp.att.com/history/milestones.html.

Austen, I. (2013, Sept. 27). BlackBerry's Future in doubt, keyboard lovers bemoan their own. *The New York Times*. http://222.nytimes.com/2013/09/28/technology/blackberry-loses-nearly-1-billion-in -quarter.html.

Baguley, R. (2013, Aug. 1). The Gadget we Miss: The Nokia 9000 Communicator. *Medium*. https://medium.com/people-gadgets/the-gadget-we-miss-the-nokia-9000-communicator-ef8e8c7047ae.

Blumberg, S.J., & Luke, J.V. (2019, June). Wireless Substitution: Early Release of Estimates from the National Health Interview Survey, July-December 2018. *National Center for Health Statistics*. https://www.cdc.gov/nchs/data/nhis/earlyrelease/wireless201906.pdf.

Bohn, Dieter. (2019). The best smartwatch to buy if you have an Android phone. *The Verge*. https://www.theverge.com/2019/11/12/20959753/best-smartwatch-android-samsung-galaxy-google-wear-os-price-features.

Carter v. AT&T, 13 F.C.C. 2d 420 (1968); also *Hush-A-Phone Corp. v. U.S.* 238 F. 2d 266, 269 (D.C. Cir., 1956).

CHM Computer History Museum (2020). *Handhelds Branch Out.* https://www.computerhistory.org/revolution/mobile-computing/18/320.

Clement, J. (2019, Aug. 1). Total global mobile app revenues 2014-2023. *Statista.* https://www.statista.com/statistics/269025/worldwide-mobile-app-revenue-forecast/.

Connors, W. (2012, March 29). Can CEO revive blackberry? *Wall Street Journal* Marketplace, B1, B4.

CTIA (2019, June 20). 2019 Annual Survey Highlights. *CTIA Report.* https://www.ctia.org/news/2019-annual-survey-highlights.

Eadicicco, L. (2019, Dec. 28). These are all the ways smartphones got even better in 2019 from triple-lens cameras to foldable screens. *Business Insider.* https://www.businessinsider.com/best-smartphone-trends-apple-samsung-oneplus-google-2019-12#lower-prices-6.

Economist. (2009). Brain scan—father of the cell phone. *The Economist* website. http://www.economist.com/node/13725793?story_id=13725793.

Farley, T. (2005, March 4). Mobile telephone history. *Telektronnik.* http://www.privateline.com/wp-content/uploads/2016/01/TelenorPage_022-034.pdf.

FCC. (2017, Sept. 27). Twentieth Annual Report and Analysis of Competitive Market Conditions with Respect to Mobile Wireless. https//apps.fcc.gov/docs_public/attachmatch/DA-15-1487A1.pdf.

FCC. (2020). The FCC's 5g FAST Plan. https://www.fcc.gov/5G.

Freier, A. (2019, June 19). A look at Gen Z mobile behaviours—64% of mobile users are always connected. *Business of Apps.com.* https://www.businessofapps.com/news/a-look-at-gen-z-mobile-behaviours-64-of-mobile-users-are-always-connected/.

Gartner. (2019, Feb. 19). Gartner says global smartphone sales stalled in the fourth quarter of 2018. Apple suffered worst quarterly decline since first quarter of 2016. https://www.gartner.com/en/newsroom/press-releases/2019-02-21-gartnew-says-global-smartphone-sales-stalled-in-the-fourth-quarter.

Grimmelmann, J. (2011). Owning the stack: The legal war to control the smartphone platform. Ars technical. http://arstechnica.com/tech-policy/news/2011/09/owning-the-stack-the-legal-war-for-control-of-the-smartphone-platform.ars.

GSMA. (2019). The Mobile Economy 2019. https://www.gsmaintelligence.com/research/?file=b9a6e6202ee1d5f787cfebb95d3639c5.

Hartmans, A. (2017a, Aug. 29). These are the 10 most used smartphone apps. *Business Insider.* http://www.businessinsider.com/most-used-smartphone-apps-2017-8/#10-pandora-1.

Hartmans, A. (2017b, Dec. 23). I tried a lot of smartwatches in 2017—here are my top picks. *Business Insider.* http://www.businessinsider.com/best-smartwatches-gift-guide-2017-12/#the-lg-watch-style-is-a-clean-minimalist-smartwatch-1.

John, R. (1999). Theodore N. Vail and the civic origins of universal service. *Business and Economic History*, 28:2, 71-81.

Juniper Research (2016, March 1). Apple & Samsung drive NFC mobile payment users to nearly $150M globally this year. JuniperResearch.com. https://www.juniperresearch.com/press/press-releases/apple-samsung-drive-nfc-mobile-payment-users.

Kastrenakes, J. (2018, June 27). Apple and Samsung settle seven-year-long patent fight over copying the iPhone. *The Verge.* https://www.theverge.com/2018/6/27/17510908/apple-samsung-settle-patent-battle-over-copying-ipnone.

Kastrenakes, J. (2019, Feb. 21). Verizon will launch 5G in 30 cities this year. *The Verge* https://www.theverge.com/2019/2/21/18234755/verizon-5g-30-cities-2019-launch.

Kolchin, J. (2019, June 25). Why the U.S. still lags behind China in mobile wallet adoption. *Forbes Los Angeles Business Council Post.* https://www.forbes.com/sites/forbeslacouncil/2019/06/25/why-the-u-s-still-lags-behind-china-in-mobile-wallet-adoption/.

Lasar, M. (2008). Any lawful device: 40 years after the Carterfone decision. Ars technical. http://arstechnica.com/tech-policy/news/2008/06/carterfone-40-years.ars.

Lee, T.B. (2011). Yes, Google stole from Apple, and that's a good thing. Forbes.com. https://www.forbes.com/sites/timothylee/2011/10/25/yes-google-stole-from-apple-and-thats-a-good-thing/.

Lemels. (2000). LEMELS N-MIT Inventor of the Week Archive—Cellular Technology. http://web.mit.edu/invent/iow/freneng.html.

Livengood, D., Lin, J., & Vaishnav, C. (2006, May 16). Public Switched Telephone Networks: An Analysis of Emerging Networks. Engineering Systems Division, Massachusetts Institute of Technology. http://ocw.mit.edu/ courses/engineering-systems-division/esd-342-advanced-system-architecture-spring-2006/projects/report_pstn.pdf.

Ludlum, N. (2019, Dec. 21). 2019: A Banner Year for Wireless. *CTIA* Blog. https://www.ctia.org/news/2019-a-banner-year-for-wireless.

McCann, J., Moore, M., & Lumb, D. (2020). 5G: Everything you need to know. *Techradar.* https://www.techradar.com/news/what-is-5G-everything-you-need-to-know.

Nield, D. (2018, Nov. 27). Mobile Technology: 5 innovations to expect from smartphones in 2019. https://newatlas.com/5-innovations-2019-smartphones/57404/.

Oehmk, T. (2000). Cell phones ruin the Opera? Meet the culprit. *New York Times* - Technology. http://www.nytimes.com/2000/01/06/technology/cell-phones-ruin-the-opera-meet-the-culprit.html.

Pai, A. (2019, Aug. 28). Remarks of FCC Chairman Ajit Pai at the University of Mississippi Tech Summit. *FCC.gov.* https://www.fcc.gov/document/chairman-pai-remarks-university-mississippi-tech-summit.

Perez, S., & Lunden, I. (2019). Cyber Monday totalled $9.4B in US online sales, smartphones accounted for a record $3B. *TechCrunch.* https://techcrunch.com/2019/12/03/cyber-monday-on-track-to-deliver-9-4b-in-u-s-online-sales/.

Priya, G. (2008, Sept. 22). T-Mobile G1, aka First 'Googlephone,' carries high expectations. *Wired.* https://www.wired.com/2008/09/since-apple-lau/.

Rasool, A. (2019, Aug. 28). Gen Z use smartphones as a constant source of entertainment, says a study. Digital Information World. https://www.digitalinformationworld.com/2019/08/gen-z-use-smartphones-as-constant-entertainment.html.

Rasool, N. (Aug. 28, 2019). Gen Z use smartphones as a constant source of entertainment, says a study. *Digital Information World.com.* https://www.digitalinformationworld.com/2019/08/gen-z-use-smartphones-as-constant-entertainment.html.

Rubin, B.F. (2015, May 28). Google's Android pay to duke it out with Apple Pay. CNET. http://www.cnet.com/news/google-refreshes-mobile-payments-effort-with-android-pay/.

Sager, I. (2012, June 29). Before iPhone and Android came Simon, the First Smartphone. *Bloomberg.com.* https://www.bloomberg.com/news/articles/2012-06-29/before-iphone-and-android-came-simon-the-first-smartphone.

Segan, S. (2020, Jan. 20). What is 5G? *PCMag.com.* https//www.pcmag.com/news/what-is-5g.

Shankland, S. (2008, Oct. 21). Google's open-source Android now actually open. *Cnet.* https://www.cnet.com/news/googles-open-source-android-now-actually-open/.

Shanklin, W. (2014, Sept. 24). Apple Watch vs. Samsung Gear S. *Gizmat.* http://www.gizmag.com/apple-watch-vs-samsung-gear-s/33960/.

Statt, N. (2019, April 22). AT&T's 5G E marketing ploy is turning out to be a disaster. *The Verge.* https://www.theverge.com/2019/4/22/18511741/att-5g-e-marketing-ploy-disaster-misleading-claims-lawsuit-confusion.

The Youth of the Nations: Global Trends Among Gen Z: An analysis conducted by GlobalWebIndex and Snap Inc. (2019, June 13). https://forbusiness.snapchat.com/blog/the-youth-of-the-nations-global-trends-among-gen-z.

Thierer, A.D. (1994). Unnatural Monopoly: Critical Moments in the Development of the Bell System Monopoly. *The Cato Journal* 14:2.

Time. (2007). Best Inventions of 2007. Best Inventors—Martin Cooper - 1926. *Time Specials.* http://www.time.com/time/specials/2007/article/0,28804,1677329_1677708_1677825,00.html.

Viswanathan, T. (1992). Telecommunication Switching Systems and Networks. New Delhi: Prentice Hall of India. http://www.certifiedeasy.com/aa.php?isbn=ISBN:8120307135&name=Telecommunication_Switching_Systems_and_Networks.

Vogelstein, F. (2008). The untold story: how the iPhone blew up the wireless industry. *Wired Magazine* 16:02—Gadgets—Wireless. http://www.wired.com/gadgets/wireless/magazine/16-02/ff_iphone?currentPage=1.

21

The Internet

Hsuan Yuan Huang, Ph.D.*

Abstract

The internet is evolving. Started as a military communication network development project, the internet has evolved to become a global information and communication tool that we can't live without. In recent decades, we have increasingly become used to the idea of living our life on the internet, so to speak, and having all of our information stored in "the cloud" with cloud computing. With our insatiable appetite for data consumption, cloud computing services aim to meet that explosion of data consumption at whatever cost. This chapter examines cloud computing services and internet companies' efforts towards building a green internet. It then discusses internet-based technology and the role it plays in facilitating socially supportive communication for individuals with a health concern and social listening/monitoring for online marketers to develop better products and services to deliver better brand experiences. This chapter concludes with an examination of the shift to increasingly mobile access and its implication for online shopping, information seeking, social communication, and emerging regulation.

Introduction

The internet serves as the nervous system of the modern global economy, supporting our financial, transportation, and communication infrastructures, and whose task it is to support the world's insatiable consumption of data, responding to every click or tap on the web. The world has come to rely on the internet—the "networks of networks"—for work, business, health, education, social interaction, and entertainment.

Started as a U.S. Department of Defense communication system development project during the Cold War, the internet now commands a whopping 4 billion users worldwide, and its popularity is expected to continue to grow. Consisting of a global networks of computer systems linked by data lines and wireless systems, the internet is used by a community of people

* Assistant Professor, School of Journalism and Graphic Communication, Florida A&M University (Tallahassee, Florida).

who share access to a globally linked system of information (DeFleur and Dennis, 2002).

The impact of the internet is more far-reaching than its creators imagined. The reach of the internet has not only made Marshall McLuhan's vision of a "Global Village" a reality, it has also fundamentally changed the way we live our lives, do business, and communicate with one another. The popularity of online shopping worldwide drove retailers that were not quick to adapt, out of business. Paying bills and getting a checking account balance is convenient through ebanking and mobile devices. Health related websites and online health forums are a significant source of health information. U.S. presidents' social media have millions of followers, and Twitter has become the most popular social media for sharing political opinions and advocating social movements.

The internet is constantly changing. Among recent innovations that marked the evolution of the internet are social media, cloud technology, and mobile connectivity. In this chapter, we begin with a review of the internet's origins and rise to popularity, followed by the recent developments on social media monitoring, social networking supportive communication, and cloud computing. We then conclude the chapter with continuing look at emerging issues including the passage of consumer data privacy laws and the rise in mobile connectivity.

Background

The internet was not invented by a single person—instead, it was the work of dozens of pioneers who developed new features and technologies that eventually merged to become the global information network we know today. The internet began as a military computer network in 1969, called the Advanced Research Projects Agency network. ARPAnet was a computer-communications development spurred by the United States Department of Defense's desire to maintain a communication system that would survive nuclear attacks by Russia. In the Cold War, the military feared that even one missile attack could destroy the network of lines and wires that the nation's communication system relied upon. The idea of an "Intergalactic Network" of computers that could talk to one another was then introduced by MIT scientist J.C.R. Licklider as a solution to this problem. Shortly after that, scientist Paul Baran developed packet switching as a faster and more secure way of sending information from one computer to another. Packet switching breaks large chucks of data down into manageable blocks—or packets—which allows each packet to travel independently over any available circuit to the final destination where the data will be reassembled.

APRAnet used packet switching—which became one of the major building blocks to the internet—to allow multiple computers to communicate on a single network. As more packet-switching enabled computer networks emerged over time, it became difficult to integrate them into one single worldwide "internet." In the late 1970s, computer scientist Vinton Cerf developed "Transmission Control Protocol" and internet Protocol (TCP/IP) as a way for computers on the world's computer networks to communicate with one another. From there, researchers began to assemble the network of networks that became the modern internet.

Rogers' (1983) Diffusion of Innovation theory notes that innovations that are easy to use are more readily adopted. Because of that, the early version of the internet that ran on text-based commands was not well accepted. The public needed a more user-friendly way to access the internet, receive the information it contained, and communicate to others that didn't involve learning text-based commands. In 1990, Tim Berners-Lee invented the World Wide Web, a graphical interface for accessing the internet using ordinary language and hyperlinks. The user-friendly nature of the World Wide Web enabled the adoption of the internet, and the innovation started to take off. Many credited the creation of the World Wide Web as the tool that help popularized the modern internet.

Today, when we surf the internet, a complex series of processes deliver content. Let's consider a simple Google search. Google.com is Google's domain name, which is the name we enter into a browser. Behind it is an Internet Protocol (IP) address, the actual location of the server that has the content we need. In the Domain Name Systems (DNS), .com is the top-level domain, and the name of the organization, for example, Google, is the second-level domain.

The Internet Corporation for Assigned Names and Numbers, or ICANN, an organization responsible for assigning domain names and address to specific websites and servers, maintains two sets of top-level domain names: generic top-level domain names that we have been familiar with such as .com, .org, .net, .edu, .gov, .mil and country codes such as .cn for China, .ca for Canada, and .ru for Russia. ICANN's board of directors reviews requests submitted by numerous organizations for new top-level domains. In early 2000, it approved a raft of new top-level domains, many of which tied a top-level domain to a particular topic such as .museum or .plumbing, including the controversial .xxx top-level domain for adult entertainment sites. ICANN also announced that it would greatly increase the number of top-level domains by allowing nearly any new top-level domain name in any language (Hall, 2020)

IP addresses are unique numeric identifiers for all digital devices that are connected to the internet ranging from smartphones and cameras to printers and laptops (ICANN, 2014). The simplest address system, IPv4, was developed when the APRAnet was invented. With rapid internet expansion in the past decades, the four billion IPv4 addresses have been almost all allocated to the Internet Service Providers (ISP) and users. Currently, there are 4.8 billion internet users worldwide (Internet World stats, 2020).

To accommodate the explosion of internet-connected devices, a more complicated system known as IPv6 was developed. Compared to IPv4's 32-bit address, IPv6 has a 128-bit address, which creates 340 undecillion (340 x 10^36) addresses, larger than the 300 billion (300x 10^9) stars in the Milky Way. That means there are a trillion trillion more IPv6 address than stars in the galaxy. With this many addresses, even considering individual consumer's increasing demand for multiple digital device addresses, IPv6 should still last longer than the 30 years that IPv4 has lasted.

IPv4 and IPv6 can exist together on the same network, known as dual stack. To illustrate this, one can look up your local internet host's IP addresses on WhatIsMyIP.com, it will give you the IPv4 and IPv6 addresses of the local internet host/internet service provider, the location, and the name of your ISP. You can also search "My IP Information," and it will show you the host's geolocation and host info (see Figure 21.2).

Figure 21.1

Most Popular Top-Level Domains Worldwide In 2019

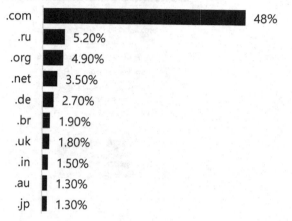

.ru = Russia, de = Germany. jp = Japan, .br = Brazil

Source: Statista

An IPv4 address usually has four numeric segments separated by dots, while IPv6 are made up of up to 8 segments of hexadecimals (a system of 16 numbers using numbers of zero to 9 and A-F to represent ten to fifteen). An IPv4 address looks like this: 68.32.24.102 while IPv6 could look like this: 2401:4c0:28f0:1d22:5949. However, in cases where IPV6 is not detected, it simply means that it was assigned an IPV4 address only.

In the example above, the internet service provider (ISP) is Comcast Cable, a cable high-speed internet access provider. When you have a computer, modem and a router for networking, you will still need to have service with an ISP to enable access to the internet. Cable broadband services offered by cable companies such as Comcast and Charter Cable have been growing while DSL (Digital Subscriber Line) offered by phone companies such as AT&T and Verizon have been on the decline because phone companies found selling cellular services with smartphone internet capabilities much more lucrative.

Figure 21.2

Dual Stack of IPv4 & IPv6 addresses

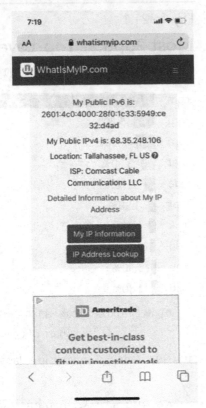

Source: H. Huang

In 2013, Google launched Google Fiber, an ultra-fast fiber optic service, for internet and TV service in Kansas City, Mo. It boasts up to 1 Gigabit (1000 mbps) speed for uploading and downloading, capacity to stream 20 devices at a time, unlimited data and 1 TB of free Cloud storage. With the service, you could download a full-length movie in just about a minute (Hobson, 2017).

Since its invention, the development of internet worldwide has generally been linked to factors such as national security, political freedom, urbanization, wealth of a country, stability of government, and telecommunication infrastructure (Hao & Kay, 2004). In 2018 and 2019, online data privacy concern compelled the Chief Executive of tech giant Facebook, Mark Zuckerberg, to testify twice in front of the U.S. Congress about Facebook's failure to protect 87 million Facebook users' personal data including names, "likes" and other personal information, allowing these data to fall into the hands of Cambridge Analytica, a political analytics firm, to improperly use Facebook users' data in targeted advertising aiming to influence the 2016 election (Romm, 2018). Lawmakers threatened to regulate the company and its tech industry peers as they cast doubt on the social giant's ability to fix the troubles on its own given this was the second time that Facebook was involved in a privacy mishap (Romm, 2018).

In Asia, Singapore is one of the exceptional countries where country wealth, stability of government, urbanization, and communication infrastructure, all contributed to the high internet penetration, despite its strict control on political freedom. In China, on the other hand, political freedom and national security play a heavy hand in its internet development. Over the past two decades, China has successfully built "a digital wall between itself and the rest of the world, a one-way barrier designed to keep out foreign companies like Facebook and Google" (Yuan, 2019). While the one-fifth of the world's internet users who live in China can't use Facebook to stay in touch with their friends, subscribe to a YouTube channel, or Google something on the internet, the Chinese government allows them to search information on China's own version of internet search engine, Baidu, watch Chinese videos on Youku, and use popular microblogging site WeChat to connect with their friends and family.

Although China maintains a closed internet, it allows its technology companies to compete with foreign rivals and expand globally. In 2019, the U.S. government decided to restrict access to American technology to Huawei, a Chinese telecommunication giant. The Trump administration accused Huawei of stealing technology, calling the firm an espionage threat (Zhong, 2019). After the announcement, Google limited the software services it provided to power Huawei's smartphones. Following the birth of the internet in the U.S. many assumed that the U.S. Would maintain a role defining the rules that governed the medium. But the emergence of other web giants, including China, have proven just how wrong this assumption was (Prodromou, 2020).

As nations apply rules and regulations that limit the internet—some protecting privacy and freedom, and some invading privacy and freedom—the internet may fragment. Former Google chief executive, Eric Schmidt, predicted that in the next 10 to 15 years, the internet will split into two—one internet led by China, and one internet led by the U.S.—with closed and open access models (Wong, 2018). China, with its "Great Firewall," resisting the western idea of a democratic, free-market model, instead created its own versions of capitalism and web properties, like Alibaba and Weibo, with closed access. While the U.S. internet model is known for its libertarian strand, with light regulation and rules, Europe adopted a more interventionist approach with heavier regulatory oversight around issues like antitrust, privacy and taxation. For example, the Europe's ever-tightened regulation on data privacy, *General Data Protection Regulation*, known commonly as GDPR, poises to govern internet data privacy and vows to give European residents greater privacy protection by mandating that

consumers have to give permission for a company to collect their data. The European model is likely to be the third internet, "which leverages the innovations and developments of America's internet and applies a greater oversight and regulatory regime" (Prodroumou, 2020). "All signs point to a future with three internets," notes an opinion essay written by the New York Times editorial board (Wong, 2018). On October 1, 2016, the U.S. Department of Commerce National Telecommunications and Information Administration (NTIA) handed control of the domain name system to a multinational non-profit organization, Internet Corporation for Assigned Names and Numbers (ICANN), whose multiple stakeholders include technical experts, business, academics, civil society, and government representatives.

ICANN oversees the address book of the internet; how those addresses are doled out, and how they are named, thanks to a contract issued by the U.S. government that expired September, 2015 (Pew, 2014). Critics such as Senator Ted Cruz argued that the handoff will "cede" American "control" of the internet to authoritarian countries that will impose regulations that change the basic open character of the internet and eventually censor internet content throughout the world. ICANN Board Chair Stephen D. Crocker said such fears were uninformed noting that ICANN is a technical organization that doesn't regulate content on the internet. It also doesn't regulate bad actors on internet or regulate internet access (the internet access is provided by the ISP). He further emphasized that a governance model defined by the inclusion of all voices, including business, academics, technical experts, civil society, governments, and many others "is the best way to assure that the internet of tomorrow remains as free, open and accessible as the internet of today" (Moyer, 2015).

ICANN's main function, as discussed earlier in this chapter, is coordinating domain name systems to ensure that when users try to access a website or send an email, they end up in the right place. The internet would break down without this kind of coordination.

Recent Developments

Computer-Mediated Supportive Communication

The internet has become an increasingly common place for the exchange of supportive communication about health issues. Considerable research noted that online social support or computer-mediated social support (CMSS) is available for individuals who share health concerns, such as weight loss (Hwang et al., 2010), breast cancer (Gustafson et al., 2001), Huntington's disease (Coulson et al., 2007), AIDS (Mo & Coulson, 2010b) or asthma (Wise et al., 2007). Members of these networks can seek, provide and receive various forms of social support such as emotional support, medical information, exercise partners, problem-solving tips, and coping strategies. One of the strongest assets of internet-based support groups is anonymity which helps people feel more comfortable with openly discussing their problems and concerns particularly with sensitive topics (Hwang et al., 2010). (For more on health applications of the internet, see Chapter 17.)

In recent years, social support groups are evolving from computer-mediated social groups to social networking sites. Facebook is the world's largest social networking site with 2.5 million active monthly users worldwide, and it is widely used by seven in ten adults in the U.S. (Pew, 2019a). Despite recent debate on whether it is tasteful or appropriate to discuss one's health on social media (Fox, 2014), social networking sites (mainly Facebook) offer a unique environment for the exchange of social support and has garnered increasing popularity in recent decades (Ellison et al., 2007; Frison et al., 2015; Lee et al., 2018; McCloskey et al., 2015).

Cloud Computing

The term "cloud computing" captures the mythical image of the internet as a "virtual" world where lives are lived and all types of memory are stored in "the cloud."

Cloud computing is the delivery of on-demand computing service or information technology—including servers, storage, database, networking, software, analytics, and intelligence—over the internet ("the cloud"), typically on a pay-as-you-go basis. Microsoft Azure, the current second market leader, gave this example: "instead of storing personal documents and photos on your personal computer's hard drive, most people now store them online: that is cloud computing" (Get to know Azure, n.d). In reality, cloud computing is not that simple.

Typically, web/online data is stored at actually physical facilities called data centers. In the data centers, there are servers, storage equipment, backups, power supplies, and high-capacity cooling systems. Major internet companies, including cloud computing services, social media, and video streaming, typically require several data centers to store all their web data and run computations. Companies like Amazon Web Services have long sited their data centers in undisclosed location due to data privacy concerns. Furthermore, those data centers consume vast amount of energy, with staggering electricity consumption (Glanz, 2012) (see the Sustainability sidebar for details on data centers energy consumption). Worldwide, digital warehouses use about 30 billion watts of electricity—the rough equivalent of energy produced by 30 nuclear power plants.

Cloud computing encompasses a wide range of services from consumer services such as Gmail and the cloud back-up of your smartphone photos to business services that allow corporations to store data and run applications in the cloud. Even software vendors such as Adobe are increasingly offering application services over the cloud on subscription basis rather than standalone products. With fast growing demand, the cloud computing industry earned total revenue of $89 billion since the third quarter of 2018, up 37% annually (Canalyst, 2019). The global cloud computing market is estimated to reach $285 billion by the end of 2025 (Bloomberg, 2019).

The top three cloud computing services providers in 2019 were Amazon Web Services (AWS), Microsoft Azure, and Google Cloud. These services typically provide three main types of computing services: infrastructure as a service (IaaS), platform as a service (PaaS), software as a service (SaaS) (Ranger, 2018).

Infrastructure as service is the most basic category of cloud computing service where computing infrastructures such as servers, storages, data centers, and networking are provided and managed over the internet. Companies save the upfront cost and the complexity of buying and managing their own physical servers and other datacenter infrastructure, and simply pay for what they use, when they use it. In other words, individual companies run their own platforms and applications within the infrastructure environment provided by the cloud services.

Platform as a service is an offering of cloud computing where the provider gives the users access to a cloud environment in which to develop, manage, and host applications. Individual users have access to a range of tools through the platform including development tools, database management systems and business analytics to support testing, development, deployment and updating of applications. The provider is responsible for the underlying infrastructure, security, operating systems, and backups (Jones, 2019).

Software as a service is a method for delivering software application over the internet. Instead of installing software applications on the user's local drives, consumers access cloud-based applications through the web on their computer, tablet, or phone—typically on a subscription basis. The service providers are responsible for hosting, managing, and upgrading the software. Some common examples are email, calendaring, and office tools (such as Microsoft Office 365).

Artificial intelligence, the internet of things (IoT) and analytics are the prime technologies for cloud vendors. Take AI for an example—with the increasing reach and popularity of artificial intelligence and machine learning, cloud computing has emerged as a system to enable offering AI as a service. AI tools are embedded in cloud applications to power a self-managing cloud, improve data management, increase the SaaS value and optimize cloud service. (For more on AI, see Chapter 25.) Furthermore, a mutli-cloud approach is becoming a common way for cloud deployment. A recent survey suggested that enterprises often pair up Amazon Web Services and Microsoft Azure, or AWS and Google Cloud (Dignan, 2019).

Social Media Monitoring

Over the past few years, internet marketing has shifted its focus towards user driven and content driven platforms such as blogs, social networks, and video-sharing sites. Similar to online information exchange and the provision of emotional and network support for individuals who share a health concern, internet-based technologies such as social media, review sites, blogs, and discussion forums are popular avenues for consumers to share their experiences, voice their opinions, and provide recommendation to all users in these social conversations. For brands, their task now is how to turn data from these conversations into business intelligence to inform them in smart decision-making.

Social media monitoring is gaining popularity because it allows "listening in" on on-going conversations on a daily basis and gathering social intelligence through consumer feedback to a product and service, measuring public opinion towards an unfolding brand crisis, or even voter's sentiments towards political candidates (Liu, 2012).

Sentiment analysis is a subset of social media monitoring. It analyzes people's opinions, sentiments, evaluations, attitudes, and emotions from written texts found on the web and social media through natural language processing (Liu, 2012). For example, Uber used sentiment analysis to discover whether users like the new features of their app. With daily monitoring of app user feedback, Uber was able to understand how users feel about the new changes they were implementing, says Krzysiek Randoszekski, Marketing Lead for central and eastern Europe at Uber. Faced with an image crisis, United Airlines analyzed the #UnitedAirlines hashtag to determine the public sentiment towards its handling of the 2017 incident that ended with a passenger bloodied and dragged out of the flight. The analysis confirmed that almost 70% of social media users expressed negative feelings towards the brand after the incident.

Current Status

In 2019, the number of people using the internet surged to almost 4.5 billion worldwide (Clement, 2020), a year-on-year increase of 298 million, or 7% (Brown, 2020), with an estimated one million people coming online for the first time each day (Kemp, 2019).

This means 58% of the world population is connected online (Clement, 2020), up from 49% in 2017 (Rainie &Anderson, 2017). In 2019, 90% of American adults used the internet, which is almost a 10% increase from 2018.

According to a Pew Research Center survey, 81% of Americans go online on a daily basis. Among them 28% said they are online almost constantly. Younger adults are at the forefront of the constantly connected: Roughly half of 18- 29-year-olds (48%) say they go online almost constantly, a 9% increase from the previous year. Another 46% of young adults go online multiple times per day. By comparison, just 7% of those 65 and older go online almost constantly, and 35% of that age group go online multiple times per day.

The share of constantly online Americans ages 50 to 64 has risen from 12% in 2015 to 19% in 2019. College-educated adults, Hispanic adults, adults who live in higher-income households, and nonrural residents, are also among those who are constantly connected.

For example, 36% of adults with a college education go online almost constantly (and 93% go online daily), compared with 23% of adults with a high school education or less. While 34% of Hispanic adults report using the internet almost constantly, only 25% of black adults and 26% of white adults reported doing so.

Income level and urban residence also plays a role in constant internet connectivity. While 34% of adults with an annual household income of $75,000 or more use the internet almost constantly, just 23% of those living in households earning less than $30,000 are constantly online. Adults who live in urban and suburban areas are also more likely to be connected online almost constantly than those who live in rural areas (Perrin & Kumar, 2019)

Worldwide, the internet has the highest penetration in the North America (95%), followed by Europe (87%), the Middle East (69%), Latin American (68.9%), Oceania/Australia (67%), Asia (54%), and Africa (39%). As the number of world internet users continue to grow, over half of the internet user population is in Asia (50.3%) with a total of 23 billion users, Europe is second (15.9%) with a total of 7 billion users, and Africa, third (11.5%) with a total of 5 billion users (Internet World Stats, 2020). Country-wise, China has the highest number of internet users with 854 million active users online, India is second with 560 million users, and the U.S., third with 292 million users (Clement, 2020).

Figure 21.3

Percent of Americans Who are Constantly Online

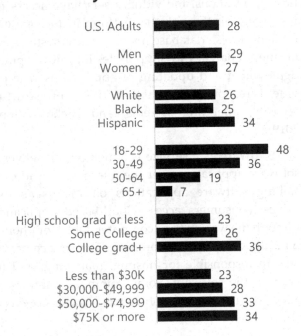

Note: White and Black include only non-Hispanics. Hispanics are of any race.

Source: Pew Research Center

Looking at time spent online, the average internet user will spend 6 hours and 43 minutes per day online in 2020—more than 100 days in total, that equals a collective 1.25 billion years globally.

Regarding online shopping, 74% of internet users aged 16 to 64 reported having purchased a product online in the past month, and over half (52%) via a mobile phone as the phone is now the top device for online shopping. A whopping $3 trillion was spent on online shopping in 2019, with shoppers spending an average of almost $500 each on consumer goods alone—a yearly increase of 9%.

When communicating online, the average American spends 2 hours and 3 minutes a day on social media—less than the global average of 2 hours 24 minutes, accounting for more than one-third of our total internet time. Globally, 70% of the population is now on social media, with 2.5 billion active users on Facebook, followed by YouTube with 2 billion active users. In the U.S., almost four in five report using YouTube each month, and three out of four report using Facebook. Interestingly, the most active social media users are in the Philippines, who average 3 hours and 53 minutes on social media every day. Overall, women only account for 45% of global social media users (We Are Social, 2020).

When using the internet for information, surprisingly, the U.S. only ranks seventh in countries concerned about misinformation despite concerns about "fake news," with two out of three (67%) expressing concern about what is real or fake on the internet. Similar percentage of U.S. adults (66%) are concerned about online personal data privacy. In addition, the Pew Internet Project survey estimated approximately 80% of internet users in the United States searched online for health information (Pew, 2010).

With the overall shift to mobile technology, 35% of American adults rely mostly on their smartphone to access the internet, among them, 10% are smartphone-only internet users. The widespread adoption of the smartphone is also increasingly cited as the reason for not having high-speed internet at home, up from 27% in 2015 to 47% in 2019 (Anderson, 2019).

This trend toward mobile internet is especially notable among younger users. Younger groups use their mobile phone more frequently than their PC, laptop, or tablet for online activities, while the older groups spent more time online on their PC, laptop, or tablet (Clement, 2019). See Figure 21.4.

In the US, although there was an increase in internet penetration across the population, it's worth noting that adoption gaps still remain considering factors such as age, education and income. Nearly all young adults 18-29 (99%) use the internet, closely followed by adults 30-49 (97%). The adoption rates drop to 88% for adults 50-64 and 73% for age 65 and above. When it comes to education, nearly all college graduates (98%) use the internet, while 75% of internet users reported having only a high school education. When comparing different income levels, high-income American households are considerably more likely to use the internet. The percentage of American households reported making $75,000 and above using the internet reaches close to saturation at 98%; 97% of households with income of $50,000-$74999 use the internet; 93% of households with income of $30,000-$49,999 use the internet; 81% of households with income less than $30,000 reported using the internet (Pew, 2019b).

Figure 21.4

Average Internet Usage by Age & Device

Average duration of daily internet usage worldwide as of first quarter 2019, by age group and device

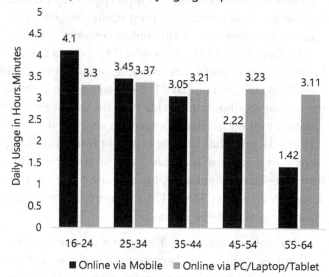

Source: Clement, 2019

In the third quarter of 2019, the global cloud market grew 37% to reach $27.5 billion, with AWS accounting for 32% of market share, followed by 17% for Microsoft Azure, and 7% for Google Cloud (Canalyst, 2019). AWS's dominance in cloud services provides substantial profit to its parent company representing roughly 13% of Amazon sales (Levy, 2019). While AWS still holds an impressive $2.3 billion quarterly growth equal to the next three competitors' combined increase, AWS actually experienced a slow growth of

35%, and its market share dropped below 33%. Microsoft Azure outpaced the market, taking home 59% growth and a 3% share increase, while Google Cloud accelerated and scored a 69% growth and a 1.5% market share compared to 2018. The cloud market was estimated to reach $100 billion in year 2019.

Factors to Watch

The *California Consumer Privacy Act* (CCPA), the toughest data privacy law in the U.S., went into effect January 1, 2020. This law requires online businesses ranging from tech giants to ordinary retailers to tell consumers what data they have collected about them, and if asked, to delete it, giving consumers a powerful tool to control their online privacy. Given there is no similar federal mandate, this Act will set the national standard for nearly two dozen other states that are copying this Act.

This is why consumers saw company privacy policies update announcements popping up in their email inbox or when they are visiting online retailers' websites particularly from businesses who do business in California (Myrow, 2019). For example, soon after CCPA went into effect on January 1, 2020, when you entered the online retailer Ann Taylor website, you were first directed to their Cookie Consent page where it specifically tells you that you have the option of making third party cookies "inactive," and if you are a California resident, you can opt out of the "sale" of personal information collected by the company by asking your personal data be deleted.

For online marketers, the economic effect of this law can be enormous. Targeted advertising will become less effective if consumers demand that their personal data be deleted. Companies like Ann Taylor will face potential sales loss if they can no longer market their products with messages tailored to consumer's personal preference or recent purchase history. Google, whose business model is based on selling targeted ads, will also likely lose revenue since generic ads commands far lower prices than targeted ads (Roberts, 2019)

Americans' concern over data privacy is evident. Auxier et al. (2019) found that the majority of Americans believe they have little control over their personal data collected by companies and the government (81% and 84%). Although many marketers claim that data driven products and services save consumers time and money, and can even lead to better health and well-being, most Americans do not believe it; 81% of Americans believe the risks outweigh benefits. It will be interesting to see if California consumers will actually assert their new right to protect their online privacy. In 2018, Europe implemented a sweeping privacy law called *General Data Protection Regulation* (GDPR), but few consumers in the European Union have taken advantage of the law (Roberts, 2019).

The Rise in Mobile connectivity

Today's Americans go online to connect with their friends and family, shop, get news search for information, apply for jobs, and get politically involved. With 81% of American now owning a smartphone up from 35% in 2011, using a smartphone as the primary method to go online as compared the stationary use of the internet has been on steady rise over the decade (Anderson, 2019).

Roughly one in five adults depend on their smartphone as the primary way of going online. This dependency is particularly high among young adults, non-white and lower-income Americans who don't have traditional broadband service at home, or have few options for online access other than their cellphones (Smith, 2015). Further data confirmed the U.S. internet usage shift to mobile access. In 2019, 54% of web traffic in the U.S. originated from mobile devices, up 33% from last year. In 2023, almost 85% of U.S. adults will access the internet via mobile phone, up from 80% in 2018 (Brown, 2020).

This trend as part of a broader shift towards mobile technology has changed how people do everything via the internet. For example, more Americans prefer using a mobile device (57%) over a desktop computer (30%) to get news (Walker, 2019). The increasing popularity of using mobile apps for a wide range of activities from paying credit card or utility bills, streaming videos, to tracking fitness results, coupled with the invention of internet of things (IoT) further contributes to the rise of mobile connectivity.

Gen Z

In 2020, the oldest of Generation Z is 23, while the majority are still in their teens and younger. Nearly all U.S. teens 13-17 (95%) reported having access to a smartphone, and almost half (45%) of those teens are on the internet "almost constantly" (Schaeffer, 2019). For those teens who are on the internet "almost constantly," nine-in-ten are reported social media users.

With the proliferation of smartphones and the rise of social media, teens who are active online have increasingly become victims of cyberbullying or harassment, and these behaviors represent a major problem for people in this age group. Teens (69%) who have experienced bullying or harassments online noted name-calling and rumor-spreading the most common forms of harassments (Anderson, 2018). Instagram is the number one social media site where teens report experiencing cyberbullying, with 42% experiencing harassment on the platform (Ditch the Label, 2017). Teens who experience cyberbullying are at a greater risk than those who are nonvictims for both self-harm and suicidal behaviors (John et.al, 2018). They were also more depressed, irritable and angry, and more likely to experience disassociation (Anunobi, 2019)

Despite the high mental and behavior health risks for cyberbullied teens, most teens think that their teachers, social media companies and politicians have failed to protect them. Parents are the only key group that has taken considerable measures in protecting their kids against online bullying (Anderson, 2018).

Sustainability

While many people may consider information and communication data consumption a "God-given" right, whether it is accessing YouTube videos, streaming movies on Netflix, checking bank balances through websites, paying bills through mobile apps, sending Google email with file attachments, purchasing holiday gifts on Amazon, posting on social media, or reading newspapers online, fewer consumers are aware of the vast amount of energy required to manufacture and power our digital devices, data centers, and related infrastructures that support the overall explosion of digital information generated by 4 billion users worldwide. When the most-watched Spanish-language music video for "Despacito" hit five billion views, as much energy as 40,000 U.S. homes use in a year was consumed (Elegant, 2019).

The foundation of the information industry is "at odds with its sleek clean and environment friendly image" (Glanz, 2012). With companies running data centers at maximum capacity around the clock, data centers can waste 90% of the electricity they pull off the grid as only 6 to 12% of the energy was used to power the servers and support computing operations. The rest is used to keep servers idling and ready in case of an unexpected surge in activity. "A single data center can take more power than a medium-size town," said Peter Cross, an industry expert (Glanz, 2012).

Faced with the danger of climate change, the internet industry is charged with building a green internet. Today, data centers consume about 2% of electricity worldwide. However, with the anticipated three-fold increase in global internet traffic, the information and technology communication (ITC) data demand—particularly driven by the streaming of video and social media—is forecast to accelerate in the 2020s, and our digital carbon footprint is likely to rise up to 8% of that global total by 2030 (Corcoran & Andrae, 2013).

In 2011, major internet companies Apple, Facebook, and Google became the first few companies to make a pledge to power their data centers with 100% renewable energy and have made impressive strides in recent years. Greenpeace's latest *Clean Click 2017*, ranked Apple, Facebook, and Google, the most aggressive internet companies in their overall efforts toward green internet. Apple powers its data center with 83% renewable energy including the deployment of a significant number of solar installations and micro-hydro projects, as well as reducing energy and carbon footprint for all of its operations (Frick, 2016). Facebook, home to four social networking sites (Facebook, Messenger, Instagram, and WhatsApp), was the first major internet company to make 100% renewable energy commitment and, in 2019, achieved 67% of its renewable energy consumption goal. Google Cloud, although lagging behind its platform competitors Apple and Facebook in energy transparency, still achieved a substantial 56% clean energy use.

Bibliography

Anderson, M. (2019). Mobile technology and home broadband 2019. Pew Research Center
   https://www.pewresearch.org/internet/2019/06/13/mobile-technology-and-home-broadband-2019/.

Anderson, M. (2018), A Majority of Teens Have Experienced Some Form of Cyberbullying. Pew Research Center.
   https://www.pewresearch.org/internet/2018/09/27/a-majority-of-teens-have-experienced-some-form-of-cyberbullying/.

Anunobi V. (2019, November 23). Cyberbullying and Mental Health. *Healthtian*. https://healthtian.com/cyberbullying-and-mental-health/.

Auxier, B., Rainie, L., Anderson, M., Perrin, A., Kumar, M., & Turner, E. (2019). Americans and privacy: concerned, confused and feeling lack of control over their personal information. Pew Research Center. https://www.pewresearch.org/internet/2019/11/15/americans-and-privacy-concerned-confused-and-feeling-lack-of-control-over-their-personal-information/.

Bloomberg. (2019, October 29). Cloud Computing Market share is Estimated to Reach US$ 285 billion by the end of 2025, with a CAGR of 29.2% - Valuates [Press release]. https://www.bloomberg.com/press-releases/2019-10-29/cloud-computing-market-share-is-estimated-to-reach-us-285-3-billion-by-the-end-of-2025-with-a-cagr-of-29-2-valuates.

Brown, E. (2020, February 27). 3 out 5 Americans use mobile first for surfing the web. ZDNET. https://www.zdnet.com/article/three-from-five-americans-use-mobile-first-for-surfing-the-web.

Canalyst. (2019, October 30). Canalyst: Global cloud market up 37%, with channels creating new growth engine. https://www.canalys.com/newsroom/global-cloud-market-Q3-2019

Clement, J. (2020, January 3). Countries with the highest number of internet users 2019. Statista. https://www.statista.com/statistics/262966/number-of-internet-users-in-selected-countries/

Clement, J. (2019, July 31). Average daily internet usage worldwide 2019, by age and device. Statista. https://www.statista.com/statistics/416850/average-duration-of-internet-use-age-device/.

Clement, J. (2020, February 3). Global digital population as of January 2020. Statista. https://www.statista.com/statistics/617136/digital-population-worldwide/.

Cohen, S., & Wills, T. A. (1985). Stress, social support, and the buffering hypothesis. *Psychological Bulletin, 98*(2), 310-357. doi: 10.1016/0022-3999(67)90010-4.

Corcoran, P., Andrae, A. (2013). Emerging Trends in Electricity Consumption for Consumer ICT. http://hdl.handle.net/10379/3563.

Coulson, N.S., Buchanan, H., & Aubeeluck, A. (2007). Social support in cyberspace: a content analysis of communication within a Huntington's disease online support group. Patient Education Council, *68*(2):173-8.

DeFleur, M. L. & Dennis, E. E. (2002) *Understanding Mass Communication: A Liberal Arts Perspective.* Houghton-Mifflin.

Dignan, L. (2019, January 24). Cloud customers pairing AWS and Microsoft Azure, according to Kentik. ZDNET. https://www.zdnet.com/article/cloud-customers-pairing-aws-microsoft-azure-more-according-to-kentik/.

Ditch the Label. (2017). *The Annual Bullying Survey 2017*. https://www.ditchthelabel.org/wp-content/uploads/2017/07/The-Annual-Bullying-Survey-2017-1.pdf.

Elegant, N. X. (2019, September 18). The internet cloud has a dirty secret. *Forbes*. https://fortune.com/2019/09/18/internet-cloud-server-data-center-energy-consumption-renewable-coal/.

Ellison, N. B., Steinfield, C., & Lampe, C. (2007) The benefits of Facebook "friends:" social capital and college students' use of online social network sites. *Journal of Computer-Mediated Communication*, 12:1143-1168.

Fox, S. (2014). The social life of health information. Pew Research Center. https://www.pewresearch.org/fact-tank/2014/01/15/the-social-life-of-health-information/.

Frick, T. (2016) *Designing for Sustainability*, O'Riley Media.

Frison, E., & Eggermont, S.(2015). The impact of daily stress on adolescent's depressed mood: The role of social support through Facebook. *Computers in Human Behavior*, 44:315-525.

Get to know Azure (n.d.). Microsoft. https://azure.microsoft.com/en-au/overview/.

Glanz, J. (2012, September 22). Power, pollution and the internet. *The New York Times*. https://www.nytimes.com/2012/09/23/technology/data-centers-waste-vast-amounts-of-energy-belying-industry-image.html

Gustafson, D. H., Hawkins, R., Pingree, S., McTavish, F., Arora, N. K., Mendenhall, J., …Salner, A. (2001). Effect of Computer Support on Younger Women with Breast Cancer. *Journal of General Internal Medicine, 16*(7), 435-445. doi: 10.1046/j.1525-1497.2001.016007435.x.

Hall, M. (2020, February 6). ICANN. https://www.britannica.com/topic/ICANN.

Hao, X., & Kay C.S. (2004). Factors affecting internet development: An Asian study. *First Monday, 9*(2). https://firstmonday.org/ojs/index.php/fm/article/view/1118/1038.

Hobson, J. (2017). How Google Fiber changed Kansas City. *Here & Now* https://www.wbur.org/hereandnow/2017/11/08/google-fiber-kansas-city.

Huang, H., & David, P. (2012). Exploring the mediating effect of social support on self-efficacy on Lose It! Facebook page. Web 2.0-5th World Congress on Social Media, Mobile Health and Web 2.0, Boston, MA.

Hwang, K. O., Ottenbacher, A. J., Green, A. P., Cannon-Diehl, M. R., Richardson, O., Bernstam,E. V., & Thomas, E. J. (2010). Social support in an internet weight loss community. *Int J Med Inform, 79*(1), 5-13.

ICANN (2014) International Corporation for Assigned Names and Numbers. (2014). IPv6 Factsheet. https://www.icann.org/en/system/files/files/factsheet-ipv6-24feb14-en.pdf.

Internet World Stats (March 15, 2020). https://internetworldstats.com/stats.htm.

John, A. Glendenning AC, Marchant A, Montgomery P, Stewart A, Wood S, Lloyd K, & Hawton K. (2018). Self-Harm, Suicidal Behaviours, and Cyberbullying in Children and Young People: Systematic Review. *Journal of Medical Internet Research, 20*(4): e129. DOI: 10.2196/jmir.9044.

Jones, E. (2020, January 13). Cloud market share – a look at the cloud ecosystem in 2020. Kinsta. https://kinsta.com/blog/cloud-market-share/.

Kemp, S. (2019). Digital 2019: Global Internet Use Accelerates. https://wearesocial.com/us/blog/2019/01/digital-2019-global-internet-use-accelerates.

Lee, E. & Cho, E. (2018). When using Facebook to avoid isolation reduces perceived social support. *Cyberpsychology, Behavior, and Social Networking*, 21:32-39.

Levy, A. (2019, April 25). Amazon Web Services revenue grew 41% in the first quarter. *CNBC*. https://www.cnbc.com/2019/04/25/aws-earnings-q1-2019.html.

Liesman, S. (2018). Nearly 60% of Americans are streaming and most with Netflix: CNBC Survey. https://www.cnbc.com/2018/03/29/nearly-60-percent-of-americans-are-streaming-and-most-with-netflix-cnbc-survey.htmlLiu (2012).

Liu, B. (2012). Sentiment analysis and opinion mining, Morgan & Claypool Publishers.

McCloskey, W., Iwanicki, S. Lauterbach, D. et al. (2015). Are Facebook "Friends" helpful? Development of a Facebook-based measure of social support and examination of relationships among depression, quality of life, and social support. *Cyberpsychology, Behavior and Social Networking* 18: 499-505.

Mo, P. K. H., & Coulson, N. S. (2010b). Living with HIV/AIDS and Use of Online Support Groups. *Journal of Health Psychology, 15*(3), 339-350. doi: 10.1177/1359105309348808.

Moyer, E. (2016, October 1). US hands internet control to ICANN. *CNET*. https://www.cnet.com/news/us-internet-control-ted-cruz-free-speech-russia-china-internet-corporation-assigned-names-numbers/.

Myrow, R. (2019, December 30). California rings in the new year with a new data privacy law. *National Public Radio*. https://www.npr.org/2019/12/30/791190150/california-rings-in-the-new-year-with-a-new-data-privacy-law.

Perrin, A. & Kumar, M. (July 25, 2019). About three-in-ten U.S. adults say they are "almost constantly" online. Pew Research Center. https://www.pewresearch.org/fact-tank/2019/07/25/americans-going-online-almost-constantly/.

Pew Research Center. (2019a). *10 tech-related trends that shaped the decade*. https://www.pewresearch.org/fact tank/2019/12/20/10-tech-related-trends-that-shaped-th-decade/.

Pew Research Center. (2019b). Mobile fact sheet. https://www.pewresearch.org/internet/fact-sheet/mobile/.

Pew Research Center. (2014). What happens to the internet after the U.S. hands off ICANN to others? https://www.pewresearch.org/fact-tank/2014/03/20/what-happens-to-the-internet- after-the-u-s-hands-off-icann-to-others/.

Pew Research Center. (2011). Health Topics. https://www.pewresearch.org/internet/2011/02/01/health-topics-4/.

Prodromou, P. (2020, January 24). How the World's Three Internets Are Changing Global Marketing. *MarTech Series* https://martechseries.com/mts-insights/guest-authors/worlds-three-internets-changing-global-marketing/.

Rainie, L. & Anderson, J. (2017, June). The internet of things connectivity binge: What are the implications? Pew Research Center. https://www.pewresearch.org/internet/wp-content/uploads/sites/9/2017/06/PI_2017.06.06_Future-of-Connectivity_FINAL.pdf.

Ranger, S. (2018, December 13). What is cloud computing? Everything you need to know about the cloud, explained. ZDNET. https://www.zdnet.com/article/what-is-cloud-computing-everything-you-need-to-know-from-public-and-private-cloud-to-software-as-a/.

Roberts, J.J. (2019, December 18). New California law giving consumers control over their datasets off a scramble. *Fortune*. https://fortune.com/2019/12/18/california-consumer-privacy-act-data-nationwide/.

Romm, T. (2018, April 11). Facebook's Zuckerberg just survived 10 hours of questioning by Congress. *The Washington Post*. https://www.washingtonpost.com/news/the-switch/wp/2018/04/11/zuckerberg-facebook-hearing-congress-house-testimony/.

Rogers, E. (1983). *Diffusion of Innovations*, 3rd Edition. Free Press

Schaeffer, K. (2019). Most U.S. teens who use cellphones do it to pass time, connect with others, learn new things. Pew Research Center. https://www.pewresearch.org/fact-tank/2019/08/23/most-u-s-teens-who-use-cellphones-do-it-to-pass-time-connect-with-others-learn-new-things/.

Smith, A. (2015). US Smartphone use 2015. Pew Research Center. https://www.pewresearch.org/internet/2015/04/01/us-smartphone-use-in-2015/.

Walker, M. (2019). Americans favor mobile devices for desktops for getting news. Pew Research Center. https://www.pewresearch.org/fact-tank/2019/11/19/americans-favor-mobile-devices-over-desktops-and-laptops-for-getting-news/.

We Are Social (2020). Digital 2020. https://wearesocial.com/digital-2020.

Wise, M., Gustafson, D. H., Sorkness, C. A., Molfenter, T., Staresinic, A., Meis, T., … Walker, N. P. (2007). Internet Telehealth for Pediatric Asthma Case Management: Integrating Computerized and Case Manager Features for Tailoring a Web-Based Asthma Education Program. *Health Promotion Practice, 8*(3), 282-291. doi: 10.1177/1524839906289983.

Wong, R. (2018, October 15). There May Soon Be Three Internets. America's Won't Necessarily Be the Best. *New York Times*. https://www.nytimes.com/2018/10/15/opinion/internet-google-china-balkanization.html?auth=link-dismiss-google1tap

Yuan, L. (2019, May 20). As Huawei loses Google, the U.S.-China tech cold war gets its iron curtain. *The New York Times*. https://www.nytimes.com/2019/05/20/business/huawei-trump-china-trade.html.

Zhong, R. (2019, May 16). Trump's Latest Move Takes Straight Shot at Huawei's Business. *The New York Times*. https://www.nytimes.com/2019/05/16/technology/huawei-ban-president-trump.html?module=inline.

22

The Internet of Things

Jeffrey S. Wilkinson, Ph.D.[*]

Abstract

The Internet of Things (IoT) refers to the growing global network of connected sensors, objects, and devices to computer networks. Active and passive chipsets are being embedded across a range of devices and applied across every sector of public life. The most notable devices are probably the home assistant (Echo, HomePod, and others), Ring doorbell, and Nest thermostat. Other applications include utilities management, inventory management, surveillance, and security. Global migration to 5G and narrow-band IoT are ushering in the next generation of smart systems. Many cities are testing largescale IoT applications such as water metering, waste disposal, and traffic and parking management.

Introduction

A familiar Hollywood trope is "our machines will destroy us." From *Frankenstein* to *2001* to *Blade Runner* and the *Terminator* franchise, humankind has been consistently warned to avoid smart machines. Yet still we persist creating a world of smart devices, and The Internet of Things (IoT) has expanded significantly since 2018. For example, most American adults now own a smartphone as well as a growing list of connected smart devices. Today, several cities in the United States report that home adoption of Alexa, Google, Siri, Cortana, Jibo, or Viv is above 50% (Stine, 2019). But the social science experiment we are playing with ourselves is fraught with unknowns. The focus of this chapter moves the IoT discussion to the practical incremental ways we are changing life on planet Earth. Smart appliances, devices, and objects continue to be developed and spontaneously launched in a relatively unstructured way. As these innovations are purchased, consumers must adjust and adapt to meet the requirements of the devices in order to integrate them into their lives. It is a struggle just to avoid being overwhelmed and controlled by the very devices we rely on to make our life simpler.

[*] Associate Professor, School of Journalism and Graphic Communication, Florida A&M University (Tallahassee, Florida).

The Internet of Things (IoT) is the umbrella term referring to the global trend of connecting as many devices as possible to computer-based networks. The range of connected items continues to expand, from clothing to appliances, and surveillance systems. Governments and institutions are working to develop smart cities, buildings, campuses and homes to monitor electric, gas, water, transportation, communication, and more. The intention is to make life better and more efficient, but we are regularly reminded of the dangers of IoT through news reports of hacking, neglect, or machine breakdown.

The Internet of Things (IoT) was officially "born" around 2009 when the ratio of connected objects-to-people became equal (Evans, 2011). According to Statista (2019), in 2020 there were more than 30 billion connected devices globally, a number that is projected to double by 2024. Eventually we expect trillions of trackable objects—thousands for each individual—to enable a connected, collective global community.

One hurdle is interoperability. Competition means there are still few coherent sets of standards—not only between companies or device manufacturers—but even nations. Governments are learning how to use IoT to monitor its citizenry and wield significant control to social movements. We are now at a time when our activities and personal and family records are increasingly collected and pulled together by companies tasked with mining these "big data" sets (see Chapter 24). The impact on our modern views of privacy suggest we have already given up in the name of standardization and efficiency. Those choosing to remain "off the grid" are increasingly finding it difficult to function with the rest of us.

Our behaviors, purchases, and even communication are increasingly being tracked, linked, scraped, mined, and archived to the extent allowed by privacy laws, which vary significantly around the globe. The boundless benefits of IoT may balance the equally dire consequences. Whether we are hurdling towards Utopia or Armageddon depends on our ability to harness ourselves. This chapter offers an overview for understanding the potential, the benefits, and the dangers this connectivity brings. IoT may be the defining technology issue of the 21st century.

Background

Since the beginning of commerce, companies have sought to be the sole provider of a good or service for every consumer. Manufacturers and service providers are competing to be part of the system that provides IoT products, from familiar giants like Amazon and Samsung to relatively unknown newcomers like Vates and Sciencesoft (Software Testing, 2019b). Each day a new product or service is being launched in the market, seeking to tap some unmet need.

How to satisfy the human desire to be served has been the subject of science fiction for generations, vividly envisioned a century ago by Jay B. Nash in *Spectatoritis* (1932):

> These mechanical slaves jump to our aid. As we step into a room, at the touch of a button a dozen light our way. Another slave sits twenty-four hours a day at our thermostat, regulating the heat of our home. Another sits night and day at our automatic refrigerator. They start our car, run our motors, shine our shoes, and curl our hair. They practically eliminate time and space by their very fleetness (p.265).

According to Press (2014), the shift from mechanical slaves to electronic devices evolved from Dick Tracy's 2-way wrist radio (1940s) and wearable computers (1955), to head mounted displays (1960) and familiar tech acronyms like ARPANET (1969), RFID (1973), and UPC (1974). As these converged with existing technologies, new applications were conceived which begat newer advances leading to today's age of ubiquitous computing and high-speed broadband networks.

Today's Internet of Things relies on four key elements; mass creation and diffusion of tagged objects (active or passive), hyper-fast machine-to-machine communication, widespread RF technologies, and an ever-increasing capacity to store and access information. The global shift from IPv4 to IPv6 (128-bit) numbering (see Chapter 21) enabled us to assign unique addresses to an almost unlimited number of objects (said to be 340 undecillion addresses which is 340 billion billion billion billion, or

340,282,366,920,938,463,463,374,607,431,768,211,456). By being tagged with a unique IP (Internet Protocol) address, an object can be identified according to parameters of time, activity, and geolocation. Furthermore, the smart objects are designated as either active or passive. A passive connection uses RFID (radio frequency identification) tags to identify and track the object. For things like clothing, toys, toiletries, or office supplies, knowing where the object is and when it was bought is the primary function of the RFID tag.

An active connection is more robust and makes a device truly "smart." Using embedded sensors and computing power allows the object to gather and process information from its own environment and/or through the internet. These objects include surveillance cameras, cars, appliances, medical devices, and anything that needs to either send or receive information.

Thanks to upgrades in networks (Chapter 20) and computing power (Chapter 11), putting billions of these passive and active objects together gives us the constantly evolving Internet of Everything. Now businesses, institutions, or governments can monitor connected objects and devices, whether it is food, clothing, shelter, or services. Each of these can be tracked and analyzed, individually or in conjunction with each other. IoT is at work almost everywhere—connecting cities, businesses, the home, and the spaces in between. Future applications are only limited by our imagination (concerns about privacy are discussed throughout this chapter).

To be an IoT object or device, it must be able to connect and transmit information using internet transfer protocols. With the implementation of 5G cellular networks (see Chapter 20), the increased bandwidth means added connectivity for more devices doing more intricate things at a much faster rate. For example, a two-hour movie takes six minutes to download using 4G; The same download using 5G takes only 3.6 seconds (Soergel, 2018).

To help connect everything seamlessly together—like having your car personal assistant call your home personal assistant to activate your smart oven—a number of platforms have emerged. These cloud platforms enable the services to connect the devices for machine to machine communication. They connect the edge hardware, access points, and data networks

to the other end which is usually the end-user application (Software Testing Help, 2019a). The top IoT platforms mix familiar names with relative unknowns; Amazon AWS, Microsoft Azure, Google Cloud, IRI Voracity, Salesforce, ThingWorx, Particle, IBM Watson, Samsung Artik, Oracle, Cisco, and Altair Smartworks.

Recent Developments

According to Meola (2016), there are three major areas rapidly developing IoT ecosystems: government, business, and consumer. Important sectors include manufacturing, transportation, defense, healthcare, food services, and dozens of others which all are adopting IoT in ways to improve efficiency, economies of scale, and security.

Government/utility ecosystems includes defense grids, electricity grids, water use, and transportation networks (airplane, train, highways, etc.). Industrial vendors focus on manufacturing and quality assurance, robotics, inventory tracking, and logistics/fleet management. Consumers are increasingly targeted for the sales and marketing of products for the home such as personal digital assistants, door cameras, smart vacuums, and more.

Government & Smart Cities

Government concerns include efficient transportation patterns, safeguarding electrical grids and improving efficiency with water supplies. In an era of sustainability, cities try to maximize limited resources. For example, parking is a never-ending headache and Palo Alto, San Francisco became a trailblazer in 2015 when they implemented smart parking (Smith, 2018; Upasana, 2019). Sensors were installed at parking spaces throughout the city and commuters could use the city's app to find the nearest available spaces. The program has been noted, but by 2019 the city admitted some tweaking was needed. Parking was still an issue, and in some neighborhoods, homeowners complained they were unable to find spaces outside their homes. Others said they would rent their driveway space to commuters using smart apps (Sheyner, 2019).

In terms of basic utilities like water and electricity, IoT is being used by countless cities to help with security as well as efficiency. The United States electrical system is managed via three distinct electrical grids; the Eastern Interconnection, the Western Interconnection, and the Texas (ERCOT) grids. Within sectors there are sub-sectors enabling the grid to stay up even when one part of the grid goes down.

These grids are routinely subjected to hacker attacks, but none had ever been successful until March 5, 2019. A first-of-its-kind cyberattack on the U.S. grid created blind spots at a grid control center and several small power generation sites in the western United States (Sobczak, 2019). The unprecedented cyber disruption did not cause any blackouts, nor signal outages longer than five minutes at the control center, but the outage was significant enough to be reported to the Department of Energy, marking it the first disruptive "cyber event" ever for the U.S. power grid. The hackers took advantage of a flaw in the firewall's interface (since patched), demonstrating the risks U.S. power utilities face as their critical control networks grow more digitized and interconnected. Similar Russian cyberattacks on Ukrainian utilities in 2015 and 2016 caused hours-long outages; by comparison the U.S. cyberthreat was simpler and far less dangerous than the hacks in Ukraine.

Water management is another area where governments are turning to IoT for help. By the year 2025, it is estimated that 70 percent of the world population will be living in the cities, placing huge strains on city water systems (John, 2019). Almost every municipality has some form of IoT to help detect and correct leaks by analyzing real-time water pressure, water levels, water temperature, the flow of water and other factors. Predictive analytics track water consumption habits and patterns, identifying and suggesting ways to minimize overall water use.

Surveillance by security cameras are becoming ubiquitous in the public square as well as the home. Facial recognition relies on ever-growing databases (including those convicted of crimes), and matched according to "degree or percentage of match." Although biometric security software is still in its beta-testing stage, it is being widely tested in businesses, in law enforcement, and even in the home. For example,

one significant flaw noted has been that the technology often errs with women and people of color (O'Brien, 2019; Singer & Metz, 2019).

Business IoT

In business, issues regarding security, inventory, and logistics are probably the main concerns (although there are undoubtedly more). In terms of security, companies are turning to a variety of ways to monitor employee activity. Besides the ubiquitous security camera and in-house facial recognition software, biometric security devices include voice analysis, palm/thumbprints/fingerprints, retina scanning, gait analysis (how you walk), and even tactile/touch patterns on screens (Horowitz, 2018).

One of the largest areas of investment is in transportation, and fleet management in particular (Torchia & Shirer, 2019). Almost all fleets—trucks, taxis, etc.—are now linked in real-time and monitored for efficiencies in terms of miles driven, time on the road, and successful deliveries. Thanks to increasingly sophisticated GPS and road monitoring, drivers are given updates and time checks, leading to efficiencies that were unheard of a generation ago.

Factories have increasingly turned to automation and robotics for standardized performance of repetitive tasks such as inventory management and tracking. Quality assurance of products is made possible by the sensors embedded in the process. According to Potoczak (2017), as "smart" automated production lines collect and send data to in-plant central servers, they can identify individual non-conformances as well as emerging trends resulting in non-conformances. This closed loop smart manufacturing provides operators, specialists, and managers accurate and timely quality metrics and alerts (Potoczak, 2017).

Consumers & the Home

The consumer world of IoT in particular continues to ramp up, and increasing numbers of smart devices are being sold to consumers. For those in older "less connected" homes, consumers tend to adopt and install IoT technology in piecemeal fashion, learning through trial, and error. Newer structures are designed to be smarter, with built-in appliances and integrated features. Leading companies offering devices, platforms, and infrastructure are Amazon, Google, Samsung, Nest (owned by Google), Philips, Ecobee, HomeSeer, and Belkin (Lee, 2019). These providers are designing fully-integrated closed systems that (with proper programming) can respond to centrally manage the home (Sorrel, 2014).

Besides Google's Nest Learning Thermostat, Apple's Homekit presents another home automation hub. Google's Nest can handle many core home maintenance tasks such as motion detection and learning user behavior (Burns, 2014). Apple's HomeKit centers around its home-speaker digital assistant called HomePod. Like the Echo and Google Home, HomePod can play music, set lighting, adjust the thermostat, and be programmed for other smart home tasks (Swanner, 2017). There are far fewer non-Apple devices available for HomeKit due to Apple's corporate policy of "MFi" (made for iPhone/iPod/iPad). But at CES 2020, global home appliance company Bosch announced "after the voice services from Amazon Alexa and Google Assistant have been integrated during the course of 2020, the Bosch Smart Home System will also be conveniently controllable via Siri, Apple's voice assistant." (Hardwick, 2020). Bosch offers a number of IoT compliant products such as shutters, light fittings, thermostats, intelligent alarms, and other accessories.

Larger appliances are also being marketed and equipped with 'smartness.' Recent trade shows see companies like Sharp and LG demonstrating fully integrated showers, kitchens, and laundry appliances. Sharp's smart kitchen suite for example includes IoT enhanced oven, microwave, refrigerator, and dishwasher (Morgan, 2019). These appliances can be controlled by either Google Home or Amazon's Echo (Gibbs, 2018). Amazon's Echo is the basic unit; a smaller version is called Dot, and a version with a screen is called Echo Show. Google still has the basic Home device and the Home Mini which can be placed in different rooms as needed.

Home Assistants/Smart Speakers

All the central managing hub devices have unique brand voices, and "Siri," "Hey Google," and "Alexa" are fast becoming household names. A sizable number of people refer to these as 'smart speakers' but all

the companies are scrambling to expand their digital assistant application outside the home. In 2019, over 111 million Americans were estimated to be using voice-assistants (Petrock, 2019), and in many American cities, the Google, Amazon, Apple platforms have been brought into the home with the digital home assistant (Stine, 2019). Top names are Amazon Alexa, Google Assistant, Apple Siri, Microsoft Cortana, and Samsung Bixby. Other significant entries in the marketplace are Viv, Nina, Jibo, Mycroft AI, Braina virtual assistant, Dragon Go (Nuance), Hound (Soundhound), and Aido (home robot). Some are working with or have been bought by the larger corporations, and some of these have been designed for specific functions like customer service (Nina) or designed as social robots designed to interact with family members in the home (Jibo, Aido)(PAT Research, n.d.).

The two most popular devices are Google Home and Amazon Echo. According to Ludlow (2019), Google is a bit easier to talk to and use while Echo seems more flexible with more capabilities. But hardware and software will undoubtedly be upgraded soon.

Amazon is trying to get Alexa into as many areas as possible in everything from motorcycle helmets to toilets, anything to make the experience seamless. They have voice-active noise reduction earbuds (echobuds), smartglasses, and smartlamps. There is the Echoflex, a hub to bring Alexa into every room in the house, the SmartOven/microwave where you can scan a food package and the device sets the appropriate temperature. As smart devices slowly integrate themselves and manage our lives, some predict we will come to rely less on the smartphone (although today the phone is the chief means for us to control smart devices).

IoT developments routinely appear almost daily in the popular press. A report from USA Today stated it costs around $2,000 to outfit your home with basic IoT features like thermostat control, smart lighting made, and home security (Graham, 2017). As mentioned earlier, one of the most popular products is the Nest smart thermostat. It can be controlled by your phone from anywhere and can learn to adjust to individual needs. The Nest retails for around $250.

Home security continues to be an issue and smart locks, smart lighting, and smart doorbells continue to be popular. Several companies offer home security and lighting. During Cyber Monday 2019, several brands of security cameras sold for between $100 and $180 each (Bradford, 2019). Companies like ADT offer 24/7 security monitoring, and Vivint offers 24/7 home security monitoring beginning at $29.99 per month (Woodall, 2019). Several off-the-shelf products can be bought at places like Best Buy, Walmart, or Target. A smart lock which is opened with your phone or via Alexa, ranges between $200-$300. So-called "smart lights" by Philips turn off and on by voice command cost about $150.

Smart doorbells have also become popular. The most popular models of smart doorbells sell for less than $200 and include the Amazon Ring, Google Nest Hello, and August Doorbell Cam Pro. These video doorbells allow you to see, hear, and speak with whomever is at the door. Smart doorbells made the news in December 2019 for catching thieves in the act of stealing packages from doorsteps (Hart, 2019).

Figure 22.1

Smart Home

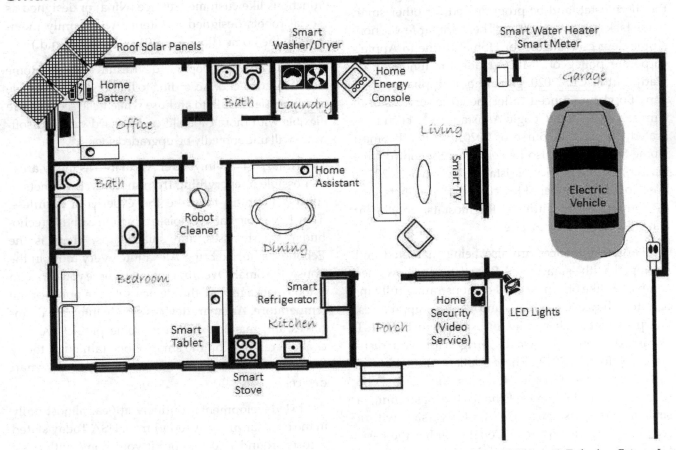

Source: J Wilkinson & Technology Futures, Inc.

Computer, OBEY"

Levels of Interaction for Personal Assistants?

August E. Grant

IoT personal assistants are rapidly improving in their capability to serve us. There are at least six stages of evolution for these automated helpers before they can respond as a person would.

The Six Stages of Personal Assistants:

1. **Simple Commands**: One clearly defined question or task.

Example: "What time is it?" [time given]. "Play Truth Hurts by Lizzo." [song plays].

2. **Multiple Commands**: A short series of clearly defined tasks.

Example: "Play three songs from Ed Sheeran, then play two songs from Billie Eilish." [tasks done].

3. **Conditional Commands**: A clearly defined task with new parameters.

Example: "Read the latest news, but don't include any sports stories." [task done].

4. **Serial Commands**: Ability to execute multiple commands at the same time, resolving conflicts between the commands.

Example: "Play that great new song from Taylor Swift about her last boyfriend, then read the latest news...Oh, and I want to see this morning's phone interview segment with the President from 'Fox and Friends'." [song played, segue to news, then cut to full interview segment].

5. **Anticipatory Commands**: Use of AI to anticipate and infer meaning to commands.

Example: "It's cold in here and it's been a rough day." The PA automatically changes the thermostat, starts playing music, or executes another command after predicting that the user might issue such a command

[cue thermostat adjustment, then play soft jazz, then refrigerator pops ice cube into prepared glass of your favorite beverage. Advanced feature might include a confidential text to best friends recommending a quick casual call.]

6. **Full Autonomy**: Use of AI to self-learn, anticipate and prioritize tasks, including the ability to execute more than one task at the same time. Extra feature: ability to predict a warp core breach or initiate photon torpedo launch with confirmation voice command from another bridge officer.

Example: "Siri, beam me up NOW!" [trouble eliminated, refrigerator pops ice cube into glass...]

But the doorbells themselves have been under scrutiny after ongoing reports of hackers using the two-way talk functions on the devices to wake people up, watch children through the internet-connected cameras, and even engage in conversations with them (Greer, 2019). An Alabama man filed a lawsuit against Ring, claiming his internet-connected camera was hacked and used to harass his children (Paul, 2019).

Another privacy concern stemmed from media reports that Amazon has hired thousands of employees to monitor what consumers say to Alexa (Valinsky, 2019). Amazon defended the practice as simply research to improve how the devices responds. The company released an upgrade with instructions on how get Alexa to 'not listen' as well as delete conversations or commands.

Even the family car is being integrated with voice assistants. Alexa has been installed in a number of models from BMW, Ford, Lexus, Toyota, and others (Charlton, 2019), but because of access issues experiments are continuing with unique inside-the-car assistants. For example, the inside-the-car assistant might recommend when to turn on high beams or when to pull over and take a rest while Alexa or Siri would be used primarily for news, hands-free phone calls, and checking home security (for more on Automotive Telematics, see chapter 12).

Current Status

According to research firm IDC, total global spending on IoT devices and appliances across all environments (work and home) grew from $646 billion in 2018 to an estimated $745 billion in 2019 (Torchia & Shirer, 2019). The industries forecast to spend the most on IoT are discrete manufacturing ($119 billion), process manufacturing ($78 billion), transportation ($71 billion), and utilities ($61 billion). IoT spending in manufacturing focus on operations and production asset management. In transportation, more than half of IoT spending goes into freight monitoring, followed by fleet management. In the utilities industry, firms continue to work with governments and municipalities to maintain and upgrade smart grids for electricity, gas, and water. The industries projected to see highest growth are insurance, federal/central government, and healthcare (Torchia & Shirer, 2019).

The continued acceleration of IoT reflects the money and innovation being poured into the next generation of smart devices. One innovation is in the area of Micro-Electro-Mechanical Systems (MEMS) or "smart dust." The term smart dust was coined by UC-Berkeley researcher Kristofer Pister in 1997 to describe a wireless array of sensor nodes. These Tiny, 3D-printed sensors work together, drawing power from subtle vibrations, even from the surrounding air (Bonderud, 2019). MEMS are ideal for highly sensitive applications. For example, MEMS are ideal for quality assurance in food processing and packaging (Bonderud, 2019). The Institute of Electrical and Electronics Engineers suggests next-generation smart dust sensors could use paper or plastic sensors to detect food freshness, then report this data via a smartphone app. MEMS can also be used where people gather—from small laboratories to giant sports facilities—to track and monitor facilities use and traffic. For example, stadiums and concert halls could identify locations where bottlenecks occur as well as where seats most often need to be replaced (proactive maintenance).

The application of MEMS can obviously extend to safety and security as well. Smart dust networks in on-campus residences and high-value facilities could be used to act as intelligent alert and alarm systems. Smart dust can be extended to ID Cards, where radio frequency identification coupled with MEMS enables individualized permission control. In school dormitories, for example, student safety would be enhanced because every student could be immediately located and identified during an emergency.

Smart dust sensors are now being used to power safety mechanisms in cars and trucks, but can be used on anything with wheels. MEMS-based accelerometers in airbags have improved performance at reduced cost. Tire pressure sensors can intelligently collect tire data using vibration as their power source.

While it would fill an entire book to report on all IoT activity, a couple of areas are worth detailing; surveillance/security cameras and home digital assistant adoption. These two have broad implications for the intersection between communication and technology.

Surveillance camera installation continues to rise exponentially worldwide. According to the Carnegie Endowment for International Peace, at least 75 out of 176 countries are actively using AI technologies for surveillance purposes (Feldstein, 2019). Governments are using surveillance in their smart city platforms (fifty-six countries), facial recognition systems (sixty-four countries), and smart policing (fifty-two countries).

At the end of 2019, it was estimated there were 350 million surveillance cameras in operation worldwide (Aamir, 2019). Although no official numbers are reported from places like Russia, India, and Brazil, it appears China leads the world in total number (200 million) followed by the U.S. (50 million). But since the U.S. has fewer people, it actually now boasts the highest ratio of cameras per 100 people (15.3 versus 14.4).

Eight of the top ten cities with the highest overall cameras-per-100 people are in China (topped by Chonqing, Shenzhen, and Shanghai are 16.8, 15.9, and 11.3, respectively). London ranks #6 (6.84) and Atlanta #10 (1.55) as the top surveilled city in the U.S., employing 7,800 cameras (Aamir, 2019).

Inside the home, Juniper Research projects that the use of voice assistants will triple from 2.5 billion at the end of 2018 to eight billion by 2023 (Perez, 2019). Most voice assistant activity is on the smartphone (see Chapter 20), and although Google announced it's now on a billion devices (Kumparak, 2019), they're mostly smartphones. Amazon's Alexa is said to be the home digital assistant leader, reaching 100 million devices sold at the end of 2019 (Matney, 2019). In fact, the smart speakers at home have been a boon for radio stations, who have adapted streams for Alexa, Google and the others. Nielsen reports that, in major markets like New York and Philadelphia, nearly 50% of homes currently have one, with 9% owners say 'they listen to a lot more radio' since owning a smart speaker (Stine, 2019).

Factors to Watch

Juniper research predicts that the device showing the fastest growth will not be smart speakers, but smart TVs (Perez, 2019). Significant growth is also predicted to hold for smart speakers and wearable technology. Juniper notes that, while Alexa is the market leader on smart speakers, they anticipate challenges from Chinese companies who will launch their own smart devices.

Juniper is also predicting that digital assistants will ultimately reduce consumer demand for apps. As consumers increasingly turn to multi-platform capable assistants, the need for standalone apps and smartphone use may decline. Simple things (like weather or news) will become outsourced to voice assistants. This will notably reduce our screen time.

As we give our devices increasingly complex tasks to perform, there is bound to be breakdowns in communication. In "the telephone game" we laughed when our friends garbled a message when whispering it from person-to-person-to-person. Yet during emergencies our smart systems may ignore us due to mispronunciation or poorly worded commands. At a more basic level, if we want the best place to buy milk AND bread, will Alexa suggest one grocer or two?

An entire IoT ecosystem is being created where companies must look beyond traditional Search Engine Optimization (SEO) to embrace Smart Device Optimization (SDO). As the saying goes, 'if you're not at the table, you're on the menu,' and competition will heat up determining which eatery Alexa will suggest when you ask for a 'good restaurant nearby.' Businesses cannot afford to be left behind, or even be listed second when Siri recommends where to go for gas.

A related area to watch is that of "Home Predictive Maintenance" (Fitzgerald, 2019). The sensors now being used in factories, airplanes, and automobiles are also being used for the devices and appliances coming into your home. As our home systems grow smarter, they will begin to inform us about things like plumbing leaks, appliance failures, and electrical problems. After that we'll see the ramping up of SDO again, and who gets recommended to come in and make the needed home repairs. Predictive Maintenance can also extend to the mundane when our smart washer one day recommends we switch to a different detergent either for cost, conservation, or cleanliness.

A final factor to watch is what happens when smart devices become dumb. What problems arise when the devices are in error. When the semantic messages in our voice commands are missed we will blame the dumb machine, but at what cost? Throughout history, our machines have broken down or ceased to work or no longer could do what was expected of them. Likewise, despite the utopian expectations of smart dust, one must wonder what happens when we buy 'fresh certified food' and find that it has some new heretofore undetected pathogen? Or when a student has human permission to enter a dorm, but the smart device signals a false positive and determines that the individual constitutes a threat.

Humankind continues the march toward self-imposed machine dependence. Whether the devices will serve us or enslave us may simply be a matter of semantics. But it is the only way to efficiently manage the vast—yet limited—resources of planet Earth as it seeks to accommodate billions of people across nations, languages, and cultures. Privacy and security will continue to be of concern across all sectors, yet devices will be adopted regardless of their vulnerabilities. There is simply too much money at stake for us to wait.

Gen Z

There are some distinctive traits associated with Gen Zers such as a general drive for simplicity, perhaps due to growing up in a 24/7 information overloaded world. In the workplace they see writing (and reading) lengthy reports as antiquated and time-consuming, and long stories or descriptions bore them quickly. This generalized impatience may make them more competitive to manage and complete projects. Some reports indicate many consider skipping university study to go directly into the workforce.

As digital natives immersed in technology, Gen Zers are comfortable with and adept at using IoT devices. For them, life is technology and anything that makes it simpler and less complicated is okay with them. Although they have grown up with smartphones, many will be able to migrate to an integrated world of IoT devices and appliances which may reduce the reliance on smartphones.

But there are also consequences to being "always-on." Critics warn that this group has far less experience developing face-to-face social skills and are therefore less adept in navigating tricky social environments. For teens and twenty-somethings, these crucial years are marked by less time with friends, less time dating, and less practice building and maintaining relationships. The resulting general lack of confidence and self-esteem can produce a recursive loop whereby social interactions are actively avoided and replaced by the comfort and safety of being online.

According to Davidson (2019), information technology has also made Gen Zers probably the most globally aware age group. The internet has opened up the world to them—the good, the bad and the ugly. They are knowledgeable about global environmental and social issues and know how to get involved. But they have also been exposed to the most graphic, inappropriate, and dangerous content online. Most Gen Zers have witnessed or experienced cyberbullying of a friend, sibling, or themselves. Being so aware and so adept at technology provides hope they will actively address and work to improve pressing world problems.

So overall, Gen Zers are perhaps best prepared to adapt to an IoT world. They are comfortable with technology and smart devices and will readily adopt IoT as they can afford it or it is provided for them in the workplace.

Global Concerns & Sustainability

A driving force behind IoT is global concerns about sustainability. As such, governments and businesses are researching, developing, and implementing IoT systems in order to manage the quality of our air, water, and soil. All these systems are evolving and subject to security breaches, and patches must be applied as they become exposed.

Numerous IoT systems are currently in place around the world and new ones are added daily. These smart systems are projected to save billions of dollars annually by reducing fuel consumption, carbon emissions, environmental damage, and time saved by travelers avoiding traffic congestion in urban areas.

In particular, IoT is now commonly used for water management. Sensors help provide tracking data on water levels and quality, helping increase efficiencies during peak usage or in the event of a flood or natural disaster (Muraleedharan, 2019). Smart irrigation systems track soil water content and soil nutrients for farming communities.

Wildlife are increasingly being tracked for overall numbers migration, mating, and feeding habits (Naveen, 2019). An online global database, eBird, shows in real-time the location and distribution of various birds around the world. In the water, ThisFish is an internet tracking network which can identity bad fishing practices, and FishPal helps local fisheries know the type and quantity of fish caught on a daily and monthly basis.

On farms, livestock are commonly fitted with chips to provide comprehensive data on the overall health and physical condition of each animal.

Significant progress continues in smart buildings, helping regulate power consumption and optimal use of renewable energy. Green designs help energy conservation by maximizing use of daylight (solar), rainwater capture, smart cooling, and ventilation systems.

Innovations in waste disposal are helping to alleviate health hazards associated with inefficient or improper practices. Trash cans and dumpsters fitted with sensors measure trash volume and send notice when it is time to be emptied. Waste companies are tracking waste disposal patterns in order to better schedule collection trucks for maximum efficiency while reducing health risks.

In the cities, advances in wastewater and sewage management track water levels, leakage, overflow, and usage patterns in order to reduce health risks and optimize efficiency.

Bibliography

Aamir, H. (2019, December 6). Report finds the US has the largest number of surveillance cameras per person in the world. Techspot.com. https://www.techspot.com/news/83061-report-finds-us-has-largest-number-surveillance-cameras.html.

Bonderud, D. (2019, September 6). Smart Dust: The Big Education Impact of IoT's Smallest Device. Edtechmagazine.com. https://edtechmagazine.com/higher/article/2019/09/smart-dust-big-education-impact-iots-smallest-device-perfcon.

Bradford, A. (2019, December 2). Cyber Monday home security camera deals: Arlo, Blink, and Nest. Digital Trends. https://www.digitaltrends.com/dtdeals/cyber-monday-home-security-camera-deals/.

Burns, M. (2014, June 23). Google Makes Its Nest At The Center Of The Smart Home Techcrunch.com. https://techcrunch.com/2014/06/23/google-makes-its-nest-at-the-center-of-the-smart-home/.

Charlton, A. (2019, April 29). Which cars have Amazon Alexa integration? Updated for 2019. GearBrain.com. https://www.gearbrain.com/which-cars-have-amazon-alexa-2525958778.html.

Davidson, B.J. (2019, February 8). Impact of Technology on Generation Z. Percentotech.com. https://percentotech.com/bobbyjdavidson/impact-of-technology-on-generation-z/.

Evans, D. (2011, April,). The Internet of Things: How the next evolution of the Internet is changing everything. Cisco Internet Business Solutions Group (IBSG) White paper. https://pdfs.sematicscholar.org/e434/2d4687233ae12aa689407f97502d87a9f27b.pdf.

Feldstein, S. (2019, September). The Global Expansion of AI Surveillance. Carnegie Endowment for International Peace. White paper. https://carnegieendowment.org/files/WP-Feldstein-AISurveillance_final1.pdf.

Fitzgerald, M. (2019, September 23). Top 14 IoT Mobile App Development Trends to Expect in 2020. Towardsdatascience.com. https://towardsdatascience.com/top-14-iot-mobile-app-development-trends-to-expect-in-2020-7fd7718155dc.

Gibbs, S. (2018, January 12). CES 2018: voice-controlled showers, non-compliant robots and smart toilets. The Guardian.com. https://www.theguardian.com/technology/2018/jan/12/ces-2018-voice-controlled-showers-robots-smart-toilets-ai.

Graham, J. (2017, November 12). The real cost of setting up a smart home. *USA Today*. https://www.usatoday.com/story/tech/talkingtech/2017/11/12/real-cost-setting-up-smart-home/844977001/.

Greer, E. (2019, December 17). Google Nest or Amazon Ring? Just reject these corporations' surveillance and a dystopic future. https://www.nbcnews.com/think/opinion/google-nest-or-amazon-ring-just-reject-these-corporations-surveillance-ncna1102741.

Hardwick T. (2020, January 8). CES 2020: Bosch Smart Home System to Support Apple HomeKit Later This Year. Macrumors.com. https://www.macrumors.com/2020/01/08/ces-2020-bosch-smart-home-system-homekit/.

Hart, G. (2019, December 5). Thieves arrested after string of package thefts caught on camera in Henderson. News3lv.com. https://news3lv.com/news/local/thieves-arrested-after-string-of-package-thefts-caught-on-camera-in-henderson.

Horowitz, B. (2018, December 6). 5 Biometric Security Measures to Keep You Safer in 2019. *PC magazine*. https://www.pcmag.com/feature/365292/5-biometric-security-measures-to-keep-you-safer-in-2019.

John, N. (2019, March 20). World water day: Using IoT to manage water. SmartEnergy.com. Smart Energy International. https://www.smart-energy.com/industry-sectors/business-finance-regulation/using-internet-of-things-for-water-management/.

Kumparak, G. (2019, January 7). Google says Assistant will be on a billion devices by the end of the month. TechCrunch.com. https://techcrunch.com/2019/01/07/google-says-assistant-will-be-on-a-billion-devices-by-the-end-of-the-month/.

Lee, M. (2019, February 23). 8 Top Smart Home Companies in the United States. DIYSmarthomehub.com. https://www.diysmarthomehub.com/smart-home-maker-company-in-united-states/.

Ludlow, D. (2019, September 10). Google Home vs Amazon Echo: Which is the best smart speaker? Trusted Reviews.com. https://www.trustedreviews.com/news/google-home-vs-amazon-echo-2945424.

Matney, L. (2019, January 4). More than 100 million Alexa devices have been sold. TechCrunch.com. https://techcrunch.com/2019/01/04/more-than-100-million-alexa-devices-have-been-sold/.

Meola, A. (2016, December 19). What is the Internet of Things (IoT)? *Business Insider*. http://www.businessinsider.com/what-is-the-internet-of-things-definition-2016-8.

Morgan, G. (2019, August 9). Sharp to unveil its first-ever complete kitchen suite at primetime show: Line includes IoT-enabled appliances, dishwashers and refrigerator. Twice.com. https://www.twice.com/industry/sharp-to-unveil-its-first-ever-complete-kitchen-suite-at-primetime-show.

Muraleedharan, S. (2019, November 15). Role of Internet of Things (IoT) in Sustainable Development. EcoMENA.org. https://www.ecomena.org/internet-of-things/.

Nash, J.B. (1932). *Spectatoritus*. New York: Sears Publishing Company. http://babel.hathitrust.org/cgi/pt?id=mdp. 39015026444193;view=1up;seq=6.

Naveen, J. (2019, September 4). How IoT And AI Can Enable Environmental Sustainability. Forbes.com. https://www.forbes.com/sites/cognitiveworld/2019/09/04/how-iot-and-ai-can-enable-environmental-sustainability/

O'Brien, M. (2019, December, 18). Why some cities and states ban facial recognition technology. *Christian Science Monitor*. Csmonitor.com. https://www.csmonitor.com/Technology/2019/1218/Why-some-cities-and-states-ban-facial-recognition-technology.

PAT Research, iManual (n.d.). Top 22 intelligent personal assistants or automated personal assistants. PAT.com. https://www.predictiveanalyticstoday.com/top-intelligent-personal-assistants-automated-personal-assistants/.

Paul, K. (2019, December 27). Ring sued by man who claims camera was hacked and used to harass his kids. *The Guardian*. https://www.theguardian.com/technology/2019/dec/27/ring-camera-lawsuit-hackers-alabama.

Pelino, M. (2018, November 6). Predictions 2019: The Internet Of Things. Forrester Research. https://go.forrester.com/blogs/predictions-2019-iot-devices/.

Perez, S. (2019, February 12). Report: Voice assistants in use to triple to 8 billion by 2023. Techcrunch.com. https://techcrunch.com/2019/02/12/report-voice-assistants-in-use-to-triple-to-8-billion-by-2023/.

Petrock, V. (2019, July 15). US Voice Assistant Users 2019: Who, What, When, Where and Why. eMarketer.com. https://www.emarketer.com/content/us-voice-assistant-users-2019.

Potoczak, E. (2017, October 17). Capitalizing on the Convergence of Manufacturing Quality, IoT and Lean. Quality Magazine. https://www.qualitymag.com/articles/94260-capitalizing-on-the-convergence-of-manufacturing-quality-iot-and-lean.

Press, G. (2014, June 18). A very short history of the Internet of Things. *Forbes*. http://www.forbes.com/sites/gilpress/2014/06/18/a-very-short-history-of-the-internet-of-things/.

Sheyner, G. (2019, May 14). Palo Alto prepares to revamp neighborhood parking programs. https://www.paloaltoonline.com/news/2019/05/14/palo-alto-prepares-to-revamp-neighborhood-parking-programs.

Singer, N., and Metz, C. (2019, December 19). Many Facial-Recognition Systems Are Biased, Says U.S. Study. *The New York Times*. https://www.nytimes.com/2019/12/19/technology/facial-recognition-bias.html.

Smith, L. (2018, October 1). Palo Alto: A Role Model for Smaller Smart Cities. *Smart City Portraits*. https://hub.beesmart.city/city-portraits/smart-city-portrait-palo-alto.

Sobczak, B. (2019, September 6). Report reveals play-by-play of first U.S. grid cyberattack. EnergyWire. E&E News.com. https://www.eenews.net/stories/1061111289.

Soergel, A. (2018, January 9). 5G: The Coming Key to Technology's Future. *U.S. News & World Report*. https://www.usnews.com/news/economy/articles/2018-01-09/5g-the-coming-key-to-technologys-smart-future.

Software Testing Help (2019a, December 19). 10 Best IoT Platforms To Watch Out In 2020. SoftwareTestingHelp.com. https://www.softwaretestinghelp.com/best-iot-platforms/.

Software Testing Help (2019b, November 22). TOP 11 Best Internet Of Things (IoT) Companies To Watch In 2019. https://www.softwaretestinghelp.com/top-iot-companies/.

Sorrell, S. (2014). Smart Homes—It's an Internet of Things Thing. Juniper Research. Smart Home Ecosystems & the Internet of Things Strategies & Forecasts 2014-2018. Hampshire, U.K.

Statista (2019). Internet of Things (IoT) connected devices installed base worldwide from 2015 to 2025 (in billions). file:///D:/IoT_%20number%20of%20connected%20devices%20worldwide%202012-2025%20_%20Statista.html.

Stine, R.J. (2019, September 9). Smart Speakers Grow in Importance: Media companies, marketers and agencies explore how to capitalize and monetize. Radioworld.com. https://www.radioworld.com/news-and-business/smart-speakers-grow-in-importance.

Swanner, N. (2017, June 8). Can Homepod make Homekit a winner for Apple? Dice.com. https://insights.dice.com/2017/06/08/can-homepod-make-homekit-winner-apple/.

Torchia, M. and Shirer, M. (2019, January 3). IDC Forecasts Worldwide Spending on the Internet of Things to Reach $745 Billion in 2019, Led by the Manufacturing, Consumer, Transportation, and Utilities Sectors. IDC.com. https://www.idc.com/getdoc.jsp.

Upasana, (2019, May 22). Real World IoT Applications in Different Domains. Edureka! Training in top technologies. Edureka.com. https://www.edureka.co/blog/iot-applications/.

Valinsky, J. (2019, April 11). Amazon reportedly employs thousands of people to listen to your Alexa conversations. CNN.com. https://www.cnn.com/2019/04/11/tech/amazon-alexa-listening/index.html.

Woodall, M. (2019, December 18). Vivint Smart Home Security Review 2020. Reviews.org. https://www.reviews.org/home-security/vivint-smart-home-security-review/.

Yehiav, G. (2019, February 19). Paving The Way For Gen Z In Tech. *Forbes*. https://www.forbes.com/sites/forbestechcouncil/2019/02/19/paving-the-way-for-gen-z-in-tech/#53dbeaf52165.

23

Social Media

Rachel Stuart, Ph.D.[*]

Abstract

Social media has come to permeate almost every aspect of both our online and real-world experiences. The first listservs like CompuServe enabled the evolution of social media to multi-modal platforms such as Facebook, Instagram, and Snapchat. Video has fast become the post of choice on social media, and augmented reality features have become commonplace. How we interact with one another, celebrities, and brands has been indelibly changed by the complete integration of social media into our lives.

Introduction

A few years ago, while visiting family, we asked our nine-year-old nephew what he wanted to be when he grew up. Without missing a beat, he said he wanted to play Minecraft while streaming on YouTube to millions of followers. Little did we know at the time, but our nephew was tapping into one of the most popular trends in social media today: the Influencer. In the three decades since its inception, social media is now a driving force behind how we connect, get news, and disseminate information. Social media has come to permeate almost every aspect of both our online- and real-world experiences. Even with so much exposure,

there is still some confusion as to what constitutes a social networking site (SNS), and which of the literally millions of web pages on the internet can be considered SNSs.

What is an SNS? According to boyd and Ellison (2008), there are three criteria that a website must meet to be considered an SNS. A website must allow users to "(1) construct a public or semi-public profile within a bounded system, (2) articulate a list of other users with whom they share a connection, and (3) view and traverse their list of connections and those made by others within the system" (boyd & Ellison, 2008, p. 211). These guidelines may seem to restrict what can be considered an SNS; however, there are still literally hundreds of vastly diverse websites that are functioning as such.

---

[*] Faculty member in the Communication Studies Department, Highline College (Des Moines, Washington).

Social networking sites have seemingly permanently cemented their place in the landscape of the internet, and as we become comfortable explorers of the online social world, the nuances between social networking and social media become clearer and more defined. For over a decade, the terms social media and social networking sites have been used almost interchangeably. However, as the sites and apps dedicated to creating connections to people become more nuanced, there does seem to be a differentiation between the two.

According to *Social Media Today*, social media are forms of "...electronic communication (as websites for social networking and microblogging) through which users create online communities to share information, ideas, personal messages, and other content (as videos)" (Schauer, 2015, par. 3), whereas, social networking sites are dedicated to "...the creation and maintenance of personal and business relationships... online" (Schauer, 2015, par. 4). One way to think about the difference between social networking sites and social media is to think about how the internet and the world wide web are delineated: If you are on the world wide web, the you are on the internet, but just because you are on the internet does not necessarily mean you are on the web. The same is true for social media versus social networking sites; if you are on a social networking site, you are on a form of social media, but just because you are on social media, does not mean you are on a social networking site. As the social media and social networking sites become more nuanced and comprehensive, the distinction between them will become ever more blurred. Rather than looking toward the future at this point, let's take a look at the background and history of social media and SNSs.

Background

Social media sites have taken on many forms during their evolution. Social networking on the internet can trace its roots back to listservs such as CompuServe, BBS, and AOL where people converged to share computer files and ideas (Nickson, 2009). CompuServe was started in 1969 by Jeff Wilkins, who wanted to help streamline his father-in-law's insurance business (Banks, 2007).

During the 1960s, computers were still prohibitively expensive; so many small, private businesses could not afford a computer of their own. During that time, it was common practice to "timeshare" computers with other companies (Banks, 2007). Timesharing, in this sense, meant that there was one central computer that allowed several different companies to share access in order to remotely use it for general computing purposes. Wilkins saw the potential in this market, and with the help of two college friends, talked the board of directors at his father-in-law's insurance company into buying a computer for timesharing purposes. With this first computer, Wilkins and his two partners, Alexander Trevor and John Goltz, started up CompuServe Networks, Inc. By taking the basic concept of timesharing already in place and improving upon it, Wilkins, Trevor, and Goltz created the first centralized site for computer networking and sharing. In 1977, as home computers started to become popular, Wilkins started designing an application that would connect those home computers to the centralized CompuServe computer. The home computer owner could use the central computer for access, for storage and, most importantly, for person-to-person communications—both public and private (Banks, 2007**).**

Another two decades would go by before the first identifiable SNS would appear on the internet. Throughout the 1980s and early 1990s, there were several different bulletin board systems (BBSs) and sites including America Online (AOL) that provided convergence points for people to meet and share online. In 1996, the first "identifiable" SNS was created—SixDegrees.com (boyd & Ellison, 2008). SixDegrees was originally based upon the concept that no two people are separated by more than six degrees of separation. The concept of the website was fairly simple: sign up, provide some personal background, and supply the email addresses of ten friends, family, or colleagues. Each person had his or her own profile, could search for friends, and for the friends of friends (Caslon Analytics, 2006). It was completely free and relatively easy to use. SixDegrees shut down in 2001 after the dot com bubble popped. What was left in its wake, however, was the beginning of SNSs as they are known today. There have been literally hundreds of different SNSs that have sprung from the footprints of SixDegrees. In the decade following the demise of SixDegrees, SNSs such as Friendster, MySpace, Facebook, Twitter LinkedIn, and Instagram became internet zeitgeists.

Friendster was created in 2002 by a former Netscape engineer, Jonathan Abrams (Milian, 2009). The website was designed for people to create profiles that included personal information—everything from gender to birth date to favorite foods—and the ability to connect with friends that they might not otherwise be able to connect to easily. The original design of Friendster was fresh and innovative, and personal privacy was an important consideration. In order to add someone as a Friendster contact, the friend requester needed to know either the last name or the email address of the requested. It was Abrams' original intention to have a website that hosted pages for close friends and family to be able to connect, not as a virtual popularity contest to see who could get the most "Friendsters" (Milian, 2009).

Shortly after the debut of Friendster, a new SNS hit the internet, MySpace. From its inception by Tom Anderson and Chris DeWolfe in 2003, MySpace was markedly different from Friendster. While Friendster focused on making and maintaining connections with people who already knew each other, MySpace was busy turning the online social networking phenomenon into a multimedia experience. It was the first SNS to allow members to customize their profiles using HyperText Markup Language (HTML). So, instead of having "cookie-cutter" profiles like Friendster offered, MySpace users could completely adapt their profiles to their own tastes, right down to the font of the page and music playing in the background. As Nickson (2009) stated, "it looked and felt hipper than the major competitor Friendster right from the start, and it conducted a campaign of sorts in the early days to show alienated Friendster users just what they were missing" (par. 15). This competition signaled trouble for Friendster, which was slow to adapt to this new form of social networking. A stroke of good fortune for MySpace also came in the form of rumors being spread that Friendster was going to start charging fees for its services. In 2005, with 22 million users, MySpace was sold to News Corp. for $580 million (BusinessWeek, 2005); News Corp. later sold it for only $35 million. MySpace, amazingly, is still a functional social media site. The site, which was at one point the pinnacle of social networking sites is now focused on providing a platform for individuals to feature their music videos and songs (Arbel, 2016).

From its humble roots as a way for Harvard students to stay connected to one another, Facebook has come a long way. Facebook was created in 2004 by Mark Zuckerberg with the help of Dustin Moskovitz, Chris Hughes, and Eduardo Saverin (Newsroom, 2020). Originally, Facebook was only open to Harvard students, however, by the end of the year, it had expanded to Yale University, Columbia University, and Stanford University, with new headquarters for the company in Palo Alto, California. In 2005, the company started providing social networking services to anyone who had a valid e-mail address ending in .edu. By 2006, Facebook was offering its website to anyone over the age of 13 who had any valid email address (Newsroom, 2020). What made Facebook unique, at the time, was that it was the first SNS to offer the "news feed" on a user's home page.

In all other social media platforms before Facebook, in order to see what friends were doing, the user would have to click to that friend's page. Facebook, instead, put a live feed of all changes users were posting—everything from relationship changes, to job changes, to updates of their status. In essence, Facebook made microblogging popular. This was a huge shift from MySpace, which had placed a tremendous amount of emphasis on traditional blogging, where people could type as much as they wanted to. Interestingly, up until July 2011, Facebook users were limited to 420-character status updates. Since November 2011, however, Facebook users have a staggering 63,206 characters to say what's on their mind (Protalinski, 2011). The length of the post on Facebook has become more irrelevant as the platform tries to highlight its usefulness in providing a multimedia experience, including a new host of emotions to communicate to our friends, such as allowing for the more appropriate frowny-face emoji than the generic "thumbs-up" like button when your best friend's dog died. In addition, Facebook has introduced more interactive features, like 360 panorama photos, live vlogging, and increased control over photographs with filters and augmented reality enhancements. Facebook, unlike Twitter, is no longer focused on the length of the post, but the quality of content you include with it—including the increase of video integration into our posts.

Twitter was formed in 2006 by three employees of podcasting company Odeo, Inc.: Jack Dorsey, Evan Williams, and Biz Stone (Beaumont, 2008). It was created out of a desire to stay in touch with friends easier than allowed by Facebook, MySpace, and LinkedIn. Originally, Twitter took the concept of the 160-character limit the first iteration of text messaging imposed on its users and shortened the message length down to 140 (to allow the extra 20 characters to be used for a user name; McArthur, 2017). This made Twitter one of the first truly mobile social media platforms because you could tweet via a texting app. The 140-character limit had been the identifying hallmark of the platform, even as it added video and photo functionality seamlessly to its mobile and desktop interfaces. On November 7, 2017, Twitter officially expanded its tweet length to 280, effectively doubling the message length you can tweet (Perez, 2017). The response to this development has been surprising. On one hand, the expansion was celebrated because it would allow people to express themselves more clearly than 140 characters would allow. On the other, the overall sentiment toward the change has been negative, with analysis from 2.7 million tweets on the matter showing that 63% were critical of the expanded length of a tweet (Perez, 2017).

Social media sites such as Twitter and Facebook clearly helped cement social media within the landscape of the internet. The last 15+ years of social media tells a story of more dynamic, specialized social media sites entering the scene and stealing the spotlight from the above-mentioned giants. Sites like LinkedIn, Snapchat, Instagram, and Reddit point toward the increasingly diversified and specialized trajectory social media is taking. LinkedIn was created in 2003 by Reid Hoffman, Allen Blue, Jean-Luc Vaillant, Eric Ly, and Konstantin Guericke (About Us, 2020). LinkedIn returned the concept of SNSs to its old CompuServe roots. According to Stross (2012), LinkedIn is unique because "among online networking sites, LinkedIn stands out as the specialized one—it's for professional connections only" (par. 1). Instead of helping the user find a long-lost friend from high school, LinkedIn helps build professional connections, which in turn could lead to better job opportunities and more productivity. In June 2016, Microsoft bought the social media platform for $26.2 billion in cash (Darrow,

2016). Growth overall for LinkedIn has been slow but steady. In April 2017, the site reported to have 500 million users worldwide (Darrow, 2017). By November 2019, the site boasts over 660 million users with the US being the largest market with 125 million unique profiles (Iqbal, 2019a).

Snapchat, which was originally released under the name Picaboo, was created by two Stanford University students, Evan Spiegel and Robert Murphy, in 2011 (Colao, 2012). The original premise of Snapchat was simple: users send friends "snaps," photographs and videos that last anywhere from one to ten seconds, and when the time expires the photos or videos disappear. In addition to the fleeting nature of the snap, if the recipient of a snap screen captures it, Snapchat will let the sender know that the person they sent it to saved it. By May 2017, Snapchat started to allow users to send snaps of any length and allowed the recipients of the snaps to keep them indefinitely (Newton, 2017). In 2018, Snapchat stepped away from exclusively hosting user-generated content and added Snapchat Originals with content produced by NBCUniversal, ABC, and BBC to name a few (Iqbal, 2019b). To say that Snapchat has become popular would be an understatement. By October 2012, one billion snaps had been sent, and the app averaged more than 20 million snaps a day (Gannes, 2012). In the first quarter of 2019, Snapchat had approximately 203 million daily users worldwide, with 97.6 million concentrated in the United States (Iqbal, 2019b).

Instagram was originally developed by Kevin Systrom and Mike Krieger in 2010 as a photo-taking application where users could apply various filters to their photographs and then upload them to the internet. In 2011, Instagram added the ability to hashtag the photographs uploaded as a way to find both users and photographs (Introducing hashtags, 2011). Facebook purchased Instagram in 2012 from Systrom and Krieger for $1 billion in cash and stock (Langer, 2013). By the time Facebook acquired the filtered photo app giant, Instagram had over 30 million users (Upbin, 2012). While many in the tech industry saw Facebook's extravagant price paid for Instagram as a bad business decision, the move has proved to be lucrative as Instagram now has one billion monthly active users, and 500 million people use Instagram Stories daily (Newberry, 2019). If Instagram was a standalone company, it would be worth more than $100 billion today (McCormick, 2018).

Trying to keep up with the shift to visually-based, mobile social media platforms, Twitter bought the video-sharing start up Vine for $30 million in 2012 (Vine, 2018). Vine was created by Dominik Hoffman, Rus Yusupov, and Colin Kroll the year before and operated as a private, invite-only application (Dave, 2013). After its acquisition by Twitter, the number of active users of Vine shot up to over 200 million (Levy, 2015). Vine did not find the same traction that sites like Instagram or Snapchat did, and in late 2016, Twitter discontinued the video platform (Foxx, 2016). One of Vine's original developers, Dominik Hoffman, is now in closed beta testing of what he calls version two of Vine, a video loop site called byte (byte, n.d.).

Perhaps the ultimate, if not most misunderstood and underrated, example of the increasingly specialization of social media is the website Reddit. The self-proclaimed "front page of the internet," Reddit is the epitome of the user generated content that is the hallmark of Web 2.0. Reddit was developed in 2005 by Alexis Ohanian and Steve Huffman when the pair wanted to create a website that aggregated the most popular and shared links on the internet. In addition to moderators posting links on the site, Reddit also allowed users to post their favorite links to the site. In its first few years, the website was competing with sites like Del.icio.us and Digg, however, in 2008, Reddit allowed the ability for users to create pages for specialized content, also known as subreddits (Fiegerman, 2014).

The creation of subreddits proved to be the major turning point for Reddit, letting users to both post their own content and to create pages within the website to share content on a specific thing. As of January 2020, there were over 1.84 million subreddits hosted on the site, up from just over 1.2 million subreddits available at the same time the year before (New Subreddits, n.d.). To demonstrate how fast subreddits are created on the site, on January 12, 2020, 1917 new Subreddits were added to the site (New Subreddits, n.d.) There are subreddits for just about every subject imaginable, and while this liberty has created issues surrounding questionable to illegal content (Fiegerman, 2014), the self-generated

and self-policing on the site has created a massive community of Redditors, participating in a global social media experiment where you are just as liable to find a plethora of cat pictures as you are a group of Redditors saving a fellow user from the temptation of suicide. Reddit is still lagging behind the other social media giants in keeping up with the evolving demands of mobile technologies, but the site is a hallmark of the social media trend of today: giving users what they want, when they want it.

Recent Developments

Social media is entering its third decade of being in the front of the collective consciousness of the internet community around the world. As the technology has matured and reached a relatively stable saturation point, there have been issues that users and platforms alike have had to navigate. These include issues of data privacy and sites "listening" (tracking our page visits online, and then using targeted marketing based on that data), have made many leery of living our lives partially online. In addition, the continued questioning of the veracity of the content we are exposed to on social media has created virtual communities of people who either uncritically consume intentionally misleading information or are suspicious of any information that is posted online. These complications in the virtual realm are divorcing us from our ability to know what is real and what is not. Even with the troubling trend of information manipulation in social media, one thing is certain: social media sites are not going away. Some of the most recent developments we have seen in social media include a shift toward video, continued integration of augmented reality into social media platforms, and the rise of "influencers" as the voice of social media.

Video is King

One of the most surprising, yet logical, developments in social media is the transition YouTube has gone through from a video hosting platform to a social media site. YouTube now qualifies as a social media site as it allows fellow viewers and posters to follow one another and send friend requests (Foster, n.d.). Because of this, YouTube is now the most popular social media site on the internet with 73% of U.S. adults using the site (Social media fact sheet, 2019). YouTube is not the only site ushering in the era of video- and GIF-based social media. Globally, relative newcomers like TikTok, QZone, and Sina Weibo have provided competition to established sites like YouTube, Instagram, Snapchat, Periscope, Facebook, and Twitter in the shift toward video-centric social media. The former three social media sites gained popularity in their native China before getting a global audience in recent years.

All of these sites have embraced a shift to platforms that support pictures, memes, GIFs, videos, and live streaming as the predominant forms of communication. Two of the most popular forms of video sharing taking social media by storm are live streaming our lives and the integration of "stories" into platforms like Snapchat, Instagram, and Facebook. The stories featured on these sites allow users to post multiple pictures and videos that are played in a slideshow format and are usually only available for 24 hours. Instagram's stories feature is a straight-up copycat of Snapchat's stories feature, including the 24-hour availability time limit, and the ability to add text, doodles, stickers, and the use of augmented reality lenses (Tillman, 2018). The purposes of the stories feature is to allow our online friends to see what we are doing in our day-to-day lives. In addition to the integration of stories into social media, you are also able to share your stories, videos and posts across platforms, so rather than having to post to multiple sites independently or using a third party site to post to multiple site immediately, sites like Instagram and Twitter give you the option to link your social media profiles together and allow you to post instantaneously to all of them.

In addition to the full-embrace of "stories" across many social media platforms, there is also an increase in live video streaming on social media. According to van Moessner (n.d.), 47% of social media consumers worldwide increased their live streaming from 2017 to 2018. In addition, Cisco Systems predicts that 82% of all internet traffic globally will be video (Cisco visual networking index, 2019). Live video streaming has become popular with not only the individual users of social media, but also among businesses and advertisers. Businesses are using live streaming to promote new products to consumers while also using it to conduct conferences and employee meetings internally (Golum,

n.d.). The film and television industries are also taking note of this trend and releasing shows like *SKAM Austin* onto Facebook Watch. The original series, *Skam*, is a popular Norwegian show that would release 6–8 minute segments on social media throughout the week and then stitch the entire episode together on Fridays. Also, during the week, the characters of the show are consistently posting to social media—not the actors, but the characters. *SKAM Austin* is hoping to capture the same popularity the original show had with a similar layout of the simultaneous release of information across platforms. The ability to simultaneously add content to all of your social media platforms has fueled the increase in user generated video. Like the stories feature on social media, users also have the ability to live stream across multiple platforms simultaneously using third party platforms like Vimeo (van Moessner, n.d.).

What is Reality?

Total Integration of Augmented Reality

Augmented reality (AR) burst into our collective consciousness with Pokémon Go. Different from virtual reality, which is the complete computer generation of an immersive environment, augmented reality programs add to the reality that we already live in (Emspak, 2018). In the few short years since then augmented reality features have found their way onto our devices through a wide range of apps. Want to know what that couch from IKEA is going to look like in your living room? Augmented reality can show you. Want to freak your cat out by turning your face into a cat face? Augmented reality can do that. Augmented reality features have become most popular on—you guessed it—social media apps. Fully 78% of social media apps on the market utilize some form of AR on their platforms (Martin, 2019).

The most popular social media app using AR is Snapchat with their ever-increasing selection of lenses and filters that modify us and our surroundings. According to Martin (2019), 80% of Snapchatters are using the AR features on their snaps at least 50% of the time. (For more on augmented reality, see Chapter 13.)

The largest group to sit up and take notice of AR in social media are businesses and marketers.

Companies are creating virtual markets on social media where you can "step into their store" and try out products. If you like what you see, you can buy it right there online—no need to ever leave your couch. This innovation is going to continue to steer business away from traditional brick and mortar stores. In addition, you can very easily create viral marketing strategies through the effective use of AR filters and lenses. If you create a fun, funky new lens for a social media platform, people will want to share the experience with their friends and followers and you will have more widespread exposure than with a lot of more traditional forms of advertising (Bullock, 2018). In the past few years, how businesses reach their customers has taken a radical shift through features like AR. One relatively new trend that has blurred the line between friends and advertising is the tidal wave of influencers that have come crashing through our social media.

Why be Famous when You can be Instafamous?

Influencers on social media

What is an influencer? An influencer "…is a user who has established credibility in a specific industry, has access to a huge audience and can persuade others to act based on their recommendations" (9 of the, 2019). Influencers online can be anyone who achieves enough reach and credibility to attract consistent traffic to their posts and profiles. Some of the most famous influencers online as of mid-2020 include Huda Kattan, a makeup expert who has over 40.7 million followers on Instagram, and Kayla Itsines, a personal trainer with 12 million followers on Instagram. Itsines posts videos and messages with health and diet tips along with inspirational quotes. Influencers have found a novel way to make money off of the posts they were already making on social media. One example of an influencer creating her own social media empire is Kylie Jenner, originally of *Keeping Up with the Kardashians* fame. Jenner has over 158 million followers and is reported to make $1 million for each of her sponsored posts on Instagram (Mejia, 2018). Kylie Jenner is the youngest self-made billionaire, with most of her money coming from her social media driven company, Kylie Cosmetics (Robehmed, 2019). When her cosmetic line finally hit physical store fronts via Ulta Beauty, Jenner was quoted as saying: "I popped up at a few stores, I did my usual social media—I did what I usually do, and it just worked" (Jenner, as quoted in Robehmed, 2019, para. 2). The big paychecks for influencers are not just for the Kardashian clan and their ilk, either. Selina Gomez makes $800,000 per sponsored post, and Christiano Ronaldo makes $750,000 per post (Mejia, 2018).

Why have influencers become, well… so influential? A big reason for the rise of influencers on social media is the desire users have to a personal connection to the businesses and brands that they are using. When someone buys a lip kit from Kylie Cosmetics online, they are not just buying the product of a cosmetic company, they are buying a product created by a trusted online "friend" (Goodwin, 2020). In addition, Influencers have begun to fundamentally change how people (especially in GenZ) expect companies to interact with them in regards to advertising and marketing. They want to feel like their interaction with the Influencer and their brand is "authentic." This is a big reason why Kylie Jenner has had more success both as an Influencer and businesswoman than her half-sister Kim Kardashian—because more authentic influencers, like Kylie, "…will share content they care about and actually use on a regular basis" (Goodwin, 2020, para. 36).

This desire for a more authentic interaction has also led to the social media teams for major companies to take a more personal and novel approach to their interaction with customers online. For example, the social media profiles of companies like Netflix, Wendy's, Moon Pie, and Burger King have all created viral campaigns based on spicy tweets. Netflix gained attention in 2019 by getting other companies involved with the following tweeted question: "what's something you can say during sex but also when you manage a brand twitter account?" (Netflix US, 2019). The tweet went viral almost immediately, and as of January 14th, 2020, it has been liked over 436,000 times, retweeted almost 111,000 times and replied to 26,500 times. Amazingly, many of the replies were from the verified accounts of other companies, and the thread gained international attention from other social media sites and more traditional news and media outlets. This shift to a much more informal interaction with customers and other brands is a result of users and consumers of social media craving a more authentic engagement online.

Fake News & Social Media

Since 2016, there has been a revolution in how we perceive the information we receive on the internet in general, and from social media in particular. Our ability to find literally anything that will validate our points have created the era of disinformation and "fake news." If we don't like or agree with information, whether it is a verified fact or not, we can just say that it is fake news. The term, "fake news," has been defined by *Cambridge Dictionary* as "false stories that appear to be news, spread on the internet or using other media, usually created to influence political views or as a joke…" (Cambridge Dictionary, n.d.). One of the biggest issues in the ever-evolving battle against fake news on social media is that it is no longer blatantly obvious what is an outright lie and what is merely misleading (Hutchinson, 2020). In an era where people skim headlines to form their political, personal, and spiritual points of view, a well-crafted—if not totally misleading—headline can be enough to perpetuate the spread of disinformation across the internet. As an example, Hutchinson (2020) points to an article with the headline: "Vegans want you to stop saying 'bring home the bacon' because it is offensive." The article itself, however, is actually about how the metaphors we use for food will most likely evolve over time (Hutchinson, 2020). However, because most people do not make it past the headline, the rumor spreads that vegans hate bacon… and fun.

While vegans have to deal with a lot of (some would say justified) hate online, the spread of disinformation and fake news can have serious consequences. As this book is going to press in spring 2020, most of you in school—from kindergarten through college—have been sent home for the foreseeable future because the world has been rocked by one of the most serious pandemics of the modern era—the spread of COVID-19. For years to come, researchers will study how the efforts to contain the virus in the United States have been hampered by a plethora of fake news proliferating across social media. One issue that is compounding the coronavirus situation is the spread of disinformation coming from the highest echelons of our government.

Fake news affected the outcome of a presidential election in 2016 (Lipowsky, 2016), our citizens' perceptions about the ethics and morality of creating concentration camps for those seeking refuge (Serwer, 2019), and whether or not climate change exists (Ward, 2018). These are just examples of fake news that is coming from political leaders who should know better. The misinformation is not limited to one group; there are numerous examples of misleading headlines and carefully framed news stories from all sectors of society. Our inability to trust the news we find on social media has a profound effect on how we gather and disseminate information that in some cases is merely funny or annoying, but in other, more severe cases, can have life or death consequences.

Current Status

Social media usage around the world continues to increase at a steady rate. Of the 7.7 billion people on the planet, 4.4 billion are active internet users, and 3.5 billion, or 79.4%, of that population are active on at least one social media platform (Kemp, 2019). Signaling the shift of social media to our mobile devices, 2.6 billion social media users are accessing their profiles via mobile devices (Kemp, 2019). Between January 2018 and January 2019 there was a 9% increase in the number of social media users worldwide. Throughout the world, social media saw an increase of 288 million new users. The worldwide saturation of social media compared to the entire population is 45%, however that number is not distributed evenly; 70% of the population in North America uses social media, whereas middle and eastern Africa have usage in the single digits (7% and 8%, respectively; Kemp, 2019). Worldwide, the most popular social media platforms are Facebook with 2.4 billion unique users and YouTube with 1.9 billion unique users (Ortiz-Ospina, 2019). YouTube reigns supreme, however, on the average amount of time spent on the site, where the average visit lasts 21 minutes 36 seconds, compared to the average time of 11 minutes 44 seconds spent on Facebook (Kemp, 2019).

As of January 2019, approximately 72% of adults access at least one or more social media site in the United States (Social Media Fact Sheet, 2019). There was only a 2% growth overall in the percentage of Americans on social media from 2015 through 2019

(Social Media Fact Sheet, 2018). Two sites that have gained in popularity, especially among the 18-29-year old demographic, are Instagram and Snapchat. Perrin & Anderson (2019) reported that among American internet users between 18-29, 67% have used Instagram and 62% have used Snapchat. In addition, there is a very clear split *within* that demographic of users between 18-24 and 25-29: Instagram (75% vs. 57%, respectively) and Snapchat (73% vs. 47%; Perrin & Anderson, 2019).

According to the Pew Research Center, social media usage remained steady at 69% amongst black users, whereas Hispanic users of social media declined from 72% in 2018 to 70% in 2019 (Social Media Fact Sheet, 2018). Women are more likely to be on social media than men, with 78% of women in the United States reporting to have one or more social media profiles, compared to 65% of men. (Social Media Fact Sheet, 2019). Looking at education, 79% of college graduates, 74% of those with some college, and 64% of those with high school or less have at least one active social media profile online in 2019 (Social Media Fact Sheet, 2019).

Figure 23.1

Percentage of U.S. adults who use at least one social media site, by race

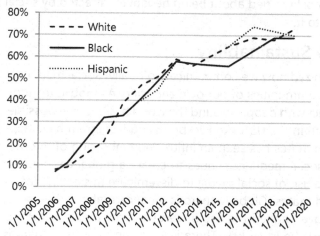

Source: Social Media Fact Sheet

As mentioned previously, YouTube has taken over Facebook as the most widely used social media platform in the United States. Facebook's saturation has remained steady at 69% from 2017 to 2020, whereas 73%

of the population uses YouTube (Social media fact sheet, 2019). Facebook still has the highest percentage of daily users at 74%, compared to YouTube's reported daily use of 51% among those with profiles. The next most popular social media sites behind Facebook are Instagram (37% of the population), Pinterest (28%), LinkedIn (27%), and Twitter (21%) (Social Media Fact Sheet, 2019). Each of these social media sites saw an increase in popularity between January 2018 and February 2019.

Figure 23.2

Percentage of U.S. adults who use at least one social media site, by education level

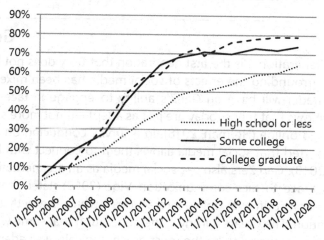

Source: Social Media Fact Sheet

Factors to Watch

The biggest trend in social media is video being the primary type of online post. Another trend that will continue to enhance our experience of social media is augmented reality. In the business world, augmented reality will start to be utilized more during meetings and conferences. Social media is also going to continue to be the primary source of news on the internet. According to Owen (2019), 50% of internet users stated they hear a breaking news story via social media before hearing it on other websites or traditional media outlets. It is going to be imperative for traditional news sources to create a more dominant social media presence to gain attention and to fight the ever-increasing issue of "fake news" online.

Adaptation to social media is not just a trend that news and information sites are going to have to get used to. Expect to see other technologies affected by and catering to our reliance on social media. One example of other communication technologies embracing social media is the Sero, Samsung's vertical TV prototype. The Sero which debuted at a pop-up store in Seoul, South Korea in mid-2019, is a TV that can very easily have its orientation changed from horizontal to vertical. Samsung admits that the conception of this television is an acknowledgement that much of the video media we are consuming online is filmed in a vertical orientation (Savov, 2019).

Finally, pay attention to Tik-Tok. As of mid-2020, this Chinese social media site is growing faster globally than other social media in 2020. The app was released worldwide in 2016, and as of 2019 already had 500 million global users (Mohsin, 2019). This number does not include the number of users of the Chinese version, Douyin, which is maintained as a different platform because of China's censorship laws. Tik-Tok was the most downloaded app on the Apple App Store in the first quarter of 2019 with 33 million individual downloads (Mohsin, 2019). What is setting Tik-Tok apart from other social media platforms is that it is being marketed as an entertainment social media app rather than a lifestyle social media app. It encourages its users to embrace their creativity and uniqueness, which is the perfect app for millennials and GenZ to be able to express themselves. Move over Facebook and Instagram, you might have finally met your match.

Gen Z

Generation Z is the first generation that truly does not know a world without social media. Much of the reporting surrounding the effects of social media has been mixed. There have been concerns about the effects that social media will have on GenZ's ability to engage in face-to-face interpersonal communication, however, research shows that teens today are just as adept—if not more so—in embracing the fluidity of living both in the real world and online (Herman, 2018). In a survey conducted on the social media habits of teenagers today, a majority reported that social media allows them to connect with both friends and more diverse groups of people (Herman, 2018). Gen Zers also see social media as an augmentation to their day-to-day communication, not a replacement of face-to-face communication. Vigo (2019) stated that in recent research done, Gen Z prefers face-to-face communication over online interactions—which is exactly the opposite from the preferences of the Millennial generation. The greatest irony is the generation that some were most worried about being negatively affected by social media, seems to be the most psychologically well adapted to it.

Global Concerns & Sustainability

The internet tore down geographical barriers and revolutionized how we communicate globally. Social media has created islands of refuge where we gather into intentional communities of our own choosing. As mobile technologies have become more sophisticated, our ability to connect with people around the world has become possible. Most of the growth of the old school social media sites born in the USA like Facebook and Instagram have happened internationally, and in the past few years, the United States has seen an influx and embracing of international social media sites like Tik-Tok and Weibo. Even as social media continues to create a global community, there are still some concerns about content and access. The use of social media to disseminate misleading information is not something that has affected only the United States. Globally there is a concern about the use of social media as a propaganda tool, and many countries have banned the use of certain social media platforms entirely—including Facebook in China and Russia, and YouTube in Turkey. There is a growing fear that there is the potential for social media to be weaponized against us, and indeed that may already be true.

Bibliography

9 of the biggest social media influencers on Instagram. (2019). *Digital Marketing Institute*. https://digitalmarketinginstitute. com/blog/9-of-the-biggest-social-media-influencers-on-instagram?utm_source=Organic+Facebook&utm_medium= Blog&utm_campaign=Facebook&utm_content=9+of+the+Biggest+Social+Media+Influencers+on+Instagram&fbclid= IwAR1wgCXCPIt7CipBR3iOFFdZNfHhQtUXlkZo3yojVjiAFgCCpH0TmuOoI6Q.

About us. (2018). *LinkedIn*. http://press.linkedin.com/about.

Arbel, T. (2016, February). Myspace still exists? Yes, and now Time Inc. owns it. *Seattle Times*. https://www.seattletimes. com/business/magazine-publisher-time-inc-buys-whats-left-of-myspace/.

Banks, M.A. (2007, January). The Internet, ARPANet, and consumer online. *All Business*. https://www.questia.com/ magazine/1G1-158091728/the-internet-arpanet-and-consumer-online.

Beaumont, C. (2008, November). The team behind Twitter: Jack Dorsey, Biz Stone and Evan Williams. *Telegraph*. http://www.telegraph.co.uk/technology/3520024/The-team-behind-Twitter-Jack-Dorsey-Biz-Stone-and-Evan-Williams.html.

boyd, d.m. & Ellison, N.B. (2008). *Social networking sites: Definition, history, and scholarship*. Journal of Computer-Mediated Communication, 13, 210-230.

Bullock, L. (2018, November). AR and social media: Is augmented reality the future of social media? *Forbes*. https://www.forbes.com/sites/lilachbullock/2018/11/16/ar-and-social-media-is-augmented-reality-the-future-of-social-media/#23e76bb9e141.

BusinessWeek. (2005, July). MySpace: WhoseSpace?. *BusinessWeek*. http://www.businessweek.com/echnology/content/ jul2005/tc20050729_0719_tc057.htm.

byte. (n.d.). byte! https://byte.co.

Cambridge Dictionary (n.d.). Fake news. https://dictionary.cambridge.org/dictionary/english/fake-news.

Caslon Analytics. (2006). Caslon Analytics social networking services. *Caslon Analytics*. http://www.caslon.com.au/ socialspacesprofile2.htm.

Cisco visual networking index. (2019, February). *Cisco*. https://www.cisco.com/c/en/us/solutions/collateral/service-provider/ visual-networking-index-vni/white-paper-c11-741490.html.

Colao, J. J. (2012, November). Snapchat: The biggest no-revenue mobile app since Instagram. *Forbes*. http://www.forbes.com/sites/jjcolao/2012/11/27/snapchat-the-biggest-no-revenue-mobile-app-since-instagram/.

Coppins, M. (2020 March 11). Trump's dangerously effective coronavirus propaganda. *The Atlantic*. https://www. theatlantic.com/politics/archive/2020/03/trump-coronavirus-threat/607825/.

Darrow, B. (2016, June). Microsoft buying LinkedIn for $26.2 billion. *Fortune*. http://fortune.com/2016/06/13/microsoft-buying-linkedin/.

Darrow, B. (2017, April). LinkedIn claims half a billion users. *Fortune*. http://fortune.com/2017/04/24/linkedin-users/

Dave, P. (2013, June). Video app Vine's popularity is spreading, six seconds at a time. *The Los Angeles Times*. http:// articles.latimes.com/2013/jun/20/business/la-fi-vine-20130620.

Emspak, J. (2018, June). What is augmented reality? *Live Science*. https://www.livescience.com/34843-augmented-reality.html

Fake News. (n.d.). *Cambridge Dictionary*. https://dictionary.cambridge.org/us/dictionary/english/fake-news.

Fiegerman, S. (2014, December). Aliens in the valley: The complete history of Reddit, the front page of the internet. *Mashable*. http://mashable.com/2014/12/03/history-of-reddit/.

Foster, T. (n.d.). Is YouTube considered a social media site? *Foster Web Marketing*. https://www.fosterwebmarketing.com/ faqs/is-youtube-considered-a-social-media-site-.cfm.

Foxx, C. (2016, October). Twitter axes Vine video service. *BBC News*. http://www.bbc.com/news/technology-37788052.

Gannes, L. (2012, October). Fast-growing photo-messaging app Snapchat launches on Android. *All Things D*. http:// allthingsd.com/20121029/fast-growing-photo-messaging-app-snapchat-launches-on-android/.

Golum, C. (n.d.). How corporations are using live video to grow and adapt. *Vimeo-Livestream*. https://livestream.com/blog/corporations-using-live-video.

Goodwin, D. (2020, January). 10 important 2020 social media trends you need to know. *Search Engine Journal*. https://www.searchenginejournal.com/2020-social-media-trends/342851/#close.

Herman, B. (2018, December). Gen Z and the impact of social media: It's complicated. *R/GA*. https://www-prd.rga.com/ futurevision/pov/gen-z-and-the-impact-of-social-media-it-s-complicated.

Hutchinson, A. (2020 January 3). What if fake news isn't the real problem on social media? *Social Media Today*. https://www.socialmediatoday.com/news/what-if-fake-news-isnt-the-real-problem-on-social-media/569711/.

Introducing hashtags on Instagram. (2011 January). *Instagram*. http://blog.instagram.com/post/8755963247/introducing-hashtags-on-instagram.

Iqbal, M. (2019a, December). LinkedIn usage and revenue statistics (2019). *Business of Apps*. https://www.businessofapps.com/data/linkedin-statistics/.

Iqbal, M. (2019b, September). Snapchat revenue and usage statistics (2019). *Business of Apps*. https://www.businessofapps.com/data/snapchat-statistics/.

Kemp, S. (2019, January). Digital 2019: Global internet use accelerates. *We Are Social*. https://wearesocial.com/blog/2019/01/digital-2019-global-internet-use-accelerates.

Langer, A. (2013, June). Six things you didn't know about the Vine app. *Yahoo! Finance*. http://finance.yahoo.com/news/six-things-didnt-know-vine-192222105.html.

Levy, A. (2015). What can Vine be for Twitter Inc? *The Motley Fool*. https://www.fool.com/investing/general/2015/11/08/what-can-vine-be-for-twitter-i nc.aspx.

Lipowsky, I. (2016 November 15). Here's how Facebook actually won Trump the presidency. *Wired*. https://www.wired.com/2016/11/facebook-won-trump-election-not-just-fake-news/.

Martin, C. (2019, December). Augmented reality dominated by social media. *Media Post*. https://www.mediapost.com/publications/article/344179/augmented-reality-dominated-by-social-media.html.

McArthur, A. (2017, November). The real history of Twitter, in brief. *Lifewire*. https://www.lifewire.com/history-of-twitter-3288854.

McCormick, E, (2018, June). Instagram is estimated to be worth more than $100 billion. *Bloomberg*. https://www.bloomberg.com/news/articles/2018-06-25/value-of-facebook-s-instagram-estimated-to-top-100-billion.

Mejia, Z. (2018, July). Kylie Jenner reportedly makes $1 million per paid Instagram post—here's how much other top influencers get. *CNBC*. https://www.cnbc.com/2018/07/31/kylie-jenner-makes-1-million-per-paid-instagram-post-hopper-hq-says.html.

Milian, M. (2009, July). Friendster founder on social networking: I invented this stuff (updated). *The Los Angeles Times*. http://latimesblogs.latimes.com/technology/2009/07/friendster-jonathan-abrams.html.

Mohsin, M. (2019, November). 10 Tik-Tok statistics that you need to know in 2020 [Infographic]. *Oberlo*. https://www.oberlo.com/blog/tiktok-statistics.

Netflix US [@netflixUS]. (2019 December 5). What's something you can say during sex but also when you manage a brand twitter account? [Tweet].

New subreddits. (n.d.). *Reddit Metrics*. https://redditmetrics.com/history.

Newsroom. (2020). *Facebook*. https://newsroom.fb.com/company-info/.

Newberry, C. (2019, October). 37 Instagram stats that matter to marketers in 2020. *Hootsuite*. https://blog.hootsuite.com/instagram-statistics/.

Newton, C. (2017, May). Snapchat adds new creative tools as its rivalry with Instagram intensifies. *The Verge*. www.theverge.com/2017/5/9/15592738/snapchat-limitless-snaps-looping-videos-magic-eraser-emoji-drawing.

Nickson, C. (2009, January). The history of social networking. *Digital Trends*. http://www.digitaltrends.com/features/the-history-of-social-networking/.

Ortiz-Ospina, E. (2019, September). The rise of social media. *Our World in Data*. https://ourworldindata.org/rise-of-social-media.

Owen, J. (2019, December). Top 20 social media future trends [2020 & beyond]. *Tech Jackie*. https://techjackie.com/social-media-future/.

Perez, S. (2017, November). Twitter officially expands its character count to 280 starting today. *TechCrunch*. https://techcrunch.com/2017/11/07/twitter-officially-expands-its-character-count-to-280-starting-today/.

Perrin, A. & Anderson, M. (2019 April). Share of U.S. adults using social media, including Facebook, is mostly unchanged since 2018. *Pew Research Center*. https://www.pewresearch.org/fact-tank/2019/04/10/share-of-u-s-adults-using-social-media-including-facebook-is-mostly-unchanged-since-2018/.

Protalinski, E. (2011, November). Facebook increases status update character limit to 63,206. *ZDNet*. http://www.zdnet.com/blog/facebook/facebook-increases-status-update-character-limit-to-63206/5754.

Robehmed, N. (2019 March 5). At 21, Kylie Jenner becomes the youngest self-made billionaire ever. *Forbes*. https://www.forbes.com/sites/natalierobehmed/2019/03/05/at-21-kylie-jenner-becomes-the-youngest-self-made-billionaire-ever/#353a920a2794.

Savov, V. (2019, April). Samsung thinks millennials want vertical TVs. *The Verge*. https://www.theverge.com/2019/4/29/18522287/samsung-sero-vertical-tv-price-release-date-millennials.

Schauer, P. (2015, June). 5 biggest differences between social media and social networking. *Social Media Today*. http://www.socialmediatoday.com/social-business/peteschauer/2015-06-28/5-biggest-differences-between-social-media-and-social.

Serwer, A. (2019 July 3). A crime by any name. *The Atlantic*. https://www.theatlantic.com/ideas/archive/2019/07/border-facilities/593239/.

Social media fact sheet. (2018, February). *Pew Research Center*. http://www.pewinternet.org/fact-sheet/social-media/.

Social media fact sheet. (2019, June). *Pew Research Center*. https://www.pewresearch.org/internet/fact-sheet/social-media/#who-uses-each-social-media-platform.

Stross, R. (2012, January). Sifting the professional from the personal. *The New York Times*. http://www.nytimes.com/2012/01/08/business/branchout-and-beknown-vie-for-linkedins-reach.html?_r=2.

Tillman, M. (2018, June). What are Instagram stories and how do they work? *Pocket Lint*. https://www.pocket-lint.com/apps/news/instagram/138416-what-is-instagram-stories-and-how-does-it-work.

Upbin, B. (2012, April). Facebook buys Instagram for $1 billion. Smart arbitrage. *Forbes*. http://www.forbes.com/sites/bruceupbin/2012/04/09/facebook-buys-instagram-for-1-billion-wheres-the-revenue/.

van Moessner, A. (n.d.). Getting started with social media streaming: what you need to know. *Vimeo-Livestream*. https://livestream.com/blog/social-media-streaming-livestream.

Vigo, J. (2019, August). Generation Z and new technology's effect on culture. *Forbes*. https://www.forbes.com/sites/julianvigo/2019/08/31/generation-z-and-new-technologys-effect-on-culture/#6a00dc355c2a.

Vine. (2018). *Crunch Base*. http://www.crunchbase.com/company/vine.

Ward, B. (2018 July 17). President Trump's fake news about climate change. *Ecologist*. https://theecologist.org/2018/jul/17/president-trumps-fake-news-about-climate-change.

24

Big Data

Tony R. DeMars, Ph.D.[*]

Abstract

Using data for business analytical purposes is nothing new; using data for business analytical purposes is a new and significant component of communication technology. Can both statements be true? In fact, they are. Big Data refers to the volume of data created, distributed and stored by way of internet connections. It is important because of its extraordinary value to business organizations, that can use Big Data analytics for everything from advertising applications to understanding customers in order to create new products, services and experiences (How Big Data Can Help, n.d.). Big data can be gathered in a variety of ways, including consumer surveys, browsing behavior, social media commenting and credit card transactions (Delgado, 2017).

In the "old" of data analytics, we recognize, for example, that original broadcasting services from the 1920s to the 1940s led to a need to analyze and use audience data, as exemplified by the work of such pioneers as Archibald Crossley, George Gallup, Elmo Roper, and Arthur C. Nielsen (Locke, 2017). In the "new" of data analytics, the internet of Things (IoT), means tens of billions of devices are connected. IoT is a major force driving Big Data analytics (Jain, et al., 2016). Anything that can be digitized has been or is being digitized, and all these things in our lives are connecting and communicating through the internet. The International Data Corporation forecasts that there will be over 41 billion connected devices generating 80 zettabytes of data in the next five years (Bhagat, 2019). Two-and-a-half quintillion bytes of data are created daily (Mills, 2018). Users and devices on the internet are directly related to the continued increase in the volume of data, with 2.5 billion worldwide internet users in 2012 and 4.1 billion in 2019 (Petrov, 2019). Analysts believe Big Data is changing the way we live, work, and think (Mayer-Schönberger & Cukier, 2013; Mills, 2018).

---

[*] Professor and Director, Radio-Television Division, Texas A&M University-Commerce (Commerce, Texas)

Introduction

Early research on Big Data was more in the Information Technology (IT) field than in communication. Despite still not being a top-of-mind topic in the communication technology realm, Big Data now deserves greater attention. The term "Big Data" only started to gain broad attention over the past decade, but it is a long-evolving term when viewed from a broader perspective.

Big Data refers to the large volume of both structured and unstructured data that belongs to different organizations, with each organization often using the data for different purposes (Top 13 Best, 2020). DeMauro, et al. (2015) note that the "fuel" of Big Data is information, adding "One of the fundamental reasons for the Big Data phenomenon to exist is the current extent to which information can be generated and made available" (p. 98). Data analytics company SAS says:

> Big data is a term that describes the large volume of data—both structured and unstructured—that inundates a business on a day-to-day basis. But it's not the amount of data that's important. It's what organizations do with the data that matters. Big data can be analyzed for insights that lead to better decision and strategic business moves (Big Data: What it is, n.d.).

Gandomi & Haider (2015), while focusing in a research project on the analytic methods used for Big Data, say a major challenge to defining the term is that the industry was ahead of academicians in recognizing Big Data's development and in becoming engaged in evaluating its emergence, recognition, and potential for the industry.

Laney (2001) is credited by some as providing an early definition that has continued to be applied to what constitutes Big Data and that is now commonly called "the three Vs," as described in more detail below. Laney's explanation connects the technological development to economic issues:

> The effect of the e-commerce surge, a rise in merger/acquisition activity, increased collaboration, and the drive for harnessing information as a competitive catalyst is driving enterprises to higher levels of consciousness about how data is managed at its most basic level (Laney, 2001, p. 1).

As defined in one instance, Big Data is "large pools of data that can be captured, communicated, aggregated, stored and analyzed" (Manyika et al, 2011, p. iv). Dewey (2014, para. 1) defines Big Data as "in-field shorthand that refers to the sheer mass of data produced daily by and within global computer networks at a pace that far exceeds the capacity of current databases and software programs to organize and process." Yee (2017) notes how remarkable it is "this new form of predictive analysis has taken hold on the way people learn and companies do business" (p. 1).

When you use Facebook, Instagram, or TikTok, you contribute to datasets underlying Big Data. When you take a picture with your smartphone, your phone—through all its sensors to collect and store data—adds to the pool of stored data, and your online actions with this digital image add to the pool of stored data. When you add a video to YouTube, when you do a search on Google, when you interact with your smart speaker, when you shop on Amazon, when you watch Netflix, when your automotive telematics communicate your location through its wireless, connected system, and when you comment on someone's online post, you contribute to the ever-growing pool of Big Data. Humans and their internet-related communication have created an extraordinary volume of data in recent years, far outdistancing the volume of data created in the entire history of humans previously (Marr, 2016; Yee, 2017).

So where is this Big Data stored? There are two options: a company may own its own hardware, do its own data storage, pay its own people to maintain the system, and by doing so maintain control of their data and equipment, or a company can outsource to a third-party—where the server is "in the cloud," connected to the internet, with the data stored and maintained by the third party (Angeles, 2013). In reality, most companies have multiple data centers in different locations in order to assure data availability and for security purposes. The decision of which option to use is based on business needs, costs, and data security. Hofacker, et al. (2016) note that Big Data became important when the cost of storing it became less costly than deleting.

Figure 24.1

Bytes & Bigger Bytes

Byte (8 bits)

1 byte	A single character
10 bytes	A single word
100 bytes	A telegram, tweet, or short SMS

Kilobyte (1,000 bytes)

1 kilobyte	A few paragraphs
2 kilobytes	Typewritten page
10 kilobytes	Encyclopedia page or static Web page
100 kilobytes	Low-resolution photograph
300 kilobytes	Average-resolution photograph (jpg) or 20 seconds of 8-bit mono audio
500 kilobytes	30-second audio/radio commercial announcement

Megabyte (1,000 kilobytes or 1 million bytes)

1 megabyte	A small novel, 1 minute of 8-bit mono audio or 1-minute stereo MP3
2 megabytes	High-resolution photograph, 7-megapixel, 2,832 x 2,128
5 megabytes	Complete works of Shakespeare, 30 seconds of TV-quality video, or 5-minute podcast
10 megabytes	One minute of uncompressed, CD-quality sound or 25 seconds of smartphone mp4 video
100 megabytes	One meter of shelved books or a two-volume encyclopedia
216 megabytes	Standard digital video, 720 x 480, 5:1 compression, 1 minute

Gigabyte (1,000 megabytes or 1 trillion bytes)

1 gigabyte	Pick-up truck filled with paper, 10 meters of shelved books, or a CD-quality symphony
5 gigabytes	One standard DVD (digital video disc or digital versatile disc)
15 gigabytes	High-definition DVD (holds 4 hours of HD video)
50 gigabytes	Floor of books
300 gigabytes	A full broadcast day, standard DV, 720 x 480, 5:1 compression

Terabyte (1,000 gigabytes)

1 terabyte	50,000 trees made into paper and printed
2 terabytes	Entire contents of an academic research library
20 terabytes	Printed collection of the U.S. Library of Congress

Petabyte (1,000 Terabytes)

3 petabytes	Total holdings, all media, U.S. Library of Congress
20 petabytes	Total production of hard-disk drives in 1995
200 petabytes	All printed material or production of digital magnetic tape in 1995

Exabyte (1,000 petabytes or 1 billion gigabytes)

5 exabytes	All words ever spoken by human beings up to the year 2000
12 exabytes	Total volume of information generated worldwide 1999
500 exabytes	Total digital content in 2009

Zettabyte (1,000 exabytes)

1.2 zettabytes	Estimated total amount of data produced in 2010

Yottabyte (1,000 zettabytes)

1 Yottabyte	Let your imagination run wild!

Source: Adapted from Wilkinson, Grant, & Fisher, 2012

Big data means we must become familiar with a few terms. While the term "growing exponentially" can start to sound cliché, in this case it is what has happened over the past two decades with data through the internet. Even though we have just started becoming accustomed to terabyte data storage devices, we must start becoming comfortable with the next terms for the size of sets of data. Mitchell (n.d.) lists Facebook, Microsoft, Amazon, and Google as the "big four data companies" and says they collectively store about 1,200 petabytes. (A terabyte is 1,000 gigabytes, and thus 1,200 petabytes is 1.2 million terabytes.) Cisco (2019) says annual global Internet Protocol traffic will reach 4.8 zettabytes per year by 2022 (396 exabytes each month), four times that of 2016, when the annual run rate for global IP traffic was 1.2 zettabytes per year.

The bottom line is that the data stored and mined by companies have value. A report by Economist Intelligence Unit says "big data analysis, or the mining of extremely large data sets to identify trends and patterns, is fast becoming standard business practice" (Marr, 2016, p. 1). Consider some other issues:

- While the business models, benefits and procedures are often the focus of studies of the impact of Big Data, some researchers recognize that as the data and its analyses grow, so will the threats of cyber security (Subroto & Apriyana, 2019).

- In contrast to cyber security risks, Big Data analytics gives increasing risk management tools to businesses, with tools that allow companies to identify, quantify, and model risks faced every day by the business (Kopanakis, 2018).

- Although Big Data has created a revolution in the IT world, one issue that has not been properly addressed is data storage location compared to where the data is generated: should data created in Germany only be stored in Germany? (Patel, 2019).

- Data is not limited only to stagnant, stored data, but also real-time data constantly being created. Organizations must be aware of how to equally use all forms of data (Vaghela, 2018).

How to benefit from the data is a major challenge. A recent study conducted by Economist Intelligence Unit (EIU) on behalf of the technology company ZS suggests the majority of business executives know how important sales and marketing analytics are, but only a small percentage believe their own analytics are working. These research findings support businesses using external firms with expertise in analyzing Big Data effectively (Stahl, 2016).

Background

The idea that Big Data is a relatively new phenomenon is based on when the term started being used. Lohr (2013) acknowledged 2012 as being the first point of wide scale recognition of the importance of Big Data. In his *New York Times* column, he further attempted to trace a history of Big Data and decided it was John Mashey, chief scientist at Silicon Graphics, who deserved recognition as first understanding and articulating the phenomenon in the mid-1990s. Mashey was also recommended as the originator of the term by data analyst Douglas Laney from the Gartner Analysts company, who, as noted above and below, has also contributed to the early evolving definition and recognition of the importance of Big Data. Diebold (2012), also discussed by Lohr, attempted to some level to take credit for the term Big Data in the econometrics field, while recognizing others in other fields. Diebold says "Big Data the term is now firmly entrenched, Big Data the phenomenon continues unabated, and Big Data the discipline is emerging," adding Big Data is "arguably the key scientific theme of our times (pp. 2-3). Others, including Roger Mougalas from O'Reilly Media, have been included as contributors to developing the term (van Rijmenam, 2013).

Although he did not specifically use the term Big Data, a pivotal point in the development of Big Data as a unique category of digital communication is the description in the early 2000s by Doug Laney of collected and organized data based on what are now commonly called "the three Vs" of Big Data: volume, velocity, and variety. The data analysis company SAS, begun as a project to analyze agricultural research at North Carolina State University, describes Laney's terms in a publicly available white paper released by SAS as a marketing tool (Big Data: What It Is, n.d.). They, plus others (What is Big Data, 2015) summarize Laney's now-mainstream definition known as the three Vs of Big Data. This model comes from the IT industry's effort to define what Big Data is and what it is not:

Volume: The amount of data is immense.

Velocity: The speed of data (always in flux) and processing (analysis of streaming data to produce near or real time results).

Variety: The different types of data, structured, (and) unstructured.

IBM suggested **Veracity** suggested as a 'fourth V of Big Data,' and since then others have suggested Big Data also is measured by **Value** and **Viability** (Powell, 2014). By 2017, analysts had listed 10 Vs of Big Data and suggested there are many more (Firican, 2017; Bresnick, J., 2017). Not all analysts list the same Vs, but Jain, et al. (2016) add, for example, validity, variability, venue, vocabulary, and vagueness.

Diebold (2012), as noted above, investigated the origins of the term Big Data in industry, academics, computer science and statistics. His research led him to claim the first use of the term Big Data in a research paper, completed in 2000 (Diebold, 2000). Lohr (2013) disagrees with this assertion, and Diebold conceded the origins of the term to others in his 2012 article.

In 2007, a major progression of Big Data was the white paper released by John F. Gantz, David Reinsel, and other researchers at IDC that was the first study to estimate and forecast the amount of digital data created and replicated each year. Gantz and Reinsel (2012) later released a video analysis that further explored the future of Big Data. Swanson and Gilder (2008) used the term exaflood to project toward what was expected in online data expansion, listing such items as movie downloads, P2P file sharing, video calling, cloud computing, internet video, gaming, virtual worlds, non-internet IPTV, business IP traffic, phone, email, photos, and music as some of the contributors to increased volumes of internet data.

Bryant et al. (2008) compared Big Data's ability to transform the way companies, the health industry, government defense and intelligence, and scientific research operate to the way search engines changed how people access information. Cukier (2010) recognized "an unimaginably vast amount of digital information which is getting ever vaster more rapidly" (para. 3). In this article, Cukier notes "Scientists and computer engineers have coined a new term for the phenomenon: "Big Data" (para. 6).

By 2011, Big Data had gained significant recognition, as researchers from the McKinsey Global Institute published an analytical paper that confirmed Big Data's importance and offered what they called seven key insights. Looking at several major sectors of society, they projected that, with the correct use of their Big Data, a retailer could improve its profit margin by 60 percent, U.S. healthcare could create more than $300 billion in value each year, and, from services enabled by personal-location data, users could capture 600 billion in consumer surplus (Manyika et al., 2011).

A final publication in the emergence of Big Data noted by Press (2013) is the 2012 article *Critical Questions for Big Data* (boyd & Crawford, 2012). These scholars describe why they think Big Data is a flawed term. Comparing, for example, the data set of any one particular topic on Twitter to all the data contained within one census, where the census data is significantly larger, they thus suggest "Big Data is less about data that is big than it is about a capacity to search, aggregate, and cross-reference large data set (p. 663)." Further, at the time the term was coined, memory and storage on computers was minimal (Abadicio, 2019). Manovich (2011) notes that the term originated in the sciences from the initial recognition that the data sets were so large so as to require supercomputers; now, desktop computers running standard software can analyze large quantities of data.

It is also important to understand that many Big Data databases are actually composed of numerous, separate databases that are linked together by some key variable—some unique identifier of a person, company, or other record. For example, to merge all of person's credit information, the key variable in the U.S. is usually a social security number, while an airline uses a frequent flier number to connect all of the information about each of its customers.

Recent Developments

Unlike many other communication technologies, Big Data is not one particular product category, set of tools, or technological development. Rather, it is the accumulated data that is in essence a byproduct of the emergence of digital communication, networking and storage over the past few decades. Some recent industry developments and uses of Big Data can help us gain a better understanding of Big Data as it applies to communication technology. In addition to media, Big Data is being used by manufacturing, retail, industry, government, healthcare, banking, and education (Big Data, n.d.; Patel, 2019).

Big Data is Changing Healthcare

Beall (n.d.) says, although healthcare was slower than other industries to effectively benefit from Big Data, benefits are now being seen from IoT connected devices and people, with Big Data and analytics dramatically changing everything from patient education to treatment. Beall quotes physician Eric Topol, executive vice president of the Scripps Research Institute, who equates Google maps' varying levels of geographical mapping to what is expected with our ability to digitize and quantify everything about the human body. Big Data is thus expected to save and improve lives, while also providing better medical care and reducing waste.

Analysts expect predictive analytics to help target interventions to patients who could most benefit (Carecloud, n.d.). Specific patient and environmental information could help improve decision-making about patient needs and services. Cooperation between government and healthcare systems could drive better access to and use of healthcare data. Interoperability of medical records could improve connections among multiple practitioners treating a single patient and improve flow of healthcare information to patients. Healthcare organizations can use analytics to mine data and improve health care and, as a result, enhance financial performance and administrative decision-making (Big Data & Healthcare, 2015).

Richards (2018) lists seven expected healthcare trends: (a) using the Internet of Things to improve patient health (citing an Allied Market Research report that predicts the IoT healthcare market will reach $136.8 billion by 2021, (b) making patient-centric care the focus, (c) reducing fraud waste and abuse through predictive analytics, (d) building better patient profiles leading to better predictive models, (e) accelerating

value and innovation in the ecosystem of healthcare, connecting the right innovation, the right care, the right provider and the right value to the patient, (f) reducing healthcare costs, reversing a 20-year upward trend, and (g) providing real time infection control.

Big Data is Changing Business Practices

Arsenault (2017) says media companies adopt Big Data applications to further their market strategies and amass capital and/or audiences. Impact on business is of course the most significant issue related to the importance of Big Data. Sarma (2017) suggests that "most organizations now understand that if they capture all the data that streams into their businesses, they can deploy Big Data analytics to get significant leverage in understanding their customers, forecasting business trends, reducing operational costs, and realizing more profits" (p. 1). Sarma further says the advanced analytics provided by Big Data can provide lean management operations, improve target segmentation, make better business decisions, improve employee hiring and retention, and increase revenue and reduce cost.

Examples of business use of Big Data are readily available (Kopanakis, 2018).

- To boost customer acquisition and retention, such as when Coca-Cola built a digital-led loyalty program in 2015.

- To solve advertising problems and build marketing insights, such as what Netflix does to communicate with suggestions to users about what to watch.

- To provide risk management insight, such as when Singapore's UOB Bank was able to reduce their time to calculate the value of business risk from eighteen hours to only a few minutes.

- To drive innovations and product development, such as Amazon's use of Big Data analytics to develop their Amazon Fresh grocery delivery service, through understanding the market from both the consumer purchasing and supplier delivery perspectives.

Morgan (2015) notes as additional business examples: (a) Bristol-Myers Squibb reduced the time it takes to run clinical trial simulations by 98% by extending its internally hosted grid environment into the Amazon Web Services Cloud, (b) Xerox used Big Data to reduce the attrition rate in its call centers by 20%, (c) Kroger personalized its direct mailer based on the shopping history of the individual customer to get a coupon return rate of 70%, compared to the average direct mail return rate of 3.7%, and (d) Avis Budget implemented an integrated strategy to increase market share, yielding hundreds of millions of dollars in additional revenue, working through its IT partner CSC (now DXC Technology). CSC applied a model to predict lifetime value to Avis Budget's customer database. Avis Budget used Big Data to forecast regional demand for fleet placements and pricing.

One of the best media-related Big Data examples comes from Netflix (Chowdhury, 2017). Did you know that, as you watch Netflix, Netflix is also watching you? With a user base of around 100 million worldwide, Netflix gathers an extraordinary amount of data. Certain data points ("events" in the world of Big Data analytics) help predict success with specific content but also as a means of recommending content to specific users. Netflix analytics evaluate such events as searches, when a user watches a show, behaviors while watching, the device used to watch, and ratings of shows watched. Netflix is said to save $1 billion a year by using data analytics (Petrov, 2019). Of course, today's media world is increasingly based on streaming distribution. Services like Hulu, CBS All Access, and Disney+ are using data analytics in the same way as Netflix.

Arsenault (2017) says Big Data is shaping media networks globally in two primary ways: (1) media content is being digitized, creating new means of distribution, and (2) media organizations are attempting to compete in this new media world by using Big Data analytics for decision-making—tailoring content to specific audiences. Wheeler (2017) suggests television programming has undergone a renaissance because of Big Data. Wheeler reports that Americans still watch more than five hours of television per day, and that over 50% of homes have subscription services like Netflix and Hulu and says data is the key to television's profitability: "Advertisers have access to more audience information than ever, while networks and content providers use data, in addition to instinct, to guide programming decisions" (p. 1).

Big Data is Changing Defense by Government

Rossino (2015) says the Defense Advanced Research Projects Agency (DARPA—the same agency that created the internet in the 1970s) is investing significantly in Big Data, resulting in the development of distributed computing as a component of networked weapons systems. Research and development funding requests from all branches of the U.S. military continue to spend more each year on Big Data analysis, allowing possible development of better defenses in areas including cyber threats and insider threats. The U.S. Department of Defense spent more than $7 billion in 2017 on artificial intelligence, Big Data, and cloud computing, a 32% increase over five years earlier. Experts suggest "the U.S. military can either lead the coming revolution, or fall victim to it" (Corrin, 2017, p. 1).

As one example, as part of improving human-systems integration (HSI), researchers are using advanced sensor technologies, Big Data, and artificial intelligence, with the expectation that quality Big Data analytics, will guide quality implementation. (Abadicio, 2019). Such activities do not necessarily mean Big Data analytics is as good as it needs to be. Research by Anton, et al. (2019) found, among other things, that, while the Department of Defense has improved its data analysis, a lack of data sharing across military departments because of cultural and security issues is still a problem, and the U.S. military needs strategic planning and long term investments in data analytics. Heckman (2019) says what is needed is to move from data providing a good snapshot of defense strategies and readiness as it currently does well, to providing good predictive analyses.

Big Data is Changing Education

Reyes (2015), says the shift from classroom to blended and online learning systems such as Blackboard, Canvas, and Moodle inevitably leads to a place for Big Data analytics in education, although there are challenges in changing from traditional to learner-centered analytics. There are technological issues to address and resolve, and there are ethical concerns, but research shows so far that Big Data will transform teaching and learning. Some, for example, envision a classroom where cameras running constantly capture data about each child, while a Fitbit-like device tracks each child's reactions and development (Herold, 2016). Schroer (2019) lists several major U.S. companies using Big Data to improve education, with services from creating lesson plans to tracking progress to helping students find colleges and majors. As just one example of the extraordinary benefit of Big Data, Texas State University enlisted the services of the company Civitas, allowing three administrators to enroll thousands of freshman students in only a few days.

Current Status

Smartphone penetration continues to grow, hundreds of billions of emails are sent every day, there are over one billion Google searches every day, and trillions of sensors monitor, track and communicate with each other, generating real-time data for the Internet of Things (Quick Facts, n.d.). IBM and Oracle are two of the companies involved in Big Data business models. IBM in 2020 is recognizing how artificial intelligence (AI) has become a critical component of Big Data analytics (Quick Facts, n.d.).

Oracle (n.d.) says machine learning is now significant because of Big Data, noting that companies now have the tools train machines instead of programming them. Woodie (2019) notes some of the current developments and expectations related to Big Data: (a) improvements in using metadata to make sense of the huge volume of data created from artificial intelligence (AI) and machine learning (ML), (b) a decline in Hadoop as companies build their own data lakes or warehouses or go directly to the cloud, (c) renewed focus on the challenges in managing the volume of Big Data, in such areas a model management and data orchestration, and (d) improvements in Data Fabric architectures, connected to advancements in smart technologies and IoT devices. Data Fabric helps coordinate data management across internal and cloud-based data storage.

Kneale (2016) provides several communication and media-related examples. The more than 3.5 billion smartphones worldwide (Holst, 2019) are considered a marketer's dream—as they track almost everything you do—watch TV, or interact with apps, websites and Facebook friends. Further, analyzing the data helps distinguish between what consumers say they do and what they really do. Instead of the old system of television programming based as much on intuition as research, programmers can now rely on the latest in machine learning and artificial intelligence. On the business decision-making side, correct use of Big Data analytics helps TV programmers know which programs will be able to generate profits from advertising revenue versus their cost compared to those that will cause a loss, no matter what.

A major player in Big Data analytics in recent years has been Apache Hadoop. Kerner (2017) says "The open source Apache Hadoop project provides the core framework on which dozens of other Big data efforts rely" (p. 1). Hortonworks describes Hadoop as an "open source framework for distributed storage and processing of large sets of data on commodity hardware," enabling businesses to analyze large amounts of structured and unstructured data. (What is Apache, n.d., para. 1). Hadoop is even a component of IBM and other companies' Big Data analytics services. However, Heller (2017), prior to its late 2017 3.0.0 update, suggested Hadoop was declining in importance, as new trends and tools in Business Intelligence (BI) and Business Analytics (BA) emerge. Woodie (2019) says Hadoop struggled in 2019 and quotes Haoyuan Li, founder and Chief Technical Officer of Alluxio, as suggesting the Hadoop Distributed File System (HDFS) is dead. Hadoop Compute, in the form of Apache Spark, used as a computational engine for Hadoop data, is still viable. As described further below, tools like Spark and Databricks continue to see new adoptions (Menon, 2019).

Google and Amazon are two other major Big Data companies. Do you think Big Data is not really a part of your life? Google proves it is. Google is transparent when it comes to disclosure, even if you may not be concerned about how they use your online behavior to target you with advertising (Ads Help, n.d.). In their online support pages, Google reminds you that, as you browse the internet, read email, shop online, and do similar online activities, you will notice advertisements for items and content based on previous searches. Google uses tools like DoubleClick and AdSense, but essentially all their apps, from Gmail to Chrome to YouTube, to track your behaviors and your interests in some way (O'Reilly, 2012). Every time you do a Google search while logged in to your Gmail account, Google keeps track of your searches, and their tools connect advertisers with your interests. Google gives users a means of controlling how data about them is used, in Google Dashboard. There, you can view almost everything Google knows about you and manage your privacy settings. George Orwell was concerned Big Brother would be watching you; more accurately today, through companies like Google, Big Data is watching you.

In 2015, Google added Google Cloud Dataflow and Google Cloud Pub/Sub to its existing BigQuery SQL-query based tools (Babcock, 2015). These services are used to analyze large data sets and data streams and put Google into full competition with Amazon Web Services (AWS). The AWS suite includes Amazon DynamoDB, Data Pipeline, Amazon Kinesis, and Hadoop-like service Elastic MapReduce (Amazon EMR). Amazon Web Services is one of the platforms most widely used by other companies for Big Data analytics and decision-making. On its AWS promotional web page, Amazon says "Amazon Web Services provides a broad and fully integrated portfolio of cloud computing services to help…build, secure, and deploy…Big Data applications…(with) no hardware to procure, and no infrastructure to maintain and scale" (What is Big Data, n.d.)

The fact is, however, as of 2020, there are dozens upon dozens of Big Data companies, most of which the average consumer has no knowledge about. Schroer (2019) lists 32 of these, and says, over the years, a large number of companies has emerged to provide data solutions, some focusing on analytics and some helping to organize datasets for clients. At a basic level, the reality comes down to how the data analyses can be done and by whom. Harris (2019) addresses the current challenge in how businesses use data visualization tools. Harris characterizes Big Data analytics within what is called the business intelligence (BI) market, noting that vendors like Tableau and Qlik have made analytics accessible to users with

limited technical knowledge, while traditional BI solutions require the technical expertise of an information technology department. Just as the graphical user interface for personal computers and the World Wide Web on the internet made technology usable by the everyday person, such applications in the Big Data analytics world has the potential to do the same.

Ultimately, however, nothing about all of the developments in Big Data matter if there is not a return on investment. Firican (2019), says Big Data has substantial value for businesses in understanding and serving their customers and improving business performance Any business' Big Data strategies must include recognition of the value to the company.

Factors to Watch

Menon (2019) suggests some of the most important trends in Big Data for 2020 are: (a) the Cloud is the new Data Lake, with the prevalent trend being hybrid storage of on-premises plus cloud, and a multi-cloud footprint, (b) more businesses will move from Hadoop and switch to Spark or Databricks, allowing better interactive processing abilities, with a user-friendly interface, (c) digital transformation, finding ways to sell the data that can generate new revenue streams, finally becomes a key factor in business' data strategies, and (d) continued evolution of machine learning and artificial intelligence, allowing increased use of automation tools. Western Digital (2020) expects, among others, (a) machine-generated data to surpass human data, (b) managing the ever-growing volume of data to be improved through new data center architectures, (c) the number of data centers will grow, both locally and regionally, and worldwide, and (d) despite their predicted demise for years, hard disc drives for storage will thrive, since they continue to meet growing data demands and continue to provide value as storage media.

Otherwise, the battle continues between the so-called NoSQL companies against relational database systems by companies like Oracle and IBM, with SQL systems making a resurgence in recent years (Kulkarni, 2019). SQL (structured query language) databases are table-based; NoSQL are document-based, where queries are based on a structure of documents.

Smart devices, e-commerce and social media create data demands on businesses that relational database systems were not thought to be able to support at the necessary speed and cost, resulting in the emergence of a variety of NoSQL databases (Clark, 2014). Once the NoSQL trend started, there was an assumption NoSQL would take over, yet the battle between the two continues. NoSQL gained traction, as relational databases were judged to be incompatible with changing business requirements and limited in providing affordable scalability (Bhatia, 2017). At the same time, Apache Spark is recognized as the leading platform for large-scale SQL and "has become one of the key Big Data distributed processing frameworks in the world…(and)…provides native bindings for the Java, Scala, Python, and R programming languages, and supports SQL, streaming data, machine learning, and graph processing" (Pointer, 2017). Apache Spark is used by major technology giants like Apple and IBM and by governments, telecommunications companies and banks. Apache Spark is expected to continue to be a significant platform in Big Data analytics.

Big data also means consumers need to know how data is being used to monitor a society; businesses need to know how to use data that customers create and that represents who the customers are, while still respecting their privacy and maintaining their trust. The practices surrounding "Big Data" research raise serious ethical and privacy questions. Who should be able to use this data, for what purposes, and what does this data mean? (boyd & Crawford, 2012).

In healthcare, for example consumers will have to accept giving up some privacy in exchange for the benefits of data pooling (McGuire et al., 2012). Companies and governments will likewise have to find the balance between privacy and data security. The recent growth of digital assistants, including Google Assistant, Alexa, and Siri create further concerns, as each uses a level of artificial intelligence to "learn" more about you as you interact with the technology (Maisto, 2016), and of course generate even more data. In relation, consumers need to be better informed about how data about them are being used and be aware of related privacy issues (Jain, et al., 2016; Manyika et al., 2011), while companies must find ways to accommodate individual privacy issues.

Analytics software company SAS acknowledges that advertisers don't really care about our personal information, they just need data to make sales. (SAS Insights Staff, n.d.). SAS lists some of typical areas of concern: (a) who owns our personal data and what can they do with it? (b) we live our life in a public space, whether we like it or not, in the internet age, (c) the combination of data breaches and aggressive hackers will continue to be a challenge, (d) in an increasingly dangerous world, how do we ensure safety, and (e) the digital age is erasing geographical borders, causing governments to seek ways to protect how their citizens' data is used.

One example: plans are developing to use product sensor data to identify precise consumer targets for post-sale services or for cross-selling. Would you object to companies monitoring how you use their products? Just as in healthcare, businesses will need to work with policy makers and consumers to find the correct balance for how data about individuals are used and the level of transparency provided to the consumer (Manyika et al, 2011). Marr (2016, p. 1) says "…customer trust is absolutely essential…people are becoming increasingly willing to hand over personal data in return for products and services that make their lives easier (but) that goodwill can evaporate in an instant if customers feel their data is being used improperly, or not effectively protected."

Finally, one of the biggest current challenges for Big Data analytics is the limited number of people in the U.S. with the necessary skills to work in the field. The people shortage expected in the coming years is for those ready to handle the statistics, machine learning, problem framing and data interpretation, analytical and managerial skills required to generate Big Data analytics with the quality of results needed for correct business decision-making. (McGuire et al., 2012). The top career areas for people with Big Data skills include (1) Big Data analyst, (2) Data Security Engineer, (3) Big Data engineer, (4) Big Data scientist, and (5) Big Data architect. Entry level salaries start at over $60,000 a year (Top 5 Most, 2019).

Gen Z

Francis & Hoefel (2018) provide insight into the presumed characteristics of Generation Z (Gen Z), suggesting overall that their influence is expanding. Recognized as the first generation of fully digital natives, Gen Z have a natural expectation of collecting information digitally, with expertise in matter-of-factly drawing on and cross-referencing a variety of online resources. These users expect to have digital relationships, expect information and entertainment to be at their demand, and seamlessly integrate their "real-world" experiences with "virtual experiences." As digital natives, almost all have smartphones, and more than half use their smartphones for more than five hours per day (Boucher, 2018). The iGen Project is beginning to document some of the characteristics that have emerged within this new generation, most of which are attributed to the technology world in which they have been reared (Katz, 2019).

Related to the kind of digital experiences a Generation Z individual sees as natural, these users also expect a reasonably free and unfettered experience online, paying only for a limited number of experiences while expecting much of their information-searching, entertainment, and social media interactions to be cost-free. This expectation is directly connected to the importance of Big Data. Data analytics of today is a direct evolution of data analysis since the beginning of media, only on a much more sophisticated level. Everything has a cost, and in the case of freely available online experiences, the cost to the user is the data about them that advertisers and marketers find has great value.

It might thus be expected that these Gen-Z individuals have less concern regarding the privacy of their information than it was for previous generations. Research shows this is not true. Sheer ID (2019) found that 75% of Gen Z users only allow location sharing for apps that require it to function and nearly two-thirds say they are more comfortable sharing personal information with brands when there is confidence the data is securely protected and stored. As much as Big Data is an integrated component of the technology world in which Gen Z individuals have grown up, greater attention is likely to be paid in coming years for how it is used and how users can have control over data about themselves.

Global Concerns & Sustainability

Implementation of the General Data Protection Regulation (GDPR) by the European Union is just one example of the worldwide impact of data and privacy issues (IBM, n.d.). The United Nations recognizes the importance of Big Data and its implications for the future, but it is within the discussion of risks that the U.N.'s information is most useful (United Nations, n.d.). There are concerns regarding the passive collection of data and the related question of whether or not removal of an individual's personal information can fully protect privacy. It may also be problematic that algorithms used to analyze multiple datasets can reconnect deleted private data. There is also concern for the classic "haves and have-nots." Many people in the world may be excluded from this new data and data analysis world, widening the gaps between least developed and most isolated countries from the rest of the digital world.

What are some of the other major global issues? The global market for Big Data is expected to grow from $2.5 billion is 2018 to $6.2 billion by 2025 (Research and markets, 2019). Telemedicine applications will grow worldwide, allowing sharing of medical research and techniques, but also leading to deep learning (a subset of artificial intelligence) and robotics applications (Hemanth & Balas, 2019). Agricultural advances will play an important role in global issues, and there is a current recognition of needing to improve data sharing and data analysis in worldwide agricultural operations and decision-making (Schaer, 2019). The assumption of this component of the impact of Big Data is based on the fact that there will always need to be a sound agriculture system to maintain the food supply, so Big Data analytics applied here can have an extraordinary worldwide impact.

But beyond some potential positives, a concern is arising of a reinvention of colonialism by the United States, in a belief that U.S. multinational corporations are increasingly exercising the same kind of control as emerged in past levels of imperialism, only this time based on the digital ecosystem. When the software, hardware, network connectivity and financial infrastructure are based on a foundation of control by mostly American companies, a concern exists for the subsequent economic domination of lesser-developed countries (Kwet, M. (2019). In this view, the centerpiece of surveillance capitalism is Big Data.

Bibliography

Abadicio, M. (2019). Big data in the military – Preparing for AI. https://emerj.com/ai-sector-overviews/big-data-military/

Ads Help (n.d.). Why you're seeing an ad. https://support.google.com/ ads/answer/1634057?hl=en.

Angeles, S. (2013, August 26). Cloud vs. data center: What's the difference? *Business News Daily*. http://www.businessnewsdaily.com/4982-cloud-vs-data-center.html.

Anton, P. S., McKernan, M., Munson, K., Kallimani, J. G., Levedahl, A., Blickstein, I., Drezner, J. A., & Newberry, S. (2019). Assessing the use of data analytics in Department of Defense acquisition. RAND Corporation. https://www.rand.org/pubs/research_briefs/RB10085.html.

Arsenault, A. H. (2017). The datafication of media: Big data and the media industries, *International Journal of Media & Cultural Politics*, 13: 1+2, 7–24. https://doi.org/10.1386/macp.13.1-2.7_1.

Babcock, C. (2015, August 13). Google adds Big Data services to cloud platform. *Information Week*. http://www.informationweek.com/big-data/big-data-analytics/google-adds-big-data-services-to-cloud-platform/d/d-id/1321743.

Beall, A. (n.d.) Big data in health care. https://www.sas.com/en_us/insights/articles/big-data/big-data-in-healthcare.html.

Bhagat, V. (2019, November 8). IoT in 2020: 5 Things you need to know. https://www.iotforall.com/iot-devices-by-2020/.

Bhatia, R. (2017). NoSQL vs SQL — Which database type is better for Big Data Applications. Analytics India Magazine. https://analyticsindiamag.com/nosql-vs-sql-database-type-better-big-data-applications/.

Big Data: What It Is and Why It Matters (n.d.). SAS. http://www.sas.com/ en_us/insights/big-data/what-is-big-data.html#mdd-about-sas.

Big Data & Healthcare Analytics Forum (2015). http://www.himss.org/Events/EventDetail.aspx?ItemNumber=42336.

Boucher, J. (2018). Top 10 Gen Z statistics from 2018. *The Center for Generational Kinetics*. https://genhq.com/top-10-ways-gen-z-is-shaping-the-future/.

boyd, D. & Crawford, K. (2012). Critical questions for Big Data. *Information, Communication & Society*,15(5), 662-679. https://doi.org:10.1080/1369118X.2012.678878.

Bresnick, J. (2017, June 5) Understanding the many Vs of healthcare Big Data analytics. https://healthitanalytics.com/news/understanding-the-many-vs-of-healthcare-big-data-analytics.

Bryant, R. E., Katz, R. H. & Lazowska, E. D. (2008, December 22). Big-Data computing: Creating revolutionary break-throughs in commerce, science and society. http://cra.org/ccc/wp-content/uploads/sites/2/2015/05/Big_Data.pdf.

Carecloud (n.d.). Healthcare will deliver on the promise of 'Big Data' in 2015. http://www.carecloud.com/healthcare-will-deliver-promise-big-data-2015/.

Chowdhury, T. (2017, June 28). How Netflix uses Big Data analytics to ensure success. http://upxacademy.com/netflix-data-analytics/.

Cisco (2019, February 27). Cisco visual networking index: Forecast and trends, 2017–2022 white paper. https://www.cisco.com/c/en/us/solutions/collateral/service-provider/visual-networking-index-vni/white-paper-c11-741490.html.

Clark, L. (2014, March). Using NoSQL databases to gain competitive advantage. *ComputerWeekly.com*. http://www.computerweekly.com/feature/NoSQL-databases-ride-horses-for-courses-to-edge-competitive-advantage.

Corrin, A. (2017, December 6). DoD spent $7.4 billion on Big Data, AI and the cloud last year. Is that enough? https://www.c4isrnet.com/it-networks/2017/12/06/dods-leaning-in-on-artificial-intelligence-will-it-be-enough/.

Cukier (2010, February 25). Data, data everywhere. *The Economist*.http://www.economist.com/node/15557443.

Delgado, R. (2017, June 16). Improved decision making comes from connecting Big Data sources. http://data-informed.com/improved-decision-making-comes-from-connecting-big-data-sources/.

DeMauro, A., Greco, M., & Grimaldi, M. (2015). What is Big Data? A consensual definition and a review of key research topics. AIP Conference Proceedings. 2015, Vol. 1644 Issue 1, 97-104.

Dewey, J. P. (2014, January). Big Data. *Salem Press Encyclopedia*.

Diebold, F.X. (2000). Big data dynamic factor models for macroeconomic measurement and forecasting. http://citeseerx.ist. psu.edu/viewdoc/download;jsesionid=3DB75BD2F8565682589A257CE3F4150C?doi=10.1.1.12.1638&rep=rep1&type=pdf

Diebold, F.X. (2012). A personal perspective on the origin(s) and development of 'Big Data': The phenomenon, the term, and the discipline, second version. http://papers.ssrn.com/sol3/papers.cfm?abstract_id=2202843.

Firican, (2017, February 9). The 10 Vs of Big Data. https://tdwi.org/articles/2017/02/08/10-vs-of-big-data.aspx.

Francis, T. & Hoefel, F. (2018, November). True Gen': Generation Z and its implications for companies. *McKinsey and Company*. https://www.mckinsey.com/ industries/consumer-packaged-goods/our-insights/true-gen-generation-z-and-its-implications-for-companies.

Gandomi, A., & Haider, M. (2015). Beyond the hype: Big data concepts, methods, and analytics. *International Journal Of Information Management, 35*(2), 137-144. https://doi.org:10.1016/j.ijinfomgt.2014.10.007.

Gantz, J. & Reinsel, D. (2012, December). The digital universe in 2020: Big data, bigger digital shadows, and biggest growth in the Far East. http://www.emc.com/leadership/digital-universe/2012iview/analyst-perspective-john-gantz-david-reinsel.htm.

Harris, D. (2019, August 15). A guide to free and open source data visualization tools. https://www.softwareadvice.com/resources/free-open-source-data-visualization-tools/.

Heckman, J. (2019, November 27). DoD overcoming culture challenges to turn data 'snapshot' into predictive analytics. Federal News Network. https://federalnewsnetwork.com/big-data/2019/11/dod-overcoming-culture-challenge-to-turn-data-snapshot-into-predictive-analytics/.

Heller (2017, August 7). 10 hot data analytics trends — and 5 going cold. https://www.cio.com/article/3213189/analytics/10-hot-data-analytics-trends-and-5-going-cold.html.

Hemanth, D. J., & Balas, V. E. (2019). Telemedicine technologies: Big data, deep learning, robotics, mobile and remote applications for global healthcare. Elsevier.

Herold, B. (2016, January 11). The future of Big Data and analytics in K-12 education. https://www.edweek.org/ew/articles/2016/01/13/the-future-of-big-data-and-analytics.html.

Hofacker, C. F., Malthouse, E.C. & Sultan, F. (2016). Big data and consumer behavior: imminent opportunities, *Journal of Consumer Marketing, 33*(2), 89-97.

Holst, A. (2019, November 11). Number of smartphone users worldwide from 2016 to 2021. Statista. https://www.statista.com/statistics/330695/number-of-smartphone-users-worldwide/.

How Big Data Can Help You Do Wonders in Your Business (n.d). https://www.simplilearn.com/how-big-data-can-help-do-wonders-in-business-rar398-article.

IBM (n.d.). Proliferating privacy regulations. https://www.ibm.com/security/privacy.

Jain, P., Gyanchandani, M. & Khare, N. (2016). Big data privacy: a technological perspective and review. *Journal of Big Data* 3(25). https://doi.org/10.1186/s40537-016-0059-y.

Katz, R. (2019, April 2). How gen z is different, according to social scientists (and young people themselves). *Pacific Standard*. https://psmag.com/ideas/how-gen-z-is-different-according-to-social-scientists.

Kerner, S. M. (2017, December 22). Apache Hadoop 3.0.0 boosts Big Data app ecosystem. http://www.enterpriseappstoday.com/data-management/apache-hadoop-3.0.0-boosts-big-data-app-ecosystem.html.

Kneale, D. (2016, January 21). Big data is everywhere—now what to do with it? New tools unlock the secrets of consumer desire. *Broadcasting and Cable*.http://www.broadcastingcable.com/news/rights-insights/big-data-dream/147166.

Kopanakis, J. (2018, June 14). 5 real-world examples of how brands are using Big Data analytics. https://www.mentionlytics.com/blog/5-real-world-examples-of-how-brands-are-using-big-data-analytics/.

Kulkarni, A. (2019, July 8). Why SQL is beating NoSQL, and what this means for the future of data. https://blog.timescale.com/blog/.

Kwet, M. (2019). Digital colonialism: US empire and the new imperialism in the Global South. *Race & Class, 60*(4), 3–26. https://doi.org/10.1177/0306396818823172.

Laney, D. (2001, February 6). 3D Data management: Controlling data volume, velocity, and variety. http://blogs.gartner.com/doug-laney/files/2012/01/ad949-3D-Data-Management-Controlling-Data-Volume-Velocity-and-Variety.pdf.

Locke, M. (2017, November 16). How to measure ghosts: Arthur C. Nielsen and the invention of Big Data. https://medium.com/s/a-brief-history-of-attention/how-to-measure-ghosts-arthur-c-nielsen-and-the-invention-of-big-data-3231cec8c0dd

Lohr, S. (2013, February 4). Searching for origins of the term 'Big Data.' *New York Times, New* York edition, B4.

Maisto, M. (2016, March 2). Siri, Cortana are listening: How 5 digital assistants use your data. https://www.informationweek.com/big-data/siri-cortana-are-listening-how-5-digital-assistants-use-your-data/d/d-id/1324507?

Manovich, L. (2011, July 15). Trending: The promises and the challenges of big social data. *Debates in the Digital Humanities*, M. K. Gold (Ed.). The University of Minnesota Press. http://manovich.net/index.php/projects/trending-the-promises-and-the-challenges-of-big-social-data.

Manyika, J. & Chui, M., Brown, B., Bughin, J., Dobbs, R., Roxburgh, C., & Byers, A.H. (2011). Big Data: The next frontier for innovation, competition, and productivity, Washington, D.C.: McKinsey Global Institute. http://www.mckinsey.com/business-functions/business-technology/our-   insights/big-data-the-next-frontier-for-innovation.

Marr, B. (2016, January 13). Big data facts: How many companies are really making money from their data? *Forbes*. http://www.forbes.com/sites/bernardmarr/ 2016/01/13/big-data-60-of-companies-are-making-money-from-it-are-you/#773b9f194387

Mayer-Schönberger, V. & Cukier, K. (2013). *Big data: A revolution that will transform how we live, work, and think*. Houghton Mifflin Harcourt.

McGuire, T., Manyika, J., Chui, M. (2012, July/August). Why Big Data is the new competitive advantage. *Ivy Business Journal*. http://iveybusinessjournal.com/ publication/why-big-data-is-the-new-competitive-advantage/.

Menon, R. (2019) Big data trends: Our predictions for 2020 plus what happened in 2019. Infoworks. https://www.infoworks.io/big-data-trends/.

Mills, T. (2018, May 16). Big data is changing the way people live their lives. https://www.forbes.com/sites/forbestechcouncil/2018/05/16/big-data-is-changing-the-way-people-live-their-lives/#6797e5b43ce6.

Mitchell, G. (n.d.) How much data is on the internet? *Science Focus*. https://www.sciencefocus.com/future-technology/how-much-data-is-on-the-internet/.

Morgan, L. (2015, May 27). Big data: 6 real-life business cases. https://www.informationweek.com/software/enterprise-applications/big-data-6-real-life-business-cases/d/d-id/1320590?

O'Reilly, D. (2012, January 30). How to prevent Google from tracking you. *CNET*. http://www.cnet.com/how-to/how-to-prevent-google-from-tracking-you/.

Oracle (n.d.) What is Big Data? https://www.oracle.com/big-data/guide/what-is-big-data.html#link1

Patel, A. (2019, March 08). How Big Data will impact different sectors. https://www.itproportal.com/features/how-big-data-will-impact-different-sectors/.

Petrov, C. (2019, March 22). Big data statistics 2020. https://techjury.net/stats-about/big-data-statistics/#gref.

Pointer, I. (2017, November 13). What is Apache Spark? The Big Data analytics platform explained. https://www.infoworld.com/article/3236869/analytics/what-is-apache-spark-the-big-data-analytics-platform-explained.html.

Powell, T. (2014, August 25). The fourth V of Big Data.http://researchaccess.com/2014/08/the-fourth-v-of-big-data/.

Press, G. (2013, May 9). A very short history of Big Data. http://www.forbes.com/ sites/gilpress/2013/05/09/a-very-short-history-of-big-data/#4a0f418b55da.

Quick Facts and Stats on Big Data (n.d.) IBM Big Data and analytics hub. http://www.ibmbigdatahub.com/gallery/quick-facts-and-stats-big-data.

Reyes, J. (2015). The skinny on Big Data in education: Learning analytics simplified. *Techtrends: Linking Research & Practice To Improve Learning, 59*(2), 75-80.

Research and markets issues report: Big data in E-commerce: Global markets. (2019). *Manufacturing Close – Up*. https://www.researchandmarkets.com/ reports/4876588/

Richards, T. (2018, May 3). Big data healthcare trends will improve outcomes. https://www.chthealthcare.com/blog/healthcare-trends.

Rossino, A. (2015, September 9). DoD's big bets on Big Data. http://www.c4isrnet.com/story/military-tech/blog/business-viewpoint/2015/08/25/dods-big-bets-big-data/32321425/.

SAS Insights Staff (n.d.). What Big Data has brought to the privacy discussion. https://www.sas.com/en_us/insights/articles/big-data/big-data-privacy.html.

Sarma, R. (2017, March 22). 5 business impacts of advanced analytics and visualization. https://resources.zaloni.com/blog/5-business-impacts-of-advanced-analytics-and-visualization.

Schaer, L. (2019, Apr 02). Ag needs better data use to aid decision-making; Big Data has tons of potential to solve global issue but is so far coming up short. *Ontario Farmer*.

Schroer, A. 2019, September 26). Big data in education: 9 companies delivering insights to the classroom. https://builtin.com/big-data/big-data-in-education.

Sheer ID (2019, May 19). Marketing to Generation Z. https://www.sheerid.com/blog/marketing-to-generation-z/.

Stahl, G. (2016, June 22). Broken links: Why analytics investments have yet to pay off. https://perspectives.eiu.com/marketing/broken-links-why-analytics-investments-have-yet-pay.

Subroto, A. & Apriyana, A. (2019) Cyber risk prediction through social media Big Data analytics and statistical machine learning. *Journal of Big Data*, 6(1):1-19. https://doi.org/10.1186/s40537-019-0216-1.

Swanson, B. & Gilder, G. (2008, January 29). Estimating the exaflood: The impact of video and rich media on the internet – A 'zettabyte' by 2015?http://www.discovery.org/a/4428.

Top 5 Most In-demand Big Data Jobs in 2020 (2019, October 29). https://robots.net/big-data/best-big-data-jobs-2020/.

Top 13 Best Big Data Companies (2020, January 29). https://www.softwaretestinghelp.com/big-data-companies/

United Nations (n.d.). Big data for sustainable development. https://www.un.org/ en/sections/issues-depth/big-data-sustainable-development/index.html.

Vaghela, Y. (2018, June 28). Four common Big Data challenges. https://www.dataversity.net/four-common-big-data-challenges/.

van Rijmenam, M. (2013, January 6). A short history of Big Data. https://datafloq.com/read/big-data-history/239.

Western Digital (2020, January 28). 2020 trends for the evolving data center, part 1. https://www.networkworld.com/article/3516394/2020-trends-for-the-evolving-data-center-part-1.html.

What is Apache Hadoop? (n.d.). Hortonworks. http://hortonworks.com/hadoop/.

What is Big Data (n.d.). https://aws.amazon.com/big-data/what-is-big-data/.

What is Big Data? And Why is it Important to Me? (2015, March 31).http://www.gennet.com/big-data/big-data-important/

Wheeler, C. (2017, June 1). The future of TV is today: How TV is using Big Data. https://www.clarityinsights.com/blog/television-big-data-analytics.

Woodie, A. (2019). Big data predictions: What 2020 will bring. https://www.datanami.com/2019/12/23/big-data-predictions-what-2020-will-editbring/.

Yee, S. (2017, February 8). The impact of Big Data: How it is changing everything, including corporate training. https://elearningindustry.com/impact-of-big-data-changing-corporate-training.

25

Artificial Intelligence

Ching-Hua Chuan, Ph.D.*

Abstract

AI technologies are now an integral part of our everyday experiences. We have robot vacuums that clean the floor at a scheduled time, smart map apps that tell us, not only the shortest or quickest route to the destination, but also when we need to leave to arrive on time, and robo-financial advisors that automatically select investments for us. For our media consumption, we read the news on Facebook and watch the movies recommended by algorithms, talk to chatbots on social media, and are constantly exposed to targeted advertisements that reflect our online behavior and preferences. Have you ever wondered how these intelligent systems are created, and what your life will be like when computer systems become smarter and smarter? This chapter will tell you what AI is and why a field started in 1960s has generated hype and controversy. This chapter provides you with fundamental knowledge of AI so you can navigate and be part of this technological revolution.

Introduction

Artificial Intelligence (AI) is a field in computer science in which researchers create computer hardware and/or software to perform intelligent tasks. What are the intelligent tasks? To answer this question, we need to think about *what intelligence is*. In English dictionaries, intelligence is defined as "the ability to acquire and apply knowledge and skills," (Oxford English Dictionary, n.d.) or "the ability to learn or understand or to deal with new or trying situations" (Merriam-Webster, n.d.).

These are broad definitions, which contribute to the broad scope of AI. Generally speaking, AI is a field that aims to create computer systems to perform intelligent tasks normally requiring *human intelligence*, such as reasoning, learning, planning, decision-making, and predicting.

The task of "mimicking human intelligence" can be conceptualized via four aspects: thinking humanly, thinking rationally, acting humanly, and acting rationally (Russell & Norvig, 2009). The four aspects reflect two dimensions—thinking versus acting, and

* Research Associate Professor, Interactive Media, University of Miami (Miami, Florida)

humanly versus rationally—which drive two vital questions in AI: (1) Do we want to create a computer that operates in the same way as humans *think*, or do we want it to focus on acting/reacting in the same way as humans *do* without requiring it to exactly follow humans' train of thoughts? (2) Do we want the computer to mimic human behaviors in every way, or we only want it to focus on our rational behaviors and ignore others such as emotional expressions or impulsive decisions? Researchers continue to debate these questions, but the prevailing view that determines the goal of AI, is to create computer systems to *act rationally*.

Fundamental Concepts in Understanding AI

To understand AI, it is important to differentiate several terms that are often associated with AI in media coverage: algorithm, program, agent, and machine.

The word *"algorithm"* means a sequence of steps to solve problems. It is an essential concept in computer science but it is not a unique term specific to AI. In fact, it originates in mathematics "from the name of Persian mathematician al-Khwārizmī, author of a ninth-century book of techniques for doing mathematics by hand" (Christian & Griffiths, 2016). Computer scientists have been developing algorithms to solve various problems, for example, finding information on the Web (searching) and presenting such information in the order of relevancy (sorting). Algorithms are then written or implemented as computer *programs* to execute a sequence of steps to solve the problem. However, not every computer program contains an algorithm.

The term "agent" refers to "anything that can be viewed as perceiving its environment through sensors and acting upon that environment through actuators" (Russell & Norvig, 2009). The agent uses its sensors to gather information as *input* to its system, running *algorithms* to make sense of the information and make decisions (i.e., solve problems), and creating *output* via actuators to carry out the decision to change the environment. For example, a AI car agent uses sensors including cameras, radar, and lidar to observe its surroundings, constantly running algorithms to determine if the car in the front is slowing down or if there is a pedestrian about to walk across the street, and if so, the car puts on the brakes to slow down. However, not every AI application has physical sensors or actuators. Therefore, *AI system* is a more general term to describe a computer program that has an AI algorithm to process input and generate output. In machine learning, a subfield of AI, the process of using algorithms to build models of concepts and relations for decision making is called training a *machine*, which does not refer to a physical machine or robot.

AI, Machine Learning (ML) & Deep Learning (DL)

If you follow news on AI, you must have heard of machine learning and deep learning. As Portland State University computer science professor Melanie Mitchell stated, it is an "unfortunate inaccuracy" that some media reports equated AI with machine learning or deep learning (2019). The relationship between these three is illustrated in Figure 25.1. Machine learning is one of the subfields in AI that focuses on creating algorithms for AI systems to learn from examples or experiences. In machine learning, there exist many methods to create such algorithms, and deep learning is one specific type of approaches that has greatly influenced AI in recent years.

Figure 25.1

Relationship Between AI, Machine Learning & Deep Learning

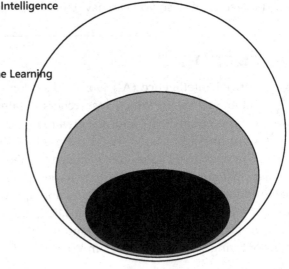

Artificial Intelligence

Machine Learning

Source: C.H. Chuan

The manner in which AI algorithms are created can be classified into two groups of approaches: rule-based and data-driven. Rule-based approaches create an AI system by first figuring out the knowledge that human experts have and then translating the knowledge into rules that the system can apply to the task. By contrast, data-driven approaches give the AI system examples of both correct and wrong answers or by setting up an environment in which the system can experiment on its own to gain experiences (i.e., to learn). Such data-driven approaches are the core of machine learning.

There are three types of machine learning: supervised, unsupervised, and reinforcement learning. Imagine you are a teacher trying to teach something; you need to decide which teaching materials you are going to use and how you are going to use them. The difference between the three types of machine learning lies in these two decisions. Assume that we are trying to teach an AI system what cats and dogs look like. For both supervised and unsupervised learning, we first collect a few pictures of cats and dogs as our teaching materials. For supervised learning, each picture is labeled with the "answer" such as cat or dog as shown on the left in Figure 25.2. Therefore, the algorithm of supervised learning knows the correct type of animals in the picture when it learns, and its job is to identify which features define most cats or dogs or features that best distinguish these two types of animals.

As for unsupervised learning, as shown on the right in Figure 25.2, we just give the algorithm all the pictures without the animal labels and tell the algorithm that these pictures represent two different types of animals—cats and dogs. You may be surprised and wonder "how can the algorithm know what's what without us telling it first?" An unsupervised algorithm learns by identifying *similarity*, assuming that objects in the same group are more similar to each other than to the ones from another group. As the teacher, we can provide some guidance by telling the algorithm what kind of features (e.g., shape, color, length of the nose) that contribute to the *similarity*. The algorithm then tries to divide the pictures into two groups so that pictures in the same groups are more similar to each other than to the ones in the other group.

Figure 25.2

Supervised versus Unsupervised Learning: Differentiating Cats from Dogs

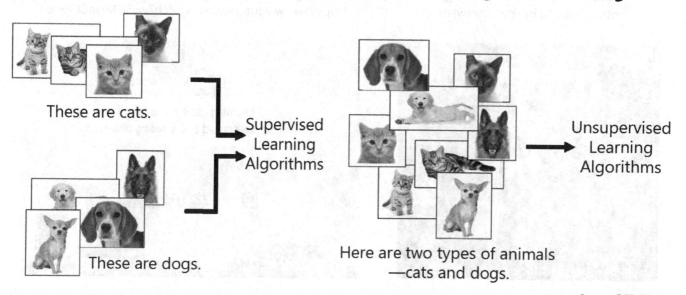

Source: C.H. Chuan

The third type of machine learning—reinforcement learning—is quite different from the previous two. Supervised and unsupervised learning is like sitting down with a teacher to learn some concepts, while reinforcement learning is more like trying to learn a skill by trial-and-error (e.g., learning how to ride a bike as a child). A reinforcement learning algorithm does not need learning materials. Instead, it is planted in the environment to figure out the right way to do things. When we were learning how to ride a bike, we often fell a lot until we learned how to balance. The key difference between us learning how to bike and reinforcement learning is that falling causes our physical pain so we will want to avoid falling from the bike. But there is no such thing as "physical pain" for the algorithm. Therefore, when we design the algorithm, we need to define rewards and/or penalties as consequences for different actions that the algorithm receives as positive or negative feedback. It is quite fun to watch the progress of reinforcement learning (follow the YouTube links in Figure 25.3): the agent does terribly in the beginning but gradually becomes much better after numerous trials.

A well-known machine learning algorithm is *neural networks* which are inspired by how human brain works. Figure 25.4 (a) shows an example of a neuron, a cell in the nerve system that transmits electrical signals to other cells. Signals propagated from neurons to neurons control brain activity and these mechanisms "form the basis for learning in the brain" (Russell & Norvig, 2009). Figure 25.4 (b) shows an artificial neuron called *perceptron* as a computational form to simulate information processing in neurons (Rosenblatt, 1958). Instead of electrical signals, a *perceptron* receives numerical values as inputs and can be activated when the arithmetic combination of the inputs exceeds a certain threshold to transmit an output. However, as a computational form, one *perceptron* can only send out signals of 1 or 0 (or on or off), which is very limited. To simulate complex activities, we need a network of multiple perceptron—a *neural network* as shown in Figure 25.4 (c)—connected together like neuros in our brain.

Figure 25.3

Applying Reinforcement Learning

Learning How to Swing
https://youtu.be/ke6xkwxwfzu

Learning How to Play the Mario Video Game
https://www.youtube.com/watch?v=5GMDbStRgoc

Source: YouTube

Figure 25.4

Neural Networks

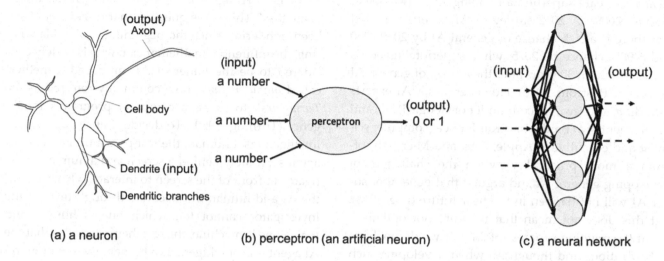

(a) a neuron   (b) perceptron (an artificial neuron)   (c) a neural network

Source: C.H. Chuan

Our brain has around 100 billion neurons (Kandel et al., 2000). Similarly, we need a neural network with many of layers of perceptron, known as "deep" network, in order to learn complicated concepts. However, the complex structure of the network come at two costs. First, more layers create much more complex calculation, which requires a lot of computing power and more efficient algorithms. Second, we need a massive volume of data to train such networks. It was in the 2010s when the two obstacles were overcome by *deep learning* which represented a quantum leap in solving many machine learning problems.

Different Types of AI

AI applications can be categorized as narrow AI or general AI. Narrow AI focuses on a specific goal or task, while general AI can perform various tasks with human-like flexible skills. For example, the Roomba vacuum by iRobot is narrow AI because it focuses on one task only–cleaning the floor. The helper robot in the movie *Robot & Frank* that can clean, cook, and plan activities to improve the human character's physical and mental health would be an example of general AI. None of the existing AI systems "could be called intelligent in any general sense" (Mitchell, 2019).

Some books label narrow AI as weak and general AI as strong. This is a mistake. As Russell and Norvig discussed (2009), weak AI means that the AI agent can

possibly act intelligently, but it only simulates the action without really intelligently "thinking." In contrast, strong AI assumes that the AI agent can actually "think." Let's use the helper robot as an example here again: if the agent is considered strong AI, it has to perform all the tasks as a result of its deliberate "thinking" instead of simply performing pre-programmed daily routines.

Beyond strong AI, there is ultra-intelligent machine (Good, 1965) or super(human) AI (Dickson, 2017). Good (1965) explained:

Let an ultraintelligent machine be defined as a machine that can far surpass all the intellectual activities of any man however clever. Since the design of machines is one of these intellectual activities, an ultraintelligent machine could design even better machines; there would then unquestionably be an 'intelligence explosion,' and the intelligence of man would be left far behind. Thus the first ultraintelligent machine is the last invention that man need ever make, provided that the machine is docile enough to tell us how to keep it under control (page 33

To use the helper bot example again, a super AI helper has to be a Mary Poppins robot!

In 1993, the computer science professor and award-winning science fiction writer, Vernor Vinge, predicted "within thirty years, we will have the technological means to create superhuman intelligence." (Russell & Norvig, 2009). A 2012 survey of AI experts reported that there is a 50% chance of general AI by 2040-2050 and a 90% chance by 2075, while superintelligence is expected within 30 years after the advent of general AI (Müller & Bostrom, 2016). However, other AI experts including Andrew Ng (co-founder of Google Brain and Chief Scientist at Baidu) and Kai-Fu Lee (computer scientist and executive at Apple, IBM, and Microsoft) offered a more practical view on the challenges of developing general AI and argued that general or super AI will not happen in the near future (Lee, 2018). But this does not mean that they are not optimistic about the ongoing progress of AI, nor we do not have to be cautious and thoughtful when developing such technologies. If you would like to learn more about what it takes for the development of general AI, you can read Urban's (2019) post on the Artificial Intelligence Revolution.

Evaluating AI: How Intelligent is it?

Human intelligence has been the conventional standard for measuring intelligence of other creatures, such as dolphins and chimpanzees. But do you remember reading news about AI beating humans in games? First in 1997, IBM's Deep Blue defeated the world chess champion Garry Kasparov. In 2011, IBM's Watson played on *Jeopardy!* and defeated human champions. More recently, in 2016, the AI program AlphaGo from DeepMind played with a Go champion Lee Sedol and won four out of the five games.

Is it possible that AI can beat humans for non-game tasks? A recent news article published by *Wall Street Journal* (Abbott, 2020) boasted a headline, "Google AI Beats Doctors at Breast Cancer Detection—Sometimes." In these "sometimes" cases, does it mean that the AI system is better than all the doctors, the majority of the doctors, or an average doctor in the field? It turns out that the conclusion this case was drawn upon were the breast cancer cases that six U.S. radiologists failed to detect but were caught by AI. Is this significant enough to declare "AI beats human doctors"? As Mitchell indicates (2019, chapter 5), such headlines should be taken with a grain of salt.

Beyond such one-on-one game settings in which AI with extraordinary computing power can quickly calculate all the possible winning options, how do we know if an AI agent is intelligent enough for a non-game task? This is the question that researchers have been pondering since the beginning of AI. Alan Turing, the influential mathematician who is widely considered to be the father of AI, proposed a method called *imitation game*, now commonly referred as the *Turing test*, to assess a computer program's intelligence (Turning, 1950). To decide if an AI agent is as intelligent as a human, the test puts a human and AI agent separately behind a screen and a human investigator in front of the screen to interactively chat with the AI and human interlocutors through texts. If the investigator cannot tell which one is human and which one is machine, the test then concludes that the AI agent is as intelligent as a human. Even though researchers are still debating about what exactly the Turing test measures, it remains to be one the most commonly used evaluation methods.

Background
A Brief History of AI

Throughout its history, AI has been developed through interdisciplinary efforts, drawing theories from disciplines such as philosophy, mathematics, economics, neuroscience, psychology, engineering, and linguistics (see Chapter 1.2, Russell & Norvig, 2009). The history of AI is characterized with a series of interchanging springs and winters. The term "artificial intelligence" was invented in a summer workshop organized by John McCarthy at Dartmouth in 1956. In the 1950s and 1960s, AI was a promising field with early enthusiasm and optimism. Researchers were excited about using rule-based AI such as logic to solve puzzles like humans do, and they anticipated that general AI would be soon realized. However, they encountered various barriers and funding agencies and industries lost interests in AI. Although progress continued, such as the breakthrough in narrow AI that solves problems in specific tasks using expert systems, the late 1960s to mid 1970s was generally considered as the first AI winter.

In the early 1980s, AI made a comeback as industries started adopting expert systems. Computers were commercially produced with specialized software and hardware for expert systems, vision systems, and robots. However, another AI winter arrived because the specialized hardware was challenged by powerful and general-purpose computers.

From mid the1980s to 2000s, many important AI algorithms were developed that paved the way to the AI spring two decades later. The invention of the web and the internet further boosted the availability of data, which became an important fuel for machine learning to flourish in 1990s. Since then, the field witnessed steady progress until deep learning brought "AI spring fever" in the 2010s (Mitchell, 2019, chapter 3).

Deep Learning and ImageNet

In the machine learning community, there are annual competitions that provide a common dataset as the benchmark for researchers around the world to fairly compare how their algorithms perform against each other. One famous competition is for object recognition, i.e., identifying the objects in an image, using the ImageNet dataset created in 2010 with 1.2 million manually labeled images (Deng et al., 2009). In the first two years of the ImageNet Challenge, the highest accuracy rate of the best team was 74%. But in 2012, the performance of the best team that used an efficient algorithm to train a deep network shot up to 85% which was an unprecedent leap for the research community. With the availability of big data and increasing computing power, deep learning has become very powerful. Since then, most participating teams switched to deep learning. At the same time, other AI communities were also creating large scale datasets and adopting the deep learning approach that has led to many breakthroughs in AI.

Recent Developments
Popular Areas of AI Applications

Powered by machine learning and especially deep learning, AI has been adopted by various industries in a transformative manner. The dramatic speed of adoption has led to the recognition of AI as the core of the fourth industrial revolution (Schwab, 2016). In his book, Rouhiainen (2018) described how AI has revolutionized various industries, including finance, retail, transportation, health care, journalism, education, and agriculture. For example, the London-based AI company DeepMind developed an AI algorithm that examines eye scans to detect early signs of eye diseases such as age-related macular degeneration (DeepMind, 2018; Kahn, 2018). Another example is Wordsmith, an AI-based platform for natural language generation. In 2016, Wordsmith used algorithms to write and publish 1.5 billion news stories, which stirred up discussions on whether human journalists will be replaced by algorithms (Miroshnichenko, 2018).

Many AI applications that made news headlines are from one of the listed areas in Table 25.1, where you can find specific research topics in the areas and some examples of such applications.

Issues of Fairness and Inclusion in AI Systems

While we enjoy the convenience of Siri or Alexa, more and more important decisions in our lives are now made by AI, including job employment (Nguyen 2014), loan approval (Lessmann et al. 2015; Vaidya 2017), and bail and sentencing decisions in criminal justice (Wang et al. 2013; Berk 2016). Unfortunately, it has been observed that AI systems can make biased decisions. For example, the AI algorithm used by Amazon "taught itself that male candidates were preferable… and penalized resumes that included the word *women's*, as in *women's chess club captain*." (Dastin, 2018). The Gender Shades project tested facial recognition algorithms from three major AI companies including IBM and Microsoft and found out that the accuracy rate for darker-skinned females can be 34% lower than that of lighter-skinned males (Buolamwini & Gebru, 2018). Google's AI chief thus warned that the immediate threat to humanity is not killer robots but rather the danger of biased algorithms (Knight 2017).

Table 25.1.

Popular AI Areas with Related Research Topics and Examples of Applications

Areas	Research Topics	Examples Of Applications
Natural Language Processing	Machine Translation, Text Mining, Text Generation, Sentiment Analysis, Topic Modeling	Google Translation, Wordsmith
Speech Recognition	Speech-To-Text, Real-Time Video/ Audio Caption Generation	Siri, Alexa, Google Home, Youtube Closed Caption
Computer Vision	Face Recognition, Image Recognition, Object Detection	Medical Image Diagnosis, Faceid, Surveillance
Robotics	Navigation, Interaction, Collaboration	Drones, Self-Driving Cars
Data Analytics	Prediction, Classification, Regression	Credit Score Prediction

Source: C.H. Chuan

Why would AI systems make such biased decisions? In the context of machine learning where developers prepare and provide "teaching" materials to the system, how the materials are gathered, labeled, and used makes an enormous difference to the algorithm. For example, there have been more male employees in the tech industry than female. If the algorithm is trained on the resumes of past and current employees who are mostly men, the algorithm naturally tends to select people who are more similar to the existing employees and ignore those who are different. In a broader context, AI algorithms are designed to simulate human's rational action in a given circumstance, i.e., doing the right thing. The task becomes complicated when an AI system is designed for problems that lack common agreement on what is "the right thing" to do. Beyond the potential misuse of AI, people are increasingly aware of the dark side of AI that the technology can be purposefully set to create damages, such as deepfakes that have been used to generate fake news videos, fake celebrity pornographic videos, and bots that are designed disseminate misinformation online.

Williams (2018) cautioned about the AI divide between those who benefit from AI advances and those who are primarily exploited by them, leading to more inequality in the future. While developing AI algorithms is the job for computer scientists and engineers, it is important for everyone to understand how AI systems are designed, such as how the training data are collected and used, monitor for potential misuse of AI, and advocate for regulations and guidelines for developing AI applications.

Current Status

According to McKinsey Analytics (Chui et al., 2019), the overall adoption of AI remains relatively low among businesses (about 20%), but many executives are looking into what AI can do for their business. Deloitte (2018) conducted a survey with 1,100 business leaders of U.S.-based companies that are already using AI and reported an increasing investment in cognitive AI technologies (e.g., machine learning and deep learning for speech recognition, computer vision, and natural language processing).

Factors to Watch

Machine learning—especially deep learning—requires big datasets. That is why a recent article in *Forbes* argues that data is the new oil (Bhageshpur, 2019). Our data has been collected by companies via various online services and platforms, mobile apps, and wearable devices. What data is collected and sold to whom, how it is used, and what it is used for are usually unknown to users. Critical issues such as data privacy, data security, and data ownership become more important than ever as AI is now omnipresent in every aspect of our lives. As of mid-2020, the General Data Protection Regulation (GDPR) is the main source for data protection and privacy as a European Union law. More countries are developing their versions of GDPR and California is the first state in the U.S. to have such a law. The website gdpr.eu provides useful resources including checklists to help companies comply with the regulation.

Gen Z

A basic understanding of AI is essential for the younger generation who grows alongside the technology. More and more universities have recognized such demands and started to offer courses about AI in an interdisciplinary setting, such as the course *Design with AI* that the author teaches in the Department of Interactive Media at University of Miami.

Similar to many new, but powerful technologies, media tend to frame AI through the competing visions of "wizard" or "prophet" that portrays the technology either as a magical solution to our problems or as an uncontrollable destructive force that threatens humanity. It is important to apply critical thinking and evaluate the benefits and risks of particular AI applications in an analytical manner. A helpful resource is organizations such as AI Now Institute which provides critical analyses on the social implications of emerging AI technologies.

Global Concerns & Sustainability

The development and uses of AI are increasingly politicized and there is now a global "AI race." For example, China aims to be the global AI leader by 2030 (Sagramsingh, 2019), while in early 2019, President Trump of the United State signed an executive order on maintaining American leadership in AI. A recent global survey published by IEEE-USA (Sagramsingh, 2019) reported that many countries have dedicated big budgets on AI but each country has a different development focus and guidelines to address issues of ethics, privacy, and security. Due to the AI race and the connection between AI and national security, more countries will start to protect their AI advances as government secrets. However, as Kai-Fu Lee (2018) wrote: "As both the creative and disruptive force of AI is felt across the world, we need to look to each other for support and inspiration."

Bibliography

Abbott, B. (2020) Google AI beats doctors at breast cancer detection—sometimes. *Wall Street Journal*.

Bhageshpur, K. (2019). Data is the new oil – And that's a good thing. *Forbes*. [News article] https://www.forbes.com/sites/forbestechcouncil/2019/11/15/data-is-the-new-oil-and-thats-a-good-thing/.

Buolamwini, J., & Gebru, T. (2018). Gender shades: Intersectional accuracy disparities in commercial gender classification. *Conference on fairness, accountability and transparency* (pp. 77-91).

Berk, R. A., Sorenson, S. B., & Barnes, G. (2016). Forecasting domestic violence: A machine learning approach to help inform arraignment decisions. *Journal of Empirical Legal Studies*, 13(1), 94-115.

Christian, B., & Griffiths, T. (2016). *Algorithms to live by: The computer science of human decisions*. Macmillan.

Chui, M., Henke, N., and Miremadi, M. (2019, March) Most of AI's business uses will be in two areas [report]. https://www.mckinsey.com/business-functions/mckinsey-analytics/our-insights/most-of-ais-business-uses-will-be-in-two-areas.

Dastin, J. (2018, October 9) Amazon scrapes secrete AI recruiting tool that showed bias against women [News article] https://www.reuters.com/article/us-amazon-com-jobs-automation-insight/amazon-scraps-secret-ai-recruiting-tool-that-showed-bias-against-women-idUSKCN1MK08G.

DeepMind. (2018, August 13) DeepMind health research and Moorfields eye hospital NHS foundation trust: A patient's story [Video file]. https://youtube/2MxtZ5XFquE.

Deloitte (2018). *State of AI in the enterprise, 2nd edition* [Survey report]. https://www2.deloitte.com/content/dam/insights/us/articles/4780_State-of-AI-in-the-enterprise/DI_State-of-AI-in-the-enterprise-2nd-ed.pdf .

Deng, J., Dong, W., Socher, R., Li, L. J., Li, K., & Fei-Fei, L. (2009, June). Imagenet: A large-scale hierarchical image database. *2009 IEEE conference on computer vision and pattern recognition* (pp. 248-255). IEEE.

Dickson, B. (2017, May 12) What is narrow, general and super artificial intelligence [Article]. https://bdtechtalks.com/2017/05/12/what-is-narrow-general-and-super-artificial-intelligence/.

Good, I. J. (1965). Speculations concerning the first ultraintelligent machine. *Advances in computers* (Vol. 6, pp. 31-88). Elsevier.

Kahn, J. (2018). Google's DeepMind to create product to spot eye disease [News article]. https://www.bloomberg.com/news/articles/2018-08-13/google-s-deepmind-to-create-product-to-spot-sight-threatening-disease.

Kandel, E. R., Schwartz, J. H., Jessell, T. M. (2000). *Principles of neural science*, 4th edition. McGraw-Hill.

Knight, W. (2017, October 3). Google's AI chief says forget Elon Musk's killer robots, and worry about bias in AI systems instead. *MIT Technology Review*.

Lee, K. F. (2018). *AI superpowers: China, Silicon Valley, and the new world order*. Houghton Mifflin Harcourt.

Lessmann, S., Baesens, B., Seow, H. V., & Thomas, L. C. (2015). Benchmarking state-of-the-art classification algorithms for credit scoring: An update of research. *European Journal of Operational Research*, 247(1), 124-136.

Merrium-Webster (n.d.) Intelligence. https://www.merriam-webster.com/dictionary/intelligence?utm_campaign=sd&utm_medium=serp&utm_source=jsonld.

Miroshnichenko, A. (2018). AI to Bypass Creativity. Will Robots Replace Journalists? (The Answer Is "Yes"). *Information*, 9(7), 183.

Mitchell, M. (2019). *Artificial intelligence: A guide for thinking humans*, Farrar, Straus and Giroux.

Müller, V. C., & Bostrom, N. (2016). Future progress in artificial intelligence: A survey of expert opinion. *Fundamental issues of artificial intelligence* (pp. 555-572). Springer, Cham.

Nguyen, L. S., Frauendorfer, D., Mast, M. S., & Gatica-Perez, D. (2014). Hire me: Computational inference of hirability in employment interviews based on nonverbal behavior. *IEEE transactions on multimedia*, 16(4), 1018-1031.

Oxford English Dictionary (n.d.) Intelligence. https://www.oed.com/view/Entry/97396?rskey=ITwaIa&result=1&isAdvanced=false#eid.

Rosenblatt, F. (1958). The perceptron: A probabilistic model for information storage and organization in the brain. *Psychological Review*, 65(6), 386-408.

Rouhiainen, L. (2018). *Artificial Intelligence: 101 Things You Must Know Today about Our Future*. Lasse Rouhiainen.

Russell, S. & Norvig, P. (2009). *Artificial Intelligence: A Modern Approach*. Pearson.

Sagramsingh, R. (2019). *AI: A global survey*. IEEE USA.

Schwab, K. (2016). *The Fourth Industrial Revolution*. World Economic Forum.

Turing, A. M. (1950). Computing machinery and intelligence. *Mind*, 49, 433-460.

Urban, T. (2019). The AI revolution: the road to superintelligence. WaitButWhy. https://waitbutwhy.com/2015/01/artificial-intelligence-revolution-1.html.

Vaidya, A. (2017). Predictive and probabilistic approach using logistic regression: Application to prediction of loan approval. *Proceedings of 2017 IEEE 8th International Conference on Computing, Communication and Networking Technologies (ICCCNT)* (pp. 1-6).

Wang, T., Rudin, C., Wagner, D., & Sevieri, R. (2013). Learning to detect patterns of crime. *Proceedings of Joint European conference on machine learning and knowledge discovery in databases* (pp. 515-530). Springer, Berlin, Heidelberg.

Williams, A. B. (2018). The potential social impact of the artificial intelligence divide. 2018 AAAI Spring Symposium Series.

Section V

Conclusions

26

Other New Technologies

Jennifer H. Meadows, Ph.D.[*]

Introduction

Every two years, the editors of this book struggle to decide which chapters to include. There are always more options than there is space because new communication technologies develop almost daily. This chapter outlines some of these technologies deemed important, but not ready for a full chapter. The 18th edition of this book may cover these technologies in full chapters just as virtual reality and artificial intelligence were once featured in an "Other New Technologies" chapter. Among the choices three technologies stand out: surveillance and biometrics, consumer robotics, and power storage. in addition, a sidebar on accessibility is presented.

Surveillance

The concept of surveillance has been around for centuries. Defined as "the act of carefully watching someone or something especially in order to prevent or detect a crime (Merriam-Webster, n.d.)," surveillance has now extended beyond preventing and detecting crime, to generating control and empowering governments and other agencies, as well as to bring a sense of safety to communities and locations. While in the past surveillance used people to spy, these days surveillance uses technology such as cameras and microphones. New technologies are being employed including artificial intelligence, facial recognition, and other biometrics such as fingerprint and retinal scanning. Surveillance technologies are employed at both private and public spaces like a local retail store or a public park.

Government surveillance began in 1919 with the creation of U.S. State Department's Cipher Bureau. Project SHAMROCK collected all telegraphic data coming in and going out of the U.S. starting in the 1940's, but was halted during the 70's when the Supreme Court ruled that surveillance for domestic threats has Fourth Amendment protection thus requiring a warrant. The 1970s Watergate scandal led to more investigations into U.S. surveillance, which eventually led to the creation of the Senate Select

[*] Professor, Department of Media Arts, Design, and Technology, California State University, Chico (Chico, California).

Committee on Intelligence and the *Foreign Intelligence Surveillance Act of 1978* (FISA). FISA allows a court to consider secret warrants for domestic spying (Factbox, 2013).

The attacks of September 11, 2001 led to Congress passing the *USA PATRIOT Act*, which extended domestic surveillance capabilities, and the Bush administration approved an NSA secret spying program which was eventually exposed for widespread warrantless surveillance. Further legislation expanded the government's authority for warrantless surveillance and gave protections to U.S. telecommunication companies who turned over data (Factbox, 2013).

Section 702 of the *FISA Amendment Act* of 2008 authorized and expanded the NSA's warrantless wiretapping program so U.S. intelligence agencies can search all wireless communications going in to and out of the United State (Yachot, 2017).

What about local governments and police departments? Increasingly local communities are using surveillance technologies to prevent and detect crime. Municipal or civic surveillance systems have been used in the U.K. since 1964. CCTV blankets Britain with one camera for every 11-14 people (Young, 2018). Interestingly, the majority of CCTV cameras in Britain are privately owned. Scotland Yard reports about 95% of murder cases use CCTV footage as evidence (Young, 2018). In the U.S. the most surveilled city is Atlanta which has 15.56 camera per 1,000 residents.

Actual surveillance using video cameras has been around for some time; increasingly facial recognition technologies are used in addition to video to identify individuals. Use of facial recognition is controversial as it can be seen as a violation of privacy. When used with general surveillance, such as publicly placed cameras, it can be used without user's consent or knowledge (Face Recognition, n.d.). Many police departments and government agencies use this technology, but facial recognition technology is also being employed by businesses. For example, Facewatch, a U.K. company, sells security systems that employ facial recognition to alert business owners to "persons of interest" entering their business (Facewatch, 2020). In Europe, companies have to have "substantial public interest" to be able to capture and store biometrics, a somewhat vague threshold to meet (Olsen, 2019).

How are facial recognition databases created? Police departments and governments can use official identification information such as driver's license photographs. However, some companies rely on user sourced photographs. Facewatch creates a database of "persons of interest" using photographs from users that are shared with other Facewatch users. What about all of the images of people on the Internet? Facebook's face recognition technology can identify the faces of your friends in your images and tag them for you. Facebook lost a class action lawsuit in Illinois over the technology because it did not ask permission and did not reveal how long it would keep the data. The company has since begin allowing users to opt out of the technology (Singer & Isaac, 2020).

No company involved in surveillance and facial recognition created as much controversy in 2019 and 2020 as Clearview AI. The company scrapes internet sites such as Facebook, Youtube, LinkedIn, and Twitter to add to a database of billions of photographs without people's permission. These photos are used for facial recognition. Over 600 law enforcement agencies use the technology including police departments, the FBI, and ICE. Private businesses also use the technology including Macys, Eventbrite, Best Buy, and the NBA (Mac, Haskins & McDonald, 2020a). The app allows users to identify anyone with their image. Clearview AI has plans for augmented reality glasses that would identify anybody in the line of sight as well as a security camera with instant identification (Hill, 2020; Mac, Haskins & McDonald, 2020b). Google, Microsoft, Twitter, and YouTube have sent cease-and-desist letters to the company, and it is the subject of lawsuits including one brought by some citizens of Illinois for violation of the state's biometrics laws (Reichert, 2020). Both Facebook and Venmo have also demanded that Clearview stop using their users' images (Bonifacic, 2020). Clearview AI itself claims on its website that they use only public information in full compliance with the law. They claim that their goal is stopping criminals and protecting the innocent (Clearview AI, n.d.)

Privacy advocates argue that technologies such as Clearview AI create a mountain of problems beyond the invasion of privacy including abuse. For example, university police could easily identify protestors.

Strangers could use the technology to identify stalking victims. Governments could use the technology to identify and target ethnic minorities. For example, in China, facial recognition and surveillance technologies are being used to identify Muslim Uighur populations and track their movement (Mozur, 2019).

Back in the States, there's another group of surveillance technologies that many homeowners are happily employing—connected security systems including security cameras, doorbells, and lighting. Industry leader Ring is most famous for its video doorbells (See Figure 26.1). While used by consumers to identify people at the door and to check out who porch pirated their Amazon shipment, Ring users can bring up the Neighbors app to see real time alerts on crime in their neighborhood. The app has also been used to find lost pets. The controversial use, though, is video sharing with law enforcement. Users can post their video to the app publicly. That video is available to law enforcement. If a crime happens and law enforcement wants to see video from doorbells/security cameras in the area, Ring will send a request to users in the area asking for permission to share the video with the law enforcement agency. Users have to opt-in to share (Ring, 2020).

Figure 26.1:

Ring Doorbell

Source: Ring

Ring changed its user dashboard and security in early 2020 due to criticism of its security and privacy protection. Early mistakes include exposing customer video and audio and hacker vulnerability. Even now there are concerns over Ring's Android app and the collection of data by third parties (Schwarz, 2020).

Consumer Robots

Back in the early 1960's people watched the animated series *The Jetsons*. An important character was Rosie the robot, the Jetson's maid. The idea of a household robot, though, didn't reach reality until iRobot, a company, which made robots for organizations such as DARPA and the National Geographic Society, released the Roomba floor vacuuming robot in 2002. Since then iRobot has released robots to clean your gutters, pool, and mop your floor (iRobot, 2020). Also used for cat transportation, the robot vacuum has gained a foothold in American households. iRobot reports it has sold over 20 million Roombas worldwide (iRobot, 2020).

Robots have been used for years in manufacturing, essentially replacing human workers in many cases. This displacement is now hitting the service industry as robots are replacing bartenders and cooks. For example, the Tipsy Robot in Las Vegas has two robot bartenders, each one capable of 120 drinks per hour. While an "attraction" now, robot bartenders don't need tips and won't over-pour (Tipsy Robot, 2020). Picnic's robot pizza makers can make up to 300 pizzas in an hour while Flippy the robot can identify and flip burgers (Dormehl, 2020; Thorne, 2019).

One way Americans may communicate with robots is through their food delivery service. Companies like Doordash are increasing their use of delivery robots to replace humans. Starship, a delivery robot company, is rolling its service out to college campuses. (See Figure 26.2) At Bowling Green State Universities, students can order food from a number of campus vendors with an app and have their food delivered by robot for just $2 (Malaska, 2020). Starship notes its robots move at a pedestrian speed and are programed to avoid objects and people. The robots can cover a four mile radius and can only be unlocked by the recipient using the app (Starship, 2020). Other delivery robot users include Doordash and Postmates.

Figure 26.2:
Starship Delivery Robot

Source, Starship

What about a personal robot? While many homes have a smart speaker such as the Amazon Echo or Google Home, these devices can't follow you around and always be there for you. Enter Samsung's Ballie. Called a "real life BB-8," Ballie is a small ball shaped personal robot (Peters, 2020). (See Figure 26.3) Part of Samsung's vision of robots as life companions, Ballie has AI capabilities and a camera that allows it to be a fitness assistant, and "seek solutions for people's changing needs" (Samsung Electronics, 2020). Still in the prototype stage, Samsung's vision of Ballie can be seen in their Waltz for Ballie video (https://youtu.be/c7N5UDZX7TQ). In the video Ballie opens the windows, helps with a yoga session and even calls the vacuum robot to clean up after the dog spills some food. Time will tell if we are ready for a personal care robot

Figure 26.3:
Samsung Ballie

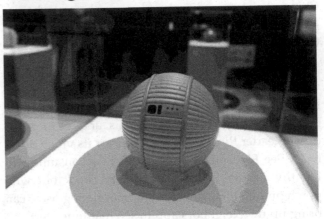

Source, Samsung

Remember Sony's robot dog? It heralded the arrival of a potential market for robot pets. At CES 2020, Elephant Robotics showed off a robot cat, MarsCat, which develops its personality over time based on interactions with the user but, just like a real cat, it does a lot of cat tasks independently like stretching, kneading, and playing (Ellis, 2020). What about a cuddly labrador puppy? Jenny, the puppy robot, was developed by Tombot after the owners' mother developed Alzheimer's disease and had to give up her real dog. Jenny was designed as a support robot for people struggling with stress, loneliness, anxiety, and depression. Jennie is a lap dog with over 50 sensors to feel how hard you are petting her, and she can respond to touch and sound. She makes dog sounds and can wag her tail. She was designed in collaboration with the Jim Henson Company, the company behind the Muppets (Collins, 2020).

Power Storage

Think of all the technologies you use on a daily basis. How many of them use a battery? Laptops, mobile phones, tablets, and earbuds use batteries and developments in battery or power storage technologies have a huge impact on the usability of technology. But batteries aren't just limited to our little technology devices. Electric cars and home electrical systems use them too.

Lithium ion batteries have shrunk in size and increased in power while dropping in price over the last decade. The development of lithium ion batteries is key to the advancement of electric vehicles. The range of an electric car is determined by its battery. BloombergNEF has predicted that 57% of passenger car sales will be electric by 2040 (Stevens, 2019).

Auto manufacturers are in a race to deploy better, smaller batteries that give longer driving range. GM showed off an improved battery pack in March, 2020 called Ultium. The batteries "could" allow for a driving range of 400 miles or more on a single charge. GMs trucks will come with 800-volt battery packs while electric cars will have half that power. The batteries also use less cobalt, which is limited in supply (Hawkins, 2020). GM and other automakers have been chasing Tesla, the leader in electric vehicle battery packs. Tesla's battery pack design allows it deliver power at a lower cost. In addition, the company's Gigafactories in Nevada and China produce batteries for Teslas as well as Toyota (LeBeau, 2020).

It's easy to think of batteries for phones and even cars, but what about utility-scale batteries? These are batteries that can store excess energy that isn't being used on the power grid. Traditionally, power generation plants (coal, nuclear, wind, gas, solar, etc.) generate electricity that is sent over the grid to be used immediately but if it isn't used it is wasted. Utility-scale batteries allow better grid management because excess energy can be stored and released later if the need arises. This is possible, in part, because the cost of lithium ion batteries has dropped (Gorrill, 2020).

The cost has dropped enough that residential solar customers can also have batteries connected with their solar systems, making them truly "off the grid."

As solar installations grow, the deployment of energy storage systems in homes has also grown. Driven in part by the California wildfire power shut offs that began in 2019, the number of installations for residential batteries in the state has almost tripled (Wesoff, 2020). According to Wesoff (2020), the US energy storage market is expected to grow to 7.3 GW by 2025 up from just 523 MW in 2019.

The batteries that make the most impact on your daily life are probably in your phone. The battery life on mobile phones has increased over time while charging has become easier with wireless systems such as Qi. For example, Apple's iPhone 11 Pro Max's battery lasts 5 hours longer than the iPhone X Max. Power banks have become a must have accessory since sometimes there isn't an outlet available. The market for power banks was $6.8 billion in 2019 and is expected to grow 18.4% by 2027 (Grand View Research, 2020).

Other Emerging Factors

Two other emerging factors, geofence warrants and accessibility, are discussed in the sidebars below.

Conclusions

This chapter provides just a brief overview of some other new technologies. Who know, maybe the next edition of this book will dedicate entire chapters to the technologies discussed here. What's important to remember, though, is that technology is always changing, and you can keep up with these changes using tools your learned in this book.

Geofence Warrants

Jeffrey S. Wilkinson, Ph.D., Florida A&M University

Since 2017, law enforcement officials across the country have been using a new investigative technique called a "geofence warrant." Geofence warrants allow investigators to cast a virtual dragnet over crime scenes, sweeping up Google location data from every users' device in a specified area during a specified period of time. Users' GPS, Bluetooth, Wi-Fi and cellular connection data are collected to help identify possible suspects (Schuppe, 2020). According to Sobel (2020), Google has not released specific numbers of requests for geofence warrants, but admitted they had a 1,500% increase in 2018 compared to 2017, and over 500% increase from 2018 to December, 2019.

There have been concerns leading to litigation regarding whether geofence warrants violate Fourth Amendment protection against unreasonable searches. For example, a defendant charged with a 2019 Virginia bank robbery took his case to the US Supreme Court (Sobel, 2020). Surveillance footage showing the perpetrator held a cell phone to his ear before entering the bank, resulting in a geofence warrant for data from every device within 150 meters of the bank 30 minutes before and 30 minutes after the robbery. The resulting data produced a pool of 19 suspects. A second round of anonymized location data narrowed that down to the individual charged for the crime. Defense lawyers argued that the location data is the private property of the user, so dragging all the information from every device in that area violates the expectation of innocent parties that they are not being scrutinized for a crime they did not commit.

Using the data to charge those who are innocent occurred during the investigation of a home burglary in Gainesville, Florida. A geofence warrant led police to focus on a biker who had passed the victim's house three times within an hour. According to Schuppe (2020), the innocent biker was using his exercise-tracking app, RunKeeper, to record his frequent loops through his neighborhood. The app used his phone's location services which fed his movements to Google. Afterward the biker said "It was a nightmare scenario...I was using an app to see how many miles I rode my bike and now it was putting me at the scene of the crime. And I was the lead suspect" (Schuppe, 2020).

Apple has said it does not have the ability to perform these types of searches. To avoid the data dragnet of geofence warrants, users are advised to turn off location tracking and history for your android devices (Naked Security by Sophos, 2019). This can be done by signing into your Google account and disabling the data location and tracking settings.

Accessibility

Norman E. Youngblood, Ph.D., Auburn University

Reaching audience members who might rely on adaptive technology when they consume media is too often an afterthought in the production process, regardless of whether creators are producing content for broadcast, film, or internet distribution. Not taking these consumers into account disadvantages those audience members, makes your content less flexible, and increases the chances of your company getting into legal trouble for not making content accessible. Vision- and hearing-related disabilities are probably the first that come to mind when you are thinking about disabilities that might keep people from consuming media. Around 2% of Americans have a vision-related disability, and almost twice as many have a hearing-related disability. Over 5% of Americans have a mobility-related disability.

In the case of video and audio products, consumers with hearing problems typically rely on captions or transcripts to get the information they might otherwise miss. Consumers with vision-related issues often rely on audio descriptions that use a separate audio track to provide a narration of what's happening on-screen. Captions, transcripts, and audio descriptions also have a lot to offer consumers who don't have issues with their hearing or vision. Need to skim the content of an audio podcast or radio news story? A transcript is a quick way to get the information. In a noisy restaurant and trying to watch a TV show? Or maybe you are trying to sneak a peek at a social media video when you are supposed to be doing something else? Captions are your friend. Most television caption users don't have a hearing problem, and it's estimated that more than 80% of Facebook users watch videos with the audio turned off. Audio descriptions offer consumers without vision problems advantages as well. Want to catch up with the latest episode of *Daredevil* while you are driving? Turn on the audio descriptions and set down the phone to get a whole new perspective on the show.

Online media, particularly web-based content, is another area that content creators need to make accessible. In addition to making the audio and video portions of the content accessible, creators need to keep in mind that consumers with disabilities may not access computers with a mouse or a regular touch interface. Many consumers with vision-related disabilities and some with motor-skill related disabilities, rely on just the keyboard to navigate a website. Consumers unable to use a physical keyboard often use a virtual keyboard controlled with eye-gaze technology, which relies on pupil fixation and/or blinks to control the cursor, or sip-and-puff controllers, which allow the consumer to control the cursor by inhaling or exhaling into a straw-like device.

The process of building an accessible website isn't complicated but, like designing an accessible building, it requires you to think about accessibility from the very beginning—it's easier to design something to be accessible than to go back and make it accessible. Some of the most common tasks are related to making the site accessible for screen-reader users. Screen-reader software reads the screen content to the user. When a screen reader comes upon an image, it looks for the image's alt attribute to describe what the image's meaning is, information that is particularly important when the image is also a hyperlink. Screen readers allow users to form mental maps of the page, allowing the user scan to essentially scan the page by getting a list of the headings on the page or a list of all the links on the page, so it's important to use heading elements in the correct order, and not just to changes sizes, and to use descriptive text for hyperlinks, avoiding vague text like, "click here." Your non-screen users will thank you for paying attention to these issues as well. Consistent headings make it easier to skim the page, and users often scan the page for hyperlinks—no one wants to find a "click here" and have to back up to figure out where the link is taking them. The World Wide Web Consortium (W3C) provides the Web Content Accessibility Guideline (WCAG) to help media creators build accessible content, and the current version of the guidelines, WCAG 2.1, serves as the national accessibility standard for more than 30 countries, including the United States.

While you should create accessible content because it's the right thing to do, it's also often the legal thing to do as well. The Americans with Disabilities Act (ADA) has been applied to the virtual world as well as the brick-and-mortar world. If you run a business, your site needs to be accessible. Section 508 of the Rehabilitation Act requires most government websites and media be accessible. Broadcasters need to be aware of accessibility requirements as well. The 21st Century Communications and Video Accessibility Act called for increased use of captions and audio descriptions by broadcasters, including when material moves from broadcast to online.

Bibliography

Bonifacic, I. (2020). Facebook and Venmo demand Clearview AI stops scraping their data. Engadget. https://www.engadget.com/2020/02/06/facebook-venmo-cease-and-desist-clearview-ai/

Clearview AI (n.d.). Clearview Facts. https://clearview.ai/

Collins, K. (2020). A snoring robot puppy stole my whole heart at CES 2020. Cnet. https://www.cnet.com/news/a-snoring-robot-labrador-puppy-stole-my-whole-heart-at-ces-2020/

Dormehl, L. (2020). Flippy the burger-flipping robot is changing the face of fast food as we know it. *Digital Trends*. https://www.digitaltrends.com/cool-tech/flippy-burger-robot-changing-fast-food/

Ellis, C. (2020). The best robots from CES 2020: the cute, the cuddly and the confusing. Techradar. Face Recognition (n.d.) https://www.techradar.com/news/the-best-robots-from-ces-2020-the-cute-the-cuddly-and-the-confusing

Facewatch (2020). https://www.facewatch.co.uk/

Factbox (2013). ACLU. https://www.aclu.org/issues/privacy-technology/surveillance-technologies/face-recognition-technology

Gorrill, L, (2020). Battery manufacturing must take utility-scale energy storage to the next level. *Electronic Design*. https://www.electronicdesign.com/power-management/whitepaper/21125735/battery-manufacturing-must-take-utilityscale-energy-storage-to-the-next-level

Grand View Research (2020). Power bank market share, size & trends analysis report. https://www.grandviewresearch.com/industry-analysis/power-banks-market

Hawkins, A. (2020). GM unveils a new electric vehicle platform and battery in bid to take on Tesla. *The Verge*. https://www.theverge.com/2020/3/4/21164513/gm-ev-platform-architecture-battery-ultium-tesla

Hill, K. (2020). The secretive company that might end privacy as we know it. *The New York Times*. https://www.
nytimes.com/2020/01/18/technology/clearview-privacy-facial-recognition.html

iRobot (2020). History. https://www.irobot.com/about-irobot/company-information/history

LeBeau, P. (2020). Tesla's competitors play catch-up on electric batteries. CNBC. https://www.cnbc.com/2020/02/10/
teslas-competitors-play-catch-up-on-electric-batteries.html

Mac, R., Haskins, C. & Mcdonald, L. (2020a). Clearview's facial recognition app has been used by the Justice Department,
ICE, Macy's, Walmart, and the NBA. Buzzfeed News. https://www.buzzfeednews.com/article/ryanmac/clearview-
ai-fbi-ice-global-law-enforcement

Mac, R., Haskins, C. & Mcdonald, L. (2020b).The facial recognition company that scraped Facebook and Instagram photos is
developing surveillance cameras. Buzzfeed News. https://www.buzzfeednews.com/article/carolinehaskins1/clearview-
facial-recognition-insight-camera-glasses

Malaska, B. (2020). Starship delivery robots roll along at BSCU. NBC24. https://nbc24.com/news/local/starship-delivery-
robots-roll-along-at-bgsu

Merriam-Webster (n.d.) Surveillance. https://www.merriam-webster.com/dictionary/surveillance

Mozur, P. (2019) One month 500,000 face scans: How China is using A.I. to profile a minority. *The New York Times*.
https://www.nytimes.com/2019/04/14/technology/china-surveillance-artificial-intelligence-racial-profiling.html

Naked Security by Sophos (2019, December 16). Police get "unprecedented" data haul from Google with geofence warrants.
Nakedsecurity.sophos.com. https://nakedsecurity.sophos.com/2019/12/16/police-get-unprecedented-data-haul-from-
google-with-geofence-warrants/

Olson, P. (2019). With Brits used to surveillance, more companies try tracking faces. *The Wall Street Journal*.
https://www.wsj.com/articles/with-brits-used-to-surveillance-more-companies-try-tracking-faces-11575369002

Peters, J. (2020). Samsung's new Ballie robot is like a real-life mini BB-8. *The Verge*. https://www.thev-
erge.com/2020/1/6/21054353/samsung-ballie-robot-ai-assistant-ces-2020

Reichert, C. (2020). Clearview AI facial recognition company faces another lawsuit. CNet. https://www.cnet.com/news/
clearview-ai-facial-recognition-company-faces-another-lawsuit/

Ring. (2020). Join the Neighborhood. https://shop.ring.com/pages/neighbors

Samsung Electornics Declairs "Age of Experience" at CES 2020 (2020). *Samsung Newsroom*. https://news.samsung.
com/us/samsung-age-of-experience-keynote-ces-2020/?utm_source=twitter&utm_medium=usnews&utm_
campaign=ces2020

Schuppe, J. (2020, March 7). Google tracked his bike ride past a burglarized home. That made him a suspect. NBCnews.com
https://www.nbcnews.com/news/us-news/google-tracked-his-bike-ride-past-burglarized-home-made-him-n1151761

Schwarz, J. (2020). Ring gets "dinged" for its video doorbell privacy. *The Hill*. https://thehill.com/opinion/cybersecu-
rity/485449-ring-gets-dinged-for-its-video-doorbell-privacy

Singer, N. & Isaac, M. (2020). Facebook to pay $550 million to settle facial recognition suit. *The New York Times*.
https://www.nytimes.com/2020/01/29/technology/facebook-privacy-lawsuit-earnings.html

Sobel, N. (2020, February 24). Do Geofence Warrants Violate the Fourth Amendment? Lawfareblog.com. https://www.law-
fareblog.com/do-geofence-warrants-violate-fourth-amendment

Starship. (2020). https://www.starship.xyz/business/

Stevens, P. (2019). The battery decade: How energy storage could revolutionize industries I the next 10 years. CNBC.
https://www.cnbc.com/2019/12/30/battery-developments-in-the-last-decade-created-a-seismic-shift-that-will-play-out-
in-the-next-10-years.html

Tipsy Robot (n.d.) http://thetipsyrobot.com/

Thorne, J. (2019). Secretive Seattle startup Picnic unveils pizza making robot – here's how it delivers 300 pies/hour.
GeekWire. https://www.geekwire.com/2019/secretive-seattle-startup-picnic-unveils-pizza-making-robot-heres-delivers-
300-pies-hour/

Wesoff, E. (2020). ESA: More records set for US energy storage install in Q4, massive growth forecasted. *PV Magazine*.
https://pv-magazine-usa.com/2020/03/10/esa-power-capacity-record-for-us-energy-storage-installs-in-q4/

Yachot, N. (2017). History shows activists should feat the surveillance state. ACLU. https://www.aclu.org/blog/national-
security/privacy-and-surveillance/history-shows-activists-should-fear-surveillance

Young, J. (2018) A history of CCTV Surveillance in Britain. SWNS. https://stories.swns.com/news/history-cctv-surveillance-
britain-93449/

27

Your Future & Communication Technologies

August E. Grant, Ph.D.*

This book has introduced you to a range of ideas on how to study communication technologies, given you the history of communication technologies, and detailed the latest developments in about two dozen technologies. Along the way, the authors have told stories about successes and failures, legal battles and regulatory limitations, and changes in lifestyle for the end user.

So, what can you do with this information? If you're entrepreneurial, you can use it to figure out how to get rich. If you're academically-inclined, you can use it to inform research and analysis of the next generation of communication technology. If you're planning a career in the media industries, you can use

it to help choose the organizations where you might work, or to find new opportunities for your employer or for yourself.

More importantly, whether you are in any of those groups or not, you are going to be surrounded by new media for the rest of your life. The cycle of innovation, introduction, and maturity of media almost always includes a cycle of decline as well. As new communication technologies are introduced and older ones disappear, your media use habits will change. What you've learned from this book should help you make decisions on when to adopt a new technology or drop an old one. Of course, those decisions depend upon your personal goals—which might be to be an

---

* J. Rion McKissick Professor of Journalism, School of Journalism and Mass Communications, University of South Carolina (Columbia, South Carolina).

innovator, to make the most efficient use of your personal resources (time and money), or to have the most relaxing lifestyle.

This chapter explores a few ways you can apply this information to improve your ability to use and understand these technologies—or simply to profit from them.

Researching Communication Technologies

The speed of change in the communication industries makes the study of the technologies discussed in this text kind of a "spectator sport," where anyone willing to pay the price of admission (simple internet access) can watch the "game," predicting winners and losers and identifying the "stars" who help lead their teams to success. There are numerous career opportunities in researching these technologies, ranging from technology journalist and professor to stock market analyst and government regulator.

As detailed in Chapter 3, there are many ways to conduct this type of research. The most basic is to provide descriptive statistics regarding users of a technology, along with statistics on how and when the technology is used.

But greater insight can usually be obtained by applying one or more theories to help understand WHY the technology is being used by those doing so. Diffusion theory helps you understand how an innovation is adopted over time among members of a social system. Effects theories help you understand how and why a technology impacts the lives of individuals, the organizations involved in the production and distribution of the technology, and the impact on social systems.

In order to do this type of research successfully, you need three sets of knowledge:

- History of communication technologies

- Understanding of major theories of technology adoption, use, and effects

- Ability to apply appropriate quantitative and qualitative research methods

With these three, you will be prepared for any of the careers mentioned above, and you will have the chance to enjoy observing the competition as you would any other game that has so much at stake with some of the most interesting players in the world!

Making Money from Communication Technologies

You have the potential to get rich from the next generation of technologies. Just conduct an analysis of a few emerging technologies using the tools in the "Fundamentals" section of the book; choose the one that has the best potential to meet an unmet demand (one that people will pay for); then create a business plan that demonstrates how your revenues will exceed your expenses from creating, producing, or distributing the technology.

Regardless of the industry you want to work in, there are five simple steps to creating a business plan:

1) **Idea:** Every new business starts with an idea for a new product or service, or a new way to provide or enhance an existing business model. Many people assume that having a good idea is the most important part of a business plan, but nothing could be further from the truth. The idea is just the first step in the process. Sometimes the most successful ideas are the most innovative, but other times the greatest success goes to the team that makes a minor improvement on someone else's idea that enables a simpler or less expensive option. Applying Rogers' five attributes of an innovation (compatibility, complexity, relative advantage, trialability, and observability, discussed in Chapter 3) can help you understand how innovative your idea is.

2) **Team:** Bringing together the right group of people to execute the idea is the next step in the process. The most important consideration in putting together a team is creating a list of competencies needed in your organization, and then making sure the management team includes one expert in each area. Make sure that you include

experts in marketing and sales as well as production and engineering. The biggest misstep in creating a team is bringing together a group of people who have the same background and similar experiences. At this stage you're looking for diversity in knowledge and experience. Once you've identified the key members of the team, you can create the organizational structure, identifying the departments and divisions within each department (if applicable) and providing descriptions for each department and responsibilities for each department head. It is also critical to estimate the number of employees in each position.

3) **Competitive Analysis:** The next step is describing your new company's primary competition, detailing the current strengths and weaknesses of competing companies and their product or service offerings. Consider the resources and management team, and agility of the company as factors that will influence how they will compete with you. In the process, this analysis needs to identify the challenges and opportunities they present for your company.

4) **Finance:** The most challenging part of most business plans is the creation of a spreadsheet that details how and where revenues will be generated, including projections for revenues during the first few years of operation. These revenues are balanced in the spreadsheet with estimates of the expected expenses (e.g., personnel, technology, facilities, and marketing). This part of your business plan also has to detail how the initial "start-up" costs will be financed. Some entrepreneurs choose to start with personal loans and assets, growing a business slowly and using initial revenues to finance growth—a process known as "bootstrapping." Others choose to attract financing from venture capitalists and angel investors who receive a substantial ownership stake in the company in return for their investment. The bootstrapping model allows an entrepreneur to keep control—and a greater share of profits—but the cost is much slower growth and more limited financial resources. Once the initial plan is created, study every expense and every revenue source to refine the plan for profitability.

5) **Marketing Plan:** The final step is creating a marketing plan for the new company that discusses the target market and how you will let those in the target know about your product or service. The marketing plan must identify the marketing and promotional strategies that will be utilized to attract consumers/audiences/users to your product or service, as well as the costs of the plan and any related sales costs. Don't forget to discuss how both traditional and new media will be employed to enhance the visibility and use of your product, then follow through with a plan for how the product or service will be distributed and priced.

Conceptually, this five-step process is deceptively easy. The difficult part is putting in the hours needed to plan for every contingency, solve problems as they crop up (or before they do), make the contacts you need in order to bring in all of the pieces to make your plan work, and then distribute the product or service to the end users. If the lessons in this book are any indication, two factors will be more important than all the others: the interpersonal relationships that lead to organizational connections—and a lot of luck!

Here are a few guidelines distilled from 30-plus years of working, studying, and consulting in the communication technology industries that might help you become an entrepreneur:

- **Ideas are not as important as execution.** If you have a good idea, chances are others will have the same idea. The ones who succeed are the ones who have the tools, vision, and willingness to work long and hard to put the ideas into action.

- **Protect your ideas.** The time and effort needed to get a patent, copyright, or even a simple non-disclosure agreement will pay off handsomely if your ideas succeed.

- **There is no substitute for hard work.** Entrepreneurs don't work 40-hour weeks, and they always have a tool nearby to record ideas and keep track of contacts.

- **There is no substitute for time away from work.** Taking one day a week away from the job gives you perspective, letting you step back and see the big picture. Plus, some of the best ideas come from bringing in completely unrelated content, so make sure you are always scanning the world around you for developments in the arts, technology, business, regulation, and culture.

- **Who you know is more important than what you know.** You can't succeed as a solo act in the communication technology field. You have to a) find and partner with or hire people who are better than you in the skill sets you don't have, and b) make contacts with people in organizations that can help your business succeed.

- **Keep learning.** Study your field, but also study the world. The technologies that you will be working with have the potential to provide you access to more information than any entrepreneur in the past has had. Use the tools to continue growing.

- **Create a set of realistic goals.** Don't limit yourself to just one goal, but don't have too many. As you achieve your goals, take time to celebrate your success.

- **Give back.** You can't be a success without relying upon the efforts of those who came before you and those who helped you out along the way. The best way to pay it back is to pay it forward.

This list was created to help entrepreneurs, but it may be equally relevant to any type of career. Just as the communication technologies explored in this book have applications that permeate industries and institutions throughout society, the tools and techniques explored in this book can be equally useful regardless of where you are or where you are going.

Keep in mind along the way that working with and studying communication technologies is not simply a means to an end, but rather a process that can provide fulfillment, insight, monetary rewards, and pure joy. Those people who have the fortune to work in the communication technology industries rarely have a chance to become bored, as there is always a new player in the game, a new device to try, and a great chance to discuss and predict the future.

In the process, be mindful of all of the factors in the communication technology ecosystem—it's not the hardware that makes the biggest difference, but all of the other elements of the communication technology ecosystem that together define this ecosystem.

The one constant you will encounter is change, and that change will be found most often at the organizational level and in the application of system-level factors that impact your business. As discussed in the first chapter of this book, new, basic technologies that have the potential to change the field come along infrequently, perhaps once every ten or twenty years. But new opportunities to apply and refine these technologies come along every day. The key is training yourself to spot those opportunities and committing yourself to the effort needed to make your ideas a success.

Index

Index

Index

Index

Index

Printed in the United States
by Baker & Taylor Publisher Services